Materials
Science

PERIODIC TABLE OF THE ELEMENTS

I	IIA	IIIA	IVA	VA	VIA	VIIA	VIII	VIII	VIII	IB	IIB	IIIB	IVB	VB	VIB	VIIB	O
1·008 H 1																	4·003 He 2
6·940 Li 3	9·012 Be 4											10·811 B 5	12·011 C 6	14·007 N 7	15·999 O 8	18·998 F 9	20·183 Ne 10
22·990 Na 11	24·312 Mg 12											26·982 Al 13	28·086 Si 14	30·974 P 15	32·064 S 16	35·453 Cl 17	39·948 A 18
39·102 K 19	40·080 Ca 20	44·956 Sc 21	47·900 Ti 22	50·942 V 23	51·996 Cr 24	54·938 Mn 25	55·847 Fe 26	58·993 Co 27	58·710 Ni 28	63·540 Cu 29	65·370 Zn 30	69·720 Ga 31	72·590 Ge 32	74·922 As 33	78·960 Se 34	79·909 Br 35	83·800 Kr 36
85·470 Rb 37	87·620 Sr 38	88·905 Y 39	91·220 Zr 40	92·906 Nb 41	95·940 Mo 42	99 Tc 43	101·070 Ru 44	102·905 Rh 45	106·400 Pd 46	107·870 Ag 47	112·400 Cd 48	114·820 In 49	118·690 Sn 50	121·750 Sb 51	127·600 Te 52	126·904 I 53	131·300 Xe 54
132·905 Cs 55	137·340 Ba 56	RARE EARTHS La-Lu	178·490 Hf 72	180·948 Ta 73	183·85 W 74	186·20 Re 75	190·2 Os 76	192·2 Ir 77	195·09 Pt 78	196·967 Au 79	200·590 Hg 80	204·370 Tl 81	207·190 Pb 82	208·980 Bi 83	210 Po 84	210 At 85	222 Rn 86
223 Fr 87	227 Ra 88	ACTINIDE SERIES Ac-Lw															

RARE EARTHS	138·92 La 57	140·13 Ce 58	140·92 Pr 59	144·24 Nd 60	145 Pm 61	150·35 Sm 62	152·0 Eu 63	157·25 Gd 64	158·92 Tb 65	162·50 Dy 66	164·93 Ho 67	167·3 Er 68	169 Tm 69	173·04 Yb 70	174·98 Lu 71
ACTINIDE SERIES	227 Ac 89	232·038 Th 90	231 Pa 91	238·030 U 92	237 Np 93	242 Pu 94	243 Am 95	247 Cm 96	249 Bk 97	251 Cf 98	254 E 99	255 Fm 100	256 Mv 101	255 No 102	257 Lw 103

The atomic number of an element appears below its symbol, while its atomic weight is above.

Materials Science

Science

Fourth edition

J.C. ANDERSON   K.D. LEAVER
R.D. RAWLINGS   J.M. ALEXANDER

Imperial College of Science, Technology and Medicine

CHAPMAN & HALL
London · New York · Tokyo · Melbourne · Madras

Published by Chapman & Hall, 2–6 Boundary Row, London SE1 8HN

Chapman & Hall, 2–6 Boundary Row, London SE1 8HN, UK

Van Nostrand Reinhold Inc., 115 5th Avenue, New York NY10003, USA

Chapman & Hall Japan, Thomson Publishing Japan, Hirakawacho Nemoto Building, 7F, 1-7-11 Hirakawa-cho, Chiyoda-ku, Tokyo 102, Japan

Chapman & Hall Australia, Thomas Nelson Australia, 102 Dodds Street, South Melbourne, Victoria 3205, Australia

Chapman & Hall India, R. Seshadri, 32 Second Main Road, CIT East, Madras 600 035, India

First edition 1969
Reprinted with amendments 1971, 1973
Second edition 1974
Reprinted 1977, 1979, 1981, 1982, 1983
Third edition 1985
Reprinted 1985, 1986, 1987
Fourth edition 1990
Reprinted 1991

© 1990 J. C. Anderson, K. D. Leaver, R. D. Rawlings and J. M. Alexander

Typeset in 10/12 Times by Colset Private Ltd, Singapore
Printed in Great Britain by St Edmundsbury Press Ltd, Bury St Edmunds, Suffolk

ISBN   0 412 34150 6   0 442 31189 3 (USA)

A catalogue record for this book is available from the British Library

Library of Congress Cataloging-in-Publication data
Materials science/J. C. Anderson . . . [et al.].—4th ed.
    p.   cm.
    Includes bibliographical references.
    ISBN 0 442 31189 3
    1. Materials.   I. Anderson, J. C. (Joseph Chapman), 1922–
TA403.M346   1990
620.1′1—dc20   90-33214
                                                         CIP

PREFACE

Since 1985, when the third edition was published, there have been rapid and significant developments in materials science and this new edition has been revised to include the most important of these.

Chapter 11 has been subject to major revision and updating to take account of the newly developed ceramics and high-strength composites.

A completely new chapter on plastics has replaced the original Chapter 12 on organic polymers. This has been written by Dr. P.S. Leevers of Imperial College and includes the basic science and technology relating to the plethora of modern plastic materials that have had such a profound influence on everyday life. The authors are most grateful to Dr. Leevers for his valuable contribution.

Chapter 13 now includes a section on superconductors and covers the new 'high-temperature' superconducting materials, discovered in 1986, for which the Nobel Prize in Physics was awarded in 1987.

Chapter 15 has been extensively revised to cover the most modern technologies adopted by the semiconductor industry, including molecular beam epitaxy and ion implantation.

Chapter 16 now includes coverage of modern permanent magnet materials and of the new 'glassy' amorphous magnetic materials.

Chapter 17 has been extended to give more details of piezo-, pyro- and ferroelectric materials, which have found increasingly wide applications in medical diagnostics, infrared imaging and communications.

Chapter 18 has been revised and extended to cover materials and techniques used in opto-electronic applications, which are increasingly important in information technology.

Finally, the authors would once again like to express their appreciation of the helpful comments and suggestions received from colleagues and of the publisher's continued enthusiam for our book.

J.C.A.
K.D.L.
R.D.R.
J.M.A.

Preface to the first edition

The study of the science of materials has become in recent years an integral part of virtually all university courses in engineering. The physicist, the chemist, and the metallurgist may, rightly, claim that they study materials scientifically, but the reason for the emergence of the 'new' subject of materials science is that it encompasses all these disciplines. It was with this in mind that the present book was written. We hope that, in addition to providing for the engineer an introductory text on the structure and properties of engineering materials, the book will assist the student of physics, chemistry, or metallurgy to comprehend the essential unity of these subjects under the all-embracing, though ill-defined, title 'Materials science'.

The text is based on the introductory materials course which was given to all engineers at Imperial College, London. One of the problems in teaching an introductory course arises from the varying amounts of background knowledge possessed by the students. We have, therefore, assumed only an elementary knowledge of chemistry and a reasonable grounding in physics, since this is the combination most frequently encountered in engineering faculties. On the other hand, the student with a good grasp of more advanced chemistry will not find the treatment familiar and therefore dull. This is because of the novel approach to the teaching of basic atomic structure, in which the ideas of wave mechanics are used, in a simplified form, from the outset. We believe that this method has several virtues: not only does it provide for a smooth development of the electronic properties of materials, but it inculcates a feeling for the uncertainty principle and for thinking in terms of probability, which are more fundamental than the deterministic picture of a particle electron moving along a specific orbit about the nucleus. We recognize that this approach is conceptually difficult, but no more so than the conventional one if one remembers the 'act of faith' which is necessary to accept the quantization condition in the Bohr theory. The success of this approach with our own students reinforces the belief that this is the right way to begin.

In view of the differences which are bound to exist between courses given in different universities and colleges, some of the more advanced material has been separated from the main body of the text and placed at the end of the appropriate chapter. These sections, which are marked with an asterisk, may, therefore, be omitted by the reader without impairing comprehension of later chapters.

In writing a book of this kind, one accumulates indebtedness to a wide range of people, not least to the authors of earlier books in the field. We particularly wish to acknowledge the help and encouragement given by our academic colleagues.

Our students have given us much welcome stimulation and the direct help of many of our graduate students is gratefully acknowledged. Finally, we wish to express our thanks to the publishers, who have been a constant source of encouragement and assistance.

J.C.A.

K.D.L.

Preface to the second edition

The goals of the first edition remain unchanged, but the need was felt to provide in the second edition a wider and more detailed coverage of the mechanical and metallurgical aspects of materials. Accordingly we have extensively rewritten the relevant chapters, which now cover mechanical properties on the basis of continuum theory as well as explaining the microscopic atomic mechanisms which underlie the macroscopic behaviour. The opportunity has also been taken to revise other chapters in the light of the many helpful comments on the first edition which we have received from colleagues around the world. We would like to thank here all who have taken the trouble to point out errors and inconsistencies, and especially our colleagues who have read the manuscript of this and the first edition. We are also indebted to Dr. D.L. Thomas and Dr. F.A.A. Crane for several micrographs which appear here for the first time.

J.C.A.

K.D.L.

J.M.A.

R.D.R.

Preface to the third edition

This edition represents a general updating and revision of the text of the second edition to take account of continuing developments in the field of materials science and the helpful comments and criticisms that we have received from our colleagues around the world.

Chapter 7, on thermodynamics, has been revised and extended to include the basic ideas of lattice waves and this is applied, in the chapter on electrical conductivity, to phonon scattering of electrons.

Chapters 8 and 9 on mechanical properties have been revised and extended and the fundamental principles of the relatively new subject of fracture toughness have been introduced; the microstructural aspects of fracture, creep and fatigue, together with the interpretation of creep and fatigue data, are dealt with in more detail.

The section on steel in Chapter 10 has been considerably extended. A completely new chapter on ceramics and composites has been added to take account of the increasing development and importance of these materials.

The chapter on semiconductors has been revised and updated by including the field effect transistor. An innovation is the introduction of a complete new chapter on semiconductor processing which is an important area of application of materials science in modern technology.

The addition of a chapter on optical properties has given us the opportunity of discussing the principles of spectroscopy, absorption and scattering, optical fibre materials and laser materials.

In general, all chapters have been reviewed and minor revision, correction and updating has been carried out where required.

J.C.A.
K.D.L.
R.D.R.
J.M.A.

SELF-ASSESSMENT QUESTIONS

A series of self-assessment questions, with answers, will be found at the end of each chapter. By using these the student can easily test his understanding of the text and identify sections that he needs to re-read. The questions are framed so that the answer is a choice between two or more alternatives and the answer is simply a letter A, B, C, etc. The correct answers are given at the end of each set of questions and it should be noted that, where a question involves more than two possibilities, there may be more than one right answer.

CONTENTS

BUILDING BLOCKS: THE ELECTRON

1·1 Introduction

Science is very much concerned with the identification of patterns, and the recognition of these patterns is the first step in a process that leads to identification of the building bricks with which the patterns are constructed. This process has all the challenge and excitement of exploration combined with the fascination of a good detective story and it lies at the heart of materials science.

At the end of the nineteenth century a pattern of chemical properties of elements had begun to emerge and this was fully recognized by Mendeléev when he constructed his periodic table. Immediately it was apparent that there must be common properties and similar types of behaviour among the atoms of the different elements and the long process of understanding atomic structure had begun. There were many wonders along the way. For instance, was it not remarkable that *only* iron, nickel, and cobalt showed the property of ferromagnetism? (Gadolinium was a fourth ferromagnetic element discovered later.) A satisfactory theory of the atom must be able to explain this apparent oddity. Not only was magnetism exclusive to these elements, but actual pieces of the materials sometimes appear magnetized and sometimes do not, depending on their history. Thus a theory that merely states that the atoms of the element are magnetic is not enough; we must consider what happens when the atoms come together to form a solid.

Similarly, we wonder at the extraordinary range–and beauty–of the shapes of crystals; here we have patterns–how can they be explained? Why are metals ductile while rocks are brittle? What rules determine the strength of a material and is there a theoretical limit to the strength? Why do metals conduct electricity while insulators do not? All these questions are to do with the properties of aggregations of atoms. Thus an understanding of the atom must be followed by an understanding of how atoms interact when they form a solid because this must be the foundation on which explanations of the properties of materials are based.

This book attempts to describe the modern theories through which many of these questions have been answered. We are not interested in tracing their history of development but prefer to present, from the beginning, the quantum-mechanical concepts that have been so successful in modern atomic theory. The pattern of the book parallels the pattern of understanding outlined above. We must start with a thorough grasp of fundamental atomic theory and go on to the theories of aggregations of atoms. As each theoretical concept emerges it is used to explain relevant observed properties. With such a foundation the electrical, mechanical, thermal, and other properties of materials can be described, discussed, and explained.

To begin at the beginning we consider the building bricks of the atoms themselves, starting with the electron.

1·2   The electron

The electron was first clearly identified as an elementary particle by J. J. Thomson in 1897. A more detailed description of his experiment is given towards the end of this chapter; here it is sufficient to say that he was able to conclude that the electron is a constituent of all matter. It was shown to have a fixed, negative charge, e, of $1·6021 \times 10^{-19}$ coulomb and a mass of $9·1085 \times 10^{-31}$ kilogramme at rest. Thomson's proof of the existence of the electron was the essential prerequisite for the subsequent theories of the structure of the atom. However, before going on to consider these we must first review some of the known facts about atoms themselves.

1·3   Avogadro's number

In the electrolysis of water, in which a voltage is applied between two electrodes immersed in the water, hydrogen is observed to be given off at the negative electrode (the cathode). This indicates that the hydrogen ion carries a positive charge and measurement shows that it takes 95,650 coulombs of electricity to liberate one gramme of hydrogen. Now if we know how many atoms of hydrogen make up one gramme it would be possible to calculate the charge per hydrogen ion. This can be done using Avogadro's number, but before defining it we must describe what is meant by a 'gramme-atom' and a 'gramme-molecule' (or 'mol').

It is known from chemistry that all substances are either elements or compounds and that compounds are made up of elements. Any quantity of an element is assumed to be made up of atoms, all of equal size and mass, and the mass of each atom when expressed in terms of the mass of the hydrogen atom was defined as its atomic weight. In 1815 Prout suggested that if the atomic weight of hydrogen were taken as unity the atomic weights of all other elements should be whole numbers. This turned out not to be quite true and internationally agreed atomic weights were based, instead, on the atomic weight of oxygen being 16, which gives hydrogen the atomic weight 1·0080. It was later discovered that an element could have different isotopes. i.e., atoms of the same atomic *number* could have differing atomic *weights*: this will be discussed in a later chapter. Mixtures of naturally occurring isotopes were the cause of the atomic weights not being exactly whole numbers. In 1962 it was internationally agreed to use the isotope ^{12}C, with an atomic weight of 12, as the basis of all atomic weights, which still gives to hydrogen the value 1·0080.

Thus a gramme-atom of a substance is the amount of substance whose mass in grammes equals its atomic weight. Similarly a gramme-molecule (or mol) of a compound is the amount whose mass in grammes equals its molecular weight which, in turn, is the sum of the atomic weights of the atoms which go to make up the molecule.

We may now define Avogadro's number, which is the number of atoms in a gramme-atom (or molecules in a gramme-molecule) of any substance and it is a universal constant.

Returning to our electrolysis experiment, the amount of electricity required to liberate a gramme-atom of hydrogen will be $95,650 \times 1\cdot0080 = 96,420$ coulombs. Now, suppose we *assume* that the charge on the hydrogen ion is equal to that on the electron, then we may calculate Avogadro's number, N.

$$N = \frac{96,420}{1\cdot602 \times 10^{-19}} = 6\cdot02 \times 10^{23} \text{ mol}^{-1}$$

Experimentally, the accurate value has been determined as $6\cdot023 \times 10^{23}$ and so it may be concluded that the charge on the hydrogen ion is equal in magnitude and opposite in sign to that on the electron.

Also using Avogadro's number we may calculate the mass of a hydrogen atom as

$$\frac{\text{Atomic weight}}{\text{Avogadro's number}} = \frac{1\cdot0080}{6\cdot023 \times 10^{23}}$$
$$= 1\cdot672 \times 10^{-24} \text{ gramme}$$

This is just 1,840 times the mass of the electron.

Such a calculation suggests that there is another constituent of atoms, apart from the electron, which is relatively much more massive and which is positively charged.

1·4 The Rutherford atom

In 1911 Rutherford made use of the *a*-particle emission from a radioactive source to make the first exploration of the structure of the atom. By passing a stream of particles through a thin gold foil and measuring the angles through which the beam of particles was scattered he was able to conclude that most of the mass of the gold atom (atomic weight 187) resided in a small volume called the nucleus which carried a positive charge. He was also able to show that the radius of the gold nucleus is not greater than $3\cdot2 \times 10^{-12}$ cm.

Since the number of gold atoms in a gramme-molecule of gold is given by Avogadro's number and, from the density (mass per unit volume) we may calculate the volume occupied by a gramme-molecule, then we may make an estimate of the volume of each atom. If we assume them to be spheres packed together as closely as possible we can deduce that the radius of the gold atom is in the region of 10^{-8} cm, which is 10,000 times larger than the radius obtained by Rutherford. Thus the atom seems to comprise mainly empty space.

To incorporate these findings and the discovery of the electron Rutherford proposed a 'planetary' model of the atom; this postulated a small dense nucleus carrying a positive charge about which orbited the negative electrons like planets round the sun. The positive charge on the nucleus was taken to be equal to the sum of the electron charges so that the atom was electrically neutral.

This proposal has an attractive simplicity. Referring to Fig. 1·1: if the electron moves in a circular orbit of radius, r, with a constant linear velocity, v, then it will be subject to two forces. Acting inwards will be the force of electrostatic attraction described by Coulomb's law:

$$F = \frac{q_1 q_2}{4\pi\epsilon_0 r^2} \tag{1·1}$$

where q_1 and q_2 are the positive and negative charges, in this case each having the value of the charge, e, on the electron and ϵ_0 is the permittivity of free space given by $10^{-9}/36\pi$.

Acting outwards there will be the usual centrifugal force given by mv^2/r, where m is the electron mass and v^2/r is its radial acceleration. The orbit of the electron will settle down to a stable value where these two forces just balance each other, that is, when

$$\frac{mv^2}{r} = \frac{e^2}{4\pi\epsilon_0 r^2} \tag{1·2}$$

Unfortunately, this model has a basic flaw. Maxwell's equations, which describe the laws of electromagnetic radiation, can be used to show that an electron undergoing a change in velocity (i.e., acceleration and deceleration) will radiate energy in the form of electromagnetic waves. This can be seen by analogy: if the electron in a circular orbit were viewed from the side it would appear to be travelling rapidly backwards and forwards. Now it carries a charge and a moving charge represents an electric current. Thus its alternating motion corresponds to an alternating electric current at a very high frequency. Just such a high-frequency alternating current is supplied to an aerial by a radio transmitter and the electromagnetic radiation from the aerial is readily detectable. Thus we must expect the electron in Rutherford's model of the atom continuously to radiate energy. More formally, since an electron moving in an orbit of radius, r, with a linear velocity, v, is, by the laws of Newtonian mechanics, subject to a continual radial acceleration of magnitude v^2/r, it must, by the laws of electromagnetic radiation, continuously radiate energy.

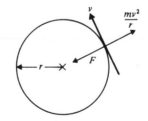

Fig. 1·1 Rutherford's planetary atomic model

The kinetic energy of the electron in its orbit is proportional to the square of its velocity and if it loses energy its velocity must diminish. The centrifugal force, mv^2/r, due to the radial component of acceleration will therefore also diminish and so the electrostatic attraction between the positive nucleus and the negative electron will pull the electron closer to the nucleus. It is not difficult to see that the electron would ultimately spiral into the nucleus.

The resolution of this difficulty was only made possible by the bringing together of a variety of observations and theories and we must now consider these briefly in turn.

1·5 Waves and particles

One of the more remarkable puzzles which gradually emerged from experimental physics as more and more physical phenomena were explored was that it appeared that light, which was normally regarded as a wave (an electromagnetic wave to be precise), sometimes behaved as if the ray of light were a stream of particles. Similarly, experimental evidence emerged suggesting that electrons, which we have so far treated as particles, may behave like waves. If this is so, the Rutherford atom, based on treating the electron as a particle having a fixed mass and charge and obeying Newtonian mechanics, is evidently too crude a model. This is obviously a crucial matter and we must examine the relevant experiments carefully.

1·5·1 Electron waves

Remember that the properties which convince us that light is a wave motion are *diffraction* and *interference*. For instance, a beam of light impinging on a very narrow slit is diffracted (i.e., spreads out behind the slit). A beam of light shining on two narrow adjacent slits, as shown in Fig. 1·2, is diffracted and if a screen is placed beyond the slits interference occurs, giving a characteristic intensity distribution as shown. Yet another pattern is produced if we use many parallel slits, the device so produced being called a *diffraction grating*. As shown in Fig. 1·3, if we take a line source of light and observe the image it produces when viewed through a diffraction grating we see a set of lines whose separation depends on the wavelength of the light and the

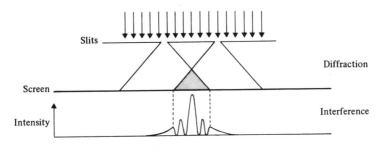

Fig. 1·2 Diffraction of light through two slits

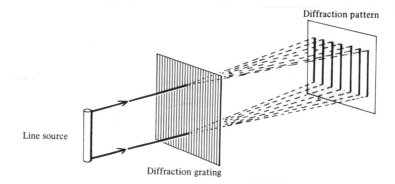

Fig. 1·3 Diffraction grating and pattern

distance between the slits in the diffraction grating. If two diffraction gratings are superimposed with their lines intersecting at right-angles a spot pattern is produced in which the distances between the spots are a function of the wavelength of the light for a given pair of gratings. For the experiment to work the dimensions of the slits in the grating *must be comparable to the wavelengths of the light*, i.e., the slits must be about 10^{-4} cm wide and separated by a similar distance.

In order to observe the wavelike behaviour of electrons, we can expect to do so with a diffraction grating of suitable dimensions; but we do not yet know the dimensions necessary. It so happens that the size required is very small indeed and, in fact, is comparable with the mean distance between atoms in a typical crystal, that is, in the region of 10^{-8} cm. Thus if we 'shine' a beam of electrons on a crystal which is thin enough for them to pass through, the crystal will act as a diffraction grating if the electrons behave as waves.

Referring to Fig. 1·4, a source of electrons, such as a hot filament, is mounted in an evacuated enclosure with suitable electrodes such that the electrons can be accelerated and collimated into a narrow beam. The beam is

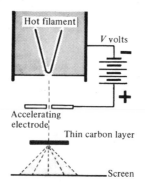
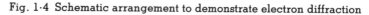

Fig. 1·4 Schematic arrangement to demonstrate electron diffraction

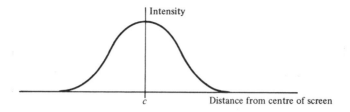

Fig. 1·5 Intensity distribution for randomly scattered light

allowed to fall upon a thin carbon layer on the other side of which is placed a fluorescent screen of the type commonly used in television tubes. Electrons arriving at the screen produce a glow of light proportional to their number and energy.

The energy imparted to the electron beam is determined by the potential, V, applied to the accelerating electrode. If we are regarding it as a charged particle the potential energy of an electron when the applied voltage is V volts is, by definition of potential, just eV, where e is its charge in coulombs. The accelerating electron increases its kinetic energy at the expense of its potential energy, and, if we treat it as a particle, it will reach a velocity, v, given by:

$$\tfrac{1}{2}mv^2 = eV \tag{1·3}$$

This is a relationship of which we shall make much use later.

If the electrons are to be regarded as waves it is not clear how we can attribute a kinetic energy of $\tfrac{1}{2}mv^2$ to them since we do not normally attribute mass to a wave. We must just say that they have gained energy and it is convenient to define this energy as the change in potential energy of the wave in passing through a potential difference of V volts. This energy will be eV joules and we therefore define a new unit of energy, the *electron volt*. This is defined as the energy which an electron acquires when it falls through a potential drop of one volt and is equal to $1·602 \times 10^{-19}$ joule.

Returning to Fig. 1·4, the electrons thus impinge upon the carbon layer with high energy. If they were particles we would expect them to be scattered randomly in all directions when they hit atoms in the carbon crystals. The intensity distribution beyond the carbon layer would then be greatest at the centre, falling off uniformly towards the edges, as shown in Fig. 1·5, and the pattern of light on the screen would be a diffuse one, shading off gradually in brightness from the centre outwards.

In fact the experiment produces a quite different result. The pattern seen on the screen consists of rings of intense light separated by regions in which no electrons are detected, as shown in Fig. 1·6.

This can only be explained by assuming that the electron is a wave. If layers of atoms in a crystal act as a set of diffraction gratings we can regard them as line gratings superimposed with their slits at right-angles and the diffraction pattern should be a set of spots, as described earlier. The pattern seen is a series of rings rather than spots because the carbon layer is made up of many tiny crystals all orientated in different directions; the rings correspond to

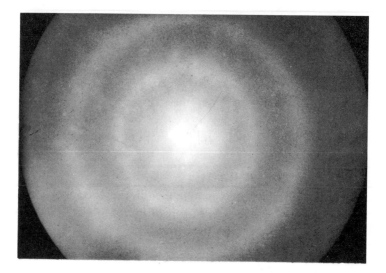

Fig. 1·6 Electron diffraction rings

many spot patterns, with differing orientations, superimposed. Each ring corresponds to diffraction from a different set of atomic planes in the crystal. Diffraction of electrons by crystals is a powerful tool in the determination of crystal structures and is dealt with in detail in more advanced textbooks (see, for instance, Azaroff and Brophy).

If in the experiment the accelerating voltage, V, is varied, we observe that the diameters of the rings change, and it is not difficult to establish that the radius, R, of any ring is inversely proportional to the square root of the accelerating voltage. That is,

$$R \propto \frac{1}{V^{1/2}}$$

To understand this we must go to yet another branch of physics–that of X-ray crystallography. X-rays are electromagnetic waves having the same nature as that of light but with much shorter wavelengths. 'Hard' X-rays have wavelengths in the range of 1 to 10 Å (one ångstrom unit, denoted by 1 Å, is 10^{-10} metres) and are therefore diffracted by the planes of atoms in a crystal, in the same way as light is diffracted by a diffraction grating. Study of X-ray diffraction by crystals led to the well-known Bragg's law relating wavelength to the angle through which a ray is diffracted. This is derived in Chapter 6 and the law can be written as

$$n\lambda = 2d \sin \theta \tag{1·4}$$

when n is an integer (the 'order' of the diffracted ray), λ the wavelength, d the distance between the planes of atoms, and 2θ is the angle between the direction of the diffracted ray and that of the emergent ray as shown in Fig. 1·7. Since the electrons appear to be behaving as waves it is reasonable to assume

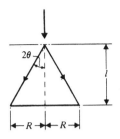

Fig. 1·7 Illustrating Bragg's law of diffraction

that we may also apply Bragg's law to them.

Referring to Fig. 1·7, in our experiment the electron beam fell normally on the carbon film and, on emergence, had been split into diffracted rays forming a cone of apical angle 4θ. From the diagram, the radius, R, of the diffraction ring is given by

$$R = l \tan 2\theta \qquad (1·5)$$

where l is the distance from the carbon film to the screen. Since θ is commonly only a few degrees we may approximate $\tan 2\theta \approx 2\theta \approx 2 \sin \theta$, so that

$$R \approx 2l \sin \theta$$

and, using Bragg's law of Eqn (1·4),

$$R \approx \frac{nl\lambda}{d} \qquad (1·6)$$

Since R and l can be measured and n is known for the 'order' of the ring ($n = 1$ for the first ring), we may determine λ from this experiment, given a knowledge of the spacing, d, between the atomic planes in carbon, which is known from X-ray diffraction work.

The importance of this result for our present purposes is that it shows R is proportional to λ, the other quantities being constants. But we have already observed that $R \propto 1/V^{1/2}$ and so we have the result that

$$\lambda \propto \frac{1}{V^{1/2}}$$

Now we earlier defined the energy of the wave, E, as the potential through which it fell, that is

$$E = eV \qquad (1·7)$$

Since e is a constant it follows that

$$\lambda \propto \frac{1}{E^{1/2}} \qquad (1·8)$$

and we have the important result that the electron wavelength is inversely proportional to the square root of the energy of the wave.

l

Now although the electron might be behaving like a wave we do know that it can also behave as if it were a particle whose energy is given by

$$E = \tfrac{1}{2}mv^2 = eV$$

and which will have a momentum, p, given by $p = mv$. Relating the kinetic energy, E, to the momentum we have

$$E = \tfrac{1}{2}mv^2 = \frac{p^2}{2m} \qquad (1\cdot9)$$

Thus, since m is a constant, $E^{1/2} \propto p$ and from the experimental result of Eqn (1·8), we may write

$$\lambda \propto \frac{1}{p} \qquad (1\cdot10)$$

Thus we are postulating that the electron wave carries a momentum which is inversely proportional to its wavelength. If the wave is to carry momentum we are implying that the wave, in some way, represents a mass, m, moving with a velocity, v, and we may refer to the electron as a 'matter wave'.

The relationship between λ and p was, in fact, proposed by de Broglie before the phenomenon of electron diffraction was discovered. Putting in a constant of proportionality, h, we obtain de Broglie's relationship,

$$\lambda = \frac{h}{p}$$

in which h is a universal constant called *Planck's constant* and has the value $6\cdot625 \times 10^{-34}$ Nm s ($=$ joule second). This is one of the most important relationships in physics and applies to all matter, from electrons to billiard balls. It implies that all moving matter will exhibit wavelike behaviour. However, because of the small value of Planck's constant, the wavelength of anything much larger than an electron will be so short that we will never see any evidence, such as diffraction, of wavelike behaviour. (See Problems at the end of this chapter.)

1·5·2   Particles of light

We have already mentioned that the properties of diffraction and interference show that light is a wave motion. But now consider the experiment shown in Fig. 1·8. Some materials (e.g. caesium) when irradiated with light will emit electrons which can be detected by attracting them to a positively charged electrode and observing the flow of current.

It can be shown from other experiments that to remove an electron from a solid it is necessary to give the electron enough kinetic energy. 'Enough' here means more than some minimum quantity, W, which represents the 'potential energy barrier' (often called just the 'potential barrier') which the electron has to surmount in order to leave the solid. Energy in excess of this amount, W, supplied by the light beam must reappear as kinetic energy, E_k, of the electron, i.e.,

$$\text{Total energy supplied by light} = E_L = W + E_k \qquad (1\cdot11)$$

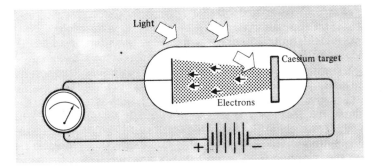

Fig. 1·8 Schematic arrangement to demonstrate photo-emission

If we reduce E_L by reducing the amount of light supplied E_k must also decrease. So reducing the *intensity* of the light should reduce E_k which is measurable.

But this does not happen! What we observe is that the *number* of electrons released by the light reduces proportionally with the light intensity but the *energy* of the electrons remains the same.

This can be explained by supposing the light beam to consist of particles of fixed energy. Reducing the intensity would reduce the number of particles flowing per unit time. If we suppose that when a particle 'collides' with an electron it can give the electron its whole energy, then reducing the rate of flow of particles would reduce the number of collisions but the energy transferred at each collision would remain the same. Thus we would reduce the rate of emission of electrons without changing their kinetic energy.

We call the light particles 'photons'. How do we know how much energy a photon has?

We can get one clue by changing the *frequency* of the light. We then find that the emitted electrons have a different energy. By trying several different wavelengths (frequencies) it is possible to show that the energy, E, of each photon is proportional to the frequency, f, that is,

$E \propto f$

This is clearly different from the case of the electron for which $E^{1/2} \propto f$ [see Eqn (1·8)]. But the constant of proportionality is again h, Planck's constant, so that

$E = hf$ (1·12)

The implications of this are profound since we are saying that the light waves behave as if they consist of a stream of particles, each particle carrying an energy hf. Thus the energy carried by the wave is apparently broken up into discrete lumps or 'quanta'. The energy of each quantum depends only on the frequency (or wavelength) of the light while the intensity (or brightness) of the light is determined by the number of quanta arriving on the illuminated surface in each second. This idea emerged just about the time of the discovery of the electron. It was in 1900 that Planck suggested that heat radiation, which is infra-red light, is emitted or absorbed in multiples of a definite amount. This definite amount was named a quantum and its energy was defined as hf, where h was a universal constant that came to be called Planck's constant. Just as wavelike behaviour can be assumed to apply to all moving matter, so all energy associated with it is *quantized* whatever its form. In most practical applications in the real world, such as in the oscillations of a pendulum, the frequency is so low that single quantum could never be observed. However, at the high frequencies of light, X-rays, and electron waves, quantum effects will be readily apparent. This general truth is the basis of the quantum theory and led to a new type of mathematics being developed for application to physical problems.

1·6 Particles or waves?

We have seen that both light and electrons behave in some ways like particles, in others like waves. How is this possible? One way of understanding this is to think about the many possible forms a wave may take. The first is a simple plane wave. If we plot amplitude vertically and distance horizontally we can draw such a wave in one dimension as in Fig. 1·9(a). This wave has a single wavelength and frequency. If we add together two plane waves with slightly different wavelengths the result is a single wave of different 'shape' as shown in Fig. 1·9(b). Its amplitude varies with position so that at regular intervals there is a maximum. Adding a third plane wave with yet another wavelength makes some of these maxima bigger than others. If enough plane waves are added together just *one* of the maxima is bigger than all the others as in Fig. 1·9(c): if they are added together in the correct phases all the other maxima disappear and the wave so formed has a noticeable amplitude only in one (relatively) small region of space as shown in Fig. 1·9(d). This type of wave is often called a *wave packet* and it clearly has some resemblance to what we call a particle, in that it is localized.

So we can regard an electron or a light wave when it acts in the manner of what we call a 'particle', as being a wave packet. It is still useful at such times to use the language which applies to particles, just as a matter of convenience.

1·7 Wave nature of larger 'particles'

We have seen that to treat the electron as a 'particle' is an approximate way of describing a wave packet. The wavelike behaviour shows up only when the electron wavelength is comparable, for example, to the diameter of an atom or the distance between atoms. This is the case, too, with light; its wave nature shows up when it passes through a slit whose width is comparable to the wavelength of the light.

Now an atom, being matter like the electron, also has a characteristic wavelength given by the relationship $\lambda = h/p$. It will thus show wavelike properties when it interacts with something of the right size: roughly, equal to its wavelength.

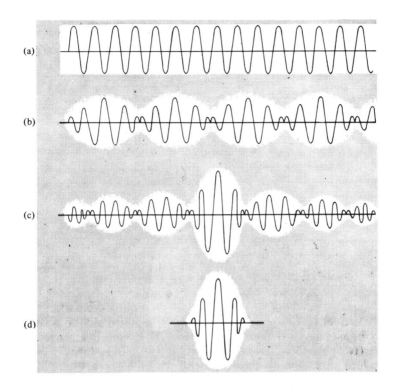

Fig. 1·9 Combination of waves of different frequency to form a wave packet

But the lightest atom is about 2,000 times heavier than the electron, so for the same velocity its momentum, p, is about 2,000 times greater and so the wavelength, λ, is correspondingly 2,000 times smaller than that of the electron. Thus its wavelike properties will almost never be observable since its wavelength is so small. The same applies, even more so, as we consider bigger and bigger 'particles'. So a dust 'particle' consisting of several million atoms virtually never displays wavelike behaviour and can always be treated as a particle. This means, in terms of physics, that Newton's laws will apply to a dust particle while they cannot be expected to apply, at any rate in the same form, to electrons. Thus we see that the difficulty in the Rutherford atom lies in treating the electron as a particle subject to Newtonian mechanics. It is necessary to develop a new mechanics–wave mechanics–to deal with the problem and this is the subject of the next chapter.

*1·8 The Thomson experiment

The following descriptions are not historically accurate but, rather, are intended to bring out the basic behaviour of electrons in the presence of electric and magnetic fields.

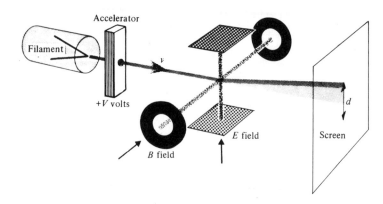

Fig. 1·10 Schematic diagram of Thomson's experiment

In the remainder of this chapter we deal in more detail with some of the ground covered earlier; these sections may be omitted on first reading.

Beginning with Thomson's type of experiment, suppose that we have the apparatus illustrated in Fig. 1·10. Electrons are emitted from a hot filament, are focused into a beam, and accelerated by a potential V to a velocity v, finally impinging on a fluorescent screen. The beam is arranged to pass between two plane parallel electrodes between which a voltage can be applied to produce a uniform electrostatic field, E volt/m, across the path of the beam. If the electrons are treated as charged particles they will be deflected by the field by virtue of the force F_ϵ newtons perpendicular to their path which it exerts on them. This force is given, from the definition of electric field, by

$$F_\epsilon = eE$$

and the resulting deflection of the beam moves its arrival point at the screen through a distance, d.

In the same position as the electrodes we also have a pair of magnetic coils which, when supplied with current, produce a uniform magnetic field of flux density B Wb/m² (or teslas) perpendicular to the path of the electron beam. Now the moving electrons, since they are charged, are equivalent to a current of value ev/l over a path of length, l. One of the fundamental laws of electromagnetism is that a wire carrying a current, i, in a uniform magnetic flux density, B, experiences a force perpendicular to both the direction of current and of the magnetic field given by Bil, where l is the length of the wire. In the present case, the electrons will therefore experience a force given by

$$F_H = Bev \qquad (1·13)$$

If now we arrange that the forces due to the electric and magnetic fields oppose each other and we adjust the fields so that the spot on the screen is returned to its original straight-through position we have

$$eE = Bev \qquad (1·14)$$

Now the velocity of the beam, as we have seen previously, is given by the relationship

$$eV = \tfrac{1}{2}mv^2 \qquad (1·15)$$

so that

$$v = \left(\frac{2eV}{m}\right)^{1/2}$$

Thus, substituting in Eqn (1·14),

$$E = B\left(\frac{2eV}{m}\right)^{1/2} \quad \text{or} \quad \frac{e}{m} = \frac{E^2}{2VB^2} \qquad (1·16)$$

As E, V, and B are known we have obtained a value of e/m for the electron.

By using different materials for the filament wire it can be established that the same value for e/m is always obtained, thus supporting the statement that the electron is a constituent of all matter.

*1·9   Millikan's experiment

Since the Thomson experiment is only able to yield the charge-to-mass ratio of the electron it is necessary to devise means for measuring either the charge or the mass separately. Thomson designed a suitable method, which was improved in Millikan's famous oil-drop experiment, which provided a value for the charge e. Both methods depend upon the speed with which a small drop of liquid falls in air under the influence of gravity. A particle falling freely under the influence of gravity will reach a velocity, v, after falling through a distance, h, given by $v^2 = 2gh$. However, if the particle is falling in air, the friction, or viscosity, of the air comes into play as the velocity increases and the particle finally reaches a constant terminal velocity. If the particle or drop is spherical with a radius a, this velocity is given by Stokes' law as

$$v = \frac{2ga^2\rho}{9\eta} \qquad (1·17)$$

where ρ is the density of the sphere and η the coefficient of viscosity of the air.

If the drops are sufficiently small, v may be only a fraction of a millimetre per second and may be measured by direct observation. It is then possible to deduce the radius, a, and hence the mass of the drop. This is the prerequisite for the oil-drop experiment.

A mist of oil is produced by spraying through a fine nozzle into a glass chamber. The drops may be observed through the side of the chamber by means of a suitable microscope with a graticule in the eyepiece. The free fall of a suitable drop is observed in order to calculate its mass as outlined above.

At the top and bottom of the chamber are plane parallel electrodes to which a potential can be applied producing a uniform electric field in the vertical direction. The drop of oil is charged by irradiating it with ultra-violet light which causes the loss of an electron so that the drop acquires a positive

charge, $+e$. The potential is adjusted so that the drop appears to be stationary when the upward force due to the field (top electrode negative) just
balances the downward force due to the weight of the drop. Thus,

$$\frac{eV}{d} = mg \qquad (1\cdot18)$$

where V is the potential, and d the distance between the electrodes, m is the
mass of the drop, and g the acceleration due to gravity.

The charging of the drop by exciting an electron out of it will be better
understood when we have dealt with atomic structure in more detail. It is not
possible in this experiment to be sure that the drop has not lost two, three, or
even more electrons. However, by observing a large number of drops the
charge always works out to be an integral multiple of $1\cdot6 \times 10^{-19}$ coulomb.
Thus the *smallest* charge the drop can acquire has this value, which we therefore assume to be the charge carried by a single electron. Since there is no *a
priori* proof that the drop may charge only by loss of an electron we are, in
fact, somewhat jumping to conclusions. However, as will be seen later from
the detailed atomic model, the only conceivable explanation is precisely the
one proposed.

*1·10   Wave analysis

Earlier in the chapter we referred to the fact that combining sinusoidal wave-
forms of different frequencies produces a wave whose amplitude, as a function of position, differs from the amplitudes of the two constituent waves. By
use of simple trigonometrical formulae it is possible to demonstrate this.

Mathematically, a plane wave whose amplitude varies with time, t, can be
represented by the expression

$$y = A \sin \omega t \qquad (1\cdot19)$$

where y is the amplitude at any instant of time, A is the maximum amplitude,
and ω is 2π times the frequency.

If the wave is travelling in a given direction at a velocity c cm/s, then it will
take a time x/c s to reach a point x cm from its point of origin, i.e., the variation of amplitude at the point x will lag, in time, behind the variation at the
origin. Thus the wave will then be represented by

$$y = A \sin \left[2\pi f \left(t - \frac{x}{c} \right) \right]$$

or

$$y = A \sin \left[2\pi \left(ft - \frac{x}{\lambda} \right) \right]$$

since $c = f\lambda$; that is

$$y = A \sin \left(\omega t - \frac{2\pi x}{\lambda} \right) \qquad (1\cdot20)$$

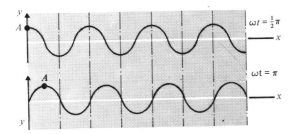

Fig. 1·11 A travelling wave

Using the formula for $\sin(A - B)$ we may write this as

$$y = A\left(\sin \omega t \cos \frac{2\pi x}{\lambda} - \cos \omega t \sin \frac{2\pi x}{\lambda}\right)$$

If we take $\omega t = \pi/2$, then $y = \cos 2\pi x/\lambda$, while, at some subsequent time such that $\omega t = \pi$, we have $y = \sin 2\pi x/\lambda$. These are illustrated in Fig. 1·11 and it is easy to see that the equation describes a wave travelling steadily through space.

If we were to combine two waves of the same frequency and amplitude travelling in opposite directions, one in the $+x$ and the other in the $-x$ directions, the individual waves would be given by

$$y_1 = A \sin\left(\omega t - \frac{2\pi x}{\lambda}\right) \quad \text{and} \quad y_2 = A \sin\left(\omega t + \frac{2\pi x}{\lambda}\right)$$

so that using the trigonometrical formulae

$$y = y_1 + y_2 = 2A \sin \omega t \cos \frac{2\pi x}{\lambda} \tag{1·21}$$

In this case the variation of the wave with distance, x, is the same for all times, t, being represented by $\cos 2\pi x/\lambda$ in every case. This is shown in Fig. 1·12 and is called a standing wave. In it there are fixed positions in space, distance $\lambda/2$ apart, where the amplitude of the wave is always zero (nodes) and positions where the wave varies, with time, between zero and a maximum value $2A$ (antinodes).

When we wish to deduce the shape of a combination of travelling waves of different frequencies we may do so by representing them by trigonometric functions having differing values of ω, λ, and amplitude A and adding them

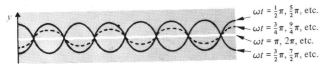

$\omega t = \frac{1}{2}\pi, \frac{5}{2}\pi,$ etc.
$\omega t = \frac{3}{4}\pi, \frac{9}{4}\pi,$ etc.
$\omega t = \pi, 2\pi,$ etc.
$\omega t = \frac{3}{2}\pi, \frac{7}{2}\pi,$ etc.

Fig. 1·12 A standing wave

up. In principle this can be done by using the trigonometric formulae to simplify the expressions and plotting a graph of the resulting equation. However, this rapidly becomes clumsy, algebraically, and the simplest method for the more straightforward cases is to draw graphs representing each of the component waves and add them up, point by point. This is, in effect, what has been done in Fig. 1·9.

There is a mathematical technique for simplifying the algebraic complexities which is known as Fourier analysis. This is based on Fourier's theorem which states that any single-valued periodic function, $y = f(t)$, having a period 2π, may be expressed in the form

$$y = C + A_1 \sin(\omega t + \phi_1) + A_2 \sin(2\omega t + \phi_2)$$
$$+ A_3 \sin(3\omega t + \phi_3) + \dots \qquad (1\cdot22)$$

Since

$$A \sin(\omega t + \phi) = A \sin \omega t \cos \phi + A \cos \omega t \sin \phi$$
$$= a \sin \omega t + b \cos \omega t$$

where $a = A \cos \phi$ and $b = A \sin \phi$, this expression may be written as

$$y = c + a_1 \sin \omega t + a_2 \sin 2\omega t + a_3 \sin 3\omega t + \dots$$
$$+ b \cos \omega t + b_2 \cos 2\omega t + b_3 \cos 3\omega t + \dots$$

that is

$$y = c + \sum_n a_n \sin n\omega t + \sum_n b_n \cos(n\omega t) \qquad (1\cdot23)$$

In order to use Eqn (1·23) to analyse a complex wave it is necessary to determine the coefficients c, a_1, a_2, a_3, $\dots$ and b_1, b_2, b_3, $\dots$ This is done by multiplying both sides of the equation by a suitable factor and integrating between limits. If the multiplying factor is correctly chosen all terms vanish except those which will give the required coefficient.

The details of this are more properly the province of a mathematical textbook. However, we give here the results, together with an example of their application.

$$c = \frac{1}{2\pi} \int_0^{2\pi} y \, d(\omega t) \qquad (1\cdot24)$$

$$a_n = \frac{1}{\pi} \int_0^{2\pi} y \sin n\omega t \, d(\omega t) \qquad (1\cdot25)$$

$$b_n = \frac{1}{\pi} \int_0^{2\pi} y \cos n\omega t \, d(\omega t) \qquad (1\cdot26)$$

where n is a positive integer.

It should be realized that for a travelling wave a graph of amplitude against time for a fixed distance will be the same as that for amplitude against distance for a fixed time. In the first instance we are standing at a fixed position and plotting the wave as it goes past and in the second case we take a look at

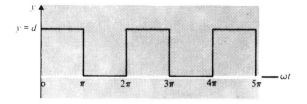

Fig. 1·13 A periodic square wave

the wave over a length of space at a fixed time; in each case the pattern is the same.

As an example we consider the series of pulses shown in Fig. 1·13 in which amplitude is plotted as a function of ωt. From $\omega t = 0$ to $\omega t = \pi$ the equation for the curve is $y = d$; from $\omega t = \pi$ to $\omega t = 2\pi$ the equation is $y = 0$ and the wave has a period 2π.

Using Eqn (1·24), we write

$$c = \frac{1}{2\pi} \int_0^{2\pi} y\, d(\omega t) = \frac{1}{2\pi} \int_0^{\pi} y\, d(\omega t) + \frac{1}{2\pi} \int_\pi^{2\pi} y\, d(\omega t)$$

$$= \frac{1}{2\pi} \int_0^{\pi} d \cdot d(\omega t) + \frac{1}{2\pi} \int_\pi^{2\pi} 0 \cdot d(\omega t)$$

$$= \frac{1}{2\pi} \cdot d\pi = \frac{d}{2}$$

This is the mean level of the wave.

Equation (1·25) gives

$$a_n = \frac{1}{\pi} \left[\int_0^{\pi} d \cdot \sin(n\omega t)\, d(\omega t) + \int_\pi^{2\pi} 0 \cdot \sin(n\omega t)\, d(\omega t) \right]$$

$$= \frac{1}{\pi} \left[-\frac{d \cos n\omega t}{n} \right]_0^{\pi}$$

$$= \frac{d}{n\pi} (1 - \cos n\pi)$$

where n is odd the term $(1 - \cos n\pi) = 2$ and when n is even then $(1 - \cos n\pi) = 0$. Thus

$$a_1 = \frac{2d}{\pi} \qquad a_3 = \frac{2d}{3\pi}, \text{ etc.}$$

and

$$a_2 = a_4 = \cdots = 0$$

From Eqn (1·26) we obtain

$$b_n = \frac{1}{\pi} \left[\int_0^{\pi} d \cos(n\omega t)\, d(\omega t) + \int_\pi^{2\pi} \cos(n\omega t)\, d(\omega t) \right]$$

$$= \frac{1}{\pi} \int \frac{d \sin n\omega t}{n} \Big|_0^\pi$$

$$= 0$$

Thus all the cosine terms are zero and the wave is represented by the expression

$$y = \frac{d}{2} + \frac{2d}{\pi}\left(\sin \omega t + \frac{1}{3} \sin 3\omega t + \frac{1}{5} \sin 5\omega t + \ldots\right) \qquad (1\cdot27)$$

Problems

1·1 What reasons are there for believing that matter consists of atoms?

1·2 Outline two experiments, one showing the electron to be a particle, the other showing it to be a wave.

1·3 The proton is a particle with a positive charge equal in magnitude to that of the electron and with a mass 1,840 times that of the electron. Calculate the energy, in electron volts, of a proton with a wavelength of $0\cdot01$ Å (1Å $= 10^{-10}$ m).

1·4 A dust particle of mass 1 μg travels in outer space at a velocity 25,000 mile/h. Calculate its wavelength.

1·5 In an electron diffraction experiment the distance between the carbon film and the screen is 10 cm. The radius of the innermost diffraction ring is $0\cdot35$ cm and the accelerating voltage for the electron beam is 10,000 V. Calculate the spacing between the atomic planes of carbon.

1·6 X-rays are produced when high-energy electrons are suddenly brought to rest by collision with a solid. If all the energy of each electron is transferred to a photon, calculate the wavelength of the X-rays when 25 keV electrons are used.

1·7 A static charge of $0\cdot1$ coulomb is placed by friction on a spherical piece of ebonite of radius 20 cm. How many electronic charges are there per square centimetre of the surface?

1·8 A parallel beam of light of wavelength $0\cdot5$ μm has an intensity of 10^{-4} W/cm^2 and is reflected at normal incidence from a mirror. Calculate how many photons are incident on a square centimetre of the mirror in each second.

*1·9 Obtain the Fourier coefficients c, a_n, and b_n, for a periodic wave whose equation is $y = \alpha t$ from 0 to π and $y = 0$ from π to 2π, where α is a constant.

Self-Assessment questions

1 The electron is a constituent of all materials.

   A) true   B) false.

2 The ratio of charge to mass for an electron is

   A) $1\cdot759 \times 10^{-11}$ C/kg   B) $1\cdot759$ C/kg   C) $1\cdot759 \times 10^{11}$ C/kg.

3 The mol or gramme-molecule is

A) the molecular weight of a substance in grammes
B) the amount of a substance whose mass is numerically equal to its molecular weight
C) the amount of a substance whose volume is equal to that of 1 gramme of hydrogen gas at s.t.p.

4 Avogadro's number is

A) the number of atoms in a gramme-atom
B) the number of molecules in a gramme-molecule
C) both (A) and (B).

5 What is the mass of a helium atom whose atomic weight is 4·003?

A) $1·672 \times 10^{-26}$ B) $6·64 \times 10^{-24}$ C) $2·41 \times 10^{24}$.

6 Rutherford's scattering experiment showed that

A) the nuclear charge is proportional to atomic number
B) electrons are small compared with the atom
C) the nucleus is small compared with the atom.

7 We know that an electron shows wavelike behaviour because

A) it is diffracted by a crystal
B) it is scattered by collision with a lattice of atoms
C) it can penetrate solid objects.

8 The distance between slits in a diffraction grating must be

A) much greater than
B) comparable to
C) much less than

the wavelength of light.

9 An electron volt is

A) the voltage required to accelerate an electron to a velocity of 1 m/s
B) the kinetic energy acquired by an electron in falling through a p.d. of 1 volt
C) the voltage required to give an electron 1 joule of energy.

10 A pattern of concentric rings on a fluorescent screen, produced by projecting a beam of electrons through a thin carbon film is evidence for the wave nature of electrons.

A) true   B) false.

11 The diameters of the rings described in the previous question are

A) inversely proportional to the electron velocity
B) inversely proportional to the square root of the electron velocity
C) proportional to the electron velocity.

12 The kinetic energy of an electron accelerated from rest to a potential of 2 kilovolts will be

A) 2,000 J   B) $3·2 \times 10^{-16}$ J   C) $4·1 \times 10^{-10}$ J.

13 What will be its velocity?

A) $2·65 \times 10^7$ m/s   B) 3×10^{10} m/s   C) $6·63 \times 10^{16}$ m/s.

14 De Broglie's relationship gives the wavelength of a hydrogen atom moving with a velocity of 10^3 m/s as

A) 1.46×10^{-9} mm   B) 3.96×10^{-10} m   C) 7.27×10^{-7} m.

15 In photo-emission of electrons the *number* of electrons emitted depends on

A) intensity   B) wavelength   C) velocity of the light.

16 In photo-emission of electrons the *energies* of electrons emitted depend on the

A) intensity   B) wavelength   C) velocity of the light.

17 The velocity of light is 3×10^{10} m/s. The photon energy of light of 6×10^{-7} m wavelength is

A) 3.97×10^{-40} J   B) 3.31×10^{-17} J   C) 1.99×10^{-23} J.

18 A wave packet is the result of adding together a large number of waves of the same frequency but different amplitudes.

A) true   B) false.

19 Newton's laws do not apply without modification to a wave packet.

A) true   B) false.

20 A wave packet has no mass.

A) true   B) false.

21 A wave packet cannot carry momentum.

A) true   B) false.

22 A wave packet cannot be formed by a wave of a single frequency.

A) true   B) false.

Answers

1 A	**2** C	**3** A	**4** C
5 B	**6** C	**7** A	**8** B
9 B	**10** A	**11** A	**12** B
13 A	**14** B	**15** A	**16** B
17 B	**18** B	**19** A	**20** A
21 B	**22** A		

WAVE MECHANICS

2·1 Introduction

In the previous chapter we saw that the properties of the electron as a particle (which is localized) and a wave (which extends over a region of space) may be brought together by treating it as a wave packet. We now require to develop a mathematical means of representing it in order that we can perform similar calculations to those for which we use Newtonian mechanics in the case of a particle.

2·2 Matter, waves, and probability

A plane wave travelling in the x direction and having a wavelength, λ, and frequency, f, can be represented by expressions of the type

$$\left. \begin{array}{l} \Psi = A \sin \left(\omega t - \dfrac{2\pi x}{\lambda} \right) \\[3mm] \Psi = A \cos \left(\omega t - \dfrac{2\pi x}{\lambda} \right) \end{array} \right\} \tag{2·1}$$

where t = time and $\omega = 2\pi f$. The difference between the sine and cosine functions is simply a shift along the horizontal axis, that is, a difference in phase. Thus we may plot the wave, at a given point in space described by $x = 0$, as shown in Fig. 2·1(a). This corresponds to standing at a particular point and plotting the variation in amplitude of the wave with time as it passes. Alternatively, we may choose to look at a length of space at a given time, described for example by $t = 0$, and plot the variation of amplitude of the wave with position along the length chosen, as in Fig. 2·1(b).

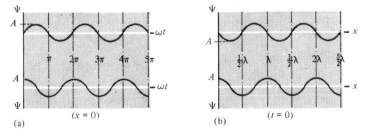

Fig. 2·1 Wave function Ψ as a function of (a) ωt for $x = 0$ and (b) position x for $\omega t = 0$

If we describe an electron by such an equation an immediate difficulty arises, namely, what does the amplitude of the wave represent? In all our experiments we can detect only that an electron is present or is not present: we can never detect only a part of it. Thus the electron must be represented by the whole wave and we must somehow relate the point-to-point variation of amplitude to this fact. It was the physicist Born who suggested that the wave should represent the probability of finding the electron at a given point in space and time, that is, Ψ is related to this probability. But Ψ itself cannot actually equal this probability since the sine and cosine functions can be negative as well as positive and there can be no such thing as a negative probability. Thus we use the intensity of the wave, which is the square of its amplitude, to define the probability. However, at this point our ideas and our mathematical representation are out of step. We can see that Ψ has no 'real' meaning and yet Eqn (2·1) defines Ψ as a mathematically real quantity. In general, Ψ will be a mathematically complex quantity with real and imaginary parts, that is, we can represent Ψ by a complex number of the type $(A + jB)$ where $j = \sqrt{-1}$.

Now the square of the modulus of a complex number is always a real number and so we can use this to represent the probability defined above. Thus we will have the wave function Ψ in the form

$$\Psi = (A + jB)$$

and the probability given by

$$|\Psi|^2 = (A^2 + B^2)$$

Traditionally–and in most physics textbooks–this is put in a different way which, nevertheless, is precisely the same thing. If $\Psi = (A + jB)$ its complex conjugate is $\Psi^* = (A - jB)$. The probability of finding an electron which is represented by the wave function Ψ at a given point in space and time is defined by the product of Ψ and its complex conjugate Ψ^*, that is:

$$
\begin{aligned}
\text{Probability} &= \Psi\Psi^* \\
&= (A + jB)(A - jB) \\
&= (A^2 + B^2) \\
&= |\Psi|^2
\end{aligned}
$$

It remains to choose a mathematical expression which will represent a complex wave. We can combine a real wave

$$\Psi' = A \cos\left(\omega t + \frac{2\pi x}{\lambda}\right)$$

with an imaginary one

$$\Psi'' = jA \sin\left(\omega t + \frac{2\pi x}{\lambda}\right)$$

to give

$$\Psi = \Psi' + \Psi'' = A\left[\cos\left(\omega t + \frac{2\pi x}{\lambda}\right) + j\sin\left(\omega t + \frac{2\pi x}{\lambda}\right)\right]$$

that is

$$\Psi = A \exp\left[j(\omega t + kx)\right] \tag{2·2}$$

by de Moivre's theorem, where $k = 2\pi/\lambda$.

We can separate the space-dependent and time-dependent parts of Eqn (2·2) by writing it as the product of two exponentials, that is:

$$\Psi = A \exp(j\omega t) \exp(jkx) \quad \text{or} \quad \Psi = \psi \exp(j\omega t) \tag{2·3}$$

where $\psi = A \exp(jkx)$ and represents the space variation of the wave. Equation (2·3) is a general expression for a complex wave in which ψ may be any function of x, not necessarily of the form $\exp(jkx)$.

Considering the probability, we have,

$$
\begin{aligned}
\Psi\Psi^* &= \psi\psi^* \exp(j\omega t) \exp(-j\omega t) \\
&= \psi\psi^* \exp(j\omega t - j\omega t) \\
&= \psi\psi^* \exp(0) \\
&= \psi\psi^*
\end{aligned}
$$

Thus if we can find ψ, and hence ψ^*, we can determine the most probable position for the electron in a given region of space. Since the electron carries a charge the function $\psi\psi^*$ gives a measure of the distribution of charge over the same region. These are the most useful applications of the theory and in the majority of cases we can forget about the time-variation term.

2·3 Wave vector, momentum, and energy

The electron has now been described as a wave by means of a mathematical expression. It must be realized that the expression developed describes a wave moving in one direction (the x direction) only. It is perfectly possible to have a wave whose amplitude and wavelength are different in different directions at the same time. For example, we can have a two-dimensional wave as shown in Fig. 2·2. In this the variation of amplitude with distance and the wavelengths are different as one proceeds along the x and along the z directions, and thus the mathematical expression for the amplitude must be a function of x and z

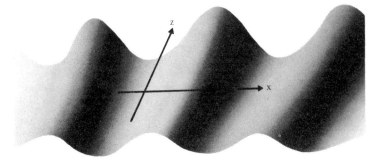

Fig. 2·2 Two-dimensional wave

as well as of time. Equally, although it is almost impossible to draw, we could clearly have a three-dimensional wave described by a mathematical expression involving three space coordinates. However, even in three dimensions, there will always be a resultant direction of travel for the wave along which its wavelength can be measured.

When we describe a particle in motion we must specify not only its speed but also the direction in which it is moving and this is done by treating its velocity as a vector, **v**. A vector is defined as a quantity having magnitude and direction and the rules for vector summation, multiplication, and so on have been set up in vector algebra.

The student is referred to mathematical textbooks for the details of vector algebra. It is sufficient for our purposes here to remember that a quantity printed in bold type, for example **v**, is a vector. This will have components of v_x, v_y, and v_z in a cartesian coordinate system as shown in Fig. 2·3. The length of the vector is called its scalar magnitude and is printed in ordinary type thus, v, or in italic thus, v. The relationship between the scalar magnitude of the components in a cartesian coordinate system is always given by

$$v = (v_x^2 + v_y^2 + v_z^2)^{1/2}$$

Although the electron is not a particle its momentum has a direction and this is the direction in which the wave itself is travelling. Now the momentum is directly proportional to $1/\lambda$, the inverse wavelength, and its direction is that along which the wavelength is measured. So we can define a vector, **k**, which has magnitude $2\pi/\lambda$ and the direction defined above and which we call the wave vector. This is, of course, related to the velocity, c, of the wave through the formula, $\lambda = c/f$, but is generally a more convenient quantity to use.

In particular, we can relate the wave vector directly to the momentum, p, carried by the wave through the de Broglie relationship, $p = h/\lambda$. Now, for a particle, $p = mv$ and since v is a vector, p must also be a vector. Similarly with the wave: since $k = 2\pi/\lambda$ we have

$$\mathbf{p} = \frac{h}{2\pi} \cdot \mathbf{k} \qquad\qquad (2·4)$$

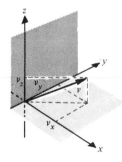

Fig. 2·3 Components of the velocity vector v

This means that the direction in which momentum is carried by the wave is given by the direction of the wave vector.

In many of our calculations the actual direction of travel of the wave may not be important and k is simply used as an ordinary scalar quantity. This will be the case when the wave is assumed to be a plane wave travelling in a single direction: however, in two-dimensional or three-dimensional problems the vector nature of k must be remembered.

Energy is, of course, a scalar quantity since it can never be said to have direction. The kinetic energy of a particle is given by $\frac{1}{2}mv^2 = p^2/2m$ where the total velocity or momentum magnitude is used without regard for its direction. Similarly for a wave, the kinetic energy, E_k, associated with it is given by

$$E_k = \frac{p^2}{2m} = \frac{h^2 k^2}{8\pi^2 m}$$ (2·5)

[It should be noted that $h/2\pi$ is written conventionally as $\hbar$, (called 'h-bar') and this is often used in textbooks. We then have $p = \hbar k$ and $E_k = \hbar^2 k^2/2m$, but this notation is not used here.]

2·4   Potential energy

By means of the wave vector we can describe the direction of travel of a wave and the energy which it carries. However, we must also be able to describe the environment in which it moves because the electron will, in general, interact with its environment. Since the electron carries a charge a convenient method of doing this is to ascribe to the electron a potential energy, E_p, which may be a function of position. Thus, for example, if the electron, which is negatively charged, is in the vicinity of a positive charge it will have a potential energy in accordance with Coulomb's law of electrostatics. This is proved in textbooks covering elementary electrostatics: it is sufficient for our purposes to define the potential energy of an electron as the work done in bringing an electron from infinity to a point distant, r, from a positive charge of strength, $+e$. This potential energy is given by

$$E_p = \frac{-e^2}{4\pi\epsilon_0 r}$$ (2·6)

where ϵ_0 is the permittivity of free space.

There will, of course, be other ways in which E_p could vary with position apart from the inverse variation of Eqn (2·2): for example, an electron wave moving through a solid such as a metal encounters positively charged atoms (ions) at regular intervals and in such a case the potential energy is a periodic function of position.

To take a simple mechanical analogy, suppose we have a mass M (kg), at a height h (m) above the floor, as illustrated in Fig. 2·4, and we define the potential energy as zero at floor level. The potential energy of the mass at the initial position will be given by the work done in raising it to that position from the floor, that is, by force × distance.

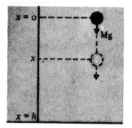

Fig. 2·4 Illustrating potential energy

If the mass is at rest in this position its kinetic energy is zero and its total energy is equal to its potential energy. When it is released it falls, converting its potential energy to kinetic energy so that, at some intermediate position, x, its total energy, E, is given by

$$E = E_k + E_p$$
$$= \tfrac{1}{2}mv_x^2 + Mg(h - x)$$

where v_x is the velocity attained at the point x. When it reaches the floor all its energy is kinetic energy and the potential energy is zero. Thus we see that the potential energy describes the position of the mass, that is, its environment.

The total energy, E, of an electron, to include the effect of its environment, must thus be expressed as

$$E = E_k + E_p \tag{2·7}$$

From Eqn (2·5), rearranging

$$k = \frac{2\pi}{h} \ \sqrt{(2mE_k)}$$

and substituting in Eqn (2·7) we have

$$k = \frac{2\pi}{h} \ \sqrt{[2m(E - E_p)]} \tag{2·8}$$

Thus we see that the inclusion of potential energy will alter the magnitude of the wave vector $\mathbf{k}$ when E_p is other than zero.

2·5   Wave equation

All these ideas are brought together in a general equation describing the wave-function, ψ, which is known as Schrödinger's wave equation.

In Eqn (2·3) we saw that the space variation of the wave-function could be expressed as

$$\psi = A \exp{(jkx)} \tag{2·9}$$

Differentiating yields

$$\frac{d\psi}{dx} = jk\, A \exp{(jkx)}$$

and differentiating again we have

$$\frac{d^2\psi}{dx^2} = -k^2\, A \exp{(jkx)} = -k^2\psi$$

Substituting for k^2 from Eqn (2·8)

$$\frac{d^2\psi}{dx^2} + \frac{8\pi^2 m}{h^2}\,(E - E_p)\psi = 0 \qquad (2\cdot10)$$

and this is the wave equation for a wave moving in the x direction. Its solution will yield information about the position of the electron in an environment described by the potential energy, E_p, which itself may be a function of position, x. Equation (2·10) has been 'derived' from the equation for a 'plane' wave (Eqn 2·9) but it does not follow that all wave-shapes will satisfy the same equation. In reality, Eqn (2·10) cannot be derived: it was put forward by Schrödinger as an *assumption*, to be tested by its ability to predict the behaviour of real electrons. A simple example involving solution of the equation for specific conditions is given at the end of the chapter. Other examples are included in the problems. In three dimensions, described by the cartesian coordinates x, y, and z, the wave equation becomes a partial differential equation.

$$\frac{\partial^2\psi}{\partial x^2} + \frac{\partial^2\psi}{\partial y^2} + \frac{\partial^2\psi}{\partial z^2} + \frac{8\pi^2 m}{h^2}\,[E - E_p(x, y, z)]\,\psi = 0 \qquad (2\cdot11)$$

The solution for ψ will be a function of x, y, and z as also is the potential energy, E_p, and the equation is the time-independent Schrödinger equation. Its solution gives the space-dependent part, ψ, of the wave function, Ψ, and is to be interpreted that the value of $|\psi|^2$ at any point is a measure of the probability of observing an electron at that point. The whole solution can therefore be regarded as a shimmering distribution of charge density of a certain shape. This nebulous solution is in contrast to the sharp solutions of classical mechanics, the definite predictions of the latter being replaced by probabilities.

2·6 Electron in a 'box'

By way of an example of solving Schrödinger's equation for a given electron environment, we consider the case in which the electron is confined to a specific region of space. In three dimensions this would mean confining it in a 'box' but we shall treat the problem as one-dimensional, with the electron confined to a line of length L. This situation can be represented by making the potential energy, V, of the electron zero within the length L and infinite outside it, as illustrated in Fig. 2·5. Since the electron can never acquire infinite

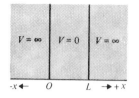

Fig. 2·5 Potential energy defining a 'box'

energy it is not able to exist where $V = \infty$, so we expect to find the solution $\psi = 0$ to the wave equation for this case. Mathematically, we put $V_p = \infty$ for $x < 0$ and $x > L$ (while $V = 0$ for $0 < x < L$). The one-dimensional wave equation is

$$\frac{d^2\psi}{dx^2} + \frac{8\pi^2 m}{h^2} (E - V)\psi = 0 \tag{2.12}$$

When V is infinity, all terms in this equation are negligible compared to the term $- V\psi$ and we can therefore straight away write

$$- V\psi = 0$$

which, since $V = \infty$, can only be true if $\psi = 0$. Thus the condition $V = \infty$ outside the length L automatically ensures that $\psi = 0$, as expected, and the probability of the electron being there is exactly zero.

Inside the length L we do not expect ψ to be zero. But at the points $x = 0$ and $x = L$ a finite value for ψ is not possible, for ψ is zero *outside L* and it cannot be discontinuous at these points. A discontinuous function has an infinite gradient, so the value of $d^2\psi/dx^2$ would be indeterminate at $x = 0$ and $x = L$, and Schrödinger's equation could not be satisfied there. We therefore conclude that $\psi = 0$ at $x = 0$ and $x = L$, and these values give the appropriate *boundary conditions* for the solution of Schrödinger's equation in the distance L.

In the next section it is shown that the solution in this region is in fact

$$\psi = j \left(\frac{2}{L} \right)^{1/2} \sin \left(\frac{n\pi x}{L} \right) \tag{2.13}$$

where n is an arbitrary integer. It is easy to show by substitution that when $x = 0$ or L then $\psi = 0$ as required. This solution is plotted in Fig. 2·6 for the three cases $n = 1, 2$, and 3, together with the function $|\psi|^2$ which is the probability distribution.

The electron energy E is also derived in Section 2·7 and is given by

$$E = \frac{h^2 n^2}{8mL^2} \tag{2.14}$$

It therefore increases with n^2. Thus, when $n = 2$ the electron has four times the energy it has when $n = 1$ and so on; the electron energy can have only certain specific values, and it is thus quantized. It will be seen from the diagram

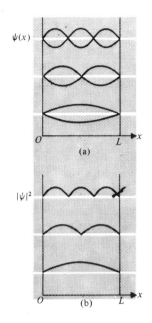

Fig. 2·6 Solutions for an 'electron in a box': (a) the wave function, and (b) the probability as a function of position for different quantum numbers

that the wavelength of the electron for the case $n = 1$ is such that $L = \lambda/2$, while for $n = 2$, $\lambda = L$ and for $n = 3$, $L = \frac{3}{2}\lambda$ and so on. This is in agreement with our picture of the electron wavelength decreasing as its kinetic energy increases [Eqn (2·8)].

The solution in Eqn (2·13), illustrated in Fig. 2·6(a), is exactly like the case of waves on a stretched string–Meldé's famous experiment. The situation of the electron is almost entirely analogous to that of a disturbance travelling along a stretched string. When the electron wave reaches a boundary at $x = L$ it is totally reflected and travels back to $x = 0$ where it is again reflected. The result is a standing electron wave with a probability distribution which is stationary in space. The analogy with a stretched string will prove very useful in Chapter 3.

Figure 2·6 shows clearly why the electron energy is quantized: there must always be an integral number of half wavelengths in the length L, so the wavelength is constrained to one of a particular set of values, and with it the energy. It is this feature which makes the results of wave mechanics quite different from those of classical mechanics.

The number n, which determines both the number of half wavelengths making up the length L and the energy, is called a *quantum number*. In subsequent chapters we shall see that in practice there is more than one such number for each electron and that these quantum numbers have great significance in the explanation of the properties of atoms.

*2·7   Solution of Schrödinger's equation

We return in this section to the problem of finding a solution of Schrödinger's equation inside the 'box' defined in the previous section. Since the potential energy is zero inside the region $0 < x < L$, Schrödinger's equation is simply

$$\frac{d^2\psi}{dx^2} + \frac{8\pi^2 m}{h^2} E\psi = 0 \qquad (2·15)$$

Since the potential energy of the electron is zero, the energy E in this equation is just the kinetic energy, E_k, which is given by Eqn (5·5). Rewriting that equation we have

$$k^2 = \frac{8\pi^2 mE}{h^2}$$

and substituting in Eqn (2·15)

$$\frac{d^2\psi}{dx^2} = -k^2\psi \qquad (2·16)$$

This is a standard form of differential equation the solution to which is described in standard mathematical textbooks and has the form

$$\psi = A \exp(jkx) + B \exp(-jkx) \qquad (2·17)$$

This may be tested by differentiation. To evaluate the constants A and B we must consider the boundary conditions of our particular problem. Since we require $\psi = 0$ for $x = 0$ we have, by substituting for ψ and x in Eqn (2·17),

$$0 = A \exp(0) + B \exp(-0)$$
$$= A + B$$

Thus $A = -B$ and Eqn (2·17) may be written as

$$\psi = A[\exp(jkx) - \exp(-jkx)]$$

But $\sin \theta = \dfrac{e^{j\theta} - e^{-j\theta}}{2j}$

so that

$$\psi = 2jA \sin kx \qquad (2·18)$$

Now we also require $\psi = 0$ at $x = L$, so that

$$0 = 2jA \sin kL$$

which can only be true if $kL = n\pi$, that is

$$k = \frac{n\pi}{L} \qquad (2·19)$$

where n is an integer; therefore

$$\psi = 2jA \sin \frac{n\pi x}{L} \tag{2·20}$$

The question now arises of finding a value for the constant, A, and this is done by what is called *normalization* of the solution. Since $|\psi|^2$ is to be interpreted as a probability and the maximum value of a probability is 1 by very definition of the term, we must arrange that the total sum of $|\psi|^2$ over all space is unity. Expressed mathematically this means that the integral of $|\psi|^2$ over all space equals 1, that is

$$\int_0^\infty |\psi|^2 d\tau = 1 \tag{2·21}$$

where $d\tau$ is an element of volume.

In the present problem, 'all space' is confined simply to the length L, so that Eqn (2·21) becomes

$$\int_0^L |\psi|^2 dx = 1$$

that is

$$\int_0^L 4A^2 \sin^2 \left(\frac{n\pi x}{L} \right) dx = 1$$

Thus,

$$4A^2 \int_0^L \frac{1}{2} \left[1 - \cos \left(\frac{2n\pi x}{L} \right) \right] dx = 1$$

$$4A^2 \left[\frac{x}{2} - \frac{L}{4n\pi} \sin \left(\frac{2n\pi x}{L} \right) \right]_0^L = 1$$

Therefore,

$$\frac{4A^2 L}{2} = 1 \quad \text{or} \quad A = \left(\frac{1}{2L} \right)^{1/2}$$

Substituting in Eqn (2·20) gives

$$\psi = j \left(\frac{2}{L} \right)^{1/2} \sin \left(\frac{n\pi x}{L} \right) \tag{2·22}$$

which is the full solution to the original equation. The energy E may now be found by using Eqns (2·5) and (2·19):

$$E = \frac{h^2 k^2}{8\pi^2 m} = \frac{h^2}{8\pi^2 m} \cdot \frac{n^2 \pi^2}{L^2} = \frac{n^2 h^2}{8mL^2} \tag{2·23}$$

Problems

2·1 Explain why it is necessary to use the *intensity* of a matter wave in order to provide a physical interpretation of the mathematical expression

$$\psi = A \exp [j(\omega t + kx)]$$

2·2 Calculate the magnitude of the wave vector of an electron moving with a velocity of 10^6 m/s.

2·3 If an electron having total energy of 10^{-21} J travels at a constant distance of 100 Å from a proton, what will be its velocity and its value of wave vector?

2·4 If an electron is in a region of constant potential V, show that one solution to the Schrödinger equation

$$\frac{d^2\psi}{dx^2} + \frac{8\pi^2 m}{h^2} (E - V)\psi = 0$$

is $\psi = Ae^{jkx}$, where A is a constant, if

$$k^2 = \frac{8\pi^2 m}{h^2} (E - \dot{V})$$

2·5 Show also that if the value of ψ at any point x is equal to the value of ψ at a point $(x + a)$, where a is a constant, then $k = 2n\pi/a$, where n is an integer.

2·6 Why is the function ψ = constant not a permissible solution to problem 2·4? (*Hint*: consider normalization.)

2·7 Assuming that at very large distances from the origin of axes a wave-function may be written in the form

$$\psi = \frac{C}{r^n}$$

where C is a constant and r is the distance from the origin, show that the wave is only a permissible solution if $n > 2$. (*Hint*: consider normalization.)

Self-Assessment questions

1 If the expression $\Psi = A \sin (\omega t - 2\pi x/\lambda)$ represents an electron as a wave, the quantity A is the amplitude of the electron wave at

A) any time t for a fixed position x
B) any position x at a fixed time t
C) a time and position such that

$$\left(\omega t - \frac{2\pi x}{\lambda}\right) = \frac{n\pi}{2}$$

where n is an integer.

2 The intensity of a wave at any time t and position x is given by

A) A^2 B) $|\Psi|^2$ C) the condition that $\left(\omega t - \dfrac{2\pi x}{\lambda}\right) = n\pi$.

3 The probability of finding an electron at a point x at a time t is given by

A) the amplitude of the wave   B) the intensity of the wave
C) the peak amplitude of the wave.

4 The charge density, due to an electron wave, at a point x at a time t is given by

A) eA^2 B) $e|\Psi|^2$ C) $e\Psi$

where e is the charge on an electron.

5 If $\Psi = (A + jB)$, then $\Psi\Psi^*$ is equal to

A) $(A + jB)^2$ B) $(A - jB)^2$ C) $(A^2 + B^2)$.

6 A complex wave, given by $\Psi = A \exp [j(\omega t + kx)]$, represents the product of space-dependent and time-dependent wave functions.

A) true   B) false.

7 De Broglie's relationship between momentum and wavelength for an electron is

A) $\lambda = \dfrac{h}{p}$ B) $p = h\lambda$ C) $p = \dfrac{\lambda}{h}$.

8 The magnitude of the wave vector is proportional to the magnitude of

A) the wavelength   B) the reciprocal of the wavelength
C) the reciprocal of the frequency of the wave.

9 The wave vector changes direction if the wave is deflected from its original path.

A) true   B) false.

10 The kinetic energy associated with a plane electron wave is given by

A) hk B) $\dfrac{1}{2}mk^2$ C) $\dfrac{h^2 k^2}{8\pi^2 m}$.

where h is Planck's constant.

11 The environment of an electron may be described by attributing a potential energy to the electron.

A) true   B) false.

12 The total energy of the electron is

A) the difference between its kinetic and potential energies
B) the sum of its kinetic and potential energies
C) the product of its kinetic and potential energies.

13 The wave vector is

A) independent of,
B) proportional to,
C) proportional to the square root of,

the total energy of the electron.

14 The total energy of an electron having potential energy depends on its position if the potential energy is

A) constant   B) zero   C) a function of position.

15 Schrödinger's equation relates the probability of an electron being at a particular point to its total energy when at that point.

A) true   B) false.

16 The solution of Schrödinger's equation allows the position of an electron to be fixed exactly.

A) true   B) false.

17 If the potential energy of an electron is zero, Schrödinger's equation does not apply.

A) true   B) false.

18 If the wavelength of an electron wave is infinite, the electron must be stationary.

A) true   B) false.

19 The energy of an electron in a deep potential well ('electron in a box') is quantized.

A) true   B) false.

20 The solutions for Schrödinger's equation for an electron in a deep potential well are standing waves.

A) true   B) false.

Answers

1 C	**2** B	**3** B	**4** B
5 C	**6** A	**7** A	**8** B
9 A	**10** C	**11** A	**12** B
13 C	**14** C	**15** A	**16** B
17 B	**18** A	**19** A	**20** A

THE SIMPLEST ATOM – HYDROGEN

3·1 Introduction

Now that we have formed a clear picture of how an electron behaves in a rather simple environment we are able to approach the more complicated problem of an electron bound to a single positively charged 'particle'. Such a combination forms a simple atom–indeed, an atom of hydrogen is exactly like this. In it a single electron moves in the vicinity of a particle called a *proton* which carries a positive charge equal in magnitude to that of the electron. Its mass, however, is 1,840 times the mass of the electron so that it behaves much more like a particle than does the electron and we can regard it as a fixed point charge around which travels the electron wave. The shape of this wave could be calculated by solving Schrödinger's equation (Eqn 2·11) with the appropriate expression for E_p. This is mathematically rather difficult so, for the present, we shall make use of the simple analogy noted at the end of Chapter 2 and treat the electron wave as if it were a real wave on, say, a length of string.

Now the electron wave clearly must be three-dimensional but in order to simplify the calculation let us see how far it is possible to go whilst limiting consideration to one dimension only. Later we shall discuss how the results thus obtained are modified by introducing the other two dimensions.

3·2 One-dimensional atomic model

The simplest model that we can take is illustrated in Fig. 3·1. The electron wave travels around a circular path whose centre is at the atomic nucleus, the proton. Using the vibrating-string analogy of Chapter 2 we can imagine a string stretched around in a circle and set in vibration. Although the path is curved the model is essentially one-dimensional since the amplitude varies only as we move along one direction, that is, the circumference of the circle.

Note that, unlike the case of a stretched string, the waves on it need not be standing waves since there are no ends at which reflections can occur. The wave thus runs around its circular path and in the corresponding electron wave this implies that the mass and charge are being carried around with it. Returning to the string analogy, and guided again by the earlier examples, we may now ask whether the wavelength is in any way restricted. The answer is yes, as can be seen from Fig. 3·2(a) where the wavelength has been arbitrarily chosen with the result that the string does not join up with itself! In electron-wave parlance, we say that the amplitude of the wave must be a smooth and

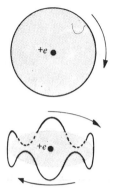

Fig. 3·1 A one-dimensional electron wave orbiting a proton

continuous function of position. Figure 3·2(b) shows a case where it is continuous but not smooth and intuitively it seems reasonable that this, too, would not be a stationary state of the system.

The requirement that the string must join up smoothly with itself forces us to the conclusion that there must be an *integral number of wavelengths* in the circumference of the circle. This result leads, not unexpectedly, to quantization of the electron energy as we shall now show.

Assume that there are *l* wavelengths in the circumference of the circle whose radius is *r*. Now *l* can have the values 1, 2, 3 . . . , ∞ (as we shall see later, the value zero is not allowed). This quantization condition may be stated mathematically thus:

$$l\lambda = 2\pi r \qquad\qquad\qquad (3·1)$$

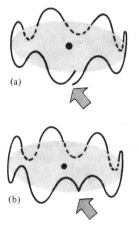

(a)

(b)

Fig. 3·2 Waves with discontinuous amplitude or slope are not permissible

From the wavelength we may obtain the electron momentum, p, using de Broglie's relationship, and combining with Eqn (3·1), we find

$$pr = \frac{lh}{2\pi} \tag{3·2}$$

The product of the linear momentum, p, with the radius, r, is the *moment of momentum* of the electron about the nucleus and is termed the *angular momentum*. We see that it can only have values which are integral multiples of $h/2\pi$, that is, it is quantized in units of $h/2\pi$.

From the momentum, p, it is easy to calculate the kinetic energy, E_k, of the electron as before:

$$E_k = \frac{p^2}{2m} = \frac{l^2h^2}{8\pi^2r^2m} \tag{3·3}$$

Unlike the examples in Chapter 2, the electron also has some potential energy since it moves in the potential distribution of the nuclear charge. This must be added to E_k to obtain the total energy. Now the potential due to the nuclear charge at a distance r is just

$$V = \frac{e}{4\pi\epsilon_0 r}$$

where ϵ_0 is the permittivity of free space.

So the potential energy, E_p, of the electron is given by

$$E_p = -eV = -\frac{e^2}{4\pi\epsilon_0 r} \tag{3·4}$$

In this equation we have used the convention that the electric potential is zero at infinity so that the potential energy of the electron is always negative.

The total energy, E, is obtained by adding Eqn (3·3) and Eqn (3·4) thus

$$E = \frac{l^2h^2}{8\pi^2r^2m} - \frac{e^2}{4\pi\epsilon_0 r} \tag{3·5}$$

From Eqn (3·5) it does not appear that the energy is quantized. However, we have so far neglected to enquire what is the value of r. We note that as r varies so does the energy but that there is a finite value of r for which the energy is least.

Now it is a universal law of nature that the energy of any system will always try to minimize itself in a given set of circumstances. For instance, a rolling ball will always come to rest at the lowest point of the hill–a position of minimum potential energy. Likewise, a ball which is rolling freely around inside a spherical saucer follows a circular path such that the *sum* of its kinetic energy, E_k and its potential energy, E_p, is a minimum (Fig. 3·3). The radius, R, of its path is thus determined by the condition

$$\frac{dE_k}{dR} + \frac{dE_p}{dR} = 0$$

Fig. 3·3 Mechanical model to illustrate energy minimization

This condition is exactly equivalent to the statement that the centrifugal force and the inward gravitational force just balance one another, that is, the net radial force on the ball is zero. This can be seen if we remember that energy (or work done) is equal to the integral of force with respect to distance R; conversely $F = dE/dR$. If the total force is to be zero, then $dE/dR = dE_k/dR + dE_p/dR$ must also be zero.

Now we may apply the same principle to the electron and minimize its energy [Eqn (3·5)] with respect to the radius, r:

$$\frac{dE}{dr} = -\frac{l^2h^2}{4\pi^2r^3m} + \frac{e^2}{4\pi\epsilon_0r^2} = 0$$

From this equation the equilibrium value of r may be determined,

$$r = \frac{\epsilon_0 l^2 h^2}{\pi e^2 m} \tag{3·6}$$

By substituting this value of r into Eqn (3·5) the energy under equilibrium conditions is obtained:

$$E = -\frac{e^2\pi m e^2}{4\pi\epsilon_0 l^2 h^2 \epsilon_0} + \frac{h^2 l^2 \pi^2 m^2 e^4}{8\pi^2 m h^4 \epsilon_0^2 l^4}$$

which simplifies to

$$E = -\frac{e^4 m}{8h^2 \epsilon_0^2} \cdot \frac{1}{l^2} \tag{3·7}$$

In exactly the same way as in the example in Chapter 2 we find that the energy is quantized and the permitted values for it are determined by putting $l = 1, 2, 3, \ldots$, in Eqn (3·7). Note that if l were allowed to be zero, the energy would be equal to minus infinity which is physically impossible. There is a useful diagrammatic way of representing *energy levels*, as the permitted values are called. Figure 3·4 shows how energy can be plotted vertically along an axis, and a horizontal line is placed at each value of energy which the electron is allowed to have.

3·3   Atom in two dimensions

Before looking too closely at the results of the previous section we should enquire whether it is to be modified when the restriction to one dimension is removed. Let us now consider the electron wave to be two-dimensional and

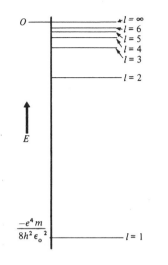

O ——
$l = \infty$
$l = 6$
$l = 5$
$l = 4$
$l = 3$

$l = 2$

E

$\dfrac{-e^4 m}{8h^2 \epsilon_o{}^2}$ —— $l = 1$

Fig. 3·4 Energy levels for the one-dimensional model of the hydrogen atom

see how the quantization condition, Eqn (3·1), is to be modified.

We have already seen in Chapter 2 that momentum is a vector quantity, that is, it has direction as well as magnitude, and that its direction in a two-dimensional wave is perpendicular to the wavefront. Since, in general, the wavefront will be curved, so will the path described by the normal to the wavefront. Many such paths may be drawn (Fig. 3·5), all crossing the wavefronts at right-angles and all indicating the direction of the momentum at each point.

The electron wave which encircles the nucleus of an atom may now be imagined in two-dimensional form. Let us concentrate our attention on any one of the many paths followed by the normal to the wavefront. This path

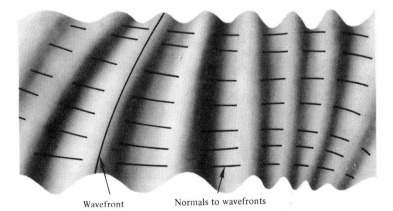

Wavefront   Normals to wavefronts

Fig. 3·5 Wavefronts and normals in two dimensions

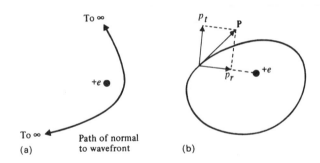

Fig. 3·6 The path followed by the wavefront must be closed, as shown in (b)

must form a closed loop, for, if it did not, the electron momentum would be carried away from the nucleus towards infinity and the wave would not represent an electron which is localized within the atom [Fig. 3·6(a)]. In Fig. 3·6(b) the electron is confined to the region of space around the nucleus, because its momentum follows a closed path. The illustration also shows the momentum, **p**, at one point, resolved into two components, a radial component, p_r, and a component, p_t, tangential to a circle, centre e.

The wavelength measured along this path must vary from point to point, but since the path is closed there must be an integral number of wavelengths along it, just as in the earlier, one-dimensional calculation. Since the wavelength varies we may express this mathematically by equating the average wavelength, $\bar{\lambda}$, to an integral sub-multiple of the path length, L, that is

$$n\bar{\lambda} = L \qquad n = 1, 2, 3, \ldots \infty$$

This, then, is our new quantization condition. To simplify the mathematics we may express L in the form $L = 2\pi\bar{r}$, where $\bar{r}$ is now an *average* radius, defined by this equation. Thus

$$n\bar{\lambda} = 2\pi\bar{r} \qquad\qquad\qquad (3·8)$$

is the condition which replaces Eqn (3·1).

If now the kinetic energy and potential energy are expressed in terms of the average values $\bar{r}$ and $\bar{\lambda}$, the total energy E is given by an expression exactly similar to Eqn (3·5).

$$E = \frac{n^2 h^2}{8\pi^2 \bar{r}^2 m} - \frac{e^2}{4\pi\epsilon_0 \bar{r}}$$

This may be minimized with respect to r exactly as before and the equilibrium energy obtained, now in terms of n instead of l:

$$E = -\frac{e^4 m}{8h^2 \epsilon_0^2} \cdot \frac{1}{n^2} \qquad\qquad\qquad (3·9)$$

As we shall shortly see, the introduction of a third dimension does not affect this result which therefore gives the energy of the electron in a hydrogen atom. The energy level diagram is exactly as in Fig. 3·4, but with l replaced by n.

The quantum number, l, is not, however, redundant. If we consider now a circular path through the wave motion, it, too, must contain an integral number of wavelengths. Indeed, this must be true of any closed path, whatever its shape, for the wave must always join up smoothly with itself on completion of a circuit. Thus we can retain the quantum number, l, which defines the number of wavelengths in any circular path and which in general is different from n, the number of wavelengths along a different path. (It is convenient to choose a circular path to define l as it simplifies discussion of the magnetic properties of the atom later on.) The quantum number, l, is related to the *tangential* component of the momentum, p_t, so that we obtain a result similar to that in Eqn (3·2):

$$p_t r = \frac{lh}{2\pi} \tag{3·10}$$

where r is the radius of the chosen circle. Equation (3·10) then shows that the angular momentum, $p_t r$, is independent of r and is quantized in units of $h/2\pi$. This equation then is an important auxiliary equation which gives further information about the shape of the wave function. It shows that there is more than one wave function with the same energy, for, if the quantum number, n, is fixed, different values of l are possible. But the range of permitted values of l is now restricted. We can see this by noting that the component, p_t, of the momentum must always be smaller than the magnitude, p, of the total momentum. This is another way of saying that the wavelength measured along the wavefront normal is the *minimum* value. Thus the path along which n is defined contains the maximum number of wavelengths, and so *l cannot be greater than n*. Moreover, l cannot equal n because in that case the radial component of momentum would be zero, and we have already seen that this is not possible.

We must conclude, then, that l must always have a value smaller than n. Note that l may take the value zero since this describes a wave with no nodes (that is, constant amplitude) around a circular path and there is nothing to prevent this. It just means that it is equally probable that the electron will be found at any point around the circumference of the circle. Observe also that when l equals zero the electron has no angular momentum–all the momentum is carried in the radial direction. This is not inconsistent with a closed path for the momentum since both the wave and the momentum flow simultaneously outward and inward along the same path, setting up a standing wave pattern.

3·4 Three-dimensional atom - magnetic quantum numbers

It might be expected that the introduction of the third dimension would involve quantization of the component of momentum perpendicular to the plane of the two-dimensional wave. This approach, however, is mathematically difficult and does not give us as much physical insight as the method we shall follow.

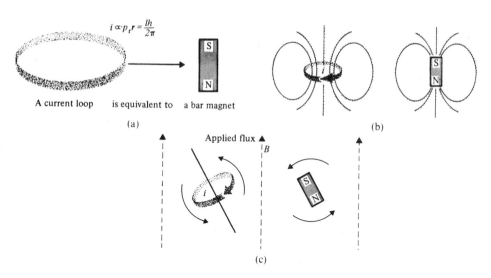

$$i \propto p_t, \quad r = \frac{lh}{2\pi}$$

A current loop   is equivalent to   a bar magnet

(a)   (b)

Applied flux ▲

(c)

Fig. 3·7 The equivalence of a current loop and a bar magnet

The electron wave possesses angular momentum about the nucleus which implies that the electron mass is circulating with the wave. The electron charge is intimately associated with its mass, so it, too, must be circulating as if there were an electric current flowing in a loop around the nucleus (Fig. 3·7). Indeed, the equivalent current is directly proportional to the angular momentum.

Such a current loop behaves exactly like a small permanent magnet. It has the same distribution of magnetic flux around it [Fig. 3·7(b)] and behaves in a similar way when in a magnetic flux applied from an external source. Thus the two-dimensional electron wave experiences a couple on it when a

Path followed by tip of spindle

Anticlockwise couple due
to weight of gyroscope and
reaction of support.

Fig. 3·8 The motion of a gyroscope under a gravitational couple is called 'precession'

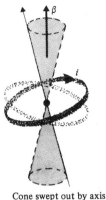

Cone swept out by axis
normal to plane of 2D wave

Fig. 3·9 The electron wave (represented here as a current loop) precesses around the field direction

magnetic flux, **B**, is applied as shown [Fig. 3·7(c)]. The couple acts in such a way as to try to turn the plane of the wave perpendicular to the flux lines. The wave is not free to so turn, however, because its circulating mass makes it behave like a gyroscope. A gyroscope under the action of the couple due to gravity (Fig. 3·8) does not fall over but *precesses* about the direction of the gravitational field. Likewise the electron wave precesses about the direction of an applied magnetic flux (Fig. 3·9).

This kind of precession must virtually always be present since it is rare that the magnetic flux in an atom is exactly zero although it may often be very weak. If there is no flux from an external source then that due to the nucleus, which is also a small magnet, is sufficient to cause precession of the electron wave.

Now this precessing wave forms a three-dimensional wave-function and, just as in the two-dimensional case, any closed path through it must contain an integral number of wavelengths. Indeed, any path may be chosen in order to define a third quantum number but the most convenient and useful one is a circle around the flux axis. The angular momentum about this axis is therefore quantized exactly–as is the total angular momentum–in units of $h/2\pi$. We shall use the symbol m_l for the new quantum number so that the angular momentum about the flux direction is $m_l h/2\pi$.

Since the total angular momentum is $lh/2\pi$, the momentum about the flux axis must be a component of it–in fact we can represent angular momentum in vectorial form as we did linear momentum. The vector representing the total angular momentum is drawn along the axis of rotation, its length proportional to the magnitude of the angular momentum. Figure 3·10 shows this and we see that this vector precesses about the flux axis with a *constant component* along that axis. This component therefore must be equal to $m_l h/2\pi$. The other component rotates about that axis and is not quantized in any simple way.

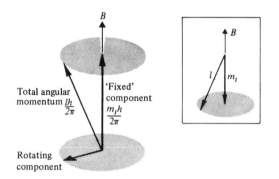

Fig. 3·10 The components of angular momentum for the precessing wave

From this diagram it is easy to see that the *magnetic quantum number*, m_l, has a maximum value equal to l, when l points along B, and the inset in Fig. 3·10 shows how the vector $m_l h/2\pi$ can point in the opposite direction to B. This is described mathematically by giving m_l a negative value.

The range of values permitted for m_l thus extends from $-l$ through zero to $+l$. If, for example, $l = 3$, then m_l can have one of the values -3, -2, -1, 0, 1, 2, 3–seven possibilities in all.

3·5 Spin of the electron

It is of interest to note that each dimension in the above problem is associated with a quantum number and that this was also true in the cases discussed in Chapter 2. It is indeed generally true that three quantum numbers are necessary to describe a wave function.

So far, however, we have not considered the possibility of motion within the electron itself. The wavelike model of the electron does not include this possibility but it has been found necessary to assume that the electron is spinning internally about an axis in order to explain many aspects of the magnetic properties of atoms. Pictorially we may represent the spinning electron as fuzzy distribution of mass rotating about its own centre of gravity rather as a planet spins on its axis while orbiting the sun. It is difficult to couple this image with that of the wave motion around the nucleus and the reader is advised not to try to do so! However, a spinning wave packet is not too difficult to imagine although more difficult to draw.

As might be expected, the angular momentum of this spinning motion is quantized, and in units of $h/2\pi$, too. To distinguish it from the *orbital* angular momentum, $lh/2\pi$, we call it the *spin* angular momentum and its magnitude is $m_s h/2\pi$, where m_s is the spin quantum number. Unlike l and m_l, though, m_s does not have integral values but can only take one of the two values $+\frac{1}{2}$ and $-\frac{1}{2}$. Thus the motion can be either clockwise or anticlockwise about the axis, the spin angular momentum being equal to $\pm\frac{1}{2}h/2\pi$. Since the vector representing this points along the axis of rotation, the two states are

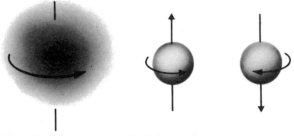

Pictorial spinning electron   Equivalent motion of a 'particle electron'

Fig. 3·11 Pictorial spinning electron (left) and equivalent motion of a 'particle electron'

often referred to as 'spin up' and 'spin down' (Fig. 3·11). An explanation of the reason for the non-integral values of m_s is beyond the scope of this book.

3·6 Electron clouds in the hydrogen atom

Since the shape of the wave-function is defined by the quantum numbers n, l, and m_l, so too is the probability density distribution for the electron. Thus we may imagine an 'electron cloud' which represents the charge distribution in a hydrogen atom just as described in Chapter 2. Such a cloud can be illustrated as in Fig. 3·12, where some examples are given which represent different states of the hydrogen atom. The quantum numbers corresponding to each state are given below the illustrations. Each cloud is shown in cross-section and the density of the image is proportional to the charge density. These clouds all have rotational symmetry about the vertical axis, which coincides with the axis along which m_l is quantized. Note that since the electron is not a standing but a running wave the probability density does not vary along a circle described around that axis since it is the time-average probability.

The $l = 0$ states are all seen to be perfectly spherical in shape, while the others are quite asymmetrical. These shapes will be referred to again when we consider atomic bonding in Chapter 4.

3·7 Energy levels and atomic spectra

One of the most direct checks on the accuracy of the expression for electron energy is to measure the energy released when the electron 'jumps' from an energy level to a lower one. In Fig. 3·13 such a transition between energy levels is represented by a vertical arrow: its length is proportional to the energy released.

Since no matter is expelled from the atom during the transition the only form which the emitted energy can take is an electromagnetic wave–a photon. The photon energy must be exactly equal to the difference in energy between

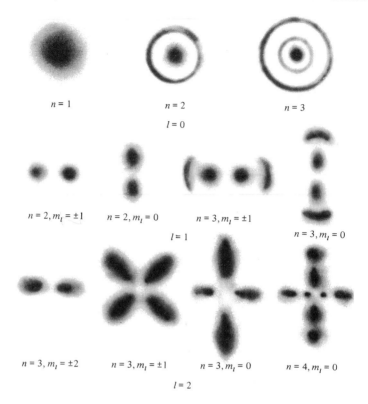

Fig. 3·12 Electron clouds for the hydrogen atom

the two electron energy levels concerned so it can be calculated from Eqn (3·9). (The possibility exists that two photons might be emitted but the chance of this is so small that it is negligible.)

If the electron makes a transition between the levels with quantum numbers n_1 and n_2 the photon frequency may thus be found from the equation

$$E_{photon} = hf = -\frac{e^4 m}{8h^2 \epsilon_0^2 n_1^2} + \frac{e^4 m}{8h^2 \epsilon_0 n_2^2}$$

that is, $f = \dfrac{e^4 m}{8h^3 \epsilon_0^2} \left(\dfrac{1}{n_2^2} - \dfrac{1}{n_1^2} \right)$ \hfill (3·11)

The frequency may thus have any one of a discrete set of values obtained by substituting different values for n_1 and n_2.

How can we arrange to observe these? The first requirement is a source of hydrogen in which there are many atoms in *excited* states, that is, states other than the one with lowest energy (called the *ground* state). Fortunately this is easily achieved by setting up an electric discharge in gaseous hydrogen. In a

$$E_{\text{photon}} = hf = -\frac{e^4 m}{8h^2 \varepsilon_0^2 n_1^2} + \frac{e^4 m}{8h^2 \varepsilon_0 n_2^2}$$

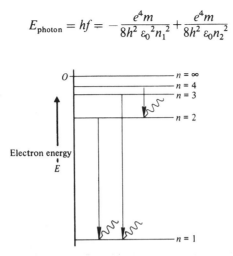

Fig. 3·13 Transitions between energy levels involve the emission of photons

discharge tube (of which the neon tube frequently used for advertisements is an example) a large potential difference is applied between two electrodes (Fig. 3·14). Once the discharge has been started the gas glows brightly and a large current flows. The current is carried across the gas by ions (that is, atoms which have lost or gained an electron and which are hence charged). The potential gradient across the tube accelerates electrons to high velocities whereupon they collide with neutral atoms and excite the latter into states of high energy. As these excited atoms return to the ground state they emit photons as described above.

The emitted light contains several wavelengths and the corresponding frequencies are given by Eqn (3·11) above. The light from a discharge tube is thus normally coloured.

The wavelengths present may be separated and measured using a spectrometer. In its simplest form, this has a glass prism to split the light into its com-

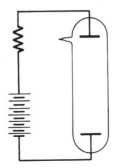

Fig. 3·14 A discharge tube and its power supply

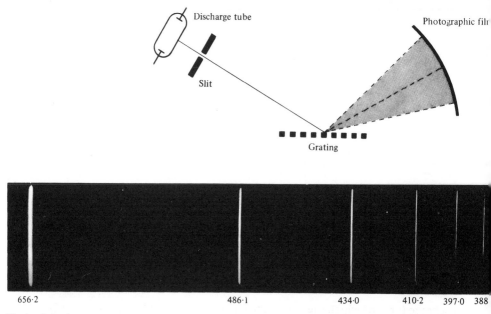

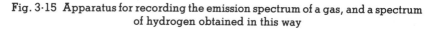

656·2   486·1   434·0   410·2   397·0   388

Wavelength (nm)

Fig. 3·15 Apparatus for recording the emission spectrum of a gas, and a spectrum of hydrogen obtained in this way

ponents but more often a diffraction grating is used. Figure 3·15 shows how the beam transmitted by a narrow slit is diffracted onto a photographic film. From a knowledge of the spacing of the grating lines and Bragg's law the wavelengths may be measured. In Fig. 3·15 we also show a photographic record of the spectrum of hydrogen obtained in this way.

It is found that Eqn (3·10) accurately predicts the measured wavelengths. (Even more accurate agreement is obtained when corrections are made for the finite mass of the nucleus and for relativistic effects.)

If the discharge tube is placed in a large magnetic field during the measurement the quantum states with different values of m_l and m_s have different energies. The highest energy occurs when the total angular momentum vector points along the field direction and the lowest when it points in the opposite direction. These shifts in the energy levels shown in Fig. 3·13 may be observed experimentally and confirm the predictions of our theory. It was through such measurements of spectral lines that the theory was originally built up, and the existence of the spin of the electron was first established.

*3·8 Bohr theory of the atom

Following the suggestion that the atom may be like a planetary system, as outlined in Section 1·4, Bohr calculated the energy levels of such a model. He

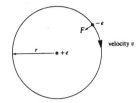

Fig. 3·16 The circular electron orbit in Bohr's theory

assumed that the electron was a particle which orbited the nucleus along a circular path (Fig. 3·16) at a distance r from it.

The attractive force, F, between the proton and the electron is just

$$F = \frac{e^2}{4\pi\epsilon_0 r^2}$$

It is this force which maintains the inward acceleration v^2/r, where v is the velocity of the electron in its orbit, so that Newton's third law of motion gives

$$\frac{e^2}{4\pi\epsilon_0 r^2} = \frac{mv^2}{r} \qquad (3·12)$$

As mentioned in Section 1·4, the accelerating electron is expected to radiate energy. In order to get round this, Bohr had to postulate that certain orbits were stable, non-radiating states. In order to obtain agreement with the wavelengths of spectral lines emitted (the equation $E = hf$ had already been established) he proposed an arbitrary quantization condition: the angular momentum of the electron should be an integral multiple of $h/2\pi$.

Thus the product of the linear momentum, mv, and the radius, r, is to be given by

$$mvr = \frac{nh}{2\pi} \qquad (3·13)$$

where $n = 1, 2, 3 \ldots \infty$ is an integer, each value of which is associated with a different orbit. There is no justification for the above quantization condition, but it is of course identical to Eqn (3·2) which resulted from the assumption that the electron is wave-like. We shall see how ignorance of the wave properties gives incorrect results, in spite of the apparently correct choice made by Bohr for the quantization condition.

Bohr then proceeded to calculate the total energy of the electron. The kinetic energy is

$$E_k = \tfrac{1}{2}mv^2$$

and the potential energy due to the presence of the nuclear charge is obtained as in Eqn (2·6) or (3·4)

$$E_p = \frac{-e^2}{4\pi\epsilon_0 r}$$

The total energy, E, is the sum of these:

$$E = \tfrac{1}{2}mv^2 - \frac{e^2}{4\pi\epsilon_0 r}$$

With the help of Eqns (3·12) and (3·13) the radius, r, and velocity, v, can be eliminated to give the result

$$E = -\frac{me^4}{8h^2\epsilon_0{}^2 n^2} \tag{3·14}$$

which is identical to Eqn (3·9) and gives the correct energy level diagram, as in Fig. 3·13.

However, this model is incomplete, for it ignores the possibility of non-circular orbits. It was Sommerfeld who extended the theory to allow for elliptical orbits, which are also possible according to Newtonian mechanics. To do this he introduced a second quantum number k, which is analogous to the angular momentum number l introduced earlier, and retained the quantum number n which is now redefined in terms of the total momentum and not simply the circumferential component of it (see Fig. 3·6). Here the lack of the wave model leads to an error in the value of the angular momentum as follows. When $k = n$, all the momentum of the electron is angular momentum, that is, the orbit is a circle as in Bohr's original theory. But it has been explained that this is not possible according to the wave theory, although in the planetary model there is no reason to exclude this case. Thus the maximum value of the angular momentum quantum number k is equal to n, while the wave theory allows it to be only as great as $(n - 1)$.

Another difficulty lies in the fact that the case $k = 0$ must be impossible if the electron is a particle. This is because $k = 0$ means that there is no angular momentum, so that the 'orbit' must be a straight line passing through the nucleus. This should be impossible for a particle-like electron, but we know from the wave theory that the angular momentum may indeed be zero, as this does not imply collision of the electron with the nucleus.

In spite of these shortcomings the Bohr-Sommerfeld theory of the atom enjoyed considerable success. Corrections were made to account for the finite mass of the nucleus (both the electron and the nucleus must rotate about their common centre of mass) and for relativistic effects which are apparent when the electron velocity is large. These changes brought the theory into very close agreement with the energy levels deduced from spectroscopic measurements.

However, it could not deal with the criticisms above nor with the fact that many of the expected spectral lines either do not appear or are extremely weak. In all these respects the wave theory has proved entirely satisfactory.

Problems

3·1 A ball of mass 1 g moves in a circular path on the inside surface of an inverted cone, in a manner similar to the example in Fig. 3·3. If the apical angle of the cone is 90°, find an expression for the energy levels of the ball, assuming its wavelength to be given by de Broglie's relation. Hence show that the quantization of its energy may be neglected for practical purposes.

3·2 Calculate the average diameter of the wave function of the hydrogen atom in its ground state. Compare it with the distance between the atoms in solid hydrogen, which may be calculated from its density, $76·3 \text{ kg/m}^3$.

3·3 Calculate the energies of the first three levels of the hydrogen atom. What are the frequencies of radiation emitted in transitions between those levels? Find the first transition for which the radiation is visible, if the shortest visible wavelength is 400 nm.

3·4 How many different values may the quantum number m_l have for an electron in each of the states $n = 3, l = 2; n = 4, l = 2; n = 4, l = 3$?

3·5 Deduce a formula for the energies of the helium ion, He^+, in which a single electron moves around a nucleus whose charge is $+2e$. (*Hint*: although the nuclear charge is doubled, the electronic charge is not, so that it is not correct to replace e by $2e$ in Eqn 3·9.)

3·6 Discuss the consequences if Planck's constant were to have the value 10^{-3} joule s.

3·7 Draw a rough plot of the amplitude of the electron wave versus distance from the nucleus for the state $n = 3, l = 0$ illustrated in Fig. 3·12. Do the same for the state $n = 3, l = 2, m_l = 0$, along the vertical axis. How many maxima are there in each case?

3·8 Explain why in Fig. 3·12, the state $n = 2, l = 1, m_l = 0$ has two maxima around a circumference, while the fact that $l = 1$ indicates that there is only *one* wavelength in the same distance.

3·9 What do you expect might happen if an electron with kinetic energy greater than 13·6 eV were to collide with a hydrogen atom?

3·10 If the space around the nucleus of a hydrogen atom were filled with a dielectric medium with relative permittivity ϵ_r, calculate the new expression for the energy levels. Hence find the energy of the level $n = 1$ when $\epsilon_r = 11·7$.

(The relevance of this question to semiconductor theory will be discussed in Chapter 14.)

Self-Assessment questions

1 In the simplest model of the hydrogen atom the path of the electron (wave) is assumed to be
   A) a straight line   B) elliptical   C) circular   D) hyperbolic.

2 The quantization condition for the electron wave is that
   A) the value of Ψ must not be discontinuous
   B) the value of $d\Psi/dx$ must not be discontinuous
   C) the values of Ψ *and* $d\Psi/dx$ must not be discontinuous.

3 The quantization condition for a closed path leads to the result that

A) the wavelength of the electron is an integral multiple of the circumference of the path

B) there is a half-integral number of wavelengths in the circumference of the path

C) there is an integral number of wavelengths in the circumference of the path.

4 The quantization condition applied to a circular path of radius r leads to an equation of the form:

A) $n\lambda = 2\pi r$ B) $\lambda = 2\pi r n$ C) $r\lambda = 2\pi n$ D) $n\lambda = 2\pi/r$.

5 The above equation leads to quantization of the angular momentum in units of

A) $2\pi/h$ B) $2\pi h$ C) $h/2\pi$ D) $\pi h/2$.

6 The kinetic energy of an electron of mass m and momentum p is

A) $\frac{1}{2}mp^2$ B) $\frac{1}{2}p^2/m$ C) $2m/p^2$.

7 The condition for equilibrium of the electron path, or orbit, is that

A) the total energy is the sum of the potential energy and the kinetic energy

B) the force of attraction to the nucleus is equal to the centrifugal force

C) the total energy is a minimum with respect to the radius of the orbit

D) the angular momentum should be quantized.

8 The total energy E of the electron in equilibrium in an orbit depends on the principal quantum number n according to

A) $E \propto - n$ B) $E \propto - 1/n$ C) $E \propto - n^2$ D) $E \propto - 1/n^2$.

9 The total energy E is inversely proportional to the angular momentum quantum number l.

A) true   B) false.

10 The energy differences between adjacent energy levels of the hydrogen atom

A) decrease with increasing energy

B) increase with increasing energy

C) are independent of energy.

11 The gyroscopic nature of the electron motion about the nucleus leads to

A) quantization of the orbital angular momentum

B) magnetic properties of the electron

C) the spin of the electron

D) precession of the electron orbit (wave-function).

12 The spin quantum number of the electron determines

A) the angular momentum about the nucleus

B) the total angular momentum of the electron

C) the angular momentum of the electron about its own centre of mass.

13 The principal quantum number n may have only the values

A) $0, 1, 2, \ldots$ B) $0, \pm1, \pm2, \pm3, \ldots$ C) $1, 2, 3, \ldots$

14 The angular momentum quantum number l may have only the values

A) $0, 1, 2, 3, \ldots (n - 1)$ B) $0, 1, 2, 3, \ldots (n)$ C) $1, 2, 3, \ldots (n)$
D) $1, 2, 3, \ldots (n - 1)$.

15 The magnetic quantum number m_l may have only the values

A) $0, \pm1, \pm2, \ldots \pm l$ B) $0, \pm1, \pm2, \ldots \pm n$
C) $0, \pm1, \pm2, \ldots \pm(l - 1)$ D) $0, \pm1, \pm2, \ldots \pm(n - 1)$.

16 The spin quantum number m_s may have only the values

A) $0, \pm\frac{1}{2}$ B) $0, \pm\frac{1}{2}, \pm1, \pm3/2 \ldots \pm(l - \frac{1}{2})$ C) $\pm\frac{1}{2}$.

17 When an electron 'jumps' from an energy level to a lower one, the energy released is usually

A) absorbed by the nucleus
B) emitted as heat
C) emitted as light
D) emitted as a continuous electromagnetic wave
E) emitted as a photon.

18 The frequency and wavelength of the emitted radiation can be found from the two equations

A) $\lambda = h/p$ B) $E = hf$ C) $E = \frac{1}{2}mc^2$ D) $c = f\lambda$.

19 In a magnetic field the energy of the electron depends additionally on the value of the quantum number

A) l B) m_l C) m_s.

20 The radiation emitted by a heated gas of hydrogen atoms contains

A) all wavelengths
B) one specific wavelength
C) a set of discrete values of wavelength.

21 The emission of radiation from a gas of atoms occurs when

A) an electron is spiralling towards the nucleus
B) an electron jumps between two energy levels
C) the wavelength of an electron changes.

22 In the emission spectrum of hydrogen the effect of a magnetic field will be

A) to increase the number of spectral lines
B) to decrease the number of spectral lines
C) to change the wavelength of the spectral lines without increasing their number.

Each of the sentences in questions 23–29 consists of an assertion followed by a reason. Answer:

A) If both assertion and reason are true statements and the reason is a correct explanation of the assertion
B) If both assertion and reason are true statements but the reason is not a correct explanation of the assertion
C) If the assertion is true but the reason contains a false statement
D) If the assertion is false but the reason contains a true statement
E) If both the assertion and reason are false statements.

23 The value of the quantum number m_l is always less than or equal to l *because* the absolute value of the orbital angular momentum must always be greater than one of its components.

24 The hydrogen atom in its ground state can emit radiation *because* the electron can make a transition to a different energy level.

25 Hydrogen gas at normal temperatures does not emit light *because* the radiation which is emitted has much shorter wavelengths.

26 The electron orbit (wave path) precesses in a magnetic field *because* the field exerts a couple on it.

27 The path of the electron in an atom must be a closed one *because* the electron is confined within the atom.

28 The electron behaves like a magnet *because* it is charged.

29 Hydrogen has the simplest atom *because* it is gaseous.

Answers

1 C (or B)	2 C	3 C	4 A
5 C	6 B	7 B, C	8 D
9 B	10 A	11 D	12 C
13 C	14 A	15 A	16 C
17 E	18 B, D	19 B, C	20 C
21 B	22 A	23 A	24 D
25 C	26 A	27 A	28 B
29 B			

ATOMS WITH MANY ELECTRONS – THE PERIODIC TABLE

4·1   Introduction – the nuclear atom

In the early years of this century the structure of atoms had to be laboriously deduced from a varied collection of facts and experimental results. It was easily discovered that atoms of most elements contained several electrons since they acquired charge in multiples of $+e$ when electrons were knocked out of the atom by collision. Because an atom is normally neutral, therefore, it must also contain a number of positive charges to balance the negative charges on the electrons. From the experiment by Rutherford mentioned in Chapter 1 it was also found possible to deduce the actual charge on the atomic nucleus by analysing the paths of the deflected a-particles. In this way Rutherford showed that the number of positive charges of magnitude e on the nucleus just equalled the atomic number–the number assigned to an element when placed with the other elements in sequence in the Periodic Table. As the reader who has studied more advanced chemistry will know, this sequence is nearly identical to that obtained by placing the elements in order of increasing atomic weight and has the merit that the relationships between elements of similar chemical behaviour are clearly displayed. This is a topic which we shall cover later in this chapter.

Now if the nucleus of an atom contains Z positive charges each equal in magnitude to the electronic charge it follows that the neutral atom must contain Z electrons. By analogy with the case of hydrogen we therefore anticipate that these electrons are in motion around the nucleus, bound to it by the mutual attraction of opposite charges. It remains only to assign appropriate quantum numbers to each electron and we shall then have a model of the atom with which it will prove possible to explain many of its properties, including chemical combination. This is the aim of this and the next few chapters.

But before proceeding, note one point that is as yet unexplained. We have already identified the proton as a fundamental particle and we expect an atomic nucleus to contain Z such protons. However, the mass of an atom is much greater than the mass of Z protons so there must be some other constituent of the nucleus. In any case, it is unreasonable to expect a group of protons, all having the same charge, to form a stable arrangement without some assistance. The extra ingredient has been identified as an electrically neutral particle of nearly the same mass as the proton, called the *neutron*. Several neutrons are found in each nucleus: the number may vary without changing the chemical properties of the atom. The only quantity which changes is the atomic weight, and this explains the existence of different isotopes of an element as mentioned in Chapter 1.

Since the proton and neutron have nearly identical masses and the electron masses may be neglected in comparison, the atomic weight A must be nearly equal to the total number of protons and neutrons. So the number of neutrons is just the integer nearest to $A-Z$. The force which binds the uncharged neutrons to the protons is a new kind of force, called *nuclear force*, which is very strong compared to the electrical repulsion between the protons.

Since the structure of the nucleus has little or no bearing on the chemical and physical properties of an element, we shall not study it further.

4·2   Pauli exclusion principle

In assigning quantum numbers to the Z electrons in an atom the first consideration must be that the atom should have the minimum possible energy. For if one electron could make a transition to a lower energy level it would do so, emitting radiation on the way. At first sight this implies that the electrons must all be in the lowest energy level, each having the same quantum numbers, that is, $n = 1, l = 0, m_l = 0$, and $m_s = +\frac{1}{2}$. But this is not so and, indeed, the truth is almost exactly the opposite. It is found that each electron has its own set of quantum numbers which is different from the set belonging to every other electron in the atom. This result has become enshrined in a universal principle named after Wolfgang Pauli, who first deduced it. In its simplest form, Pauli's exclusion principle states that:

no more than one electron in a given atom can have a given set of the four quantum numbers.

To emphasize this important principle we state it again in a different way: no two electrons in an atom may have all four quantum numbers the same.

The origin of this principle may be partly understood from the following argument. The three quantum numbers n, l, and m_l completely determine the shape of the wave function and hence also the charge distribution. If two electrons have the same values for these quantum numbers their charge distributions are thus directly superimposed on one another, or in other words the electrons are 'in the same place'. Now this is a very unfavourable situation, since the two electrons must repel one another very strongly. It occurs in practice only if the electron spins are opposed: their quantum numbers, m_s, are then different. It seems that two electrons with counter-rotating spins do not repel one another very strongly. On the other hand, if their spins are parallel the repulsion is so strong that they cannot have the same position (charge distribution) and therefore they have a different set of values for n, l, and m_l.

4·3   Electron states in multi-electron atoms

We may now use the exclusion principle to assign quantum numbers for the first few atoms in the table of elements and we shall find that we are led quite naturally to the construction of a table which reflects the chemical similarities and differences between the elements.

The element with atomic number, Z, of two is helium. It contains two electrons, and at least one of the quantum numbers of the second electron must differ from those of the first. On the other hand both electrons must have the lowest possible energy. Both these requirements are met if the electrons have the following quantum numbers.

1st electron: $n = 1$ $l = 0$ $m_l = 0$ $m_s = +\frac{1}{2}$

2nd electron: $n = 1$ $l = 0$ $m_l = 0$ $m_s = -\frac{1}{2}$

This places both electrons in states with the same wave function, but with opposite spins.

Atomic number $Z = 3$ corresponds to lithium. Two of the three electrons may have the sets of quantum numbers given above for helium. The third electron must have

$$n = 2 \quad l = 0 \quad m_l = 0 \quad m_s = +\frac{1}{2}$$

Here we note that the lowest energy for this electron is the level $n = 2$, which is higher than the energy level with $n = 1$. This should mean that this electron can be removed more readily from the atom than either of the electrons in the helium atom, since less energy is needed to take the electron away to an infinite distance. This is, indeed, the case, for while helium is a noble gas, lithium is metallic, and we know that metals readily emit electrons when heated in a vacuum while helium certainly does not.

In assigning the value $l = 0$ to this third electron we have made use of another rule–*in a multi-electron atom the levels with lowest l values fill up first.* This may be stated in another way: the energy of an electron increases with l as well as with n. Remember that this was not so in the hydrogen atom; the expression for the energy [Eqn (3·9)] contained only the quantum

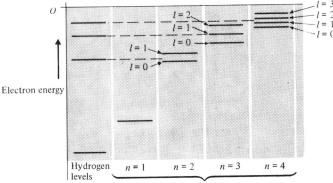

Fig. 4·1 A comparison of the energy levels in hydrogen and in atoms with more than one electron

number, n. Now, however, the electrical repulsion between the electrons alters the energy of each level as indicated in Fig. 4·1, splitting what was a single level in the hydrogen atom into a series of levels each with a different value for the angular momentum quantum number, l.

4·4   Notation for quantum states

Before going further we shall introduce a shorthand notation for the quantum numbers and their values. The principal quantum number, n, defines a series of levels (often called 'shells' because each level corresponds to a different mean radius, $\bar{r}$, of the wave function), each shell corresponding to one value of n. Each of these shells is assigned a letter according to the scheme:

$$n : \quad 1 \quad 2 \quad 3 \quad 4 \ldots$$
$$\text{Letter}: \quad K \quad L \quad M \quad N \ldots$$

Similarly, the values of l are characterized by another series of letters, which, like the above, derive from the early days of spectroscopy:

$$l : \quad 0 \quad 1 \quad 2 \quad 3 \ldots$$
$$\text{Letter}: \quad s \quad p \quad d \quad f \ldots$$

All energy levels belonging to given values of n and l are said to form a subshell and these are labelled by the number corresponding to the value of n and the appropriate letter for the value of l. Thus the subshell with $n = 3$, $l = 2$ is denoted by $3d$, that with $n = 2$, $l = 0$ is denoted by $2s$.

Table 4·1
Electrons in each subshell for the first 18 elements

Atomic Weight (A)	Atomic Number (Z)	Element	K	L		M		
			$1s$	$2s$	$2p$	$3s$	$3p$	$3d$
1·008	1	H	1					
4·003	2	He	2					
6·94	3	Li	2	1				
9·01	4	Be	2	2				
10·81	5	B	2	2	1			
12·01	6	C	2	2	2			
14·01	7	N	2	2	3			
16·00	8	O	2	2	4			
19·00	9	F	2	2	5			
20·18	10	Ne	2	2	6			
22·99	11	Na	2	2	6	1		
24·31	12	Mg	2	2	6	2		
26·98	13	Al	2	2	6	2	1	
28·09	14	Si	2	2	6	2	2	
30·97	15	P	2	2	6	2	3	
32·06	16	S	2	2	6	2	4	
35·43	17	Cl	2	2	6	2	5	
39·95	18	A	2	2	6	2	6	

We now continue to build up the electronic structure of the elements assuming that the levels are filled sequentially with rising values of n and l. Table 4·1 shows the number of electrons in each subshell for each of the first 18 elements.

Note that there is a maximum number of electrons that can be put into each subshell, and by looking at the details of quantum numbers we can see why this is.

For instance in the K (or $1s$) shell, we have $n = 1$, $l = 0$, $m_l = 0$. There are two possibilities for m_s, these are $+\frac{1}{2}$ and $-\frac{1}{2}$. Hence only two electrons are permitted in the K shell.

In the L shell, where $n = 2$, the possible values for l, m_l, and m_s are $l = 0, 1$; $m_l = -1, 0, +1$; $m_s = \pm\frac{1}{2}$. So the possible combinations of these label the various 'states' into which electrons may go:

$$
\left.
\begin{array}{lll}
l = 0, & m_l = 0, & m_s = +\frac{1}{2} \\
l = 0, & m_l = 0, & m_s = -\frac{1}{2}
\end{array}
\right\} \text{2 states in the } 2s \text{ subshell}
$$

$$
\left.
\begin{array}{lll}
l = 1, & m_l = 0, & m_s = \pm\frac{1}{2} \\
l = 1, & m_l = 1, & m_s = \pm\frac{1}{2} \\
l = 1, & m_l = -1, & m_s = \pm\frac{1}{2}
\end{array}
\right\} \text{6 states}
$$

making six states in the $2p$ subshell, and a grand total of eight in the L shell.

It is easy to derive the corresponding numbers for subsequent shells from the rules:

for each value of n there are n possible values of l;
for each value of l there are $(2l + 1)$ possible values of m_l;
for each value of m_l there are two possible values of m_s.

Thus the total numbers of electrons which can be accommodated in each succeeding shell are found to be

$$K : 2 \qquad L : 8 \qquad M : 18 \qquad N : 32$$

4·5 Periodic Table

Certain interesting features arise in Table 4·1. The elements with atomic numbers 2, 10, and 18 are the rare gases which are not merely stable but are exceedingly inert chemically. Thus we may equate inert characteristics with completely filled shells of electrons. It will be noted that the outer shell contains two electrons in helium and eight in both neon and argon. This is the first indication, as we move down the list of elements, that similar arrangements of outer electrons appear periodically and that this periodic variation is reflected in the chemical properties. In fact the periodicity of chemical properties was noted before the electronic structure of the elements was known and as long ago as 1870 the chemist Mendeléev devised a way of tabulating the elements which demonstrated it very effectively. This is now known as the Periodic Table of the Elements.

To construct the Periodic Table we take the elements listed in Table 4·1 and place them in order in horizontal rows so that the outer electron structure changes stepwise as we proceed along the row. We begin a new row whenever a p shell has just been filled with electrons. Ignoring for the moment elements one and two this means that each row finishes with a rare gas, for example, neon with its full L shell or argon with full $2s$ and $2p$ shells. In this way elements with identical numbers of electrons in their outermost shells appear directly beneath one another. Thus lithium, sodium, and potassium, each with one electron outside full shells (usually called *closed* shells) appear in the first column. More remarkably, these elements all display very similar chemical behaviour–they are all very reactive, are metals with a valency of unity, and they form similar compounds with, for example, fluorine or chlorine. We may amplify this last point by noting the properties of the chlorides: they all form transparent, insulating crystals, which are readily cleaved to form regular and similar shapes. They all dissolve, to a greater or lesser extent, in water and they all have fairly high melting points (about 700°C).

We note corresponding similarities between the elements in the second column: beryllium, magnesium, and calcium are all light metals with a valency of two; they form very stable oxides which have even higher melting points than the alkali halides mentioned above and tend to be reactive although not to the degree shown by the alkali metals.

It would be possible to fill a book by listing all the properties which are shared by elements in the same column but by now the reader should be able to recognize that this unity of chemical behaviour is common to all the columns (or *Groups*, as they are called) of the Periodic Table. Moreover, it is reasonable to associate similar chemical behaviour with a similarity in the occupancy of the outermost electron shells. The importance of these outer electrons is brought home further when we observe that the principal valency of each Group is related to the number of such electrons. Thus in Groups I, II, and III the valencies are identical to the numbers of outer electrons, while in Groups V, VI, and VII the valencies are equal to the *number of electrons which would have to be added to complete the outermost shell*. We shall return to this feature later, for this is our first glimpse of the role of electrons in chemical combination.

4·6 Transition elements

The reader will observe that we have so far been very careful to limit discussion to the first three rows of the Periodic Table plus potassium and calcium. The reason is that while the periodicity of properties continues beyond this point it is not exemplified in such a simple fashion.

Let us study the filling of energy levels in the elements of the fourth row. These are shown in Table 4·2 (note the new notation), and immediately an anomaly is apparent. The rule concerning the order of filling the various levels has been broken! Instead of the outermost electrons of potassium (K) and calcium (Ca) entering the $3d$ shell, they go into the $4s$ shell. Only when the $4s$ shell is full do electrons begin to enter the $3d$ shell–there is one $3d$ electron in scandium, two in titanium, and so on.

Table 4·2
Arrangement of electrons for elements 19 to 29

Z	Element	Electron Configuration†				
19	K	(Filled K and L shells)	$3s^2$	$3p^6$	$3d^0$	$4s^1$
20	Ca	(Filled K and L shells)	$3s^2$	$3p^6$	$3d^0$	$4s^2$
21	Sc	(Filled K and L shells)	$3s^2$	$3p^6$	$3d^1$	$4s^2$
22	Ti	(Filled K and L shells)	$3s^2$	$3p^6$	$3d^2$	$4s^2$
23	V	(Filled K and L shells)	$3s^2$	$3p^6$	$3d^3$	$4s^2$
24	Cr	(Filled K and L shells)	$3s^2$	$3p^6$	$3d^5$	$4s^1$
25	Mn	(Filled K and L shells)	$3s^2$	$3p^6$	$3d^5$	$4s^2$
26	Fe	(Filled K and L shells)	$3s^2$	$3p^6$	$3d^6$	$4s^2$
27	Co	(Filled K and L shells)	$3s^2$	$3p^6$	$3d^7$	$4s^2$
28	Ni	(Filled K and L shells)	$3s^2$	$3p^6$	$3d^8$	$4s^2$
29	Cu	(Filled K and L shells)	$3s^2$	$3p^6$	$3d^{10}$	$4s^1$

†In the notation used here the superscript indicates the number of electrons which occupy the subshell.

The reason for this oddity is that the 4s levels have *lower* energy than the 3d levels so that they fill first in keeping with the minimum energy principle. This is illustrated in the energy diagram in Fig. 4·1, where the positions of the energy levels in the hydrogen atom are shown for comparison. The shift in the relative energies in a multi-electron atom is another example of the way the interactions between electrons modify the wave-functions and their energies. The more electrons there are competing for space around the nucleus, the more important these interactions become. So, while the interactions have been small up to this point, from now on we shall find more and more examples of interchanged energy levels.

Returning to the filling of the 3d shell in elements 21 to 29 we note the irregularities at 24 chromium (Cr) and 29 copper (Cu), each of which contains one 4s electron instead of two. This is due to the fact that the exactly half-filled 3d shell and the filled 3d shell are particularly stable configurations (that is, they have lower energy) compared to the neighbouring occupancies of four and nine electrons respectively.

How do we assign these elements to their correct Group in Periodic Table?

Table 4·3
Arrangement of electrons for elements 29 to 36

Z	Element	Electron Configuration†		
29	Cu	(Filled K, L, M shells)	$4s^1$	
30	Zn	(Filled K, L, M shells)	$4s^2$	
31	Ga	(Filled K, L, M shells)	$4s^2$	$4p^1$
32	Ge	(Filled K, L, M shells)	$4s^2$	$4p^2$
33	As	(Filled K, L, M shells)	$4s^2$	$4p^3$
34	Se	(Filled K, L, M shells)	$4s^2$	$4p^4$
35	Br	(Filled K, L, M shells)	$4s^2$	$4p^5$
36	Kr	(Filled K, L, M shells)	$4s^2$	$4p^6$

†In the notation used here the superscript indicates the number of electrons which occupy the subshell.

In spite of the complications, it is clear that the first two elements, potassium (K) and calcium (Ca), fall respectively into Groups I and II both on grounds of chemical similarity and because of the similarity of the 'core' of electrons remaining after removal of the valence electrons. Assignment of subsequent elements to their groups is easier if we first discuss 29 copper and succeeding elements up to number 36 where we arrive at krypton, another inert gas, with a stable octet of outer electrons. The intervening elements, 29 to 35, all have complete inner shells and unfilled outer shells; they thus fall naturally into Groups I to VII in sequence. However, copper does not fit quite so naturally with the very reactive lithium (Li) and sodium (Na) because removal of an electron leaves not a stable core but the configuration $3d^9 4s^1$ of nickel (Ni). We therefore divide Group I (and, for similar reasons, Group II) into subgroups labelled A and B, putting Cu and Zn into the latter.

Let us now return to the elements 21 scandium (Sc) to 28 nickel (Ni). This set is called the first series of *transition elements*, which, because of the presence of the $4s$ electrons, all have similar properties (they are all metals). However, the first five fit quite naturally into Groups II–VII, though again we find it appropriate to divide the groups into subgroups, putting the transition metals into the A subgroups. But for the elements Fe, Co, and Ni there are no precedents in the table, and these we assign to a new Group (VIII).

Having dealt with the first series of transition elements, we are not surprised to find another series in the fifth row of the table, and also in the sixth. The latter, however, is more complicated owing to the filling of two inner subshells ($4f$ and $5d$) before the transition is complete and the $6p$ shell begins to fill. Here, the series of elements 57 to 71, in which the $4f$ shell is being filled, have almost identical chemical properties–as a Group they are called the *rare earth metals*. The reason for their chemical similarity is that these elements differ only in the number of electrons in a shell which is far removed from the outermost electrons while it is the latter which determine chemical behaviour. Their almost indistinguishable qualities place them all into one pigeon-hole in the Periodic Table, in Group IIIA.

The subsequent filling of the $5d$ shell beyond its solitary occupancy in 71 lutecium (Lu) marks the continuation of the Periodic Table, for the $5d$ electrons, like the $3d$ electrons in the first transition series, are of chemical importance, so that the properties of these elements differ.

Another group of elements similar to the rare earths is found in the seventh periodic row, although the majority of these do not occur in nature because their nuclei are unstable–they have, however, been manufactured artificially. The elements in this last series in Group IIIA are often referred to as the *actinide elements* (they all behave like actinium, the first of the series) just as the rare earths are sometimes called *lanthanide elements*.

4·7 Valency and chemical combination

It is necessary to clarify some points concerning valency, particularly in the transition elements. We have already hinted that valency is determined by the ease with which an atom may become ionized, that is, either lose or gain electrons in its outer shells. Thus magnesium (Mg) and calcium (Ca) are divalent because the two s electrons, known as the *valence* electrons, are readily removed from the neutral atoms. Removal of a third electron would involve destruction of the completed and very stable octet of electrons in the next lowest shell. The stability of closed s and p shells is thus displayed in all atoms, not just in the rare gases which happen to possess closed shells in the electrically neutral state.

By contrast, a transition element deprived of its outermost s electrons (for example, the two $4s$ electrons in iron) is not nearly as stable. It may easily lose yet more electrons, and can consequently have more than one valency. Thus we find ions (atoms with more or less than the normal number of electrons) such as Fe^{3+} and Fe^{2+} (the superscript gives the resultant charge on the ion). The Fe^{2+} ions would be found in a compound such as ferrous oxide (FeO) or ferrous bromide ($FeBr_2$) whilst Fe^{3+} occurs in ferric oxide (Fe_2O_3) and ferric bromide ($FeBr_3$).

In these examples the oxygen and bromine atoms have valencies of two and one respectively because they require the addition of this number of electrons to attain a stable configuration with closed outer shells. Since in this stable configuration the outer shell contains eight electrons the number of electrons to be added is $8-N$, where N is the group number. Thus the principal valency of the elements in Groups IV–VII is given by the so-called $8-N$ rule.

Hydrogen is a unique case since the nearest stable arrangement contains two electrons (a filled K shell), or alternatively no electrons. Hydrogen atoms can either lose or gain electrons so permitting combination with halogens to form acids (HF, HCl, HBr, HI) and also with metals, giving hydrides (NaH, ZrH, CaH, and so on).

The noble gases are assigned to Group zero: their valency is nominally zero and they were thought until quite recently to be completely inert. It is now known that they form a few simple compounds and that they can also combine with themselves to form solids but only at very low temperatures.

It is of interest to note at this point that although the individual wave-functions of electrons in a subshell have quite complex shapes (cf. Fig. 3·12), in a completed subshell the total charge density is quite simply spherically symmetric. The electron cloud of a filled subshell is thus like that of one of the s states in Fig. 3·12 and a noble gas atom may therefore be pictured roughly as a small sphere.

Problems

4·1 Discuss the chemical similarities which lead us to place the elements of sub-groups IIA and IIB together in Group II. Discuss also the differences between these subgroups.

4·2 Plot the melting points and densities against atomic number for the first three rows of the Periodic Table. Notice how the changes reflect the periodicity represented by the Table. (Data in Appendix III.)

4·3 Look up the densities of the elements in Group IIA and explain why they depend on atomic number.

4·4 What differences would there be in the Periodic Table if (a) the $5p$ levels had lower energy than the $4d$ levels; (b) the $5d$ levels had lower energy than the $4f$ levels?

4·5 In which elements do the nuclei contain the following numbers of neutrons?

     10   14   22

Suggest how the existence of several isotopes might explain why chlorine apparently has the non-integral atomic weight 35·4.

4·6 Write down the quantum numbers of each electron in the M shell and hence show that it may accommodate only 18 electrons.

4·7 Why do the oxides of the Group II metals have higher melting points than the alkali halides? Why are the Group II metals less reactive than those in Group I?

4·8 Without consulting Table 4·1, write down the electronic configuration of the elements with atomic numbers 4, 7, 10, 15. By counting the valence electrons, decide to which Group each element belongs.

4·9 Copper has a single valence electron in the $4s$ shell and belongs to Group I. What is the electronic structure of the Cu^+ ion? Why does copper not behave chemically in the same way as sodium or potassium?

4·10 The outermost shells of all the inert gases are filled s and p subshells. Silicon and germanium have four electrons outside closed s and p shells, and so have titanium and zirconium. Why, then, are the chemical characteristics of all these elements not identical?

Self-Assessment questions

1 If the atomic number of an element is Z and its atomic weight is A, the number of protons in the nucleus is

     A) Z B) $A-Z$ C) A D) $Z \cdot A$.

2 The difference between A and Z is due to the presence of

     A) electrons   B) protons   C) photons   D) neutrons

in the nucleus.

3 Pauli's exclusion principle states that, within one atom

     A) no more than two electrons may have the same energy

     B) the spins of the electrons interact so as to become parallel if possible

     C) no two electrons may have the same four quantum numbers

     D) there are only two values for the quantum number m_s.

4 In atoms containing many electrons the shells are filled in order of

     A) increasing n and l B) decreasing n and l C) increasing energy.

5 The maximum number of electrons in the L shell ($n = 2$) is

     A) 4   B) 6   C) 8   D) 14.

6 In the notation $2p^6$, $3s^2$, etc. the meanings of the symbols are

     A) the first number is the value of l, the letter gives the value of n, and the superscript is the number of electrons in the subshell.

B) The first number is the number of electrons in the subshell, the letter gives the value of l and the superscript the value of n.

C) The first number is the value of n, the letter gives the value of l and the superscript is the number of electrons.

7 $3f^6$ denotes a subshell containing 6 electrons for which $n = 3$ and $l = 3$.

A) true   B) false.

8 The maximum number of electrons allowed in the $4d$ subshell is

A) 14   B) 10   C) 8   D) 4.

9 The lithium atom, which contains 3 electrons has the structure

A) $1s^2 2s^1$ B) $1s^2 2p^1$ C) $1s^1 2p^2$ D) $2s^2 2p^1$.

10 The atomic number of the element whose outermost electron fills the $3s$ shell exactly is

A) 13   B) 8   C) 10   D) 12.

11 The characteristic feature of the transition elements is

A) a partly filled valence shell   B) an empty inner shell
C) an unfilled outer shell   D) a partly filled inner shell.

12 The first series of transition elements, in which the $3d$ shell is gradually filled, begins at atomic number

A) 19   B) 21   C) 11   D) 13.

13 The element with electronic structure $1s^2 2s^2 2p^6 3s^2 3p^6 3d^8 4s^2$ is a transition element.

A) true   B) false.

14 In the Periodic Table, the elements are arranged in order of increasing

A) atomic weight   B) chemical equivalent weight
C) molecular weight   D) atomic number.

15 The polonium atom has the electronic structure

$1s^2 2s^2 2p^6 3s^2 3p^6 3d^{10} 4s^2 4p^6 4d^{10} 4f^{14} 5s^2 5p^6 5d^{10} 6s^2 6p^4$

Its group number is

A) II   B) IV   C) VI   D) VIII.

16 The manganese atom has the electronic structure

$1s^2 2s^2 2p^6 3s^2 3p^6 3d^5 4s^2$

Its group number is therefore

A) I   B) VII   C) III   D) VI.

17 The outer electron configuration which gives the noble gases their extreme inertness is

A) $s^2 p^6$ B) $s^2 p^4$ C) $p^4 d^6$ D) s^4.

18 Zinc, atomic number 30, is put into Group IIB rather than Group IIA because

A) removal of two electrons leaves a stable octet in the outermost remaining shell
B) removal of three electrons is nearly as easy as removing two
C) the electronic structure of Zn^{2+} is not particularly stable.

19 Transition elements have several valencies because

A) they contain several electrons in the outermost shell
B) removal of the electrons in the outermost shell does not leave a stable octet
C) the electrons in the valence shell can readily be removed singly from the atom.

20 The principal valency of the elements in Groups I–IV of the Periodic Table is equal to

A) 6 minus the group number   B) the group number
C) 8 minus the group number.

21 The valency of the element with structure $1s^2\, 2s^2$ is

A) 2   B) 4   C) 6   D) 0.

22 The principal valency of the elements in Groups V–VII of the Periodic Table is

A) 9 minus the group number   B) the group number
C) 8 minus the group number   D) the group number minus 4.

23 Boron (Group IIIB) and fluorine (Group VIIB) may form a compound whose formula is

A) B_2F_3 B) B_3F C) BF_3 D) B_3F_2.

Each of the sentences in questions 24–31 consists of an assertion followed by a reason. Answer:

A) If both assertion and reason are true statements and the reason is a correct explanation of the assertion
B) If both assertion and reason are true statements but the reason is *not* a correct explanation of the assertion
C) If the assertion is true but the reason contains a false statement
D) If the assertion is false but the reason contains a true statement
E) If both the assertion and reason contain false statements.

24 The ground state of the helium atom is $2s^2$ *because* this is the lowest energy state.

25 The electronic structure $1s^2\, 2s^2\, 2p^6\, 3s^2\, 3p^6\, 3d^{10}\, 4s^2\, 4p^5\, 4d^2$ does not normally occur in a real atom *because* the shells have been filled in the wrong order.

26 The rare gases do not form compounds *because* they are in Group 0 of the Periodic Table.

27 The elements in any one group of the Periodic Table are chemically similar *because* they all contain the same number of electrons in the outermost shell.

28 There are nine groups in the Periodic Table *because* there are nine stable arrangements of electrons in the s and p subshells.

29 Copper (Group IB) is chemically not so similar to lithium (Group IA) as is sodium (Group IA) *because* its atoms do not contain the same number of electrons in the outermost shell.

30 The first series of transition elements are all placed together in Group VIII *because* of their chemical similarity.

31 Iron and oxygen can combine to form more than one oxide: e.g. FeO and Fe_2O_3 *because* iron and oxygen may have more than one valency.

Answers

1 A	2 D	3 C	4 C
5 C	6 C	7 B (no such subshell exists)	
8 B	9 A	10 D	11 D
12 B	13 A	14 D	15 C
16 B	17 A	18 B, C	19 B
20 B	21 A	22 C	23 C
24 E	25 A	26 B	27 C
28 A	29 C	30 E	31 C

MOLECULES AND BONDING

5·1 Molecular bond

When a gas of neutral atoms condenses to form a solid the atoms are held rigidly together by mutual attraction. The pull between them is much stronger than a mere gravitational force and we say that atoms form *bonds* between one another. In a solid each atom forms a bond with all of its neighbours but in a diatomic molecule the single bond may be studied in isolation. We therefore commence our discussion of the mechanism of bond formation with a study of the molecular bond.

5·2 Formation of a molecule

Just as the simplest atom to study is hydrogen so the simplest molecule is formed by the combination of two hydrogen atoms–the diatomic hydrogen molecule.

This combination of two protons and two electrons is not so easy to treat mathematically as is the single atom. However, a qualitative picture may easily be obtained by imagining the atoms H_a and H_b to approach one another as depicted in Fig. 5·1. As their separation decreases each electron feels two new competing forces: an attraction towards the other nucleus and a repulsion from the other electron. In practice the attraction is the stronger, for the repulsion is small if the electrons have opposite spins. This can be seen from Pauli's exclusion principle which allows electrons to have the same wave functions, that is, occupy the same region of space, only if their spins are antiparallel.

The atoms are thus pulled together until each electron cloud surrounds both nuclei at which stage the two electrons share the same wave function [Fig. 5·1(c)]. Closer approach is impossible because the repulsion between the two nuclei would be too great. At the equilibrium separation, however, there is an exact balance between the attraction of the electrons and the repulsion of the nuclei. This is shown in Fig. 5·2 where each force is plotted against the interatomic distance. As the atoms approach both attraction and repulsion forces build up, the latter more slowly at first. On close approach the repulsive force increases rapidly until it equals the attraction and the molecule is in equilibrium. In hydrogen this occurs with a separation d_0 of 0·74 Å. Thus it is possible to calculate the equilibrium separation if the dependence of the forces on separation is known.

Note that bonding results in a marked overlap of the two electron clouds.

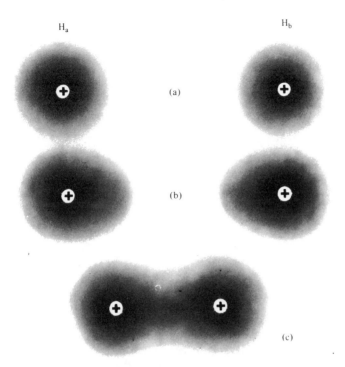

Fig. 5·1 (a) Isolated atoms; a weak attractive force is present; (b) at closer approach the electron clouds are distorted by the attractive forces, the two electron spins are antiparallel; (c) in equilibrium the electrons share the same wave function

This is a universal characteristic and it means that a pair of atoms form a bond only when they approach close enough for the valence electron clouds to overlap one another.

When two heavier atoms combine there is one small difference from the case discussed.

Consider the approach of two sodium atoms each of which we may picture as a core of negative charge (the completed K and L shells) outside of which moves the valence electron cloud (Fig. 5·3). We have already seen that the closed shells of the core are spherical in shape so that they are represented as spheres in the illustration. The bonding force arises exactly as in the hydrogen molecule but the repulsive force is now augmented at short separations by the reluctance of the closed inner shells to overlap. They could overlap only if the electron spins were opposed (see the discussion of Pauli's principle, Chapter 4) but since each closed shell contains electrons with spins already opposed, overlap is not allowed. Hence the equilibrium separation is determined largely by the radius of the inner core and the molecule is very like a pair of nearly rigid spheres glued together. This simple model is of great utility in considering the way atoms pack together in crystals (cf. Chapter 6).

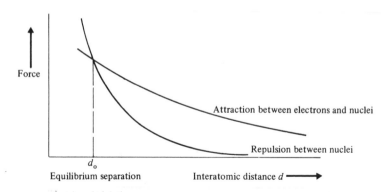

Fig. 5·2 Forces between two atoms plotted against interatomic distance

5·3 Bonds in solids

We have seen how the overlapping of the valence electron clouds of two atoms can lead to an overall attraction and hence to the formation of a molecule. The formation of a solid body may be described in similar terms but with attraction between each atom and all of its neighbours simultaneously instead of with just one or two other atoms. The question of how many neighbours surround each atom and in what arrangement is left to the next chapter. Here we shall look primarily at the bond formed between one

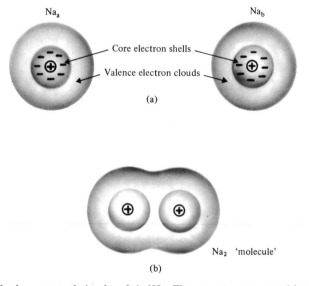

Fig. 5·3 The formation of a 'molecule' of Na. This structure is not stable in practice; however, the addition of further atoms produces a stable crystal of sodium – see later

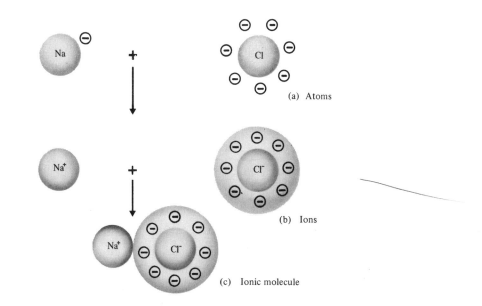

(a) Atoms

(b) Ions

(c) Ionic molecule

Fig. 5·4 Schematic representation of the formation of an ionic 'molecule' of sodium chloride

pair of atoms in a solid but we must remember that both atoms form more than just this one bond.

Bonds between atoms may vary in nature according to the electronic structure of the atoms concerned. One would not expect the closed-shell structure of the inert gases to behave in the same way as the alkali metals, each of which can so easily lose its outer electron. Indeed, we know from experiment that

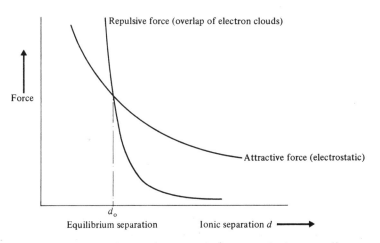

Fig. 5·5 Dependence of inter-ionic forces on ionic separation

while these metals readily form compounds with other elements, the inert gases do not.

There are actually five classes into which bonds are conventionally divided, although the boundaries between these categories are not too well defined, as we shall see.

5·4 Ionic bonding

As remarked above, the valence electron is very easily removed from an alkali metal leaving behind a very stable structure resembling an inert gas but with an extra positive nuclear charge.

On the other hand an element from Group VII (fluorine, chlorine, bromine, or iodine–the halides) is but one electron short of an inert gas structure. Since the inert gas structure is so stable we might expect that a halide atom would readily accept an extra electron and might even be reluctant to lose it given suitable conditions.

Taking these two tendencies together we can understand what happens when, for example, sodium and chlorine atoms are brought together in equal numbers. It is easy for the valence electron of each sodium atom to be transferred to a chlorine atom, making both more stable. But now there must be an electrostatic attraction between the ions so formed for each sodium ion carries a positive charge (Fig. 5·4) and each chlorine ion a negative one. The attraction brings them closer and closer together until the inner electron clouds begin to overlap. At this point a strong repulsion force is manifested exactly as for the case of two sodium atoms described earlier and the two

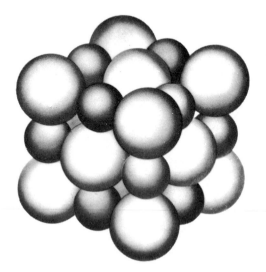

Fig. 5·6 Part of the crystal of NaCl modelled as a close packed arrangement of spheres

forces just balance one another. The way in which these forces depend upon the separation, d, is shown in Fig. 5·5, and we see that it is very similar to the case of the hydrogen molecule, Fig. 5·2.

Naturally there is also an electrostatic repulsion between ions of the same charge so that in solid sodium chloride (NaCl) we do not expect to find like ions side by side but rather alternating as shown in Fig. 5·6. In this way the attraction of unlike ions overcomes the repulsion of like ions and a stable structure is formed. We shall meet this and other kinds of atomic packing in Chapter 6.

In Fig. 5·6 we represent the ions by spheres for we have already seen (Fig. 5·3) that this is quite a realistic way of picturing the inner core of electrons.

It is to be noted that while the structure shown in Fig. 5·6 is stable, a single pair of ions do not form a stable molecule—the molecule of sodium chloride does not normally exist by itself.

Magnesium oxide (MgO) has a similar structure to that of sodium chloride but in this case two electrons are transferred from each magnesium atom to an oxygen atom, again giving each ion a stable group of eight outer electrons (Fig. 5·7). Since these ions are doubly ionized and hence carry two electronic charges it is understandable that the interatomic cohesion should be much stronger than in sodium chloride. This accounts for the much higher melting point of magnesia (2,800°C compared with 800°C for sodium chloride). The relationships between melting point and bond strength will be discussed later in this chapter but it suffices here to note that the stronger a bond the higher is the temperature needed to break it.

Yet other examples of ionic bonding occur in the compounds cupric oxide, chromous oxide, and molybdenum fluoride (CuO, CrO$_2$, MoF$_2$) showing that the metallic element need not be from Groups I or II but that any metal may readily become ionized by losing its valence electrons.

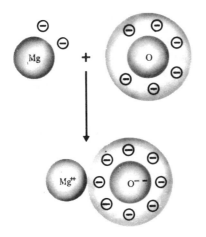

Fig. 5·7 Representation of the formation of an ionic molecule of MgO

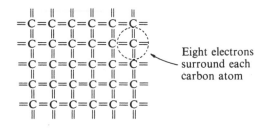

Fig. 5·8 By sharing electrons (shown by dashes) with its neighbours, a carbon atom forms four bonds, and a repetitive structure can be created

5·5 Covalent bonding

Elements from the central groups of the Periodic Table–notably Group IV–are not readily ionized. The energy required to remove all the valence electrons is too large for ionic bonding to be possible. It is still possible for each atom to complete its outer shell, however, by sharing electrons with its neighbours.

We saw earlier how the hydrogen molecule is formed by sharing electrons in this way but we now show that this bond is more universal and can be present in solids.

We may take carbon as an example. It has a filled K shell and four electrons in the L shell, that is, $1s^2 2s^2 2p^2$. Four more electrons are required to fill the L shell and these may be acquired by sharing an electron with each of four neighbours when carbon is in its solid form. One way in which this could be done is shown in Fig. 5·8 although this arrangement does not occur in practice because of the directional nature of the bonds.

To understand this we must look at the shape of the electron clouds in a carbon atom. The four electrons in the L shell interact rather strongly with one another, modifying the shapes of the s and p wave functions until they are virtually identical. The electrostatic repulsion is so strong that each electron's 'cloud' concentrates itself away from those of the other three. As illustrated in Fig. 5·9, this means that each cloud is sausage-like and points away from the nucleus, the four arranging themselves with the largest possible angle, 109·5°, between each pair. It is easy to see that they point towards the corners of an imaginary tetrahedron.

Because of the strong repulsion this arrangement is difficult to distort and the carbon atoms join up as shown in Fig. 5·9 with the electron clouds of neighbouring atoms pointing towards one another. In this way each carbon atom is surrounded by eight electrons which is a stable arrangement. The structure so formed is very strong and rigid–it is the structure of diamond, the crystalline form of carbon. The term crystal (from the Greek *kristallos*–clear ice) until recently referred to a solid which can be cleaved to form a regular geometrical shape, or which occurs naturally in such a shape. The modern definition of a crystal will be given in Chapter 6.

Just as in the case of sodium chloride, no molecule can be distinguished

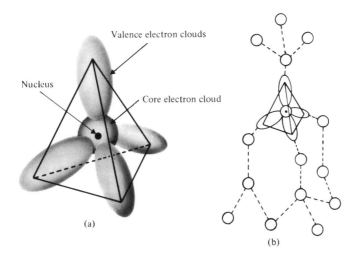

Fig. 5·9 (a) The lobe-shaped valence electron clouds in the carbon atom; (b) the carbon atom in its crystalline surroundings has a similarly shaped electron cloud

here but the solid is rather like one huge molecule (a so-called macro-molecule) since it is a never-ending structure. One can always add more carbon atoms.

The other Group IV elements, silicon and germanium, also crystallize in the same structure and compounds of an element of Group III with one of Group V form the related zinc blende structure which we shall meet in Chapter 6. In these compounds each atom is again tetrahedrally coordinated with four others even though the isolated Group III and Group V atoms have three and five valence electrons respectively.

Here we should note how the valency of covalent elements may be deduced. In forming the covalent bond the atom acquires electrons by shar-ing until it has a stable outer shell. The valency thus equals the number of electrons lacking from the outer shell; if there are N electrons present in the outer shell of the neutral atom then the valency is $8-N$.

5·6 Metallic bonding

As its name implies, metallic bonding is confined to metals and near-metals many of which are to be found in Groups I, II, and III of the Periodic Table. If we take as an example copper, we see that shells K, L, and M are full, while there is just one 4s electron in the N shell. In solid copper the outer electron is readily released from the parent atom and all the valence electrons can move about freely between the copper ions. The positively charged ions are held together by their attraction to the cloud of negative electrons in which they are embedded (Fig. 5·10) rather like ball bearings in a liquid glue.

In some respects this arrangement is like an ionically bonded solid but

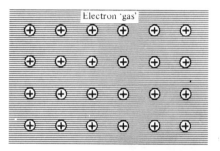

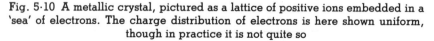

Fig. 5·10 A metallic crystal, pictured as a lattice of positive ions embedded in a 'sea' of electrons. The charge distribution of electrons is here shown uniform, though in practice it is not quite so

instead of the negative 'ions' being like rigid spheres they fill all the space between the positive ions. One might say that a metallic bond is a sort of ionic bond in which the free electron is donated to all the other atoms in the solid. (Although this is the easiest way to understand the metallic bond, it is often useful and legitimate to view it as a kind of covalent bond in which the electrons form a temporary covalent bond between a pair of atoms and then move on to another pair.) Thus the bonds are not directional and their most important characteristic is the freedom of the valence electrons to move. In later chapters we shall see how this mobility of the electrons is responsible for the high electrical and thermal conductivities of metals.

Because the ions are bonded to the valence electrons and only indirectly to each other it is possible to form a metallic solid from a mixture of two or more metallic elements, for example, copper and gold. Moreover, it is not necessary for the two constituents to be present in any fixed ratio in order that the solid be stable. The composition may thus be anywhere between that of gold with a small proportion of copper added and of copper containing a small proportion of gold. It is rather as if one metal were soluble in the other and, indeed, we often refer to this type of material as a solid solution. More commonly the term *alloy* is used and in Chapter 10 we shall learn more about alloys, some of which are of great importance because of their great strength.

5·7 Van der Waals bonding

The inert gases condense to form solids at sufficiently low temperatures although one would not expect them to form bonds of any of the above kinds. Similarly, valency requirements are fulfilled in molecules of the gases methane, carbon dioxide, hydrogen, and so on (CH_4, CO_2, H_2, etc.) and no spare electrons are available for forming bonds, yet they too can solidify. It is clear that there must be another kind of bond, often called a secondary bond, in which no major modification of the electronic structure of atoms takes place.

We know that an atom with closed electron shells consists essentially of a

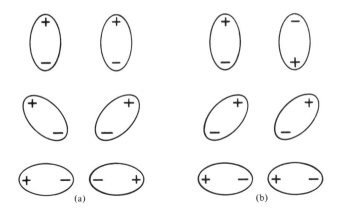

Fig. 5·11 Illustrating the origins of van der Waals force

positive nucleus surrounded by a spherical cloud of negative charge. If the electron clouds of two such atoms were static and undeformable no force would exist between them. Actually, however, the electron clouds result from the motion of the various electrons and although it is not possible to follow their trajectories the electrons must be regarded as in motion around the nucleus. Thus whilst an atom can on average have no electrical dipole moment it has a rapidly fluctuating dipole moment. At any instant the centre of the negative charge distribution does not coincide with the nucleus but rapidly fluctuates about it.

When, therefore, two atoms approach one another the rapidly fluctuating dipole moment of each affects the motion of the electrons in the other atom and a lower energy (that is, an attraction) is produced if the fluctuations occur in sympathy with one another. This can be seen by considering the alternative situations shown in Fig. 5·11, where each atom is shown as a small electric dipole. In Fig. 5·11(a) we see several situations in which little or no attraction is produced, while in Fig. 5·11(b) all cases display a marked attractive force. The fluctuations occur in such a way that the situations like those of Fig. 5·11(b) occur most frequently so that a net attractive force exists.

Clearly this force acts even over large distances for the electron clouds do not need to overlap one another. As a result it makes itself felt even in the gaseous state, and causes deviations from the perfect gas laws. The well-known gas equation suggested by van der Waals incorporates its effects and the force is thus known after him.

We shall come across many substances which are bonded by van der Waals forces: they include plastics, graphite, and the paraffins.

5·8 Hydrogen bond

There is another secondary bond which can be made between atomic groups which have no electrons to spare. It bears a close resemblance to the van der

Waals bond which, as we saw, is due to the attraction between dipoles, but occurs only when a hydrogen atom is present.

Consider an ionic molecule formed by an atom of hydrogen and some other atom or group of atoms which we shall call X. We may describe the structure by the symbol X^-——H^+ (The signs $+$ and $-$ indicate the charge on the ions.) This molecule is a small dipole. The positive charge on the hydrogen ion can attract the negative charge on a third ion or group of atoms, which may even be part of another molecule. Thus we have a weak dipole-dipole attraction exactly as in the van der Waals bond. But the H ion is very small since it has lost its valence electron and the third ionic group, Y, may approach very closely, forming a slightly stronger bond than otherwise. We may denote this structure symbolically

$$X^-——H^+——Y^-$$

Note that, since the bond with Y is not a true electronic bond, Y cannot approach as closely as X to the hydrogen ion. However, it would be quite possible to have the structure

$$X^-————H^+——Y^-$$

in which the ionic bond appears between Y and H while the X——H bond is now weak. In practice the hydrogen atom hops backwards and forwards between the two possible positions (near to X and near to Y) so the bond must be described as a mixture of the two forms. The net attractive force between X and Y is rather weak since it cannot be stronger than its weakest part (the dipolar attraction). Modified forms of this bond occur when hydrogen is present in both of the molecules which form the bond.

The hydrogen bond, as it is called, occurs frequently in organic materials in which hydrogen often plays a major role. We shall meet it in Chapter 12 for it can make an important contribution to the strength of plastics (polymers). As might be expected, it also appears in water and ice and it explains the occurrence of such hydrated inorganic salts as nickel sulphate ($NiSO_4.6H_2O$).

5·9   Comparison of bonds – mixed bonds

Both the van der Waals bond and the hydrogen bond differ in a quite distinctive way from the three primary bonds, and they give quite different properties to a solid. In the following sections we shall compare the properties of the five bonds, taking some examples to illustrate them.

Although at first sight the three primary bonds seem quite different the bonding in many substances does not fit easily into just one or other of these categories.

Let us consider a series of molecules which contain the same total number of electrons, and which are hence called *isolectronic*: methane, ammonia, water, and hydrogen fluoride.

In methane (CH_4) the carbon atom can form four covalent bonds, one with each hydrogen atom (Fig. 5·12). In hydrogen fluoride (HF) at the other extreme, the hydrogen donates its electron to the fluorine atom to complete

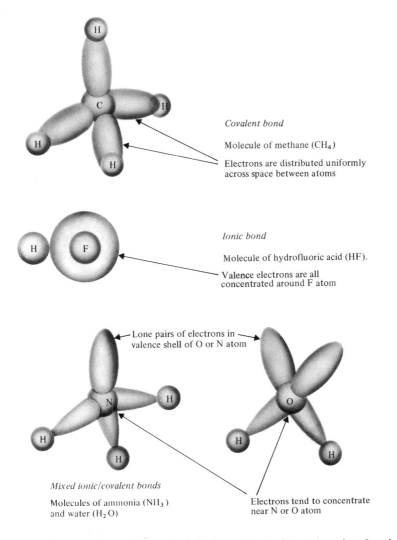

Covalent bond

Molecule of methane (CH₄)

Electrons are distributed uniformly across space between atoms

Ionic bond

Molecule of hydrofluoric acid (HF).

Valence electrons are all concentrated around F atom

Lone pairs of electrons in valence shell of O or N atom

Mixed ionic/covalent bonds

Molecules of ammonia (NH₃) and water (H₂O)

Electrons tend to concentrate near N or O atom

Fig. 5·12 Electron clouds for covalent, ionic, and mixed ionic/covalent bonds

the L shell of the latter, and the atoms are bonded ionically together. In the intervening compounds, ammonia and water (NH_3 and H_2O), neither of these pictures is quite correct in itself but a mixture of the two applies. In such cases a proportion of an electronic charge may be considered to be transferred while the rest of it is shared equally to form a partial covalent bond. In reality this means merely that the electron clouds are distorted roughly as shown in Fig. 5·12 so that there is a predominance of electronic charge density away from the hydrogen atom. This shift of the 'centre of charge' increases as we go through the series from the C–H bond to the F–H bond.

The carbon, nitrogen, oxygen, and fluorine atoms are arranged in order of increasing 'electronegativity' which is defined as the degree to which an atom can attract electrons to itself. In general, the electronegativity increases steadily across the Periodic Table from Group I to Group VIII. It also varies within a Group, decreasing with increasing atomic number except in the case of transition elements.

If we are given the electronegative series, we can thus form rules about bonding:

Two atoms of similar electronegativity form either a *metallic* bond or *covalent* bond, according to whether they can release or accept electrons.

When the electronegativities differ the bond is partially ionic, the ionic character increasing with the difference in electronegativity.

Since the covalent bond is directional, while the ionic bond is not, the degree of directionality changes with the bond character. Such changes have a marked influence on the crystal structure as we shall see in Chapter 6.

A further example of mixed ionic/covalent bonding occurs in an important class of substances based upon the dioxide of silicon (SiO_2). This compound can form both a regular, ordered structure–in which form it is called quartz–and an irregular structure called quartz glass or silica. Many of the glasses and various ceramic materials, for example porcelain, are complex structures containing silicon dioxide and other compounds in which regular structures are embedded in an amorphous matrix. [*Amorphous*-having no form or structure (from the Greek *morphe*-form). Amorphous solids do not form crystals with regular shapes because there is no regularity in the way in which their atoms are packed together. Note that not all solids which do not form regular geometrical crystals are amorphous.] Such materials will be discussed in more detail in Chapter 11.

Just as bonds occur with mixed ionic/covalent natures, so we find that there is a continuous change in bonding character in a series of alloys of metals such as Cu–Ni, Cu–Zn, Cu–Ga, Cu–As, and Cu–Se. In fact, the last few of these tend to form definite compounds (that is, the elements combine in simple ratios), indicating the presence of a bond other than metallic. However, at the same time they retain certain characteristically metallic properties.

5·10   Bond strength

Obviously, because there is a smooth transition between covalent and ionic and metallic bonds there is also a smooth variation in bonding strength. However, if we confine our attention to pure covalent, ionic, hydrogen, and van der Waals bonds, some characteristic distinctions emerge.

The strength of a bond is best measured by the energy required to break it, that is, the amount of heat which must be supplied to vaporize the solid and hence separate the constituent atoms. If we measure the heat of vaporization of one gramme mol this must be the energy used in breaking N_o bonds (N_o = Avogadro's number). Figures derived in this way are tabulated in Table 5·1 together with the type of bonding for the solid concerned. As might be expected, the weakest are the van der Waals bond and the hydrogen bond. Next comes the metallic bond, followed by the ionic and covalent bonds whose strengths are nearly comparable.

Table 5·1

Bond type	Material	Heat of vaporization H (kJ/mole)
van der Waals	He	0·08
	N_2	7·79
	CH_4	10·05
Hydrogen:		
O—H——O	Phenol	18·4
F—H——F	HF	28·5
Metallic	Na	109
	Fe	394
	Zn	117
Ionic	NaCl	641†
	LiI	544†
	NaI	507†
	MgO	1013†
Covalent	C (Diamond)	712
	SiC	1185
	SiO_2	1696
	Si	356

†Not strictly the heat of vaporization, since the solid does not break down into its constituent monatomic gases when vaporized. These figures are thus obtained by adding those for a sequence of several changes of state with the required end result.

5·11 Bond strength and melting point

Melting points are also indicative of bond strengths for the following reason. The atoms in a solid are in constant vibration at normal temperatures. In fact the heat which is stored in the material when its temperature is raised is stored in the vibrational energy of the atoms which is proportional to the (absolute) temperature. Thus melting occurs when the thermal vibration becomes so great that the bonds are broken and the atoms become mobile.

By comparing melting points it is therefore possible to illustrate some interesting variations of bond strength. For instance we have already remarked upon the extra strength conferred on the ionic bond by high valency, and how this results in the high melting point of magnesium oxide compared to sodium chloride. It is also clear from the following list of melting points of Group IV elements how bond strength decreases with increasing atomic number.

Material	Atomic number	Melting point (°C)
Carbon (diamond)	6	3,750
Silicon	14	1,421
Germanium	32	937
Tin	50	232

We conclude from these and other data that electrons of higher principal quantum numbers form weaker bonds.

Bond strength is clearly of importance in determining the strength of materials although the form of the crystal structure has at least as much, if not more, influence on strength. Thus we shall leave a discussion of this to a later chapter.

In some solids two kinds of bonding are separately present: the solid forms of the gases are a case in point. Nitrogen (N_2), carbon dioxide (CO_2), oxygen (O_2), and nitrous oxide (NO_2) all have covalent or mixed ionic-covalent bonds within the molecule. In solid and in liquid form the molecules are held together by van der Waals forces. This explains how it is that they solidify only at very low temperatures since the van der Waals bond is easily broken by thermal agitation. On the other hand these gases break down into their constituent atoms (that is, dissociate) only at very high temperatures because of the strong intramolecular bonds. Similar behaviour is found in polymers (see Chapter 12) and in many organic materials.

Problems

5·1 Using the data in Table 5·1, calculate the energy in eV required to break a single bond in MgO.

5·2 Calculate the angles between the tetrahedral bonds of carbon.

5·3 The distance between the atoms in the HF molecule is 0·917 Å. Calculate the maximum attractive force between two HF molecules separated by a distance of 10 Å and compare this with the attractive force between the H^+ ion and the F^- ion when they are separated by 10 Å.

5·4 What kind of bonding do you expect in the following materials: GdO, GdTe, SO_2, RbI, FeC, C_6H_6, InAs, AgCl, UH_3, GaSb, CaS, BN, Cu–Fe?

5·5 The attractive force between ions with unlike charges is $e^2/4\pi\epsilon_0 r^2$ while the repulsive force may be written Ce^2/r^n, where the exponent n is about 10 and C is a constant.

Obtain an expression for the equilibrium distance r_0 between the ions in terms of C and n. Hence deduce an expression for the energy E required to separate the ions to an infinite distance apart, and show that it may be written in the form

$$E = \frac{e^2}{4\pi\epsilon_0 r_0}\left(1 - \frac{1}{n-1}\right)$$

(*Hint*: energy = integral of force with respect to distance.)

5.6 To remove an electron to infinity from a Na atom, 5·13 eV of energy must be used, while adding an electron from infinity to a Cl atom to form the Cl⁻ ion liberates 3·8 eV of energy. What is the energy expenditure in transferring an electron from an isolated sodium atom to an isolated chlorine atom?

Using the expression given in Problem 5·5, calculate also the energy change when the two isolated atoms are brought together to form an NaCl molecule, given that the ionic separation is 2·7 Å at equilibrium, and that the exponent n in the expression for the energy is equal to 10. Compare your result with the figure given in Table 5·1 for the heat of sublimation. Comment on the discrepancy between the two figures.

5.7 Does the decrease in melting point with increasing atomic number which is displayed by the elements in Group IV appear in all the Groups? (Data in Appendix III.)

Self-Assessment questions

1 When the atoms in a solid are separated by their equilibrium distance

A) the potential energy of the solid is least
B) the force of attraction between the atoms is a maximum
C) the force of repulsion is zero.

2 The repulsive component in the force between two atoms which bond together is due primarily to

A) electrostatic repulsion between their nuclei
B) electrostatic repulsion between the ion cores
C) repulsion between the overlapping electron clouds of inner filled shells
D) repulsion between the overlapping electron clouds of the valence electrons.

3 The ionic bond is formed by the mixing of electrons from two atoms.

A) true   B) false.

4 The electrostatic nature of the ionic bond makes it

A) non-directional   B) weak
C) applicable only to Group I and II elements.

5 The bond between the atoms in the hydrogen molecule is called the hydrogen bond.

A) true   B) false.

6 The figure illustrates the formation of a

A) covalent
B) ionic
C) metallic
D) van der Waals bond.

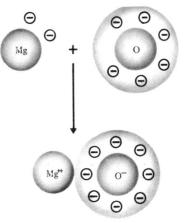

7 The covalent bond between two divalent atoms is formed by

A) two electrons from each atom
B) one electron from each atom
C) an octet of electrons.

8 The covalent bond is formed only between elements in Group IV of the Periodic Table

A) true   B) false.

9 Metallic solids are held together by

A) the attraction between the ion cores
B) the attraction between ion cores and the electrons
C) electrons shared between adjacent pairs of atoms.

10 Metallic bonds are directional.

A) true   B) false.

11 The van der Waals bond is prominent

A) between ionic molecules   B) between covalent molecules
C) when there are no valence electrons available to form primary bonds.

12 The van der Waals bond is due to

A) attraction between magnetic dipoles
B) attraction between saturated covalent bonds
C) attraction between transitory electrostatic dipoles.

13 When a hydrogen atom shares its electron with *two* chlorine atoms, a hydrogen bond is formed.

A) true   B) false.

14 Hydrogen bonds play a part in bonding

A) water of hydration to a salt   B) solid methane (CH_4)   C) ice.

15 In germanium the bonding is

A) covalent   B) tetravalent   C) semi-metallic   D) ionic.

16 Boron (Group IIIB) and oxygen (Group VIB) combine by the formation of

A) ionic   B) metallic
C) mixed ionic-covalent   D) covalent bonds.

17 Silicon (Group IVB) and carbon (Group IVB) combine by the formation of

A) metallic   B) covalent   C) ionic
D) mixed ionic-covalent   E) van der Waals bonds.

18 Caesium (Group IA) and chlorine (Group VIB) combine by the formation of

A) ionic   B) mixed ionic-covalent
C) covalent   D) metallic
E) van der Waals bonds.

19 Solid CO_2 has a low sublimation point because

A) the carbon-oxygen bonds are covalent
B) the intermolecular bonds are van der Waals
C) there are no spare electrons for bonding the molecules to one another.

20 (i) Chlorine solidifies at a temperature in the range

    A) 0–300 K   B) 300 K–1,000 K   C) above 1,000 K

    (ii) because of the presence of

    A) ionic   B) covalent   C) metallic
    D) mixed ionic-covalent   E) van der Waals bonds.

21 The electronegativity of an element is a measure of

    A) the excess of electrons over protons
    B) the number of electrons in the valence shell
    C) the strength with which electrons are attracted to the atom.

Each of the sentences in question 22–27 consists of an assertion followed by a reason. Answer:

    A) If both assertion and reason are true statements and the reason is a correct explanation of the assertion
    B) If both assertion and reason are true statements but the reason is *not* a correct explanation of the assertion
    C) If the assertion is true but the reason contains a false statement
    D) If the assertion is false but the reason contains a true statement
    E) If both the assertion and reason are false statements.

22 Magnesium oxide is bonded ionically *because* magnesium and oxygen are both divalent.

23 Diamond is bonded covalently *because* carbon lies in Group IV of the Periodic Table.

24 The van der Waals bond is weak *because* it bonds only molecules together.

25 Liquid carbon tetrachloride cannot conduct electrically *because* it does not contain C^{4+} and Cl^- ions.

26 A substance whose melting point is –154°C must consist of small covalent molecules *because* only van der Waals forces could give rise to such a low melting point.

27 CO_2 and SiO_2, whose melting points are respectively −56°C and 1,700°C, both form small covalent molecules *because* carbon and silicon are both in Group IV of the Periodic Table.

Answers

1 A	2 C	3 B	4 A
5 B	6 B	7 B	8 B
9 B	10 B	11 B, C	12 C
13 B	14 A, C	15 A	16 C
17 B	18 A	19 B, C	20 (i) A (ii) E
21 C	22 B	23 A	24 C
25 A	26 D	27 D	

CRYSTAL STRUCTURE

6·1 Introduction

It was mentioned in the last chapter that the properties of a solid depend as much upon the arrangement of atoms as on the strength of the bond between them. We met there some examples of such arrangements, diamond and rock salt (NaCl), and also an irregular solid, glass. We know that the regular structures often lead to regular crystalline solid shapes, as in the precious stones, and that amorphous solids do not form crystalline shapes.

Between these extremes lie other categories such as the metals which show regularity in the way their atoms pack together but which do not normally take on crystalline shapes. The concept we use to distinguish the regularity or otherwise of the atomic packing is that of *order*. Thus in a diamond crystal (Fig. 5·9) there is perfect order because each atom is in the same position relative to its neighbours, irrespective of where in the crystal it lies. We say that the crystal has *long range order*.

By contrast, an amorphous solid displays no order–it is completely disordered.

In yet other cases the long range order exists over distances of only 0·1 mm or so. This is still very long compared to the atomic spacing (2 Å). So the term 'long range' still applies. We illustrate this kind of structure in two dimensions in Fig. 6·1. Each region of long range order is called a crystallite or grain and the solid is termed polycrystalline. Metals, some ceramics, and many ionic and covalent solids are often found in this form.

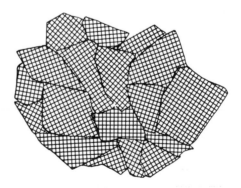

Fig. 6·1 A polycrystalline solid is composed of many grains oriented at random and separated by grain boundaries

30μm

Fig. 6·2 A photomicrograph of an etched polycrystalline specimen of copper showing grain boundaries. Twins (see Chapter 8) are also present.

The boundary between crystallites is called a grain boundary, and it is at this surface that the regularity 'changes direction'. Grain boundaries can be shown up by etching a metal with a suitable acid–the metal dissolves more readily at the boundary–and a micrograph of an etched polycrystalline

Fig. 6·3 X-ray diffraction pattern obtained photographically from a single crystal of Cr_2O_3. The dark lines and the central disc are not produced by the X-rays

specimen of copper is shown in Fig. 6·2. The dimensions of the grains are normally anything upwards of about 1 μm across.

To distinguish those solids in which the order extends right through we call them either single crystals or monocrystalline solids.

Yet another situation arises in materials like glass in which order extends over distances embracing a few atoms only. This we call short range order and it is also found in liquids. Indeed, many solids which display short range order are actually super-cooled liquids and are not strictly in equilibrium in this state. Glass will sometimes actually crystallize if given sufficient time (several hundred years) or if heat treated (see Chapter 11).

Polymers also display order but in a unique way and we leave discussion of their structure to a later chapter.

6·2 Single crystals

Now that we have met the four kinds of solid–monocrystalline, polycrystalline, glassy, and amorphous–we can go on to study the different forms of

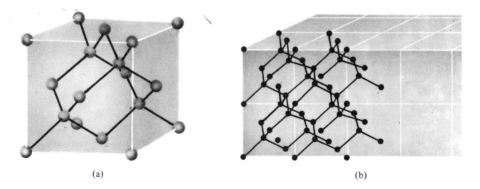

(a)   (b)

Fig. 6·4 (a) The unit cell of the diamond structure; (b) the crystal of diamond is built by stacking many unit cells together

crystalline order. Historically, crystallography began in about 1700 with the study of the external form or morphology of crystals but it was only in the first years of this century that the concept of a crystal as a regular array of atoms was established. Then in 1912 the diffraction of X-rays by crystals was discovered. X-rays are like light waves of very short wavelength, about a few ångstroms. Since this is comparable to the interatomic spacing in a crystal the X-ray is readily diffracted and this property can be used for studying the arrangement of atoms and their spacing. The wavelength of the X-rays can be independently measured by recording their diffraction by a ruled grating.

In Fig. 6·3 we show a diffraction photograph taken using the arrangement shown in Fig. 6·15 of a single crystal of chromium oxide (Cr_2O_3). The exposed dots on the photographic plate are at the intersections of the diffracted beams with the plate. The analysis of diffraction patterns to determine the crystal structure is very complex but the techniques for doing this are now well established. Later in this chapter we shall describe the experimental method and prove Bragg's law. In the meantime we consider only the results showing atomic arrangements which have been deduced from much patient analysis of X-ray photographs. We shall emphasize the way in which the choice of crystal structure follows from consideration of the bonding requirements.

First we must define more carefully what is meant by crystalline order. In a crystal it is possible to choose a small group of atoms which may be imagined to be contained in a regular sided box. The smallest such unit is called a unit cell [Fig. 6·4(a)]. If many such boxes are stacked together, all in the same orientation, like bricks [Fig. 6·4(b)], the atoms are automatically placed in their correct positions in the crystal. The example shown is diamond which may not look familiar as it has been drawn in a different orientation. This unit cell contains 8 atoms but in other solids the unit cell may contain more or fewer atoms and the cell may be other than cubic in shape. The formal definition of the unit cell is the smallest group of atoms which, by repeated translation in three dimensions, builds up the whole crystal.

Table 6·1
Ionic radii

Ion	Radius (Å)	Ion	Radius (Å)
Li⁺	0·78	U⁴⁺	1·05
Na⁺	0·98		
K⁺	1·33	Cl⁻	1·81
Rb⁺	1·49	Br⁻	1·96
Cu⁺	0·96	I⁻	2·20
Be²⁺	0·34	O²⁻	1·32
Mg²⁺	0·78	S²⁻	1·74

6·3   Interatomic distances and ionic radii

It was mentioned in Chapter 5 that the closed electron shell which is created when an ionic or metallic bond is formed can be regarded roughly as a rigid sphere so that adjacent atoms in a solid pack together like solid balls. The radius of the equivalent rigid sphere is termed the ionic radius of the element, and some figures are given in Table 6·1. These have been deduced from the interatomic distances in ionic and metallic solids, measured using X-ray diffraction. It is also possible to calculate them from the diameter of the electronic wave functions obtained by solving Schrödinger's equation. The agreement with measured values is reasonably good in view of the approximations which must be made in performing the complex calculations.

An obvious starting point for the study of crystal structures is therefore the consideration of the kinds of structures which can be built by packing identical spheres together as closely as possible. Later on structures composed of spheres of two different sizes will be discussed.

6·4   Close packed structures of identical spheres

Figure 6·5(a) shows balls packed together on a plane surface. Each is surrounded by six neighbours which just touch it. If we place two such planes in contact each ball in the second layer rests on three from the lower layer. In Fig. 6·5(b) the centres of the spheres in the lower layer are marked A while those of the second layer are marked B. When we come to add a third layer, the centres of the spheres in it can be placed over the gaps in the second layer marked C or over those above the A's, in which case they are directly above the spheres in the first layer. We show a side view of this arrangement in Fig. 6·5(c): the sequence of positioning the layers may be represented as A-B-A-B-A . . . etc. However, in the other case the third layer does not sit over the first so it is positioned differently from either layer A or B. Since the fourth layer must begin to repeat the sequence we obtain [Fig. 6·5(d)] the pattern A-B-C-A-B-C . . . etc.

These two closest packed arrangements are shown again in Fig. 6·6 where the relative atomic positions are seen more clearly and the unit cell is shown in each case. The two structures are known as *face-centred cubic* (f.c.c.) and

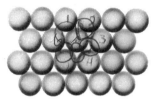

(a) Plan view of spheres resting
     on a plane surface

(b) A second layer of spheres (shown dashed)
     is placed on the first (plan view)

(c) Side elevation of several layers stacked
     A-B-A-B-A

(d) Side elevation of several layers stacked
     A-B-C-A-B-C-A-

Fig. 6·5 The formation of the two close packed structures of spheres

close-packed hexagonal (c.p.h.) respectively. In the former structure the cube shown is not the smallest possible 'building brick' but it is usually referred to as the unit cell.

These close-packed structures should be preferred by solids with van der Waals or metallic bonding because the atoms or ions have closed shells and therefore may be treated as hard spheres. Many metals do indeed crystallize in the face-centred cubic structure and so do all the noble gases excepting only helium, which chooses the close-packed hexagonal structure. Among the metals the c.p.h. structure is exemplified by cobalt, zinc, and several others.

Since ionic solids do not normally consist of atoms of identical size they do not form these close-packed structures (see, however, Section 6·5). Similarly, covalent substances do not normally seek closest packing: the main consideration is that the direction of their bonds be maintained since these are rather rigid.

These two are the only ways in which spheres can be closely packed together but there are some arrangements with looser packing which appear in nature. One of these is shown in Fig. 6·7, and is called body-centred cubic (b.c.c.). As can be seen by inspecting the atom in the centre of the cube, each atom has only eight nearest neighbours as compared to twelve in both of the close-packed structures. This is indicative of the loose packing, in spite of which many metals crystallize in the b.c.c. form. They include all the alkali metals and, most important, iron. The number of nearest neighbours, eight

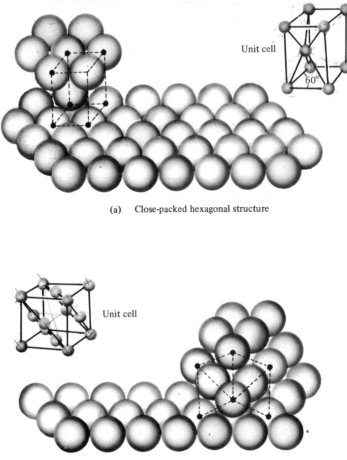

(a)   Close-packed hexagonal structure

(b)   Face-centred cubic structure

Fig. 6·6 Isometric views of the c.p.h. and f.c.c. structures, showing the unit cells

in this case, is called the *coordination number* of the structure.

The simplest cubic arrangement possible, also shown in Fig. 6·7, is called simple cubic. This is a very loosely packed structure and is not of much importance since it is exhibited only by the metal polonium. Each atom has only six nearest neighbours.

6·5 Ionic crystals

It was mentioned earlier that the ratio of the ionic radii is important in determining the structure of an ionic solid. To understand this question it helps to study the problem in two dimensions by arranging coins on a table. With

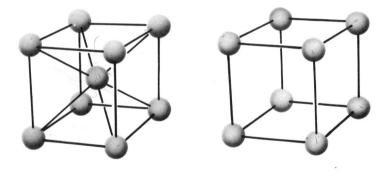

Fig. 6·7 The unit cells of the b.c.c. and s.c. structures

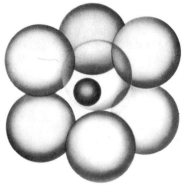

(a)

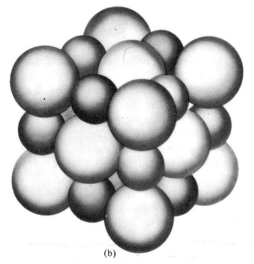

(b)

Fig. 6·8 (a) The CsCl unit cell, showing that if the central cation is too small it cannot touch the anions; (b) the rock salt unit cell

identical coins, it is possible to make each touch six neighbours. Now take a five pence piece and some ten pence pieces the ratio of whose radii is 0·83. It is possible to place five 10p coins in contact with one 5p coin but one cannot make an extended lattice in this way. However, an arrangement with four nearest neighbours, making a square lattice, can be repeated indefinitely in every direction.

The three-dimensional case is illustrated in Fig. 6·8(a) where we show how a positive ion (a *cation*) placed in the body-centred position in a simple cubic structure of *anions* (negative ions) does not touch the anions if it is too small compared to them. Because of this, the forces of attraction and repulsion between the ions do not balance and the structure is not stable. It is easy to show (Problem 6·1) that the critical ratio of radii for which the b.c.c.-like structure is just stable is $r^+/r^- = 0·732$. If the ratio is smaller than this it is possible that the arrangement shown in Fig. 6·8(b) will be stable: with the same radius ratio as in Fig. 6·8(a) the cation, placed in the body-centred position in an f.c.c. arrangement of anions, just touches the six face-centred anions giving a stable arrangement. Figure 6·9(a) and Fig. 6·9(b) show the extended lattices corresponding to those in Fig. 6·8. We see that the lattice in Fig. 6·9(a) can be regarded as two interpenetrating simple cubic lattices, one for each kind of ion. This structure is called after the ionic compound caesium chloride (CsCl) which crystallizes in this way. Each ion is surrounded by eight nearest neighbours and eight is hence the coordination number of this structure.

We have met the structure in Fig. 6·9(b) before in NaCl and it takes the name of the natural crystalline form of this material, rock salt. It can be represented as two interpenetrating f.c.c. lattices, one of anions and the other of cations, shifted with respect to one another by half a cube edge. (Because of this, the structure is basically f.c.c., since it may be regarded as a single f.c.c. lattice in which each lattice point is associated with an Na–Cl atom pair. Note, however, that it is not close-packed in the sense of the previous section.) Since each ion touches six neighbours the coordination number is six and the packing is yet looser than in the body-centred arrangement. The coordination number is actually the same as that of the simple cubic lattice and indeed the two lattices are identical if one ignores the difference between the positive and negative ions.

It is clear that in the rock salt structure also there is a limit to how small the cation may be. This is easily found (Problem 6·2) to be 0·414 times the anion size.

For radius ratios below this we might expect the coordination number to reduce yet further. Since arrangements with five neighbours cannot fit together to give an extended space lattice, four is the next lowest coordination number. Once again we have met this already when we considered carbon–the diamond structure (Fig. 6·4). It should be emphasized that carbon itself has this structure because of the directional nature of its four covalent bonds and not because of its radius ratio (its radius ratio is unity).

The ionic relative of the diamond structure is shown in Fig. 6·9(c). Like rock salt and caesium chloride it can be represented as two interpenetrating lattices, both f.c.c. in this case. It is named after zinc blende (ZnS) which

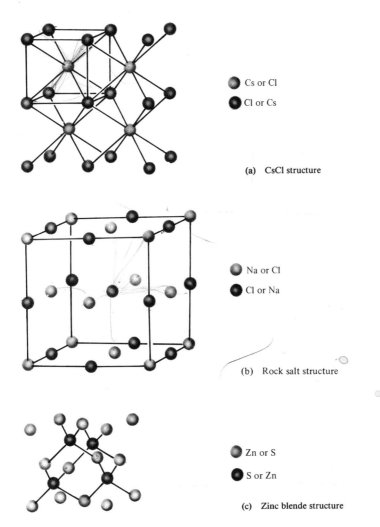

Cs or Cl
Cl or Cs

(a) CsCl structure

Na or Cl
Cl or Na

(b) Rock salt structure

Zn or S
S or Zn

(c) Zinc blende structure

Fig. 6·9 (a) Part of the extended lattice of CsCl, showing how it is built of unit cells; (b) the rock salt structure – a single unit cell is shown here; (c) the unit cell of the zinc blende structure

itself is not wholly ionic but which is the commonest compound which crystallizes in this arrangement.

As an example of a predominantly ionic solid with the zinc blende structure we may take beryllium oxide (BeO) although the bonds here are probably also partly covalent. However, this structure is favoured for the ionic bond since the radii of the two ions are 0·31 Å and 1·70 Å, that is, the radius ratio is 0·222.

It should be noted, however, that the radius ratio acts only as a rough guide to the structure of an ionic compound. There are many crystals which do not follow the rule, like lithium iodide (LiI) with a radius ratio of 0·28 which takes the rock salt structure in spite of the calculated radius ratio limit of 0·414. At the other end of the scale, potassium fluoride, radius ratio 0·98, also crystallizes in the rock salt structure. This is probably because the difference between the stabilities of the caesium chloride and sodium chloride structures is rather small and minor effects such as van der Waals forces can just tip the balance.

To summarize, we present the rules which govern the crystal structures of ionic solids, and finally in Table 6·2 we show how well (or how badly) they are followed.

(i) The non-directional ionic bond favours closest possible packing consistent with geometrical considerations.

(ii) If anion and cation cannot touch one another in a given arrangement, it is generally not stable.

(iii) The radius ratio of the ions roughly determines the most likely crystal structure.

Table 6·2
Ionic structures

Compound	Radius ratio	Predicted structure	Observed structure	Coordination number
NaCl	0·52	NaCl	NaCl	6
AgF	0·93	CsCl	NaCl	6
KBr	0·68	NaCl	NaCl	6
RbCl	0·82	CsCl	NaCl	6
CsCl	0·93	CsCl	CsCl	8
CsBr	0·87	CsCl	CsCl	8
RbCl	0·82	CsCl	CsCl†	8
BeO‡	0·22	ZnS	ZnS‡	4
BeS‡	0·17	ZnS	ZnS‡	4
CuF‡	0·71	ZnS	ZnS‡	4
LiI‡	0·28	ZnS	NaCl	6

†Structure at high pressure.
‡Partially covalent bonding.

6·6 Covalent crystals

It may be thought that the previous section deals with solids of no practical interest. Unfortunately, because of the mixed ionic-covalent bonding in so many solids, the underlying principles of ionic bonding are only clearly exemplified in the alkali halides. Among the covalent solids, however, we have a number of practical importance, mostly in the electronics industry. Carbon, silicon, and germanium all crystallize in the diamond structure as we have already seen. The main requirement in these covalent structures is that the direction of the bonds be appropriate to those of the wave functions of the bonding electrons. The compounds between Group III atoms and Group V atoms (e.g., GaAs and InSb) form the related zinc blende structure. This retains the fourfold coordination of diamond with the nearest neighbours at the corners of an imaginary tetrahedron–an arrangement called tetrahedral bonding. The preference for this structure has nothing to do with atomic radii but is determined by the directionality of the covalent bonds.

Several compounds between Group II and Group VI elements also choose this structure, e.g., ZnS (an obvious case) and BeS, while others like CdS, CdSe are normally found in a modified form of the zinc blende structure called wurtzite. Naturally there is more than an element of ionic bonding present in these compounds. In other sulphides, notably PbS and MgS, the ionic nature of the bond predominates and the rock salt structure results. This is also true of many oxides of Group II elements such as MgO, CaO, although BeO as we saw takes the zinc blende structure on account of the smallness of the Be ion. Among the covalent oxides is ZnO which has the wurtzite structure and CuO, PdO, and PtO which all form yet another kind of structure with fourfold coordination.

Because oxygen has a lower atomic number than sulphur it is more electronegative and hence tends to form ionic bonds more readily. Thus the covalent oxides mentioned above are in the minority.

6·7 Crystals with mixed bonding

We also find some covalently bonded solids with a marked metallic character. Although not really in this category, graphite (an allotrope of carbon) conducts electricity and is a soft material quite unlike diamond. The structure (Fig. 6·10) explains this: it consists of sheets of covalently bonded carbon atoms with threefold coordination, loosely bonded by van der Waals forces to neighbouring sheets. The large distance between sheets is a reflection of the weakness of the van der Waals bond and this allows the sheets to

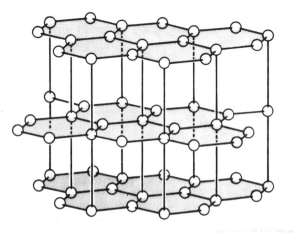

Fig. 6·10 The structure of graphite

slide over one another rather readily. This gives graphite its lubricating property, which is of such great importance since it is one of the few lubricants capable of withstanding high temperatures without vaporizing. In addition there is one free electron per atom since only three covalent bonds are formed. These free electrons cannot easily cross from sheet to sheet because of the large distance involved, so the conducting property is directional, and the resistance is low only along the sheets of atoms.

Although graphite is not strictly metallic, antimony and bismuth certainly are, in spite of their position in Group VB of the Periodic Table. Their structure is related to that of graphite, although the sheets of atoms are not flat, but the important difference is that the distance between sheets is much smaller. This is because the inter-sheet bonding is predominantly metallic and therefore stronger. As a result the conductivity is much higher and less directional than in graphite. Note that the coordination number is much lower than is expected of a metal because of the covalent bonds. However, in liquid form this influence of the covalent bonds is lost and the atoms pack more closely together, so that Sb and Bi both contract on melting. It is this property which makes them important in typecasting where the expansion as they solidify enables them to fill all the interstices in a mould.

6·8 Polymorphism

The impression given so far in this chapter is that, with the exception of carbon, each substance has but one structure which it possesses under all conditions. That this is not the case can be seen from a variety of examples–sulphur, tin, caesium chloride, zinc sulphide, and iron and many others all display allotropic forms which are generally stable over different ranges of temperature and pressure. Thus iron changes from a body-centred cubic form to a face-centred structure at a temperature of about 910°C and a pressure of one atmosphere, and back again to the body-centred structure at 1,400°C and the same pressure. We say that such solids are *polymorphic*, that is, they may have several different forms.

Changes which come about due to the action of heat cannot be discussed simply in terms of bond types and strengths as in this chapter, and we shall defer consideration of them until the nature of heat energy in solids has been explained.

6·9 Miller indices of atomic planes

Sometimes we wish to refer to planes of atoms in a lattice, such as the cube faces in the f.c.c. structure, or the sheets of atoms in graphite. A shorthand notation has been devised for this purpose, comprising a set of three numbers, called *Miller indices* which identify a given group of parallel planes. Since there are many parallel planes, all containing a similar arrangement of atoms, it is clearly unnecessary to distinguish between them.

The Miller indices for a plane may be derived by reference to Fig. 6·11. We

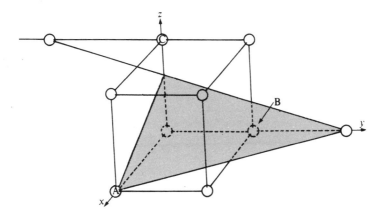

Fig. 6·11 Construction to determine the Miller indices of a plane (shown shaded) in a cubic lattice

first set up three axes along three adjacent edges of the unit cell, which in the case illustrated is cubic. We choose to measure distances along x, y, and z as multiples of (or fractions of) the length of the corresponding edge of the unit cell. Thus the three atoms labelled A, B, and C are all at unit distance from the origin, and this would apply even if the cell were not cubic.

Suppose that we wish to describe a plane, such as the one shown shaded in the figure, which passes through the centres of particular atoms in the lattice. If it intersects the axes at distances x_1, y_1, and z_1 from the origin then these three intercepts are characteristic of this one plane. In the case shown their magnitudes are respectively 1, 2, and 2/3. As these numbers are unique to the plane shown and are not the same for all similar planes, we take their reciprocals, and express them as similar fractions; thus the reciprocals 1, 1/2 and 3/2 may be expressed as 2/2, 1/2, and 3/2. The numerators of these fractions are the *Miller indices* of the plane. The indices are normally enclosed in brackets, so that the plane in question is denoted a (213) plane (pronounced 'two one three plane').

To show that a similar plane to the one chosen has identical Miller indices, we shall consider a plane in Fig. 6·11 which is parallel to the one shaded, and which passes through the atom marked B. The intercepts this plane makes on the x, y, and z axes are 1/2, 1, and 1/3 respectively. Their reciprocals are 2, 1, and 3 which, since none are fractions, are the Miller indices of this plane also. A little thought will show that all similar planes are (213) planes.

In a finite crystal it is useful to be able to distinguish opposite faces, so we allow any intercept and its corresponding Miller index to be negative. A negative sign is denoted by placing a bar above the index. A (21$\bar{3}$) plane may thus make intercepts $+1/2$, $+1$, and $-1/3$ on the axes.

If a plane is parallel to an axis, the corresponding intercept is of infinite length and the Miller index is therefore zero.

Examples. The cube face parallel to both the x and z axes has intercepts ∞, 1, ∞ [see Fig. 6·12(a)] and hence is denoted by (010). The plane having equal

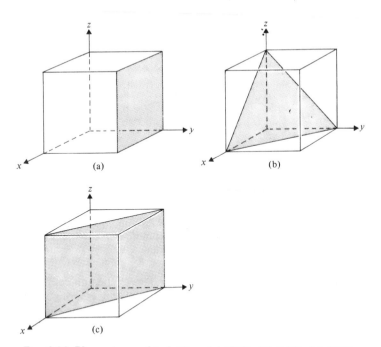

Fig. 6·12 Planes in a cubic lattice: (a) (010), (b) (111), (c) (110)

intercepts on the axes is denoted by (111) and that containing two face diagonals is denoted (110). These are all illustrated in Fig. 6·12.

Note that there are families of planes of the same character with different indices, for example (110), (101), (011), ($\bar{1}$10), ($\bar{1}$0$\bar{1}$), etc. Since one often wishes to refer to such a set of planes, the following notation has been adopted: {110} denotes a family of planes of the same basic type as the (110) planes, and would include all those listed above; similarly, there are families of {100} planes and {111} planes. These three are the types you will meet most frequently in this book.

6·10   Crystallographic directions

The direction of a line of atoms may be described using a similar notation. Here we take a parallel line which passes through the origin (Fig. 6·13) and note the length of the projections of this line on the x, y, and z axes; say they are x_0, y_0, and z_0. Then we reduce these to the smallest integers which bear the same ratio to one another. So if the intercepts were $0·75a$, $0·25a$, and a, the indices of the direction are 3, 1, and 4 since the relation 3:1:4 is the same as $0·75:0·25:1$. This direction is denoted the [314] direction–note the square brackets to distinguish it from the (314) plane.

Just like the principal planes of importance, the directions with which we shall be primarily concerned are [110], [100], and [111]. These are, respec-

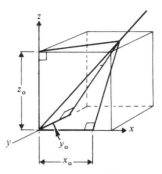

Fig. 6·13 Construction to determine the Miller indices of a direction in a cubic lattice

tively, a cube face diagonal, a cube edge, and a body diagonal. Note that the fact that a plane and its normal have the same indices is a peculiarity of the cubic lattice and does not apply to other lattice types. Families of directions are labelled by special brackets as are families of planes. Thus ⟨100⟩ denotes a family of directions which includes [100], [010], [001], [Ī00], [0Ī0], and [00Ī].

6·11 Theory of crystal structures

In this chapter the atomic arrangements, or lattices, have been discussed only in a qualitative way. An alternative approach is to study them in a formal, abstract way just as if they are geometrical structures. If this is done one finds that only a limited number of arrangements is possible. We have already met one of the reasons for this when we considered the possibility of fivefold coordination. It is just not possible to build a regular, extended lattice with such an arrangement. Similarly, many other polyhedra (that is, arrangements with different coordination numbers) cannot be stacked to form space lattices. There is another example of the packing of coins which demonstrates this: the reader may care to try building a two-dimensional lattice with the seven-sided 50 pence piece. (He is advised not to spend too long trying.) In this kind of analogy the seven-sided coin represents the unit cell of the lattice. As a result of such considerations Bravais discovered that there are only 14 different kinds of unit cell which can form an extended lattice. They are known as Bravais lattices, after their discoverer, and we show them in Fig. 6·14. Each lattice point (represented by a circle in the figure) may accommodate more than a single atom and in many crystals it does so. In consequence there may be many more than just fourteen kinds of crystal structure. These variations may be classified according to the degree of symmetry exhibited by the group of atoms situated at each lattice point. Since there is a limited number of kinds of symmetry the total number of arrangements is restricted to 230.

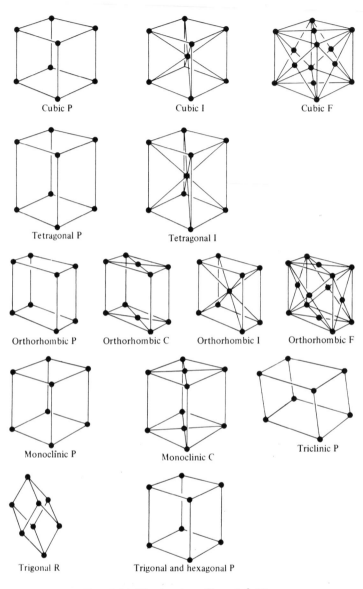

Fig. 6·14 The fourteen Bravais lattices

It is important to note that each of these 230 arrangements may be realized in many different ways using different atoms or molecules so that there is virtually no restriction on the range of different materials which may be made.

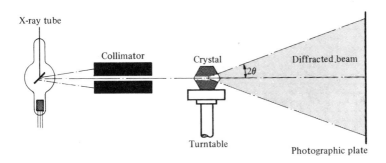

Fig. 6·15 Apparatus for the determination of crystal structure by X-ray diffraction

6·12   Measurements on crystals

Earlier in the chapter it was mentioned that the distances between atomic planes can be measured by using the diffraction of X-rays. The basic requirements are (i) a narrow beam of X-rays of known wavelength, (ii) a means of recording the direction of the diffracted beam (often a photographic film or plate), and (iii) a turntable on which the crystal is mounted so that it may be suitably rotated with respect to the beam. These elements are shown in Fig. 6·15 but the details are beyond the scope of this book and the reader is referred to the list of Further Reading at the end of the book. In this way the dimensions of the unit cell of every common material have been measured. The dimensions quoted in reference books are usually the lengths of the edges of the unit cell, and these are called the *lattice constants*. Naturally, in a cubic material, only one such constant is required to specify the size of the cell.

6·13   Bragg's law

We conclude with a derivation of the law which relates the wavelength of an X-ray to the spacing of the atomic planes–Bragg's law. We begin by assuming that each plane of atoms partially reflects the wave as might a half-silvered mirror. Thus the diffracted wave shown in Fig. 6·16 is assumed to make the same angle, θ, with the atomic planes as does the incident wave.

Now the criterion for the existence of the diffracted wave is that the reflected rays should all be in phase across a wavefront such as BB'. For this to be so, the path lengths between AA' and BB' for the two rays shown must differ by exactly an integral number n of wavelengths λ.

Thus δ, the path difference, is given by $\delta = n\lambda$, ($n = 1, 2, 3 \ldots$). Now since lines CC' and CD are also wavefronts we have

$$\delta = 2EC'$$
$$= 2CE \sin \theta$$
$$= 2d \sin \theta$$

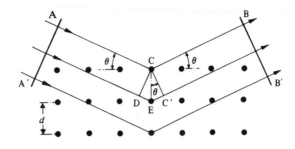

Fig. 6·16 The proof of Bragg's law

Thus

$$2d \sin \theta = n\lambda \ (n = 1, 2, 3 \ldots)$$

This is just Bragg's law.

The assumption that each atomic plane reflects like a mirror may be justified by a similar argument to that above assuming that each atom in the plane is the source of a secondary Huyghen's wavelet (Fig. 6·17). The scattered waves all add up in phase as shown to form the reflected ray.

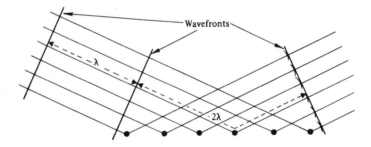

Fig. 6·17 Reflection of X-rays from a plane of atoms

Problems

6·1 In the CsCl structure shown in Fig. 6·9(a) let the ionic radii be r_1 and $r_2 \ (r_2 > r_1)$. Assuming that the anions just touch the cations, calculate the length of the body diagonal, and hence derive the value of the radius ratio at which the structure just becomes unstable.

Using the ionic radii listed in Table 6·1, find the size of the largest impurity ion which can be accommodated interstitially (i.e., between the host ions) in the CsCl lattice with its centre at the point $\frac{1}{2}, \frac{1}{2}, 0$ in the unit cell.

6·2 If the radii of the ions in the NaCl structure are r_1 and $r_2 \ (r_2 > r_1)$, calculate the length of the face diagonal in the unit cell and hence show that the structure is stable only when $r_1 > 0·414 \, r_2$.

6·3 What crystal structure do you expect to find in the following solids: BaO, RbBr, SiC, GaAs, Cu, NH_3, BN, SnTe, Ni? (Use the ionic radii in Table 6·1; the radius of Ba^{2+} is 1·35 Å.)

6·4 Calculate the angles for first-order diffraction (i.e., $n = 1$) from the (100) and (110) planes of a simple cubic lattice of side 3 Å when the wavelength is 1·0 Å. If the lattice were b.c.c., would you expect to find diffracted beams at the same angles?

6·5 How many atoms are there in the unit cell of (a) the b.c.c. lattice (b) the f.c.c. lattice?

6·6 Determine the Miller indices of a plane containing three atoms which are nearest neighbours in the diamond lattice. [Use Fig. 6·4(a).]

6·7 Calculate the density of graphite if the interplanar distance is 3·4 Å and the interatomic distance within the plane is 1·4 Å. (Note: density = mass of unit cell ÷ volume of unit cell.)

6·8 The density of solid copper is $8·9 \times 10^3$ kg/m³. Calculate the number of atoms per cubic metre.

In an X-ray diffraction experiment the unit cell of copper is found to be face-centred cubic, and the lattice constant (the length of an edge of the unit cell cube) is 3·61 Å. Deduce another figure for the number of atoms per cubic metre and compare the two results. What factors might give rise to a discrepancy between them?

6·9 Place the materials listed in Problem 6·3 in the expected order of increasing melting points. Compare with the melting points given in a standard reference book.

6·10 The lattice constants of the metals Na, K, Cu, and Ag are respectively 4·24 Å, 4·62 Å, 3·61 Å, and 4·08 Å. Both Na and K have b.c.c. structures, while Cu and Ag are f.c.c. Calculate in each case the distance between the centres of neighbouring atoms, and hence deduce the atomic radius of each metal.

6·11 What are the Miller indices of the close-packed planes of atoms in the f.c.c. lattice (see Fig. 6·6)?

Self-Assessment questions

1 A solid in which all similar atoms are in similar positions relative to their neighbours is said to

   A) have long range order   B) be crystalline
   C) be amorphous.

2 A polycrystalline material always contains

   A) crystals of different chemical composition
   B) crystallites of the same composition but different structures
   C) crystallites with different orientations.

3 A solid with long range order is ordered over distances large compared with the size of

   A) an electron   B) an atom   C) a crystallite.

4 A material in which the ordered regions contain only a few atoms is termed

   A) amorphous   B) glassy   C) polycrystalline.

5 A unit cell is

   A) the smallest group of atoms which when regularly repeated forms the crystal
   B) a group of atoms which form a cubic arrangement
   C) a unit cube containing the smallest number of atoms
   D) the smallest group of atoms which will diffract X-rays.

6 Which of the following are close-packed structures?

   A) c.p.h.   B) tetragonal   C) b.c.c.   D) f.c.c.
   E) diamond.

7 Close-packed structures are chosen by elements in which the bonding is

   A) directional   B) non-directional   C) metallic.

8 The structure of an ionic crystal is determined primarily by

   A) the relative diameters of the constituent ions
   B) the nature of the chemical bond
   C) the valency of the ions
   D) the coordination number.

9 The number of atoms per unit cell in the f.c.c. structure is

   A) 4   B) 2   C) 14.

10 The number of atoms per unit cell in the b.c.c. structure is

   A) 6   B) 9   C) 2.

11 From the diagram it can be seen that the coordination number of the CsCl structure is

   A) 4   B) 6   C) 8   D) 10   E) 12.

⬤ Cs or Cl
⬤ Cl or Cs

CsCl structure

12 The coordination number of the rock salt structure is

   A) 4   B) 6   C) 8   D) 10   E) 12.

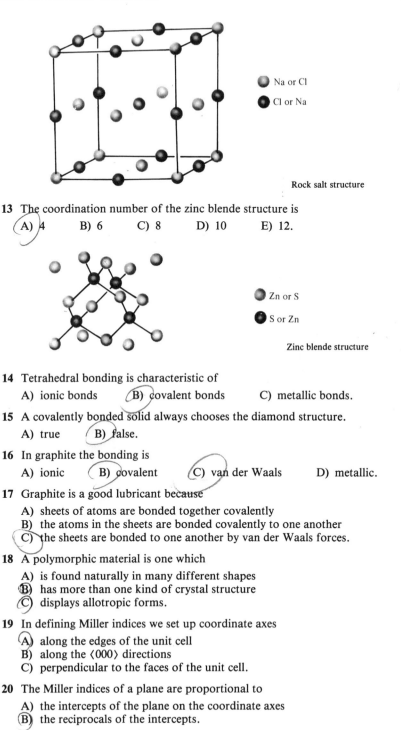

Na or Cl

Cl or Na

Rock salt structure

13 The coordination number of the zinc blende structure is

    A) 4   B) 6   C) 8   D) 10   E) 12.

Zn or S

S or Zn

Zinc blende structure

14 Tetrahedral bonding is characteristic of

    A) ionic bonds   B) covalent bonds   C) metallic bonds.

15 A covalently bonded solid always chooses the diamond structure.

    A) true   B) false.

16 In graphite the bonding is

    A) ionic   B) covalent   C) van der Waals   D) metallic.

17 Graphite is a good lubricant because

    A) sheets of atoms are bonded together covalently
    B) the atoms in the sheets are bonded covalently to one another
    C) the sheets are bonded to one another by van der Waals forces.

18 A polymorphic material is one which

    A) is found naturally in many different shapes
    B) has more than one kind of crystal structure
    C) displays allotropic forms.

19 In defining Miller indices we set up coordinate axes

    A) along the edges of the unit cell
    B) along the ⟨000⟩ directions
    C) perpendicular to the faces of the unit cell.

20 The Miller indices of a plane are proportional to

    A) the intercepts of the plane on the coordinate axes
    B) the reciprocals of the intercepts.

21 In a cubic structure the [111] direction and the (111) plane are parallel.

A) true   B) false.

22 The figure shows a cubic unit cell; the planes shown shaded in (a), (b) and (c) are

A) (110)   B) (111)   C) (102)   D) (010)   E) (100)

F) (001).

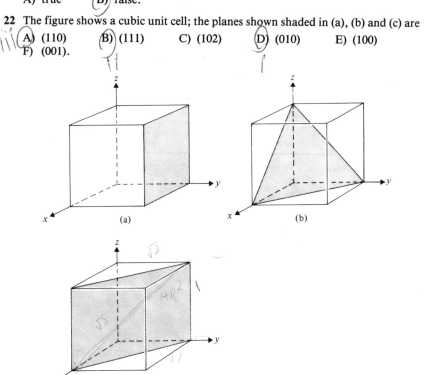

23 The Miller indices of all directions parallel to one another are identical.

A) true   B) false.

In the next two questions, R is the atomic radius and the material is an element.

24 The area of the (100) plane in a f.c.c. unit cell is

A) $8R^2$ B) $\sqrt{5}R^2$ C) $\dfrac{16}{3}R^2$.

25 The area of the (101) plane in a b.c.c. unit cell is

A) $4\sqrt{5}R^2$ B) $\dfrac{16}{3}R^2$ C) $\dfrac{16\sqrt{2}}{3}R^2$.

26 The [110] direction in a cubic unit cell is parallel to

A) the diagonal of one face of the cell

B) the body diagonal of the cell

C) one edge of the cube.

27 X-ray diffraction can be used to deduce

A) the regularity of atomic stacking in a crystal

B) the magnitude of the interatomic spacing
C) the orientation of particular atomic planes relative to the crystal faces
D) the electronic structure of the atoms.

28 In the equation $n\lambda = 2d \sin \theta$ known as Bragg's law, the angle θ is

A) the angle between the incident and diffracted X-ray beams
B) the angle between the incident beam and the normal to the diffracting planes
C) the angle between the incident beam and the diffracting planes.

Each of the sentences in questions 29–34 consists of an assertion followed by a reason. Answer:

A) If both assertion and reason are true statements and the reason is a correct explanation of the assertion
B) If both assertion and reason are true statements but the reason is *not* a correct explanation of the assertion
C) If the assertion is true but the reason contains a false statement
D) If the assertion is false but the reason contains a true statement
E) If both the assertion and reason are false statements.

29 An ion in a solid can be assigned a fairly well defined diameter *because* the interionic distance is independent of the kind of ion with which it bonds.

30 Nickel has a close-packed structure *because* its bonding is metallic.

31 Caesium chloride does not exhibit the rock salt structure *because* its radius ratio is unsuitable for closest packing.

32 The coordination number in a diamond crystal is equal to the valency of carbon *because* the valency controls the number of bonds which can be formed.

33 Graphite remains a good lubricant up to high temperatures *because* the bonding between planes of atoms is van der Waals.

34 Graphite does not conduct electricity as well as does a metal *because* electrons cannot move freely from one sheet of atoms to the next.

Answers

1 A, B	2 C	3 B	4 B
5 A	6 A, D	7 B, C	8 A
9 A	10 C	11 C	12 B
13 A	14 B	15 B	16 B, C
17 C	18 B, C	19 A	20 B
21 B	22 (a) D (b) B	23 A	24 A
	(c) A		
25 C	26 A	27 A, B, C	28 C
29 C	30 A	31 C	32 C
33 B	34 A		

7

THERMAL PHENOMENA IN MATERIALS

7·1 Introduction

The perfect crystal we have met in Chapter 6 is unfortunately an ideal which is rarely attained. In subsequent chapters we shall learn of the many ways in which real materials differ from the ideal, regular structure proposed for the perfect crystal. In this chapter we propose to concern ourselves with a perturbation which has already been mentioned–the atomic vibration which is associated with the storage of thermal energy in a solid. By considering this motion we can learn something of the relationship between thermal energy and temperature, and hence of specific heat and thermal expansion.

More important for the purpose of this book, we shall pave the way for an understanding of solid state phase changes, such as the transformation of b.c.c. iron into f.c.c. iron at 910°C, which was mentioned in Chapter 6. Such changes are basic to a study of the microstructure of alloys like steel and brass, which in turn leads to an understanding of many of their properties.

Finally, two further effects related to heat energy will be discussed–the influence of temperature on chemical reactions (because of their importance in corrosion of materials) and on atomic diffusion.

7·2 'Hard sphere' model of a crystal

In the earlier chapters of this book great stress was laid on the quantum-mechanical aspects of the electrons in an atom. In particular, the idea of the particulate electron confined to a specific orbit was shown to be an inadequate approximation for some purposes, and the electron distribution around the atom was best thought of in terms of charge clouds. Nevertheless, when considering crystal structures, especially of metallic and ionic solids, we have seen that the arrangements of the atoms are consistent with a picture of them as 'hard' spheres, closely packed together. This is justified quantum-mechanically by (a) the spherical symmetry of closed electron shells and (b) the repulsive force between overlapping electron clouds, which rises very rapidly as the overlap increases. In this and subsequent chapters on thermal and mechanical properties we shall make the tentative assumption that no serious error is involved in treating atoms and ions in the solid state as hard spheres.

It must be borne in mind that this approximation may break down, in which case the observed behaviour of solids will differ slightly from that which we might expect. Such deviations are, in fact, quite small and are rarely important.

7·3 The nature of thermal energy

How do we know that heat energy is associated with the motion of atoms, and what meaning can be attached to our mental picture of vibrating atoms?

The first direct piece of evidence for the nature of heat applies not to solids but to gases, and is the so-called 'Brownian motion' named after the botanist Brown. In 1828 he observed that, when still air containing specks of pollen dust is viewed in a microscope, the dust can be seen in continuous but irregular motion. The cause suggested for this is that the particles are being bombarded from random directions by gas atoms which are themselves continuously moving about because they have thermal energy. Confirmation of this view comes from the success of the kinetic theory of gases, which is based on this idea of continuous motion of the atoms, and which explains in considerable detail the gas laws of Boyle and Charles and also many of the deviations from them.

7·4 Summary of the kinetic theory of monatomic gases

The starting point for the kinetic theory of gases is the idea that the pressure exerted by a gas on the walls of the containing vessel arises from continual bombardment by the gas atoms. The atoms of a monatomic gas are regarded for the purpose of calculation as small elastic spheres–a viewpoint discussed above. The rate of change of the momentum of the gas atoms as they bounce off the wall can be calculated, and is equal to the force which they exert on the wall. The latter is just the pressure, P, times the area. The momentum change depends on the speed, c, of the gas atoms (we use the term speed to represent the magnitude of the velocity, independently of its direction). The calculation, which will be familiar to many readers, is to be found in any book on the kinetic theory of gases (see the book by Tabor mentioned in the Further reading). The result is

$$P = \frac{1}{3} \frac{N}{V} m\overline{c^2} \tag{7·1}$$

where $\overline{c^2}$ is the mean square speed of the gas atoms, N is the number of atoms in the volume V and m the mass of an atom.

Now the total kinetic energy, E, of the atoms is just

$$E = \tfrac{1}{2}Nm\overline{c^2} \tag{7·2}$$

so by combining Eqns (7·1) and (7·2) we find that

$$PV = \tfrac{2}{3}E \tag{7·3}$$

By comparison with the gas laws we see that the kinetic energy of the gas atoms is proportional to temperature. In fact, absolute temperature on the ideal gas scale of temperature may be *defined* by the relationship

$$\tfrac{1}{2}m\overline{c^2} = \tfrac{3}{2}kT \tag{7·4}$$

where k is called Boltzmann's constant and has the value $1\cdot380 \times 10^{-23}$ joules per kelvin. It is a *universal* constant, being the same for all ideal gases.

Substituting from Eqn (7·4) into Eqn (7·3) we obtain

$$PV = NkT$$

which is the *equation of state* of an ideal gas. If the volume V contains N_0 atoms, where N_0 is Avogadro's number, the equation of state is usually written as $PV = RT$, where R is the gas constant for a gramme molecule.

We may obtain some idea of the magnitude of the molecular speed from Eqn (7·1) if we note that the expression Nm/V is just the total mass divided by the volume–that is, the density ρ of the gas. So the mean square velocity is given by

$$\overline{c^2} = 3P/\rho \tag{7·5}$$

As an example we may take hydrogen, whose density at standard temperature and pressure is $0\cdot09$ kg/m³. Since atmospheric pressure is approximately 10^5 N/m², this gives a root mean square velocity of $1\cdot84$ km/s (about 4,000 miles per hour!). A similar calculation for carbon dioxide gives a velocity of $0\cdot393$ km/s at 0°C.

The average *energy* is also of interest. From Eqn (7·4) and the value of k we find that the average energy per molecule at 0°C is $5\cdot65 \times 10^{-21}$ J, or $3\cdot53 \times 10^{-2}$ eV. This energy is of course very small compared with the energy required to excite electrons to higher states within the atoms. The number of atoms containing electrons in excited states is therefore entirely negligible and we conclude that the energy $E = \tfrac{1}{2}Nm\overline{c^2}$ given in Eqn (7·2) represents the total heat energy of the gas atoms. The specific heat of a monatomic gas must, therefore, be a constant, independent of temperature, since the amount of heat which must be added to raise the temperature by one degree is the same whatever the temperature. The specific heat can be defined as the differential of heat energy with respect to temperature and so is constant when the internal energy, E, is proportional to temperature.

7·5   Energy distributions

The definition of temperature quoted above, Eqn (7·4), might lead one to expect the heat energy of all matter to be proportional to temperature. But this is manifestly not so, for the specific heats of solids, and polyatomic gases are not by any means constant and independent of temperature. Even monatomic gases deviate from the ideal behaviour in this respect as they approach the point of condensation to the liquid or solid state.

We naturally ask, then, what *is* the same in two substances which are at the same temperature? And the answer lies, not in the equality of total energy, but in the way in which that energy is distributed among all the atoms in the substance, as we shall now explain.

The picture of a gas in the kinetic theory is one of molecules which are moving in all possible directions at very high speeds. Naturally the molecules will frequently collide with each other and may gain or lose velocity at each collision, although the sum of the energies of the two colliding atoms must remain constant. Thus their speeds and their kinetic energies may vary wildly about an average value but, provided that the gas is in a condition of thermal equilibrium, it is possible to calculate the distribution of the molecular population among the various possible velocities by determining the number having each particular value of velocity.

Details of the calculation can be found in the book by Tabor mentioned in the Further reading; here we simply quote the results.

Since we wish to express the distribution of the particles among all possible velocities from zero to infinity we can do this by stating the number of particles with a particular velocity. However, if all velocities are possible we must decide on a minimum separation between the velocities at which we, as it were, count the number of particles. This can be expressed mathematically by letting $f(c)\mathrm{d}c$ be the number of particles having velocities between c and $(c + \mathrm{d}c)$ where $\mathrm{d}c$ is a small increment of velocity. The quantity $f(c)$ is then the number of particles per unit velocity range centred about the value c. It is argued in Section 7·7 that this is given by

$$f(c) = A \exp(-\tfrac{1}{2}\beta mc^2) \tag{7·6}$$

where A and β are constants, and m is the mass of each particle.

It will be remembered that a velocity has a magnitude and a direction and Eqn (7·6) gives the number of particles having a particular velocity in a given direction. In calculating quantities like the total kinetic energy of the particles, we need to know the total number of particles per unit volume having a velocity in the range c to $(c + \mathrm{d}c)$ irrespective of direction. We define this as $N(c)\mathrm{d}c$ and, by a process of integration, we can show it to be given by

$$N(c)\mathrm{d}c = 4\pi c^2 N \left(\frac{m}{2\pi kT}\right)^{3/2} \exp\left(-\frac{mc^2}{2kT}\right) \mathrm{d}c \tag{7·7}$$

where m is the mass of the particle, N is the number of particles per unit volume, T is absolute temperature, and k is Boltzmann's constant. This is the famous Maxwell-Boltzmann distribution law and describes, in mathematical terms, how an assembly of fast-moving particles distribute themselves over a range of velocities when they are constantly colliding with each other. We may express the result in the form of graphs of $N(c)$ against velocity, c, as in Fig. 7·1. This shows, for the case of hydrogen, how $N(c)$ varies with velocity at different temperatures and it will be seen that, as temperature rises, the velocities 'spread out' over a wider range. This is to be expected since, if on average they move faster, they will collide more often. By definition of $N(c)$ it follows that $\int_0^\infty N(c)\mathrm{d}c$ is equal to N, the total number of particles per unit volume. But $\int_0^\infty N(c)\mathrm{d}c$ is the area under the graph of $N(c)$ against c, so that the area under each of the curves of Fig. 7·1 is the same.

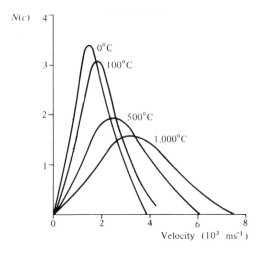

Fig. 7·1 Maxwell–Boltzmann distribution of velocities for hydrogen at different temperatures

In a monatomic gas the expression $mc^2/2$ is just equal to the thermal energy, E, of an atom, so that Eqn (7·7) can be expressed in terms of E. In doing this it must be remembered that the range of energies, dE, corresponding to the range, dc, of speed is the differential of $mc^2/2$ with respect to c. Accordingly, Eqn (7·7) can be rewritten

$$N(E)\,dE = \frac{2\pi NE^{1/2}}{(\pi kT)^{3/2}} \exp\left(-E/kT\right)dE \tag{7·8}$$

where $N(E)\,dE$ is defined as the number of atoms having an energy between E and $(E + dE)$.

Although Eqn (7·8) applies only to a monatomic gas, a corresponding expression can be derived for polyatomic gases in which some of the molecular energy is due to rotation of the molecule or vibration of its atoms with respect to one another. In each case the expression for $N(E)$ contains the factor $\exp\left(-E/kT\right)$, where E is now the total molecular energy.

In general we may write

$$N(E) = C\exp\left(-E/kT\right) \tag{7·9}$$

where C is a factor which depends on the particular system to which the equation applies. In most cases, C is a function of energy E and temperature T which varies only slowly with E and T when compared to the exponential.

Equation (7·9) applies to any assembly of atoms in which the effects of energy quantization are negligible (in its derivation quantum effects are entirely ignored). The exponential factor is therefore a very general one–it is known as the Boltzmann factor, and it arises from the random nature of all thermal energy distributions, a feature common to solids, gases, and liquids. The Boltzmann factor can, in fact, be deduced from much more general statistical considerations than those of gas kinetics alone. On the other hand, the factor C in Eqn (7·9) is dependent on the detailed distribution of available energy levels and, unlike the exponential, is not governed by the randomness of thermal processes.

It is possible to interpret Eqn (7·9) in a very simple way if we first remember that in all real situations the energy is quantized, even though the available quantum states may often be so closely spaced that for practical purposes energy may be treated as a continuous variable. In this case, the factor C may be written as the product of two quantities. One of these may be written dS/dE, where dS is the number of quantum states (i.e., states with distinguishable sets of quantum numbers) available in the energy interval dE. This factor, usually called the *density of states* is therefore the number of states per unit energy interval at the energy E, and in general it depends on E itself. The other factor we shall call A.

We can thus rewrite Eqn (7·9) in the form

$$N(E)dE = \frac{dS}{dE} \ A \exp(-E/kT)dE \tag{7·10}$$

where $N(E)dE$ is, of course, the number of *atoms* having energies due to thermal vibrations within the interval dE. The factor $A \exp(-E/kT)$ therefore represents the *fraction* of quantum states which are actually occupied. It consequently also represents the *probability*, $p(E)$, that an individual quantum state of energy E is occupied. Furthermore, because the number of occupied states at energy E is equal to the number of atoms having energy E, the probability that an individual *atom* has an energy E is proportional to the same factor.

If this probability is small compared to unity, then A does not depend upon the energy E and we can say that the relative numbers of atoms in states of differing energies E_1, E_2, E_3, etc. are in the ratios $\exp(-E_1/kT)$: $\exp(-E_2/kT)$: $\exp(-E_3/kT)$, etc. This very simple result will prove useful in discussing many atomic processes in this book.

It is worth remarking that these ratios do not *necessarily* give the relative numbers of atoms with energy E_1, E_2, E_3, etc., since there may be different numbers of quantum states at these energies. In this case recourse to Eqn (7·10) gives the correct result.

7·6 Some other energy distributions

We have implied above that, when the number of atoms in a quantum state becomes comparable to unity, the 'constant' A in Eqn (7·10) depends on the energy E. We can see that this must be so since, if there is a high probability of a particle being in a particular state, then the probability of its existing in other states is correspondingly low. This cannot be the case unless A depends upon energy.

The correct expression for the probability $p(E)$ can be deduced with the aid of statistical theory, with the result

$$p(E) = \frac{1}{(1/A) \exp (E/kT) - 1} \tag{7.11a}$$

where the quantity A is now *independent* of energy.

The reason for retaining the symbol A can be seen by noticing that, when it is very small the 1 in the denominator may be neglected. Equation (7.11a) then becomes approximately the same as Eqn (7.10); Eqn (7.11a) is often referred to as the Bose–Einstein distribution.

At this point it is worth remarking that Eqns (7.10) and (7.11a) need not apply to assemblies of atoms alone, but to any system of 'particles' which possess thermal energy and are in thermal equilibrium. The only exceptions are those particles, like electrons, to which Pauli's principle applies, for then only particles with different spin are allowed in each quantum state. This leads to the Fermi–Dirac probability $f(E)$, the derivation of which will be found in more advanced texts on solid state physics (see Further Reading). The probability that a quantum state, E, will be occupied by an electron is given by

$$f(E) = \frac{1}{1 + \exp [(E - E_F)/kT]} \tag{7.11b}$$

where E_F is a constant called the Fermi energy and is discussed in Chapter 13. Unless $(E - E_F)$ is of the order of, say, two times kT or less, the exponential term will be much greater than 1 so the expression can be approximated to $f(E) \simeq \exp [-(E - E_F)/kT] = \exp (E_F/kT) \exp (-E/kT)$ and we again have the Boltzmann distribution of the type of Eqn (7.10).

In this book we shall be interested mainly in situations in which the occupation number p of each quantum state is fairly small compared to unity.

In summary of this subject, we may say that:

(a) In most cases of interest in which atoms are involved, the Boltzmann distribution, Eqn (7.9) or (7.10), is applicable. A particularly useful form of it states that the numbers of atoms in quantum states of energies E_1, E_2, E_3, etc. are in the ratios $\exp (-E_1/kT)$: $\exp (-E_2/kT)$: $\exp (-E_3/kT)$, etc.

(b) In cases where the probability that a state is occupied is not small, the Bose–Einstein distribution is applicable.

(c) For electrons, yet another distribution must be used, since they obey Pauli's principle. This is the Fermi–Dirac probability distribution. .

7.7 Thermal equilibrium

We are now in a better position to understand the meaning of temperature. Since the Boltzmann factor is applicable to assemblies of atoms at all temperatures except those very close to absolute zero, it can be seen that this is the common factor between two bodies which are in thermal equilibrium. Hence, the probability that a quantum state of a given energy is filled is *the same in both bodies* when they are at the same temperature.

The significance of this fact can be better understood by considering how heat is exchanged between two bodies A and B when they are placed in contact. The atoms at the common surface can 'collide' with one another because they are in thermal motion, and these collisions involve the transfer of energy.

Let us suppose that both A and B are at the same temperature and consider the atoms in body A in a quantum state of energy E_1 and those in body B in a quantum state E_2. The probability that an atom in A with energy E_1 can collide with a B atom having energy E_2 is the probability that both states will be simultaneously occupied; it is therefore the product $f_1 F_2$ of the individual probabilities f_1 and F_2 that E_1 and E_2 be separately occupied. For the moment we assume that f (which refers to body A) and F (body B) depend upon energy E in different ways, as illustrated in Fig. 7·2.

After the collision, the two atoms may have any energy provided that the sum of their energies is $E_1 + E_2$. One atom can therefore have an energy E_3 anywhere in the range 0 to $E_1 + E_2$ while the other atom must have the remainder, say E_4. This is illustrated schematically by the arrows on the graph in Fig. 7·2.

The probability that the A atom moves to a state with energy E_3 depends on the number of states it has to choose from–in fact it is equal to the reciprocal of that number. The total number of available states is the number lying between the energies 0 and $E_1 + E_2$, and we shall denote it by N_a. The collision probability is then proportional to

$$\frac{f_1 F_2}{N_a}$$

If the two bodies are in equilibrium, this expression must equal the collision probability for the reverse process: that is, when an A atom with

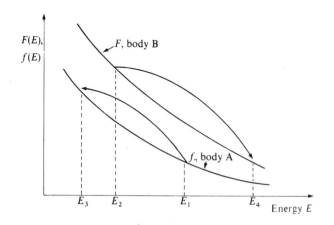

Fig. 7·2 Effect of an atomic collision on the Boltzmann distribution $f(E)$ and $F(E)$ for two bodies A and B

energy E_3 and a B atom with energy E_4 collide and end up with the respective energies E_1 and E_2. This probability is proportional to

$$\frac{f_3 F_4}{N_a'}$$

where N_a' is the number of states lying between the energies 0 and $E_3 + E_4$.

The reason why the two probabilities must be equal is that, were they not, atoms would accumulate in, say, levels E_3 and E_4, while the population in levels E_1 and E_2 diminished. In this situation the probability distributions would change when a body were placed in contact with another at the same temperature, which is clearly unreasonable.

It follows that

$$\frac{f_1 F_2}{N_a} = \frac{f_3 F_4}{N_a'}$$

But by the laws of conservation of energy, $E_1 + E_2 = E_3 + E_4$, so that the numbers N_a and N_a' must, according to their definitions, be equal. They therefore cancel in the above equation, leaving the result

$$\frac{f_1}{f_3} = \frac{F_4}{F_2}$$

This is a statement of the condition of equilibrium between the two bodies. It can only be satisfied for all energies if f and F depend on energy in the same way, and have the same value for the same energy. Thus if $f = \exp(-E/kT_a)$ and $F = \exp(-E/kT_b)$, where E is the energy and T_a and T_b are the temperatures of A and B respectively, then for equilibrium we require that $T_a = T_b$, i.e., that the bodies should be at the same temperature.

It is of interest to note that, if A and B are just two halves of the same body, the condition of equilibrium is

$$f_1 f_2 = f_3 f_4$$

Taking logarithms, we have

$$\log f_1 + \log f_2 = \log f_3 + \log f_4$$

which may be compared with the equation

$$E_1 + E_2 = E_3 + E_4$$

The comparison suggests that $E_1 \propto \log f_1$, $E_2 \propto \log f_2$, etc. The comparison also suggests a solution for f of the type

$$\log f \propto E$$

or

$$f(E) = A \exp \beta E \qquad (7 \cdot 12)$$

This is the nearest we can get to a simple justification of the Boltzmann probability factor, in which of course $\beta = -1/kT$. It leads, in particular, to the result quoted in Eqn (7·6).

7·8 Kinetic theory of solids – interatomic forces

We now return to the problem of heat energy in solids. It has already been noted that the specific heat of a solid varies considerably with temperature, and in fact it tends to zero at the absolute zero of temperature. Figure 7·3 gives an example of the form of the temperature dependence for a pure solid which undergoes no change in crystal structure in the temperature range of interest.

In view of this marked difference from the behaviour displayed by monatomic gases whose specific heat is constant, we must investigate more closely the way in which heat energy is stored in solids.

With our picture of the solid as a lattice of atoms held together by bonds which are not perfectly rigid, we can imagine that the atoms can be displaced from their equilibrium positions as if they were mounted on springs and, since they have mass, they can oscillate about a mean position with an amplitude which depends upon the amount of heat energy possessed by the solid (see Fig. 7·5).

Let us make this picture more precise. In Chapter 5 we saw that a bond between two atoms or ions provides a net attractive force which depends upon their separation and which is balanced at the equilibrium distance by a repulsion due to overlap of the electronic charge clouds.

We can express this behaviour for a pair of atoms by assigning to them a potential energy, V, which is a function of the distance, r, between them. Thus we write

$$V = -\frac{C_1}{r^n} + \frac{C_2}{r^m} \qquad (7·13)$$

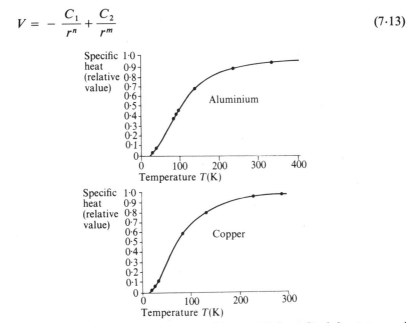

Fig. 7·3 The dependence on temperature of the specific heat C_p of aluminium and copper, measured at a constant pressure of 1 atmosphere

where, following the usual convention for potential energy, the term arising from an *attractive* force is given a negative sign and that from a repulsive force a positive sign. In this expression C_1 and C_2 are constants and n and m are to be determined; in the case of electrostatic attraction between unlike charges (the case of the ionic bond) for example, n would be unity.

Again, by definition, the net force, F, between the atoms will be given by

$$F = -\frac{dV}{dr} = \frac{-nC_1}{r^{n+1}} + \frac{mC_2}{r^{m+1}} \tag{7.14}$$

and, if the atoms are in equilibrium, this force will be zero at some critical separation r_0, so that

$$0 = -\frac{nC_1}{r_0^{n+1}} + \frac{mC_2}{r_0^{m+1}}$$

These equations are illustrated graphically for arbitrary values of n and m in Fig. 7.4; they are known as the Condon–Morse curves after their originators.

While Fig. 7.4 is constructed for a pair of atoms, similar behaviour will obviously occur between each pair of atoms in a crystal lattice. The actual values of n, m, C_1 and C_2 will depend upon the nature of the crystal bonding forces and on the crystal structure itself. They have, for example, been calculated for ionic crystals such as rock salt, in which n is unity, as would be expected for electrostatic attraction, and m is in the region of ten.

Now imagine it is possible to push one atom of the crystal aside from its equilibrium position, from point a to point b in the curve of Fig. 7.4, while keeping the rest of the lattice undisturbed. Figure 7.4(b) shows that there is an attractive force, ΔF, which is proportional to the displacement, Δr, if the latter is very small (it is exaggerated in the figure). An equal and opposite dis-

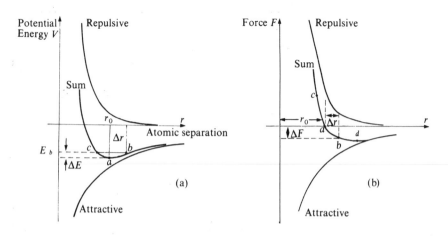

Fig. 7.4 The Condon–Morse curves: (a) potential energy V and (b) force F as a function of interatomic spacing r

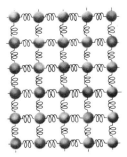

Fig. 7·5 Mechanical model of an elastic crystal lattice

placement, to point c, produces an (approximately) equal and opposite force so that, remembering the finite atomic mass, we perceive that the conditions for simple harmonic motion are met.

During the atom's motion its total energy remains constant, for the law of conservation of energy must be obeyed. This energy is made up partly of kinetic, partly of potential energy. At the turning points of its motion (points a and b in Fig. 7·4) the atom is momentarily stationary and its kinetic energy is necessarily zero. At these points, then, its potential energy–given by the curve–is equal to its total energy. The horizontal line bc in Fig. 7·4(a) therefore represents a plot of the total energy against position, and the vertical distance between this line and the potential energy curve gives the kinetic energy of the atom. The maximum value of this, shown as ΔE in Fig. 7·4(a), is the energy which would be given up if the atom were to come permanently to rest.

Now from the kinetic theory of gases, all the atoms in a gas would be at rest in the limit of zero temperature. This can be seen from Eqn (7·4). It is reasonable to suppose that the same applies to solids and therefore the energy ΔE in the above discussion represents the heat energy possessed by the atom in question.

Naturally, this hypothetical situation does not occur in nature; a single atom cannot vibrate in isolation, for in doing so it exerts forces on its neighbours and sets them moving also. We may imagine the crystal to be rather like the model depicted in Fig. 7·5, where the 'atoms' are joined by springs and the motion will obviously be complicated. We consider this, more realistic, model in Section 7·10.

7·9   Thermal expansion and the kinetic theory

Before pursuing the subject, it is of interest to note that the asymmetry of the Condon–Morse curves of Fig. 7·4(a) means that expansion will occur when a solid is heated. In Fig. 7·6 is shown the same curve with two different total energy lines superimposed on it. It is clear that the average separation between the atoms (point a) increases with the total energy and hence the temperature, because of the asymmetry of the potential energy curve.

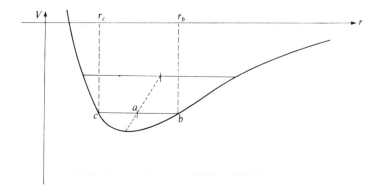

Fig. 7·6 Relationship of thermal expansion to the Condon–Morse curve for a solid

This is the first direct evidence of the correctness of the model we have chosen. In addition to the fact of expansion itself, it is possible to demonstrate experimentally that the length of a solid bar fluctuates randomly by a very small amount simply because of the randomness of the thermal vibrations of its atoms. The relative motion of two ends of a bar can be magnified until it can be visually observed, in a manner analogous to the observation of Brownian motion.

7·10   Lattice waves and phonons

The model of the motion of atoms about their idealized equilibrium positions in a solid is that each atom vibrates in a potential well of the force fields of its neighbours. Applying the kinetic theory of gases, the vibrations of all atoms are treated as independent (referred to as the Einstein model) and this is a good enough approximation in some cases, especially at high temperatures. However, if one or more atoms move in unison, the forces between them tending to restore them to equilibrium positions are reduced, and this must be taken into account.

A direct consequence of this cooperative behaviour of the atoms is that only certain elastic vibrations can exist in a crystal. In order to specify these elastic vibrations of the atoms it is convenient to describe them as waves which may be designated by wavelength, direction of propagation, frequency of vibration, and direction of polarization. There are two distinct types of waves called acoustical and optical. Acoustical waves occur in all crystalline materials and are produced by neighbouring atoms (or ions) moving in the same direction, but by slightly differing amounts, so that the spacing between the atoms remains relatively constant. This is shown graphically in Fig. 7·7(a) in which the displacements of the atoms in a linear chain from their ideal rest positions are plotted. However, the neighbouring ions in an ionically bonded crystal may move in opposite directions; this leads to a change in the spacing between the ions so producing electric dipole moments as shown in Fig.

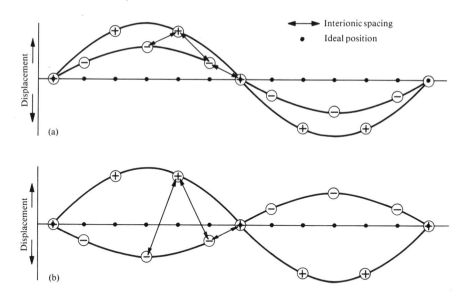

Fig. 7·7 The displacement of ions, on an exaggerated scale, in an ionic crystal
giving (a) acoustic and (b) optical waves

7·7(b). The resulting waves are called optical waves and are of higher frequency than the acoustical waves. These are the modes responsible for optical
scattering in a crystal and are further discussed in Chapter 18 in connection
with the optical properties of solids. There are in general three modes, or
polarizations, of the atomic vibrations. In the simplest case these correspond
to one longitudinal (atoms moving along the line of atoms), and two transverse modes (atoms moving perpendicular to the line of atoms) (see Fig. 7·8).

If classical mechanics were valid, the energy of the vibrational waves could
be arbitrary. However, quantum mechanics impose restrictions which limit
the energy of the waves to discrete values, that is the oscillations are
quantized. The discrete energy values that are allowed are the same as those
of any simple harmonic oscillator and are given by

$$E_n = (n + \tfrac{1}{2})h\nu \qquad\qquad\qquad (7·15)$$

where h is Planck's constant, ν is the frequency of vibration and the quantum
number n is any positive integer including zero (n is zero for all modes at
absolute zero of temperature). (Note that frequency is conventionally
denoted by ν rather than f in this book.)

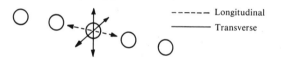

Fig. 7·8 The three modes of atomic vibrations

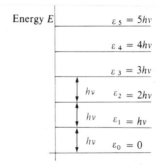

Fig. 7·9 Energy levels of atoms vibrating at a single frequency ν

The energy change accompanying a transition from one energy state with the quantum number n_1, to the next highest energy state with the quantum number n_2, is

$$\Delta E = E_{n_2} - E_{n_1} = (n_2 + \tfrac{1}{2})\, h\nu - (n_1 + \tfrac{1}{2})\, h\nu$$
$$= (n_2 - n_1)\, h\nu \qquad (7·16)$$

Since n_1 and n_2 differ by unity $\Delta E = h\nu$ that is, the transition is accompanied by the absorption of one quantum of thermal energy $h\nu$. The reverse transition from a higher energy state to one of lower energy is accompanied by the emission of $h\nu$ of thermal energy. The spectrum of energy levels is illustrated in Fig. 7·9.

The expression for the average energy of a quantum oscillator at frequency ν is different from the average energy, kT, of a vibrating atom in classical theory. In Section 7·6, we saw that the numbers of atoms in quantum states E_1 and E_2 are in the ratios $\exp(-E_1/kT)$ to $\exp(-E_2/kT)$, i.e.

$$\frac{N_1}{N_2} = \frac{\exp(-E_1/kT)}{\exp(-E_2/kT)} = \exp\left(-\frac{(E_1 - E_2)}{kT}\right)$$

But, from Eqn (7·16), $E_1 - E_2 = h\nu$ for adjacent energy levels and so the ratio of populations in the n and $(n+1)$th levels is $\exp(-h\nu/kT)$.

The *average* energy of the oscillators will be given by

$$\overline{E} = \frac{\text{Sum of energies of all oscillators}}{\text{Number of oscillators}}$$

$$= \frac{\displaystyle\sum_{n=0}^{\infty} nh\nu \exp(-nh\nu/kT)}{\displaystyle\sum_{n=0}^{\infty} \exp(-nh\nu/kT)}$$

$$= \frac{h\nu\,(e^{-h\nu/kT} + 2e^{-2h\nu/kT} + \ldots)}{(1 + e^{-h\nu/kT} + e^{-2h\nu/kT} + \ldots)}$$

Writing $x = -h\nu/kT$, the above can be written as

$$\overline{E} = h\nu \ \frac{d}{dx} \ \ln \ (1 + e^x + e^{2x} + \ldots)$$

$$= h\nu \ \frac{d}{dx} \ \ln \ \left(\frac{1}{1-e^x}\right)$$

$$= \frac{h\nu}{e^{-x} - 1}$$

Thus the average energy of a quantum oscillator of frequency ν is given by

$$\overline{E} = \frac{h\nu}{\exp \ (h\nu/kT) \ - \ 1} \tag{7·17}$$

Translating this in terms of lattice waves we can say that the average energy in a lattice wave of frequency ν is given by Eqn (7·17).

If the wave is regarded as a stream of 'particles', each of energy $h\nu$, then the average energy would be $h\nu$ times number of particles. Thus, from Eqn (7·17), the number of particles in a mode of frequency ν is given by

$$n_q = \frac{1}{\exp \ (h\nu/kT) \ - \ 1} \tag{7·18}$$

The name phonon has been given to these 'particles' each of which is a quantum of thermal energy, by analogy with the transitions involving the absorption or emission of photons (quanta of electromagnetic energy). The phonon is the most fundamental of the imperfections found in crystalline materials.

At any temperature a crystal will contain a wide spectrum of phonons as the atoms will be vibrating at many frequencies which may range from low frequency fundamental acoustical, to frequencies of 10^{13} Hz. As the atomic vibrations are not independent of each other the frequencies are quantized according to

$$\nu = \frac{v}{2} \sqrt{\left(\frac{n_x}{X}\right)^2 + \left(\frac{n_y}{Y}\right)^2 + \left(\frac{n_z}{Z}\right)^2} \tag{7·19}$$

where v is the velocity of propagation, and n_x, n_y, and n_z are integers in the direction of the three orthogonal axes, x, y, and z. X, Y, and Z are the dimensions of the crystal in the directions x, y, and z. Thus, only those phonons whose energies, or frequencies, are given by this equation can be absorbed by the crystal. The total number of allowed frequencies is $3N$ where N is the number of atoms in the crystal. The number of allowed frequencies dN in a frequency range $d\nu$ varies markedly with frequency. For a crystal with a primitive cubic structure dN is given by

$$dN = 4\pi\nu d\nu \ \left(\frac{1}{v_l^3} + \frac{2}{v_t^3}\right) \ V \tag{7·20}$$

where v_l and v_t are the velocities of propagation of the longitudinal and transverse vibrations respectively, and V is the volume of the crystal, i.e., $V = XYZ$. In general, v_l will be the velocity of propagation of a longitudinal

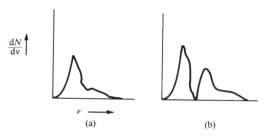

Fig. 7·10 The frequency density of allowed states for (a) primitive cubic structure and (b) lithium chloride [After A.M. Karo and J.B. Hardy, *Phys. Rev.*, **129**, pp. 2024–2036 (1963)]

sound wave through the crystal. Equation (7·20) shows that the frequency density of allowed states $dN/d\nu$ is directly proportional to the crystal volume. Thus, increasing the volume of a crystal increases the number of allowed frequencies, but does not change the distribution of these frequencies. Figure 7·10 shows the frequency density of allowed states according to Eqn (7·20), assuming $\nu_l = \nu_t$. Figure 7·10 is for a primitive cubic structure, but similar complex density of allowed states distributions are found for all structures. Indeed, in certain cases it is even possible to obtain a forbidden frequency range within the spectrum as is the case, shown in Fig. 7·10(b), for lithium chloride.

Figure 7·10 shows examples of the distribution of allowed frequencies, but it must be appreciated that not all these frequencies need exist in a crystal in a given situation. In fact, the occupation of the allowed states varies markedly with temperature, as illustrated in Fig. 7·11. At very low temperatures kT, which is a measure of the available thermal energy, is so small compared to $h\nu$ of most of the allowed waves that very few phonons exist and these will be of low frequency. At somewhat higher temperatures there are many more low-frequency phonons and also some of higher frequencies. At elevated temperature kT becomes greater than $h\nu$ for even the highest frequency waves, and so a large number of phonons with a spectrum of frequencies is found.

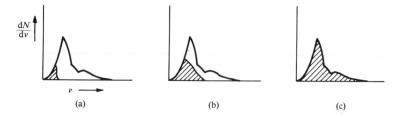

Fig. 7·11 The occupation of the allowed states at (a) a low temperature, (b) an intermediate temperature and (c) an elevated temperature

7·11 Specific heats of solids

We consider now the heat capacity of a solid, for which we shall use the ideas about phonons described above.

For simplicity, let us suppose that the atoms all vibrate at a frequency v. Since each atom can vibrate in three orthogonal directions, each atom can be in one of three quantum states for each energy and so the total number of states for N atoms is $3N$. Then the total heat energy, Q, which the lattice can hold, i.e., the heat capacity of the solid, will be given by the number of states times the mean energy of the phonons associated with them. Thus, using Eqn (7·17)

$$Q = \frac{3Nhv}{\exp(hv/kT) - 1} \qquad (7·21)$$

The specific heat at constant volume, C_v, can now be found by differentiating this with respect to T, giving

$$C_v = \frac{dQ}{dT} = 3Nk \left(\frac{hv}{kT}\right)^2 \frac{\exp(hv/kT)}{[\exp(hv/kT) - 1]^2} \qquad (7·22)$$

where N is now to be taken as the number of atoms in unit mass of the solid.

This result is illustrated graphically in Fig. 7·12 and, except at very low temperatures, it agrees with measured data for many materials fairly well if an appropriate value of v is taken. (To obtain better agreement with experiment at low temperatures it is necessary to take account of the fact that more than one frequency of oscillation is allowed for the atoms.)

At very low temperatures the atoms are mostly at rest, and few can be excited into the first energy level, because of the small amount of heat energy available. As the temperature rises, the number of excited atoms rises exponentially, so that the specific heat increases rapidly. At higher temperatures still, the mean thermal energy becomes large compared to the spacing of the energy levels, and the latter can be regarded as nearly continuous. It is therefore not surprising that the solid behaves much like a gas, so that the specific heat tends to a constant value of $3N_0k$ per mole (N_0 is Avogadro's

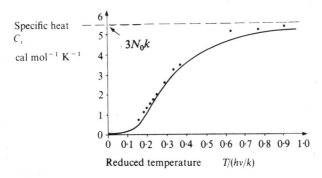

Fig. 7·12 Dependence of C_v on temperature according to Eqn (7·22), for diamond

number) (see Problem 7·5). This result is an expression of the empirical law discovered in 1819 by Dulong and Petit which stated that, at ordinary temperatures, the specific heats of all solids per mole are approximately the same. It can be seen that the law holds accurately only in the limit of very high temperature, and the several notable exceptions to Dulong and Petit's law result from the fact that 'normal' temperatures are not high enough.

Note that, except in the limit of high temperature, C_v depends upon the frequency ν. Hence the specific heat is expected to vary from solid to solid. In particular, from Eqn 7·22 it can be seen that a higher value of ν results in a *lower* specific heat. This is because, as we add the same quantity of heat to two solids, thus raising their temperatures, the atoms must be excited to higher energies in the solid with higher vibrational frequencies, since the spacing of the levels is larger.

This dependence of specific heat on frequency is of utmost importance, since it leads to the possibility of observing allotropic states of solids, as we shall see in Section 7·13. Further, the existence of such allotropes is fundamental to the microstructure and the mechanical behaviour of many structural materials such as steel.

The final point to note in this context is that the specific heat of allotropic forms of a substance must differ. Since their structures differ, the Condon–Morse curves must also differ in shape, and this affects the atomic vibration frequency. Inspection of Fig. 7·4 shows that if the force constant is increased, then the potential energy minimum displays more curvature. This is simply an expression of the mathematical relation

$$\frac{dF}{dr} = -\frac{d^2V}{dr^2}$$

A higher force constant gives a higher vibrational frequency (it is as if we made the springs joining the atoms stiffer) and this in turn gives rise to a *lower* specific heat. Hitherto, we have discussed only the specific heats at constant volume, C_v, whereas it will shortly be necessary to discuss the specific heat at constant pressure, C_p. The latter is the more easily measured quantity, so that C_v is usually obtained from C_p by making a theoretically derived correction. The correction for most solids at normal temperatures is quite small, so that the general temperature dependence of C_p is similar to that of C_v, except at very low temperatures. As might be anticipated, though, C_p tends to zero as does C_v when the temperature tends to the absolute zero.

7·12 Specific heats of polyatomic gases

Before moving on to other topics, it is appropriate to add a brief postscript on specific heats in polyatomic gases, since these form the majority of gases in nature, including those which are commonly used in heat exchangers and whose thermal properties are therefore of practical interest.

A polyatomic molecule has means of motion other than purely translational. For instance, in a diatomic molecule the atomic bond can stretch and each atom can vibrate back and forth with respect to the other, without influencing its flight. In addition, the molecule may rotate about an axis perpendicular to the axis of the bond. The rotational velocity has two orthogonal components with both of which can be associated a mean thermal energy of $\frac{1}{2}kT$, and the vibrational energy can also be $\frac{1}{2}kT$ per molecule.

But at low temperatures the vibrational and rotational motions may not be excited, as their energies are quantized–just as in a solid at low temperatures. Hence at low temperatures a diatomic gas (provided it does not liquefy) behaves like a monatomic one. As the temperature rises the rotational states begin to participate and the specific heat rises accordingly. Since the interatomic bond is normally very stiff, the vibrational mode is the last to be excited. However, the full range of behaviour is not shown by any one gas, as liquefaction and dissociation of the molecules limit the temperature range in which measurements can be made.

7·13 Allotropic phase changes

We have already noted that some solids change their structure when their temperature is raised or lowered. For example, iron changes from b.c.c. to f.c.c. at 910°C, and tin from the diamond structure to a related, tetragonal, structure at 13°C. These allotropic forms are often referred to as different *phases* of the solid material, and the transformation from one to another is very similar to the phase changes of melting and vaporization. Latent heat is usually evolved when the high temperature phase is cooled through the transition point, and the specific heat changes abruptly at the same temperature. It may well be asked, why is it that the change occurs at all? For if latent heat is evolved when iron cools through 910°C, then the b.c.c. form (commonly called α-iron) to which it transforms has lower energy and it should, according to earlier precepts, be the more stable of the two forms. This energy difference must be reflected in the Condon–Morse curves of atomic potential energy vs. atomic separation for the two structures, and we might imagine them to be as illustrated in Fig. 7·13. The deeper minimum of the α-iron curve represents a lower energy in the equilibrium state, when the atomic separation is exactly r_0.

But in fact at normal temperatures the average energy per atom is higher than the minimum of the Condon–Morse curve, since the solid contains a quantity of heat energy. The amount of energy added to a solid which is heated from 0K to a temperature T is just

$$H = \int_0^T C_p \, dT$$

where C_p is the specific heat at constant pressure.

We note immediately that solids with different values of C_p contain different amounts of heat energy at the same temperature. This, however, does not explain the phase change in iron since at and above 910°C extra heat must be given to α-iron before it will change to γ-iron (the f.c.c. form). The answer to the problem lies in the way the heat energy is distributed among the atoms.

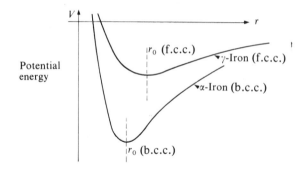

Fig. 7·13 Schematic Condon–Morse curves for α-iron and γ-iron

Let us consider the energy diagrams for the atoms in the two structures of iron. We saw in the previous section that the vibrational energy levels of the atoms in a solid may be reasonably assumed to be equally spaced by an energy $h\nu$, and that the frequency ν is related to the degree of curvature of the Condon–Morse curve at its minimum. The higher the curvature, the higher is ν, so that from Fig. 7·13 we see that the deeper minimum, which has the higher curvature, gives a higher frequency. Thus, the b.c.c. iron lattice has the higher frequency and its vibrational energy levels are therefore more widely spaced than those of f.c.c. iron.

To see more clearly what effect this has on the relative stabilities of the two phases let us consider an extreme case. We suppose that the two phases of the same material have such widely differing frequencies ν that one, phase B, has a nearly continuous band of energy levels where the other, phase A, has but two. This is illustrated in Fig. 7·14. We further suppose for simplicity that only these two lowest levels of phase A are populated at the temperature of

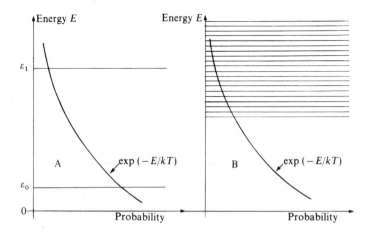

Fig. 7·14 Energy levels and distribution functions for two phases A and B of a solid held at the same temperature

interest. Since the higher frequency phase is the stable one at low temperatures, its lowest energy level must lie below that of phase B. Superimposed on each energy diagram is the Boltzmann probability distribution. It has the identical shape for both phases because they share the same temperature.

Now let the two phases be in contact with one another, so that at the interface the atoms may 'choose' one structure or the other.

In 'collision' with one another, the atoms will exchange energy, as explained earlier. Consider a 'collision' at the interface between a phase A atom of energy ϵ_1 and a phase B atom of any energy E_2. The amount of energy exchanged between them, ΔE, is a purely random quantity owing to the random nature of such collisions. The energy of the A atom subsequent to the collision is $\epsilon_1 \pm \Delta E$, which in most cases is different from both ϵ_0 and ϵ_1. It is consequently impossible for this atom to remain in phase A, since there is no suitable energy level available to it. However, it may move to a phase B energy level as these are plentiful. The atom therefore has a very high chance of making the transition from phase A to phase B.

At the same time the atom which began life in phase B is similarly unlikely to acquire exactly the correct amount of energy to allow it to make the transition to phase A, so that it has a very small chance of making that transition. There is thus a net transfer of atoms from phase A to phase B.

It can be seen that the basic cause of the phase change is the difference in the number of energy levels available in the two phases. For all the atoms in phase B to get into phase A, their purely random 'collisions' must simultaneously put them each into one of the two energy levels available to phase A. The chance of this happening is roughly like that of putting all the snooker balls on a snooker table into two pockets, by hitting them randomly with the cue. Since there are many other places in which the balls may settle, the probability is low, like that of the phase transition B to A.

On the other hand the probability of getting the snooker balls *out* of their pockets by hitting them with the cue (although against the rules of snooker) is high–as before, there are plenty of alternative resting places for them.

In summary we see that, because of the random nature of heat energy, the criterion for stability of one phase *vis-à-vis* a second is not simply that it must have lower energy, but also that a transition to the second phase must be less probable than the reverse transition.

From the above description it may be thought that the transition from phase B to phase A will never occur, but this is not so. If the temperature is lowered so that fewer atoms are present in the excited state of phase A, then the rate of transfer of atoms from phase A to phase B is lowered and may eventually fall to equality with the reverse transition. This temperature is then what is normally called the transition temperature and, below it, transitions to the lower energy phase become the more probable.

In summary of this description, the effect of *rates* of transition on thermal equilibrium may be put into two general statements.

(a) When two phases of a substance are in equilibrium at the transition temperature, there are equal and opposite rates of exchange of atoms between the two phases.

(b) If the two opposite rates become unequal through a change in temperature for example, one or other of the two phases is the more stable and hence predominates.

We have deliberately avoided quantifying the concept of stability, as it is a rather complicated one. It is possible, though, to define a quantity called the *thermodynamic probability* which measures the tendency of materials to choose a structure with a larger number of energy levels rather than one with lower thermal energy. Thermodynamic probability is a measure of the number of different ways in which the atoms can be distributed among their quantum states without changing the energy distribution–thus the larger the number of populated levels, the larger the thermodynamic probability.

7·14   Latent heat and specific heat

The two imaginary phases of the previous section must have different specific heats, according to the arguments of Section 7·11. In fact it is apparent that, were the two phases to have the same specific heat, the phase change would not occur, because the distribution of the energy levels would be the same in both phases.

The normal transition temperature, T_t, between two phases is governed by the distribution of the energy levels. The latent heat at this temperature is governed by the specific heats of the two phases. The latent heat is equal to the total energy difference between the phases at T_t. If the energy difference is H_0 at the absolute zero of temperature (the difference between the minima of the respective Condon–Morse curves), then the energy difference ΔH at the temperature T_t is

$$\Delta H = H_0 + \int_0^{T_t} C_p^a \, dT - \int_0^{T_t} C_p^b \, dT$$

where C_p^a and C_p^b are the specific heats at temperature T of the two phases. Normally the C_p of the high-temperature phase (b) cannot be measured below T_t, but it can be calculated theoretically as described earlier (Section 7·11) and values of latent heat calculated in this way confirm the theories we have been discussing in this chapter.

7·15   Melting

Allotropes will be observable where the two possible structures have Condon–Morse curves of the kind illustrated in Fig. 7·13 providing, of course, that the low-temperature phase does not melt first! But melting is also a phase transition, and can be discussed in exactly the same way. The only difference is that the high-temperature phase does not have a regular structure, a fact which can be understood if we note that at some temperature the amplitude of vibration of the atoms in a solid must become comparable to the atomic size. Thus the lattice can no longer be regarded as rigid, for as a pair of atoms move apart, another may insert itself between them, destroying the basic structure of the unit cell. Such a structure must be liquid-like, for the *mean* atomic separation will still be close to that of the solid, but rigidity is totally lost.

· It is instructive to note that melting is not a gradual process, as might be expected from the previous paragraph. A solid does not become progressively less rigid as the temperature rises until complete fluidity exists: in fact the transition occurs abruptly at a well-defined temperature. This is because the liquid state has quite a distinct *structure*, and therefore has quite a different specific heat from the solid. As the temperature is raised through the melting point, the stability of the liquid structure first becomes equal to and then greater than the stability of the solid state, exactly as for two solid structures. The transition is therefore abrupt, and there are *no stable intermediate states* between the solid and liquid phases.

7·16 Thermodynamics

In this chapter, the thermal behaviour of solids has been discussed in relation to their structure, with the emphasis on the use of atomic models to explain various phenomena. This is not the only way of treating thermal properties; historically, the measurement of heat and work and the changes they effect in matter came first, and the subject of *thermodynamics* was then built up to coordinate all the observed phenomena. Its basic tenets are that heat and work are interchangeable and that energy cannot be created or destroyed. These are enshrined in the *first law* of thermodynamics which states that if heat dQ is put into a body and work dw is done on it, their sum is equal to the change in internal energy dU. By convention, the work dw is given a negative sign while work done *by* the body on its surroundings is given a positive sign. Thus

$$dU = dQ - dw$$

The total heat content of the body at a temperature T is thus $\int_0^T (dU + dw)$, which might be thought to be the same for all bodies in thermal equilibrium. We have seen, however, that equilibrium is also governed by *thermodynamic probability*, a concept which is alien to thermodynamics and can only be appreciated by recourse to atomic models. It was therefore found necessary to 'invent' another energy term which substitutes for the concept of probability, so that phase changes could be ascribed to differences in energy of the phases concerned. The invented quantity S, called *entropy*, when multiplied by the temperature T gives an energy term which increases as a body is heated. Differences in thermodynamic probability are equivalent to differences in the product TS. The change in 'free energy' ΔG when a phase change occurs is then the total heat change minus $T\Delta S$, i.e.,

$$\Delta G = \Delta U + \Delta w - T\Delta S \tag{7·23}$$

where ΔU, Δw and ΔS are the changes in the respective quantities. Two phases of an element are in equilibrium at constant pressure if $\Delta G = 0$, so that their free energies are exactly the same. Otherwise the phase with lower free energy is the equilibrium one. In this way the minimum-energy concept was retained in the scheme. The condition for equilibrium in a two-component system like an *alloy* is slightly different and is discussed in Chapter 10. It is frequently convenient to lump ΔU and Δw into one term, ΔH, called the *enthalpy or total heat*, in which case Eqn (7·23) reads

$$\Delta G = \Delta H - T\Delta S$$

Not surprisingly, a connection between entropy, S, and thermodynamic probability, W, was established at a later date–the one increases with the other according to the equation

$$S = k \ln W \tag{7·24}$$

where k is Boltzmann's constant. This means that the higher the probability, the lower is the free energy, since the entropy term subtracts from the 'true' energy (the enthalpy) [Eqn (7·23)]. Thus the subject of thermodynamics was made self-consistent, and provides a powerful tool for discussing all kinds of changes which occur in materials. Some examples will occur in subsequent chapters.

For a more thorough treatment of thermodynamics the reader is referred to the book by Swalin which appears in the Further reading at the end of this book.

A limitation of thermodynamics is that in spite of its name it cannot deal with *rates* of phase changes and this makes it unable to explain non-equilibrium or metastable structures which constitute a number of engineering materials. It is with such questions that the next three sections deal, in the same terms as those used in earlier sections.

7·17   Multiphase solids

In Chapter 10 materials will be discussed which possess many phases. In some of these the phases exist in equilibrium at room temperature, but we shall find others exhibiting phases which are not in true equilibrium with one another. That is, the two or more phases exist side by side at room temperature, although the temperature at which they are truly in equilibrium is much higher. Although we now can understand why different phases may exist, our description so far makes their simultaneous co-existence seem impossible at temperatures other than the one at which they are in equilibrium. For if one is more stable than the other at all temperatures bar one, then the more stable one should exist alone. To explain the apparent anomaly it must be supposed that either there is some barrier to the establishment of the more stable phase, or that the transition occurs so slowly that it cannot be observed in times of less than hundreds of years. In fact both of these explanations are involved, as we shall now discover by investigating what controls the *rate* of a phase change.

7·18 Rate theory of phase changes

Until now we have discussed only differences in energy between the initial and final states of a system undergoing a phase change. However, during the change the structure must go *through* intermediate stages, though obviously none of them are stable or they would be observable over a finite temperature range. Indeed, it is important to understand that the only stable structures lie at the two end-points of the process. This implies that the structure goes through a stage (the *activated* phase) having *higher energy* than both the low-temperature and the high-temperature structures, for stable structures must have a minimum energy with respect to small changes of structure. Confirmation of this comes from the fact that in most phase changes the high-temperature state can be supercooled–it can exist slightly *below* the transition temperature if mechanical disturbances are removed–and the low-temperature phase can similarly exist at slightly higher temperatures. It is not surprising that mechanical vibration can, for instance, cause supercooled f.c.c. iron to revert to the b.c.c. form, for the lattice actually has to change shape during the process.

We can make this idea more concrete by plotting the potential energy of the structure against some convenient dimension z of the lattice which increases monotonically during the transition. A hypothetical plot is shown in Fig. 7·15, where the values z_A and z_B of the lattice dimension are the equilibrium values in the initial and final states at the absolute zero of temperature. Figure 7·15 is therefore simply a Condon–Morse curve of a different kind to those we have considered hitherto.

Because the atoms have thermal energies, their total energy is higher than E_1 or E_2, and we shall suppose that the mean total energy per atom is E_A in one phase and E_B in the other, when the temperature of both phases is the same. The differences $(E_A - E_1)$ and $(E_B - E_2)$ therefore represent the average thermal energies of atoms in either state. The points x, x' and y, y' on the potential energy curves at the energies E_A and E_B correspond to the turning points of vibrating atoms with the energies E_A and E_B respectively.

How, then, can the phase change occur, if a state of higher energy (up to

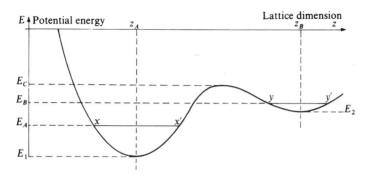

Fig. 7·15 Hypothetical plot of atomic potential energy against a lattice dimension for a solid undergoing a phase change

E_C) exists between states A and B? The clue lies in the reminder, above, that E_A and E_B represent only *mean* energies and that, if the barrier E_C is not too high, there may be many atoms with enough energy to surmount it and hence make the transition to the state B. Once there, they may lose energy by collision with neighbours so that they are trapped there in the potential energy 'well'. Note that this picture does not rule out the possibility of atoms making the reverse transition–as we have discussed earlier (Section 7·13), the *rate* of the latter change is lower, for reasons unconnected with Fig. 7·15.

We describe such a change as this as an *activated process*–the idea being that the solid in phase A must be *activated* before the change can occur, and that the activated state (state C of Fig. 7·15) has a higher energy E_C. The energy difference $E_C - E_1$ in the figure is referred to as the *activation energy* of the change. The reason why this energy difference rather than $E_C - E_A$ is used to characterize the activated state will be clear when we consider how it affects the *rate* at which the change occurs. This rate can be roughly estimated as follows. The net transition rate is the difference between two flows in opposite directions–one of atoms changing from state A to state B and the other making the reverse transition. The former rate r_A may be defined as the number of atoms which make the A–B transition in unit time, and this number must be proportional to the following factors:

(a)   the number n_A of atoms in phase A with energies higher than E_C;
(b)   the frequency, f, with which the atoms approach the barrier;
(c)   the relative thermodynamic probability of phase B compared to phase A.

We shall now discuss each of these in turn. The number n_A is given by the Boltzmann expression, Eqn (7·9), integrated over the energy range above E_C. Thus

$$n_A = \int_{E_C}^{\infty} C \exp -\left(\frac{E - E_1}{kT}\right) \, dE$$

where E_1 appears because it is the energy at which atoms possess no kinetic energy in phase A. The factor C in the above equation depends weakly upon E, so for our present purposes we shall assume it is a constant. Performing the integration thus gives the result

$$n_A = \frac{C}{kT} \exp -\left(\frac{E_C - E_1}{kT}\right)$$

The factor C can be found by extending the range of integration down to E_1, for then the number obtained is equal to the total number of atoms N_A in state A. Thus

$$N_A = \frac{C}{kT} \quad \text{and hence} \quad n_A = N_A \exp -\left(\frac{E_C - E_1}{kT}\right)$$

The second quantity, f, the frequency of approaching the barrier, may be taken to be the frequency of vibration ν_A of atoms in phase A. In most solids this is around 10^{13} Hz. The final term to be evaluated is the thermodynamic probability of phase B *relative* to phase A. This quantity has been discussed but briefly in this book, and here it will simply be assumed that the thermodynamic probabilities W_A and W_B of phases A and B can be defined, and that their ratio W_B/W_A gives the required relative probability. This ratio has the right form, since it is still finite when $W_A = W_B$, while the difference ($W_B - W_A$) would not satisfy this requirement.

The rate, r_A, being proportional to each of the above factors, is therefore proportional to their product. Thus

$$r_A = K\nu_A \frac{W_B}{W_A} N_A \exp - \left(\frac{E_C - E_1}{kT} \right) \tag{7·25}$$

where K is a constant dependent only on the geometry of the interface between phases A and B.

But Eqn (7·25) does not give the net transition rate since the reverse transition also occurs. Its rate, r_B, is given by an exactly similar expression:

$$r_B = K\nu_B \frac{W_A}{W_B} N_B \exp - \left(\frac{E_C - E_2}{kT} \right) \tag{7·26}$$

The net transition rate r is just $r_A - r_B$, so that

$$r = K\nu_A \frac{W_B}{W_A} N_A \exp - \left(\frac{E_C - E_1}{kT} \right) - K\nu_B \frac{W_A}{W_B} N_B \exp - \left(\frac{E_C - E_2}{kT} \right) \tag{7·27}$$

Since only the temperature dependence of r is of interest at present it is convenient to replace the coefficients of the two exponential terms by the symbols F and G, giving

$$r = F \exp - \left(\frac{E_C - E_1}{kT} \right) - G \exp - \left(\frac{E_C - E_2}{kT} \right) \tag{7·28}$$

This equation may be used to discuss the temperature dependence of r if we concede that all factors other than the exponential depend rather weakly on temperature, at least for small temperature changes. Let us therefore investigate how the rate of transition varies according to this theory for temperatures close to the normal transition temperature.

At the transition temperature T_t the two phases can exist side by side–they are in equilibrium and the net rate is zero. Thus

$$F \exp - \left(\frac{E_C - E_1}{kT_t} \right) - G \exp - \left(\frac{E_C - E_2}{kT_t} \right) = 0 \tag{7·29}$$

The dependence of r on T at temperatures close to T_t may be found by differentiation of Eqn (7·28) as follows

$$\frac{dr}{dT} = F \frac{(E_C - E_1)}{kT^2} \exp - \left(\frac{E_C - E_1}{kT} \right)$$

$$- G \frac{(E_C - E_2)}{kT^2} \exp - \left(\frac{E_C - E_2}{kT} \right)$$

$$= \frac{(E_C - E_1)}{kT^2} \left\{ F \exp - \left(\frac{E_C - E_1}{kT} \right) \right.$$
$$\left. - G \left(\frac{E_C - E_2}{E_C - E_1} \right) \exp - \left(\frac{E_C - E_2}{kT} \right) \right\}$$

Since $(E_C - E_2) < (E_C - E_1)$, a comparison of this with Eqn (7·29) shows immediately that the second term is smaller than the first when $T = T_t$, and hence dr/dT is positive. An increase in temperature therefore causes the transition from phase A to phase B to proceed, while a reduction has the reverse effect.

7·19   Metastable phases

Suppose that the solid discussed above is cooled instantaneously from above T_t to a very low temperature. Initially, the number of atoms in phase A will be negligible, so that transition from this phase may be neglected. The net transition rate is then equal to r_B, given by Eqn (7·26). That is,

$$r = r_B = K \nu_B \ \frac{W_A}{W_B} \ N_B \exp - \left(\frac{E_C - E_2}{kT} \right) \tag{7·30}$$

If the new, low temperature is small enough so that $E_C - E_2 \gg kT$ then this rate is so small that for practical purposes it can often be regarded as zero. Thus, by supercooling a body very rapidly to a low temperature before the transition can begin, it can effectively be halted. The number of atoms able to surmount the energy barrier is reduced to the extent that the (normal) high-temperature phase appears to be stable.

We shall encounter several examples of such metastable phases in later chapters. The most widely quoted instance is that of glass, which is a super-cooled liquid. Its structure will be discussed in more detail in Chapter 11.

7·20   Other applications of the rate theory

The above equations describe the rate of *any* process of change between two states separated by an energy barrier. One very large class of such changes is the class of chemical reactions. The rates of chemical reactions are of fundamental importance to a discussion of the corrosion in engineering materials so let us consider now the application of the ideas of the previous section to this problem.

In general, corrosion of a solid is a two-stage process. The formation, for instance, of an oxide layer on a metal obviously constitutes a chemical reaction between the metal and oxygen (also often involving water), but once the oxide is formed it covers the surface of the metal so that the latter is no longer in direct contact with the atmosphere. For the chemical reaction to proceed further it is necessary for the atoms either of the metal or of the atmosphere to migrate through the oxide layer until they meet one another. The process of migration is called *diffusion* and, since it occurs via thermal atomic motion, it is affected by temperature.

Diffusion, therefore, controls the rate at which atoms in a chemical reaction can become close enough to one another for the reaction to proceed. In some cases, diffusion may occur fast enough for the overall rate to be controlled by the rate of the reaction itself, while in others, diffusion may be slow enough to be important in determining the overall rate of the process. We shall consider diffusion later.

7·21 Chemical reactions

Let us take as a simple example the reaction of hydrogen and chlorine to form hydrochloric acid, usually written

$$H_2 + Cl_2 = 2HCl$$

This equation can, in fact, proceed in either direction, but for the present we shall regard the two acid molecules as the final state of the reaction and the two molecules of hydrogen and chlorine as the initial state. [According to the discussion in Section 7·16 on classical thermodynamics, the final state (2HCl) must have lower free energy than the initial state ($H_2 + Cl_2$).]

Now the reaction of hydrogen with chlorine is exothermic at normal temperatures—that is, heat is evolved in the production of hydrochloric acid. We may therefore conclude that the energy of the final state (the acid molecules) is lower than that of the initial state, unlike the case of the phase change we discussed in Section 7·18. [In the language of thermodynamics, this energy should be called *enthalpy* (see Section 7·16).] There is, however, an energy barrier between them, for otherwise hydrogen and chlorine would react spontaneously whenever they came into contact with one another. A mixture of the two gases does in fact appear quite stable—the reaction rate is so low that it can be ignored for practical purposes, except in the presence of sunlight.

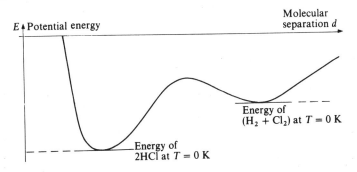

Fig. 7·16 Hypothetical plot of potential energy vs separation of the H_2 and Cl_2 molecules in the reaction $H_2 + Cl_2 \rightleftharpoons 2HCl$

An energy diagram can now be drawn (Fig. 7·16), analogous to Fig. 7·15, in which energy is plotted against the separation of the molecules taking part in the reaction. The rate of the reaction is just the number of molecules crossing the barrier in Fig. 7·16 in unit time. The process is almost exactly like that of a phase change as discussed in Section 7·18, and it will therefore obey the same equations. There will thus be a temperature at which the reaction is in equilibrium–the rate of dissociation of HCl into H_2 and Cl_2 is equal to the rate of combination at this temperature. At temperatures well removed from the equilibrium value, one or other of the two rates will be negligible and the net reaction rate, r, is therefore given by an equation similar to Eqn (7·30) having the form

$$r = A \exp -\Delta E/kT$$

where ΔE is the activation energy, as before, and A is a constant.

The rate of a chemical reaction should therefore increase rapidly with the temperature. This is indeed the case: a flame applied to a mixture of H_2 and Cl_2 gases raises the local temperature, so that molecules are more rapidly excited over the energy barrier in Fig. 7·16. The reaction goes to completion because the energy released is transmitted to other molecules, exciting them over the energy barrier, and so on. (Incidentally, the action of sunlight can also be understood if we assume that the atoms absorb photons and then have enough energy to surmount the barrier.)

In this instance, it is quite clear that the constituents of the reaction, being gases, can mix readily so that the rate of diffusion is high and has a negligible effect on the overall reaction rate.

7·22   Diffusion

The example of diffusion which is easiest to comprehend is that in which two gases, initially separated as in Fig. 7·17, diffuse into one another when the separating wall is removed. However, it was remarked earlier in connection with corrosion that diffusion can occur through solids, unlikely as this may seem. If this were not so, then only the surface layer of atoms on a metal could oxidize, while in practice the layer of oxide which forms is many

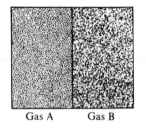

Gas A   Gas B

Fig. 7·17 Two gases, initially separated by a wall, will diffuse into one another when the common wall is removed

molecules thick. A further example is the diffusion of gases through glass, which causes vacuum flasks, thermionic valves (tubes) and cathode ray tubes to go 'soft' after many years of service, even though the glass envelope remains intact.

In all the above cases, atomic motion occurs from a region of high concentration of the diffusing species to one of low concentration. We call this process *chemical diffusion*. In contrast, *self-diffusion*, which is the random diffusion through a material of *its own* atoms does not need a concentration gradient. We shall leave self-diffusion until Chapter 8, and concentrate here on chemical diffusion. Its characteristic features are (a) that there is a net transfer of atoms across a material, and (b) that it occurs as a result of a gradient in the atomic concentration.

Knowing this, it is clear that the easiest case to understand should be one in which the concentration gradient is both uniform and controllable. Thus, measurements on gases diffusing through porous solids show that the flux, F, of atoms is simply proportional to the concentration gradient, i.e.,

$$F = - D \frac{dn}{dx} \qquad (7·31)$$

where F is the number of atoms or molecules flowing across a unit area in the y–z plane (Fig. 7·18) in one second, while the gradient in the concentration, n, is in the x-direction and is *negative*, making F positive. The constant D is known as the *diffusion coefficient*, which depends on the nature both of the diffusing species and of the medium in which diffusion occurs. Equation (7·31) is known as Fick's law.

A theory can be constructed from which this equation may be deduced, if some simple assumptions are made about the diffusion process. These are that the individual molecules move in a series of individual 'jumps' which occur at a fixed average frequency and in random directions. In gaseous diffusion, the 'jumps' would be the periods of free flight between collisions. In the next chapter we shall learn how diffusion occurs in a solid, where the jumps necessarily occur across an energy barrier. In this case, the frequency with which the jumps occur must depend upon the barrier height E_0. The

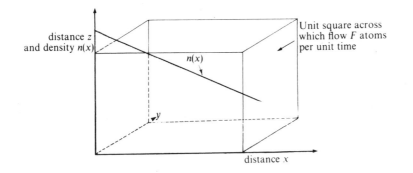

Fig. 7·18 Illustrating Fick's law

mathematics developed earlier for reaction rates is applicable here, for the diffusing atoms must be 'activated', i.e., they must gain enough energy to cross the energy barrier. Since at any one time only a fraction, approximately $\exp(-E_0/kT)$, of them have this energy, the average frequency of performing the jumps is reduced by this factor, with the outcome that the diffusion coefficient D is proportional to the same factor. Thus

$$D = D_0 \exp(-E_0/kT) \tag{7.32}$$

where D_0 is a constant.

Diffusion in solids is therefore strongly temperature-dependent while in gases and liquids, where no energy barrier exists, diffusion coefficients are generally independent of temperature. At normal temperatures, diffusion in solids is therefore relatively slow and may have a controlling effect on the rate of a chemical reaction. Thus, if the surface of a metal normally oxidizes fairly rapidly it is because diffusion of oxygen through the surface oxide layer is easy. It can be slowed down by overcoating the metal with a protective layer in which oxygen diffuses only slowly. The choice of suitable coatings for the prevention of corrosion is an important branch of materials science.

Diffusion of impurity atoms into semiconductor crystals is an important application of diffusion in solids and this is dealt with in more detail in Chapter 15.

Problems

7.1 List as many pieces of experimental evidence as you can which support the view that atoms are in constant motion.

7.2 Using Eqn (7.7) find the most probable speed for hydrogen gas molecules at 0°C. Explain with the aid of Fig. 7.1 why your result is lower than the root mean square speed quoted on p. 114.

7.3 A gas of hydrogen atoms is heated to 10,000K so that collisions between them result in electrons being excited into higher energy levels. Calculate the number of atoms in the $2s$ and $2p$ states as a proportion of the number in the ground state. (*Hint*: There are six $2p$ electron states having the same energy, but only one $1s$ state.)

7.4 Show that in Eqn (7.14) F is approximately proportional to $(r - r_0)$, if the extension of the interatomic distance is kept small enough. What implication does this have for the change in potential energy of the pair of atoms with $(r - r_0)$?

7.5 Show that, in the limit of very high temperature, Eqn (7.22) becomes approximately

$$C_v = 3Nk$$

Show also that this result is identical to that for a monatomic gas.

7.6 Find the value of C_v from Eqn (7.22) at the temperature $T_D = h\nu/k$, as a fraction of its value at $T \to \infty$.

The value of this temperature is given for some materials below. Find some of the corresponding values of ν, and explain why high values of this temperature occur in the stiffer and lighter materials.

Material	Pb	Au	NaCl	Fe	Se	Diamond
T_D (K)	95	170	280	360	650	1850

7·7 How many degrees of translational, rotational, and vibrational freedom has a triatomic molecule? Hence determine the maximum possible specific heat of a gas of such molecules at constant volume.

7·8 Describe the process of vaporization of a liquid from the point of view of the kinetic theory. Explain the reason for the existence of a latent heat of vaporization.

7·9 Explain what is meant by an *activated process*. The logarithm of the rate of an activated process is plotted against $1/T$ (T = absolute temperature). How can the *activation energy* be found from the graph? (The rate of the reverse process can be neglected.)

7·10 The diffusion coefficient of copper atoms in aluminium is found to be $1·28 \times 10^{-22}\,m^2/s$ at $T = 400K$ and $5·75 \times 10^{-19}\,m^2/s$ at $T = 500K$. Find the temperature at which its value is $10^{-16}\,m^2/s$.

Copper atoms diffuse under steady state conditions at this temperature through 0·1 mm thick aluminium foil. If the concentration of copper atoms is maintained at $10^{29}\,m^{-3}$ on one side of the foil and negligible on the other, what is the mass rate of flow of copper through the foil?

Self-Assessment questions

1 The kinetic theory of gases shows that
   A) the mean kinetic energy of a gas molecule is proportional to temperature
   B) Boyle's law is a natural consequence of the theory
   C) the pressure exerted by a gas is proportional to the mean square speed of its atoms.

2 The amount of energy possessed by a monatomic gas molecule for *each* cartesian direction of motion is
   A) $\frac{3}{2}kT$ B) $\frac{1}{2}kT$ C) $2kT^2$.

3 If the molecules in a gas may rotate and or vibrate then
   A) the average thermal energy for each component of the motion is $\frac{1}{2}kT$
   B) the specific heat in the high temperature limit is increased
   C) the material vaporizes more readily
   D) the mean square speed of the molecules at a given temperature is raised.

4 The amount of heat energy in all matter is proportional to the absolute temperature, only the constant of proportionality differing from substance to substance.
   A) true   B) false.

5 The differential of heat content with respect to temperature is called
   A) enthalpy   B) internal energy   C) specific heat of the substance.

6 The Maxwell–Boltzmann distribution law shows that
   A) the mean square speed of the molecules in a gas is proportional to temperature
   B) there is an upper limit to the velocity of a molecule at any temperature
   C) the most probable velocity increases with temperature.

7 The energy distribution law for atoms with quantized energy levels which are closely spaced can be written in the form

$$N(E) = \frac{dS}{dE}\, A \exp\left(-\frac{E}{kT}\right)$$

In this equation

A) $N(E)$ is the fraction of atoms having an energy E
B) dS/dE is the number of quantum states per unit energy interval at the energy E
C) $A \exp(-E/kT)$ is the probability that an atom lies in a quantum state of energy E.

8 In an atomic system in which the population of each quantum state is *small*, the numbers of atoms in states of differing energies E_1, E_2, E_3 are in the ratios

A) $E_1:E_2:E_3$
B) $\exp(E_1/kT):\exp(E_2/kT):\exp(E_3/kT)$
C) $\exp(-E_1/kT):\exp(-E_2/kT):\exp(-E_3/kT)$
D) $\dfrac{1}{\exp(E_1/kT)-1} : \dfrac{1}{\exp(E_2/kT)-1} : \dfrac{1}{\exp(E_3/kT)-1}.$

9 Which of the answers to question 8 is correct when the average population of some states approaches unity? If none, answer E.

10 Two bodies in contact are in thermal equilibrium when:

A) there is no net flow of heat between them
B) there is no exchange of energy between their atoms
C) the probabilities of atomic collision at the interface are the same in both bodies
D) the distribution of energy among the atoms is the same in both bodies.

11 A lattice wave can be represented as a stream of particles called phonons each of energy proportional to

A) the velocity of propagation of the wave
B) the heat content of the body
C) the volume of the crystal.

12 Figure (a) below is a graph of the forces between two atoms separated by a dis-

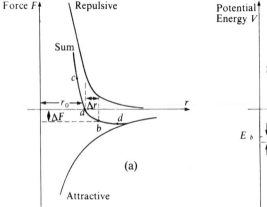

(a)

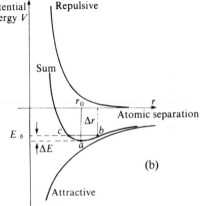

(b)

tance r. The attractive and repulsive forces can each be closely approximated by an expression of the form

A) $\pm CR^n$ B) $\pm CR^{1/n}$ C) $\pm Cr^{-n}$ D) $\pm Ce^{-nr}$

13 Figure (b) above is a graph of the potential energy of an atom versus displacement from its rest position.

(i) The rest position is at
A) a B) b C) c.

(ii) If the turning points of the motion of the atom are at b and c, the straight line bc represents a plot of
A) the kinetic energy   B) the potential energy
C) the total energy of the atom versus position.

(iii) The energy ΔE represents
A) the value of the kinetic energy of the atom
B) the energy given to the atom to set it in motion
C) the excess potential energy of the atom in motion.

(iv) The mean position of the atom when in motion is the point
A) a B) b C) c D) midway between b and c.

(v) The change in this mean position with ΔE provides an explanation of
A) thermal expansion
B) the change of specific heat with temperature
C) the thermal fluctuations in the length of a solid bar.

14 To calculate the specific heat of a body we can begin by assuming that

(i) all the atoms vibrate at a frequency ν
A) true   B) false.

(ii) the energy in lattice wave is quantized in units of
A) hc/λ B) $h\nu$ C) $\exp(-h\nu/kT)$.

(iii) the number of phonons in a mode of frequency ν is proportional to

A) $h\nu/kT$ B) $\dfrac{1}{1 - \exp(h\nu/kT)}$ C) $1/[\exp(h\nu/kT) - 1]$.

15 As a result of the calculation referred to in question 14, it is found that the total heat content Q of a body containing N atoms is

$$Q = \frac{3Nh\nu}{\exp(h\nu/kT) - 1}$$

From this equation it can easily be shown that the specific heat C_v at constant volume is equal to

A) $\dfrac{3Nh\nu}{T[\exp(h\nu/kT) - 1]}$ B) $\dfrac{3Nh^2\nu^2 \exp(h\nu/kT)}{kT^2[\exp(h\nu/kT) - 1]^2}$

C) $\dfrac{3N[\exp(h\nu/kT)(1 - h\nu/kT) - 1]}{[\exp(h\nu/kT) - 1]^2}$.

16 Dulong and Petit's Law states that the specific heats of all substances have nearly the same value when expressed per

A) mol   B) unit volume   C) gramme

of the substance.

17 Dulong and Petit's Law is confirmed by the expression in the answer to question 15, in the limit of

A) high   B) low

temperatures.

18 The term 'phase change' is used to denote changes such as

A) changes of lattice structure of a solid as the temperature is changed
B) the melting of a solid
C) the change of specific heat with temperature
D) the change of the amplitude and vibration of atoms with temperature
E) the change in volume of a gas with pressure
F) the vaporization of a liquid
G) a chemical reaction.

19 A phase change *usually* occurs

A) at a specific temperature and pressure
B) with the evolution or absorption of heat
C) over a range of temperature at a given pressure
D) spontaneously, without the absorption or evolution of heat
E) without passing through stable intermediate states.

20 When two phases of a single substance are in equilibrium with one another at a given pressure

A) the heat content (enthalpy) in each phase is the same
B) their temperatures are the same
C) their thermodynamic probabilities are the same
D) their free energies are the same.

21 The latent heat associated with a change of phase is equal to the difference in

A) enthalpy (heat content)   B) thermodynamic probability
C) free energy

between two phases.

22 The thermodynamic probability of a phase is a measure of

A) the tendency to change to another phase
B) the number of ways its atoms can be distributed among the available quantum states
C) the tendency of its atoms to acquire more than the average amount of thermal energy.

23 The quantity $\int_0^{T_1} C_p(\alpha)\,dT$ is equal to

A) the free energy   B) the heat content, or enthalpy
C) the internal energy

of a phase α at a temperature T_1.

24 The first law of thermodynamics relates the amount of heat ΔQ put into a body, the work ΔW done by it on its surroundings and the change in its internal energy ΔU, according to the equation

A) $\Delta U = \Delta Q + \Delta W$ B) $\Delta Q = \Delta U + \Delta W$
C) $\Delta W = \Delta U + \Delta Q$.

25 Two phases which coexist but are not in equilibrium at a given temperature

A) must eventually revert to equilibrium, albeit slowly

B) can remain indefinitely in the non-equilibrium state if there is a barrier to change.

26 The barrier to change referred to in question 25 may be due to

A) an intermediate state of higher energy
B) insufficient energy per atom to promote the change
C) insufficient time for the atoms to rearrange themselves.

27 In the figure, the potential energy of atoms in a two-phase material is plotted against some convenient dimension z which changes with the phase. The change between states with average energies E_A and E_B is possible because

A) atoms can 'tunnel' through the barrier
B) the height of the barrier fluctuates with time
C) atoms can 'hop' over the barrier by acquiring extra thermal energy.

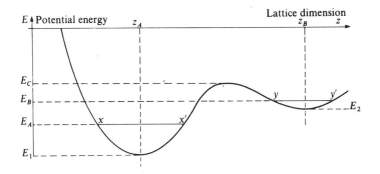

28 An activated state is

A) one which results from a phase change over an energy barrier
B) the unstable state of higher energy which constitutes the barrier
C) any state with more than average thermal energy.

29 The activation energy of the phase change illustrated in the figure in question 27 is equal to

A) $E_C - E_1$ B) $E_A - E_1$ C) $E_B - E_A$ D) $E_C - E_A$.

30 The rate of an activated phase change depends upon

A) the number of atoms with sufficient energy to cross the barrier
B) the frequency with which the atoms approach the barrier
C) the relative thermodynamic probabilities of the two phases
D) the height of the barrier.

31 A metastable phase is

A) fully stable against all disturbances
B) unstable to any disturbance
C) stable only against small disturbances.

32 The following are examples of metastable phases in air at room temperature

A) glass   B) uncoated steel   C) water vapour   D) b.c.c. iron.

33 Aluminium does not corrode in a normal atmosphere because

A) no reaction with oxygen can occur
B) the reaction with oxygen has too high an activation energy
C) atmospheric oxygen can only diffuse very slowly through the oxide layer which is formed.

34 Chemical reaction rates obey the same rate equations as phase changes.

A) true   B) false.

35 The slow passage of air through the walls of a sealed vacuum vessel is called

A) thermal activation   B) self-diffusion
C) chemical diffusion   D) thermal transpiration.

36 In chemical diffusion the flux of atoms is directly proportional to

A) the temperature   B) the concentration gradient
C) the number of atoms per unit volume.

37 If activated diffusion occurs via an energy barrier of height E, the diffusion rate

A) increases with E B) decreases with increasing E
C) is independent of E.

38 When diffusion is activated, the dependence of the diffusion coefficient on temperature is

A) $\exp - E/kT$ B) $\exp E/kT$ C) T D) $1/T$ E) zero

where E is a constant.

39 When diffusion is not activated the temperature dependence of the diffusion coefficient is

A) T B) $1/T$ C) $\exp E/kT$ D) $\exp - E/kT$ E) zero.

40 Diffusion is normally an activated process in

A) solids   B) liquids   C) gases.

Answers

1 B, C	2 B	3 A, B	4 B
5 C	6 C	7 B, C	8 C, D
9 D	10 A, D	11 A	12 C
13 (i) A	14 (i) A	15 B	16 A
(ii) C	(ii) A, B		
(iii) B	(iii) C		
(iv) D			
(v) A			
17 A	18 A, B, F	19 A, B, E	20 B, D
21 A	22 B	23 B	24 B
25 A	26 A	27 C	28 B
29 A	30 A, B, C, D	31 C	32 A, B
33 C	34 A	35 C	36 B
37 B	38 A	39 E	40 A

CRYSTAL DEFECTS

8·1 Introduction

If a perfect crystal could be obtained, atoms would only exist on lattice sites, every lattice site would be occupied by an atom, each atom would have its full quota of electrons in the lowest energy levels, and the atoms would be stationary. We have already seen that in practice the atoms vibrate about their lattice sites, and later, in Chapter 13, we will show how electrons can be excited into higher energy levels. In terms of the ideal perfect crystal, atomic vibrations and excited electrons can be regarded as defects. There are many other types of defect in crystals and all defects are of great importance to the materials scientist as their presence affects the bulk properties of materials.

Defects are classified according to their size. Those mentioned above are the smallest and are termed sub-atomic defects. The defects with which we are concerned in this chapter are, in ascending order of size, *point*, *line*, and *planar* defects.

8·2 Point defects

Figure 8·1 shows the various point defects that can exist in a crystal. All of these are characterized by having atomic dimensions in all directions.

We see from this diagram that a foreign atom (the solute), whether it be an impurity atom or a deliberate alloying addition, can occupy one of two distinct positions in a crystal. If the solute substitutes for an atom of the parent material (the solvent) it is said to be a *substitutional atom*, while if it occupies a hole or interstice in the parent lattice it is called an *interstitial atom*. Which of these two alternatives a solute atom takes depends on its size relative to the solvent atoms–the smaller the solute atom the more likely is it to exist as an interstitial. As far as metals are concerned the most important interstitials are carbon, nitrogen, and oxygen which are small atoms of less than 0·8 Å radius. We will come back to interstitial and substitutional atoms in later chapters where such factors as the amount of solute that can be accommodated by a solvent and the effect these defects have on the properties of the solvent will be discussed.

Now let us turn our attention to the vacant lattice site or *vacancy*. It is important to remember that, at all temperatures above absolute zero, the atoms in a solid are subject to thermal vibrations. This means that they continuously vibrate about their equilibrium positions in the lattice with an average amplitude of vibration that increases with increasing temperature.

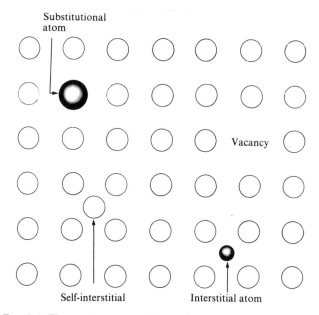

Fig. 8·1 The various point defects that can exist in a crystal

At a given temperature there is always a wide spectrum of vibration amplitudes, consequently occasionally in a localized region the vibrations may be
so intense that an atom is displaced from its lattice site and a vacancy is
formed. The displaced atom can either move into an interstice, in which case
it is called a *self-interstitial*, or on to a surface lattice site. The vacancy self-
interstitial is known as a *Frenkel defect* and the simple vacancy itself as a
Schottky defect (Fig. 8·2).

Experimentally it has been found that at a given temperature there is an
equilibrium concentration of Frenkel and Schottky defects. This means that

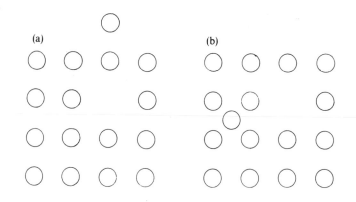

Fig. 8·2 The formation of (a) a Schottky defect and (b) a Frenkel defect

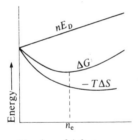

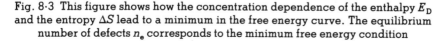

Number of defects n ⟶

Fig. 8·3 This figure shows how the concentration dependence of the enthalpy E_D and the entropy ΔS lead to a minimum in the free energy curve. The equilibrium number of defects n_e corresponds to the minimum free energy condition

the imperfect crystal must have a lower free energy than does the perfect crystal. Reference to the previous chapter reminds us that a change in the free energy ΔG of a system is related to the changes in enthalpy ΔH and entropy ΔS by

$$\Delta G = \Delta H - T\Delta S$$

The energy E_D to form a defect, i.e., the thermal energy required to displace an atom from its lattice site and into an interstice (Frenkel defect) or on to a surface site (Schottky defect), is identified with ΔH in this equation. E_D is positive and therefore corresponds to an increase in the free energy of the crystal. On the other hand, the presence of defects increases the degree of disorder of the crystal and so raises the entropy, which favours a decrease in free energy. The relative contributions of these two terms to the free energy of the crystal are given as a function of the number of defects, n, in Fig. 8·3. We can see that as the number of defects increases the free energy falls to a minimum and then increases; the equilibrium number of defects n_e corresponds to the minimum free energy condition. By calculating the minimum free energy condition as a function of temperature, Boltzmann obtained the following expression for the equilibrium number, n_e, and thus the equilibrium concentration, c_e, of defects

$$\frac{n_e}{N} = c_e = A \exp\left(\frac{-E_D}{kT}\right) \tag{8·1}$$

where N is the number of atoms in the crystal, A is a constant which is often taken as unity, T is the absolute temperature, and k is Boltzmann's constant. Values of c_e calculated from this equation with $A = 1$ and $E_D = 1$ eV/atom ($1·6 \times 10^{-19}$ J/atom) are listed in Table 8·1.

The energies of formation, and hence the equilibrium concentration at a given temperature, of Frenkel and Schottky defects differ from material to material. If we look at metals we find that E_D for Schottky defects is relatively low. The value of 1 eV/atom used to calculate the concentrations given in Table 8·1 is typical of silver, gold, and copper, which have melting points of 1,234K, 1,336K, and 1,356K respectively. With this information we conclude

Table 8·1

The equilibrium concentration of defects as a function of temperature according to $C_e = A \exp(-E_D/kT)$ with $E_D = 1$ eV/atom ($1 \cdot 6 \times 10^{-19}$ J/atom) and $A = 1$

Temperature K	c_e
0	$10^{-\infty}$
200	$6 \cdot 5 \times 10^{-26}$
400	$2 \cdot 5 \times 10^{-13}$
600	$4 \cdot 2 \times 10^{-9}$
800	$5 \cdot 0 \times 10^{-7}$
1000	$9 \cdot 2 \times 10^{-6}$
1200	$6 \cdot 1 \times 10^{-5}$

that, although the equilibrium Schottky defect concentration increases rapidly as the temperature is raised, even at temperatures approaching the melting point only about one lattice site in 10^5–10^4 is vacant. The energy of formation of Frenkel defects in a metal is much greater than 1 eV/atom and therefore the concentration of these defects under equilibrium conditions is even less than for Schottky defects and is generally negligible.

In contrast to metals, in ionic crystals there may be a preponderance of either Frenkel or Schottky defects depending on which has the smaller energy of formation. Frenkel defects are more likely to be important in crystals with open lattice structures which can accommodate interstitials without much distortion. These crystals have structures with low coordination numbers, for example the zinc blende and zinc oxide structures. In contrast we would expect to find Schottky defects in crystals with high coordination numbers such as the sodium chloride structure (Table 8·2). Where there is a large difference in size between the cations (the positive ions) and the anions (the negative ions) Frenkel defects are usually formed by the displacement of the smaller ion, which is most often the cation.

Table 8·2

Point defects in ionic crystals

Crystal	Structure	Dominant defect	Energy of formation (eV/atom)	(kJ/mole)
CdTe	ZnS	Frenkel	1·04	100
AgI	ZnO	Frenkel	0·69	67
NaCl	NaCl	Schottky	2·08	201
NaBr	NaCl	Schottky	1·69	163

Many important solid state processes to be discussed in later chapters such as recovery, recrystallization, and precipitation proceed by atoms moving (diffusing) from one lattice site to another in the crystal. One of the reasons for our interest in the equilibrium concentration of vacancies is that diffusion both of atoms from the host material and of substitutional atoms occurs by a *vacancy mechanism*, i.e., atoms diffuse by jumping into vacant lattice sites (Fig. 8·4). In fact the activation energy for diffusion, E_0 [see Eqn (7·32)],

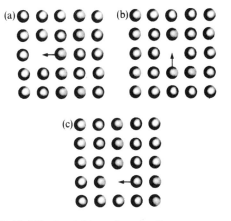

Fig. 8·4 Self-diffusion taking place by the vacancy mechanism

consists of two terms, the energy of formation E_D and the energy of motion E_m of a vacancy:

$$E_0 = E_D + E_m$$

8·3 Line defects

Line defects are long in one direction, while measuring only a few atomic diameters at right angles to their length. The only defect of this class is the *dislocation*. Knowledge of dislocations has allowed us to reconcile the rate of crystal growth from a vapour predicted by classical theory with the significantly faster growth rates obtained experimentally. It also enables us to account for the discrepancy between the stress theoretically needed to deform a perfect crystal plastically and the actual stresses measured on 'ordinary', i.e., imperfect crystals, which are much lower. In fact dislocations were first postulated in order to account for the latter; we will follow a similar path in our study of dislocations, so first we must look at the macroscopic aspects of plastic deformation.

8·3·1 Slip planes and slip directions

Consider Fig. 8·5(a) which shows a crystal loaded as in a tensile test. It must be appreciated that the applied tensile force can be resolved into components normal and parallel to any plane, such as XX', in the specimen. This is illustrated in Fig. 8·5(b) where the applied force, P, is resolved into a normal, tensile, component P_T and parallel component P_S (i.e. a shear force) on the plane XX'. This means that even a simple tensile force produces *shear forces*, and hence *shear stresses*, in a specimen. The value of the shear stress varies with the orientation of the plane and is a maximum for the planes at 45° to the tensile axis.

If we deform a single crystal plastically, that is, we increase the tensile force

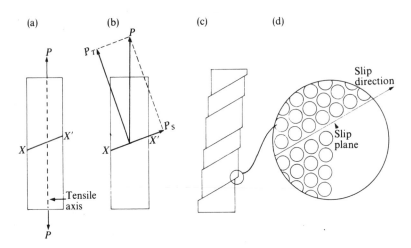

Fig. 8·5 Block slip mechanism for plastic deformation: (a) crystal before testing; (b) resolution of the applied force; (c) crystal after testing, i.e., plastically deformed; (d) a slip line (step) magnified showing the slip direction and the slip plane

until there is a permanent shape change, and then closely examine the surface under a microscope we see a series of parallel black lines such as those shown in Fig. 8·6. These lines are small steps on the surface of the crystal, called *slip lines* or slip steps. On increasing the tensile force still further, so causing more plastic strain, the number and or height of the slip lines increases. It is as if whole blocks of the crystal were slipping over one another under the action of the resolved shear stresses as shown in Fig. 8·5(c). The fact that the slip lines are parallel and that they do not necessarily correspond to slip along the planes of maximum shear indicates that the slip process is closely connected with the crystallography of the crystal. This is indeed the case as shown in the magnified diagram of Fig. 8·5(d); slip only occurs on preferred crystallographic planes, *the slip planes*, and in certain directions, *the slip directions*. The combination of a slip plane and a slip direction constitutes a *slip system*. As we will see in the following discussion on the slip systems in the f.c.c., c.p.h., and b.c.c. crystal structures, the slip direction is usually the direction in which the atoms are most closely packed and the slip plane the most densely packed plane containing that direction.

(a) *FCC*. The ⟨110⟩ directions are the closest packed directions in the face-centred cubic structure and are therefore the slip directions. The ⟨110⟩ directions lie in the most densely packed planes, which are the close-packed {111} planes (Fig. 8·7). Thus, the four {111} planes are the slip planes, and as each plane contains three ⟨110⟩ directions there are altogether $3 \times 4 = 12$ slip systems.

(b) *CPH*. The slip directions are the closest packed directions. Three such directions lie in the close-packed basal plane, giving only $3 \times 1 = 3$ slip systems. The three slip directions are shown in Fig. 8·8.

25μm

Fig. 8·6 Slip lines on the surface of a deformed specimen

(c) *BCC*. The most closely packed directions in the b.c.c. structure are the ⟨111⟩ which are therefore the slip directions. However, unlike the f.c.c. and c.p.h. structures, the b.c.c. structure is not constructed from close-packed planes. Nevertheless, there are three planes in the b.c.c. structure with relatively high packing densities. These are, in descending order of packing density, the {110}, {100}, and {112} planes. However, the {100} planes do not contain the ⟨111⟩ directions, so only the {110} and the {112} are slip

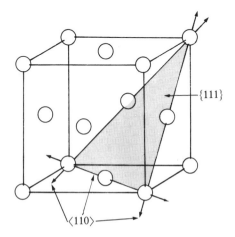

Fig. 8·7 One of the four {111} slip planes with its three ⟨110⟩ slip directions in the f.c.c. structure

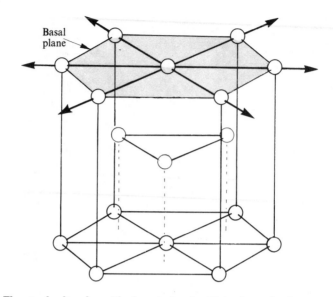

Fig. 8·8 The single slip plane (the basal plane) with its three slip directions giving three slip systems in the c.p.h. structure

planes. There are six {110} planes with two ⟨111⟩ directions in each giving 6 × 2 = 12 slip systems of the type {110} ⟨111⟩ (Fig. 8·9). There are also twelve systems of the type {112} ⟨111⟩ made up from one ⟨111⟩ direction in each of the twelve {112} planes.

So far we have considered plastic deformation to take place by the shearing

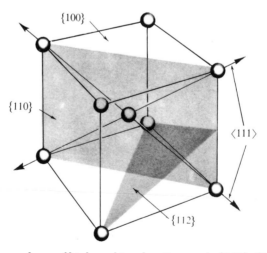

Fig. 8·9 The three planes of high packing density namely { 100}, {110}, and {112} in the b.c.c. structure. The two ⟨111⟩ slip directions for the {110} slip plane are marked

of complete blocks of a crystal on the slip planes and in the slip directions. From a knowledge of the bonding energies between atoms it is possible to estimate the magnitude of the shear stress necessary to cause plastic flow by this mechanism. These estimates give stresses of the order of $G/30$, where G is the shear modulus. In practice, shear stresses of only about $G/1000$ are required to produce plastic flow in ordinary crystalline materials so the mechanism must be different from that of simple block shearing of complete planes over one another. In the next section we will see how plastic deformation really occurs by the movement of dislocations.

8·3·2   Geometry and Burgers vector of dislocations

An extra half-plane of atoms inserted between the planes of atoms in a crystal can be accommodated as shown in Fig. 8·10. If the top of the crystal, above the line XX′, is viewed in isolation it appears perfect. Similarly, the bottom of the crystal below the line XX′ is perfect. Only when the two halves of the crystal are together is there a defect, the defect running in a line perpendicular to the page through the symbol ⊥. This line defect is an *edge dislocation* and by convention is represented by the symbol ⊥. The edge dislocation is able to move through a crystal causing slip at much lower stresses than those required for the block slip mechanism. One can see this in a general way by

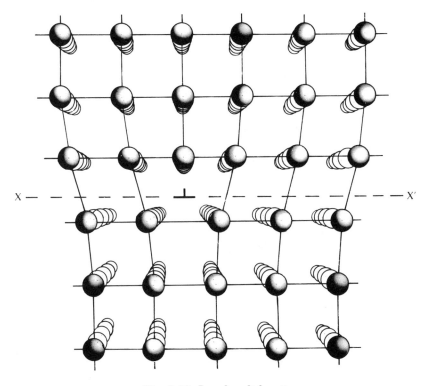

Fig. 8·10 An edge dislocation

referring to Fig. 8·11. We are going to concentrate on the atoms marked 1, 2, 3, 4, and 5. These atoms are displaced from their equilibrium positions and therefore their bond energies are altered (see Fig. 7·4). In the unstressed condition the atom pairs 1–2 and 2–3 are closer together and the pairs 1–5 and 3–4 are further apart than the normal equilibrium spacing [Fig. 8·11(a)]. On applying a shear stress [Fig. 8·11(b)] the distances apart of the atoms change; atom 2 moves a little to the left so that it is nearer atom 3 and further from atom 1, at the same time the distance between atoms 1–5 is reduced while between atoms 3–4 it is increased. The result is that there is an increase in energy of the 2–3 and 3–4 bonds (the interatomic distances between these atoms have become further removed from the equilibrium value) and a decrease in the 1–2 and 1–5 bond energies (the interatomic distances have become nearer the equilibrium value). The increase in energy from the former is greater than the decrease in energy due to the latter effect, the difference being supplied by the shear stress. On increasing the stress further, atom 2 continues to move to the left until eventually it bonds more strongly with atom 4 than does atom 3 and the dislocation has effectively moved one interatomic distance to the left [Fig. 8·11(c)]. This procedure is repeated until the dislocation leaves the crystal giving a step of height b [Fig. 8·11(d)]. If a number of dislocations move along the same slip plane the step height increases to give a configuration such as that shown in Fig. 8·5(d), i.e., a visible slip line. From this discussion it is clear that the movement of disloca-

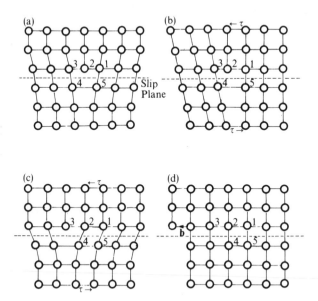

Fig. 8·11 Movement of an edge dislocation under the influence of a shear stress: (a) unstressed; (b) stress applied and the dislocation moves slightly to the left; (c) stress increased and the dislocation moves one atomic spacing to the left; (d) this procedure is repeated until the dislocation leaves the crystal forming a step of height b

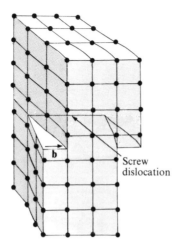

Fig. 8·12 A screw dislocation. Note that the Burgers vector is parallel to the dislocation

tions is similar in many ways to block slip (e.g., dislocations move on the slip plane and give slip lines) but requires smaller stresses as only a few bonds are being altered or broken at any one time.

The magnitude and direction of the slip resulting from the motion of a single dislocation, *b* in Fig. 8·11, is called the *Burgers vector*. We see that for the edge dislocation of Fig. 8·11 the Burgers vector is perpendicular to the line of the dislocation. This is a special case and dislocations exist with the Burgers vector at all angles to the dislocation. Another special case is when *b* is parallel to the dislocation; this is the *screw dislocation* and is illustrated in Fig. 8·12. As for the edge dislocation, under the influence of a shear stress the screw dislocation moves in the slip plane but this time the direction of motion of the dislocation is perpendicular to the Burgers vector. The motions of an edge, screw, and an arbitrary dislocation loop are compared in Fig. 8·13. The arbitrary dislocation loop has an edge component (*b* perpendicular to the dislocation), a screw component (*b* parallel to the dislocation), and a large mixed component (angle between *b* and the dislocation greater than 0° but less than 90°).

So far we have defined the Burgers vector of a dislocation in terms of the atomic displacement caused during slip. This is rather restrictive because, as pointed out earlier, dislocations play an important role in processes other than plastic deformation. A general method for obtaining the magnitude and direction of the Burgers vector is to determine the discontinuity due to the presence of the dislocation by drawing what is called a Burgers circuit. First a closed circuit is drawn around the dislocation by jumping from atom to atom in a clockwise direction [Fig. 8·14(a) and (b)]. The same number of jumps is then made in the same directions in a perfect crystal. This time the circuit does not close, that is, it does not return to the atom at which it started, and the vector needed to complete the circuit is the Burgers vector of the dislocation

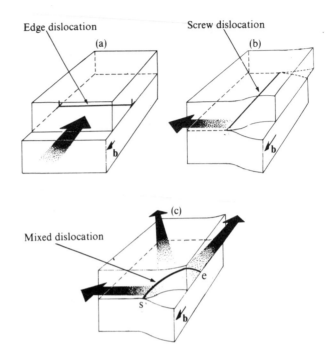

Fig. 8·13 The motion of (a) an edge dislocation; (b) a screw dislocation; (c) a mixed dislocation; s and e are screw and edge components respectively

[Fig. 8·14(c) and (d)]. It will be noted, of course, that this gives the magnitude and direction of the Burgers vector but, like all vectors, it has no fixed position. The position of the arrow in the diagram is not significant.

All dislocations are line defects and therefore only measure a few atomic diameters at right angles to their length. We define the *width*, *w*, of a dislocation as the distance over which the displacements of atoms is greater than $b/4$. When *w* is one or two atomic spacings the dislocation is said to be narrow. On the other hand a wide dislocation has a width of several atomic spacings (Fig. 8·15). Typically metals have wide dislocations, whereas covalently bonded materials have narrow dislocations.

We have already noted that the stress to move a dislocation is several orders of magnitude less than that for block slip. The reader will learn in the next chapter how other defects and second phase particles hinder dislocation motion. It follows that the minimum stress for the type of motion shown in Fig. 8·13 is the stress to move a dislocation in an otherwise perfect crystal. This intrinsic lattice friction stress is known as the *Peierls–Nabarro* stress. The significance of the width of a dislocation is that it determines the magnitude of the Peierls–Nabarro stress, τ_{PN}, as can be seen from the following equation:

$$\tau_{PN} \sim G \exp\left(-2\pi w/b\right) \tag{8.2}$$

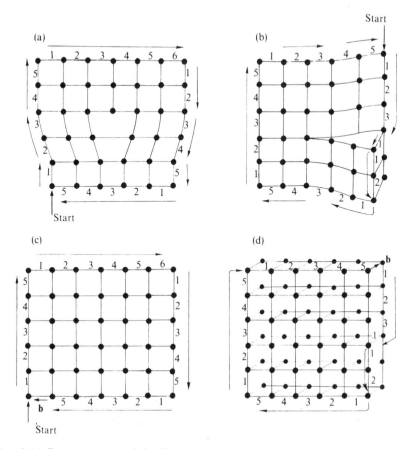

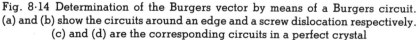

Fig. 8·14 Determination of the Burgers vector by means of a Burgers circuit. (a) and (b) show the circuits around an edge and a screw dislocation respectively. (c) and (d) are the corresponding circuits in a perfect crystal

τ_{PN} is very sensitive to w as w appears in an exponential term. The wider the dislocation the lower is τ_{PN} and the easier it is to move the dislocation. We can also see from this equation that a small Burgers vector, b, leads to a low τ_{PN} value. This is one of the reasons why the dislocations responsible for plastic deformation often have the smallest b compatible with the crystal structure, or in terms of macroscopic slip, why the slip direction is generally the closest packed direction.

8·3·3   Cross-slip and climb

The dislocation motion described in the previous section and illustrated in Figs. 8·11 and 8·13 is usually referred to as *glide*. Glide may be defined as the motion of a dislocation along a slip plane containing both the dislocation and its Burgers vector. In the case of an edge dislocation the Burgers vector is perpendicular to the dislocation, thereby specifying a single plane. It

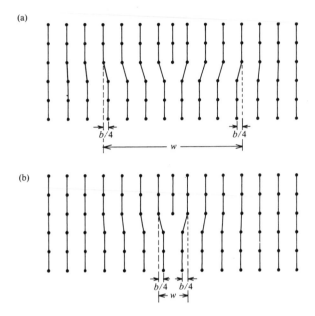

Fig. 8·15 Diagram illustrating the width of a dislocation: (a) a wide dislocation; (b) a narrow dislocation

follows that an edge dislocation is only able to glide on this specified slip plane. In contrast, the Burgers vector is parallel to a screw dislocation and so a unique plane is not defined. This means that a screw dislocation can glide in any slip planes that have a common slip direction. This gliding from one slip plane to another, which is known as *cross-slip*, enables screw dislocations to bypass obstacles to their motion (Fig 8·16).

Although edge dislocations cannot avoid obstacles by cross-slip they can avoid them by moving normal to their slip planes by the process of *climb*. The most common climb mechanism is the diffusion of vacancies to an edge dislocation and this is illustrated in Fig. 8·17. The first vacancy to reach the dis-

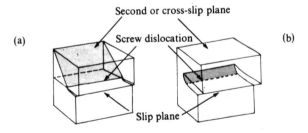

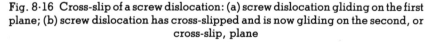

Fig. 8·16 Cross-slip of a screw dislocation: (a) screw dislocation gliding on the first plane; (b) screw dislocation has cross-slipped and is now gliding on the second, or cross-slip, plane

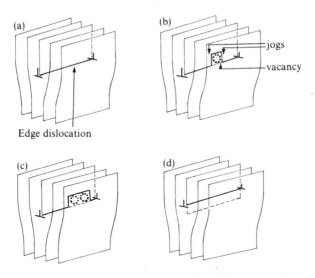

Fig. 8·17 Climb of an edge dislocation: (a) the edge dislocation before climb; (b) a vacancy diffuses to the dislocation producing jogs; (c) the jogs move apart as more vacancies arrive at the dislocation; (d) the dislocation has climbed one atomic spacing

location produces a step, the sides of which are termed *jogs* [Fig. 8·17(b)]. As more vacancies diffuse to the dislocation the jogs move apart [Fig. 8·17(c)] until eventually the whole dislocation has climbed one atomic spacing [Fig. 8·17(d)]. This process can be repeated many times so the edge dislocation can climb a number of atomic spacings. Climb is assisted by elevated temperatures which speed up the rate of vacancy diffusion.

8·3·4 Dislocation energy

The regularity of a crystal lattice results from each atom taking up a position which minimizes its potential energy; it follows that a dislocation must represent a raising of the potential energies of all the atoms whose positions are affected by its presence. Thus an energy may be ascribed to a dislocation

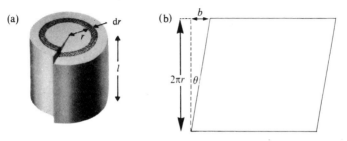

Fig. 8·18 (a) Model for the calculation of the energy of a screw dislocation. (b) Opening out the cylinder we can see that the shear strain $\gamma = \tan \theta = b/2\pi r$

which, physically, is the strain energy built into the crystal structure by displacement of the atoms from their regular positions.

We may estimate this energy by considering a screw dislocation in a cylinder of crystal of length l, as shown in Fig. 8·18(a). At a radius, r, the deformation in a thin annulus of thickness dr is given by the magnitude, b, of the Burgers vector so that the shear strain, γ, is $b/2\pi r$, see Fig. 8·18(b), and using Hooke's law for shear, the average stress, τ, will be given by $Gb/2\pi r$, where G is the elastic shear modulus.

It is interesting to note from this that the stress field of a dislocation is inversely proportional to distance from the dislocation and may, therefore, be described as long range, compared with the interatomic forces. Thus dislocations will interact with each other at quite large distances apart on the atomic scale. In fact at a distance of 10,000 atoms from the core the stress field may still have a value of the order of $10^{-5}\,G$.

The elastic strain energy dw of a small volume element dv is given by $\frac{1}{2}\tau\gamma\,dv$ so we have

$$dw = \tfrac{1}{2}\tau\gamma\,dv = \tfrac{1}{2}\,\frac{Gb}{2\pi r}\cdot\frac{b}{2\pi r}\,dv$$

$$= \tfrac{1}{2}G\left(\frac{b}{2\pi r}\right)^2 dv$$

The volume of the annular element is $2\pi r l dr$, hence

$$dw = \frac{Gb^2 l}{4\pi}\,\frac{dr}{r}$$

The elastic strain energy contained within a cylinder of radius R around the dislocation is obtained by integrating this equation up to the limit R. However, the lower limit of integration is not taken as zero for two reasons: (1) the calculation assumes the material is an isotropic continuum, but in the region near to the dislocation this assumption is unrealistic and it is necessary to consider the displacements of, and the forces between, individual atoms; and (2) the strains near the dislocation are large and Hooke's law, on which the calculation is based, fails at large strains. Thus the lower limit of integration is taken as some small value, r_0 (which is often taken to be equal to b). The region inside r_0 is referred to as the *dislocation core*. We therefore have

$$E = \frac{Gb^2 l}{4\pi}\int_{r_0}^{R}\frac{dr}{r} + E_c l$$

where $E_c l$ is the strain energy within the radius 0 to r_0, i.e., the core energy. The calculation for the core energy is complex but a reasonable estimate is $Gb^2 l/10$. Thus, on integrating to obtain the elastic strain energy we get for the total strain energy

$$E = \frac{Gb^2 l}{4\pi}\ln\frac{R}{r_0} + \frac{Gb^2 l}{10}$$

Since generally $R \gg r_0$, the logarithmic term in this expression only varies slowly with R/r_0. As an approximation, $\ln(R/r_0)$ may be taken as 4π, then

$$E \approx Gb^2l + \frac{Gb^2l}{10} \tag{8·3}$$

Two important features of this formula are: (1) the total energy is proportional to b^2; and (2) the core energy is only one-tenth of the elastic strain energy.

For typical metals such as chromium or copper $G \approx 5 \times 10^{10} \text{ N/m}^2$, and taking $b = 2·5 \times 10^{-10}$ m we obtain for the energy of a screw dislocation a value in the region of 5·5 eV/atom. An equation similar to Eqn (8·3) can be derived for an edge dislocation and would give for the energy of an edge dislocation about 8 eV/atom. These energies are considerably greater than the formation energies of Schottky defects in metals. We have seen how Schottky defects are an equilibrium feature of a crystal because the increase in entropy associated with them outweighs the increase in enthalpy and leads to a decrease in free energy of the crystal (Fig. 8·3). Dislocations also increase the entropy, but in this case the entropy change does not outweigh the very large enthalpy increase of 5·5–8 eV/atom, and consequently the free energy of the crystal is increased. Therefore, dislocations are non-equilibrium defects and if possible a crystal will decrease its dislocation density in order that its free energy may be reduced.

8·3·5 Super and partial dislocations

An established nomenclature for describing dislocations has been developed in terms of the Burgers vector. The dislocations considered so far have had Burgers vectors equal to one lattice spacing, see for example Fig. 8·14, and are called *unit* dislocations. A unit dislocation is a specific case of a *perfect* dislocation, which is a dislocation whose Burgers vector is an integral number of lattice spacings. Conversely, a dislocation is *imperfect* if its Burgers vector is not an integral number of lattice spacings. A *partial* dislocation is an imperfect dislocation with its Burgers vector less than unity, while a *super* dislocation can be perfect or imperfect with a Burgers vector greater than unity.

Sometimes a dislocation of Burgers vector b_1 splits into two, or more, dislocations of Burgers vectors b_2, b_3, b_4, etc. The reverse can also occur, namely dislocations of Burgers vectors b_2, b_3, b_4, etc., can combine to form a single dislocation of Burgers vector b_1. This behaviour can be explained in terms of the energies of the dislocations involved. We know that the energy of a dislocation is proportional to the Burgers vector squared [Eqn (8·3)], therefore a dislocation will split, if it is geometrically possible, when

$$b_1^2 > b_2^2 + b_3^2 + \ldots \tag{8·4}$$

Alternatively, a number of dislocations will combine if

$$b_2^2 + b_3^2 + \ldots > b_1^2 \tag{8·5}$$

Relationships (8·4) and (8·5) are known as *Frank's rule*.

Let us use Frank's rule to examine some examples of super and partial dislocations in more detail. First consider a super perfect dislocation of Burgers vector nb where $n = 2, 3, 4. \ldots$ Generally it is geometrically possible for such a dislocation to split into n unit dislocations each of Burgers vector b, but is this splitting energetically favourable? Applying Frank's rule we see that it is, as

$$n^2 b^2 > n b^2$$

Therefore super dislocations are rarely found in crystals as they split into unit dislocations.

The example of a partial dislocation is taken from the f.c.c. structure. The stacking sequence of the close-packed {111} planes, which are also the slip planes, is ABCABC. This is illustrated in Fig. 8·19(a) where the lower layer comprises atoms marked A, the next layer has atoms marked B, and the top layer is in the C-position. Figure 8·19(b) shows that the unit dislocation that would cause a displacement in the slip direction would move atom B_1 to B_2. In order for B_1 to move to B_2 it would have to rise over atom A_1 in the plane below. This is a difficult process and it is geometrically easier for slip to take place in two stages. The first stage is to move B_1 down the 'valley' between the A atoms to a C-position, C_1, and the second stage is to move from C_1, again down the 'valley' between the A atoms, to B_2. This zig-zag motion could be produced by partial dislocations, so let us see if it is energetically possible for suitable partial dislocations to exist in the f.c.c. structure. The Burgers vectors of the unit dislocation and the possible partial dislocations are given in Fig. 8·19(c). From the vector diagram we have for the Burgers vectors:

$$b_2 = \frac{b_1}{2 \cos 30°} = \frac{b_1}{\sqrt{3}}$$

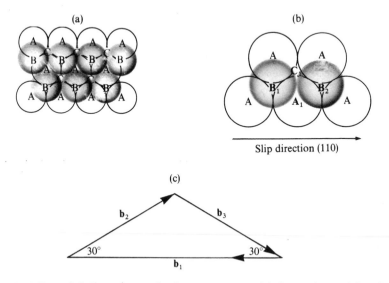

Fig. 8·19 Partial dislocations in the f.c.c. structure: (a) the packing of the close-packed {111} planes; (b) comparison of slip by a unit dislocation ($B_1 \to B_2$) to that by partial dislocations ($B_1 \to C_1 \to B_2$); (c) vector diagram for a unit dislocation b_1 and the two partial dislocations b_2 and b_3

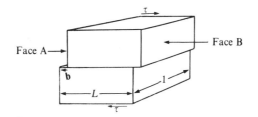

Fig. 8·20 Model for the calculation of the force on a dislocation; an edge dislocation has moved from face A to face B under the action of a shear stress τ

Similarly $b_3 = b_1/\sqrt{3}$. Therefore the partial dislocations are energetically
favourable, since

$$b_1^2 > (b_2^2 + b_3^2) = \left(\frac{b_1}{\sqrt{3}}\right)^2 + \left(\frac{b_1}{\sqrt{3}}\right)^2 = \frac{2b_1^2}{3}$$

and are responsible for slip in the f.c.c. structure. This combination of two
partial dislocations, the leading partial dislocation producing a stacking fault
and the following partial dislocation recreating the normal f.c.c. stacking
sequence, is known as an extended dislocation.

8·3·6 Forces on dislocations

The force on a dislocation due to an applied stress is simply related to the
Burgers vector of the dislocation. We refer to Fig. 8·20 which shows a crystal
of unit thickness that has been sheared by the movement of a single edge dislocation from face A to face B. If F is the force per unit length of dislocation,
then the work done when a dislocation moves from face A to face B is FL.
Now, the force on the slip plane due to the applied shear stress τ is $\tau \times$ area $=$
$\tau \times L \times 1 = \tau L$, and this force does work in producing a displacement b. The
amount of work is given by $\tau L b$ and this must equal the work done in moving
the dislocation, so we have

$$FL = \tau L b$$

Therefore,

$$F = \tau b \tag{8·6}$$

that is, the force on a dislocation is simply the product of the shear stress and
the Burgers vector.

Dislocations are not always straight, see for example Fig. 8·13(c), and
using Eqn (8·6) we can calculate the shear stress required to bow a dislocation. To do this the concept of *line tension* has to be introduced. Just as a soap
bubble has a surface energy and a related surface tension, so a dislocation has
a strain energy and an association line tension. It can be shown that the line
tension, T, per unit length is approximately equal to the elastic strain energy
per unit length, that is

$$T \simeq Gb^2 \tag{8·7}$$

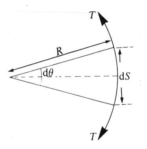

Fig. 8·21 The line tension of a dislocation tends to straighten the dislocation

Figure 8·21 shows a dislocation bowed to a radius of curvature R. The line tension will produce a force tending to straighten the dislocation and it will only remain bowed if a shear stress τ_0, produces an equal and opposite force. Resolving the forces acting on the dislocation segment dS horizontally,

$$\tau_0 b dS = 2T \sin \left(\frac{d\theta}{2} \right)$$

For small values of $d\theta$ we can take $\sin(d\theta/2)$ to be equal to $d\theta/2$, and as $d\theta/dS = 1/R$ we obtain, for the stress required to keep the dislocation bowed,

$$\tau_0 = \frac{T}{bR} = \frac{Gb}{R} \tag{8·8}$$

The significance of this equation will become apparent in the following section on dislocation multiplication and in the later discussions on precipitation hardening.

8·3·7 Dislocation multiplication

The density, ρ, of dislocations in a crystal is strictly expressed as length of dislocation line per unit volume, that is, as metres per cubic metre. This is equivalent (although not exactly) to the number of dislocations piercing a unit area anywhere in the crystal, so that ρ is usually quoted as the number per unit area and, for convenience, is usually taken as the number per square centimetre. The most perfect bulk single crystal of metal that can be produced will usually have a dislocation density of 10^2 to 10^3 cm^{-2} while a normal polycrystal has a density in the region 10^7 to 10^8 cm^{-2}. These dislocations are built in 'naturally' in the fabrication processes, but when the specimen has undergone severe plastic deformation the density is found to increase to 10^{11} to 10^{12} cm^{-2}. We must conclude, therefore, that dislocations somehow multiply in the process of deformation. A possible multiplication mechanism is called the *Frank–Read* source after its originators.

Suppose that a dislocation is pinned at two points P and P′ a distance l apart, Fig. 8·22(a). Under the action of an applied stress the dislocation will bow out, the radius of curvature R being related to the stress by Eqn (8·8) [Fig. 8·22(b)]. Increasing the stress further causes more bowing, i.e., R

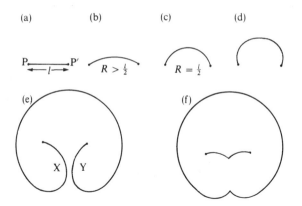

Fig. 8·22 The Frank–Read source. The stress on the dislocation increases to a maximum corresponding to minimum radius R [configuration (c)]. The sequence of events represented by (d), (e), and (f) can occur at lower stresses

decreases, until a minimum value of R is reached which is equal to $l/2$, Fig. 8·22(c). As the stress is inversely proportional to R, this minimum R configuration must correspond to the maximum stress. Consequently, the dislocation configurations shown in Fig. 8·22(d), (e), and (f) occur spontaneously at stresses less than that required for $R = l/2$. The sequence of events is that the dislocation expands into a loop, Fig. 8·22(d), the segments X and Y meet, Fig. 8·22(e), and annihilate each other forming a complete loop and reforming the pinned dislocation. The loop expands and moves away from the source PP', and the pinned dislocation can continue to multiply.

8·4 Planar defects

Planar defects are atomic in one dimension and large in the others. They may be classified as either external or internal defects. The external planar defects are the solid–gas, solid–liquid and liquid–gas interfaces. As the name suggests, internal planar defects are found in the interior of crystalline materials and the most common are grain boundaries and twin interfaces. The structure and energy of the various planar defects differ and it is particularly important to have an appreciation of the relative magnitudes of the energies. Approximate energies as a fraction of the solid–gas interface energy, E(S–G), are: E(solid–liquid) $\sim \frac{1}{12} E$(S–G), E(liquid–gas) $\sim \frac{5}{6} E$(S–G), E(high-angle grain boundary) $\sim \frac{1}{3} E$(S–G) and E(twin) $\sim \frac{1}{20} E$(S–G).

8·4·1 Grain boundaries

Grain boundaries may be defined as the interface between crystals that differ in crystallographic orientation, composition or dimensions of the crystal lattice (in some cases in two or all of these). Grain boundaries between crystals of different orientation have been described in an earlier chapter and are schematically represented in Fig. 6·1 and may be seen in the micrograph

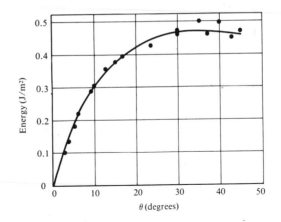

Fig. 8·23 Variation of grain boundary energy with misorientation for copper [the data are from N.A. Gjostein and F.N. Rhines, *Acta Met.*, **7**, 319 (1959) and are for a particular type of grain boundary called a simple twist boundary]

of Fig. 6·2. The energy E of these boundaries varies with the misorientation θ across the boundary according to:

$$E = E_o \theta (A - \ln \theta)$$

where E_o and A are constants. The variation in energy with misorientation is considerable at low misorientations of less than about 12°, but at high mis-

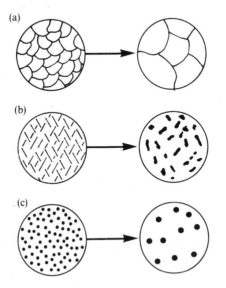

Fig. 8·24 Three processes that occur because they lead to a reduction in the total grain boundary energy: (a) grain growth in a single-phase material; (b) spheroidization in a two-phase material; (c) particle coarsening in a two-phase material

orientations the energy is nearly constant (Fig. 8·23). The grain boundaries seen in micrographs of single-phase materials, such as those of Fig. 6·2, are normally associated with large misorientation and are called high-angle grain boundaries. The other two types of grain boundaries need no explaining, and many examples of the boundaries between crystals of different compositions may be found in the micrographs in the following two chapters. The energy of these boundaries depends on the mismatch of the crystal lattices and the composition of the two crystals.

All grain boundaries have a relatively high energy, typically 0·2–1·0 J/m², and as a result are non-equilibrium defects. Thus, a crystal will always try to reduce the total grain boundary area in order to reduce its free energy. This is why the processes illustrated in Fig. 8·24, namely grain growth, spheroidization, and particle coarsening occur in materials held at elevated temperatures.

8·4·2   Twin interfaces

There are other internal planar defects other than grain boundaries, the most familiar being the *twin* interface. A crystal is twinned when one portion of the lattice is a mirror image of the neighbouring portion, the mirror being the twinning plane as shown in Fig. 8·25. This may be formed during the growth of the crystal or may be the result of dislocation motion caused by an applied stress. The straight lines traversing the grains in Fig. 6·2 are where the former type of twin interface intersects the specimen surface; these are called *annealing* twins and are formed in many f.c.c. metals and alloys during recrystallization (cf. Chapter 9).

The other type of twin, the *deformation* twin, is produced by the movement of specific dislocations. It can be seen from Fig. 8·25 that the atomic displacements required to form a twin are not integral numbers of lattice spacings, therefore we conclude that the dislocations involved are imperfect dislocations. They are in fact partial dislocations and have to move in a certain complex manner in order to produce twins.

The stress to twin a crystal tends to be higher than that required for slip, hence except under certain conditions slip is the normal deformation mechanism. For example, the twinning stress is less sensitive to temperature than the stress for slip, consequently twinning becomes more favourable as the deformation temperature becomes lower. This is the case when b.c.c. iron and its alloys are rapidly loaded at low temperatures where thin, lamellar twins appear, called *Neumann bands*. Twinning may also play a significant

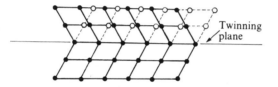

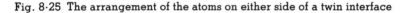

Fig. 8·25 The arrangement of the atoms on either side of a twin interface

25μm

Fig. 8·26 Deformation twins in polycrystalline magnesium. Slip lines are also visible

role in the deformation of c.p.h. materials, especially when in the polycrystalline condition. This is because the c.p.h. structure only has three slip systems, therefore slip is restricted. The twins in Fig. 8·26 were produced by deforming polycrystalline magnesium; the presence of slip lines indicates that slip has also occurred.

Problems

8·1 The energies of formation of Schottky and Frenkel defects in a material are 1 eV/atom ($1·6 \times 10^{-19}$ J/atom) and 4 eV/atom ($6·4 \times 10^{-19}$ J/atom) respectively. What are the equilibrium concentrations of these defects at 1000K? Assume that the pre-exponential constant, A, is equal to unity.

8·2 The equilibrium vacancy concentrations of Schottky defects in copper are 4×10^{-6} and 15×10^{-5} at temperatures of 1000K and 1325K respectively. What is the energy of formation of these defects?

8·3 Distinguish between edge and screw dislocations. With the aid of sketches, show that the normal glide motion of both types of dislocation produces slip lines on the surface of a crystal.

8·4 Draw a Burgers circuit for an edge dislocation in a primitive cubic crystal. In a certain crystal the Burgers vector for an edge dislocation is $2·5 \times 10^{-10}$ m. Calculate the force per unit length on the dislocation when a shear stress of 350 N/m² is applied.

8·5 A two-phase material has precipitate particles of radius $1·25 \times 10^{-8}$ m and of concentration 10^{20} m^{-3}. The energy of the grain boundary between the particle and the surrounding material (the matrix) is $0·5$ J/m^2. After heat-treatment at an elevated temperature the particles were found to have coarsened to 3×10^{-8} m radius and their concentration decreased to $7·2 \times 10^{18}$ m^{-3}. What is the reduction in the total grain boundary energy due to the coarsening?

Self-Assessment questions

1 Point defects have atomic dimensions in all directions.

    A) true   B) false.

2 Substitutional atoms occupy interstices in the parent lattice.

    A) true   B) false.

3 Many different point defects exist in crystals, for example,

    A) excited electrons   B) self-interstitials
    C) slip lines   D) vacancies
    E) precipitate particles   F) substitutional atoms.

4 This figure shows the formation of a

    A) Schottky defect   B) grain boundary
    C) Frenkel defect   D) substitutional atom
    E) line defect   F) point defect.

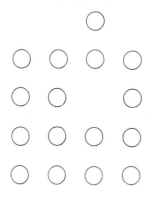

5 This graph refers to the formation of Schottky or Frenkel defects. It shows that

    A) the entropy term favours an increase in free energy
    B) the enthalpy term favours an increase in free energy
    C) a crystal will always try to minimize the point defect concentration
    D) the free energy of a crystal is a minimum when the equilibrium number of point defects is present.

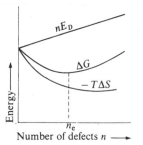

Number of defects n ⟶

6 The equilibrium concentration of Schottky defects c_e, at any given temperature T, may be obtained using the expression

A) $c_e = AT^Q$ B) $c_e = A [\ln T - \ln Q]$
C) $c_e = A \exp(-Q/kT)$ D) $c_e = kT \ln Q$

where A is a constant, Q is the energy of formation and k is Boltzmann's constant.

7 In metals the energy of formation of a Schottky defect is of the order of

A) 10^{-8} eV   B) 10^{-3} eV   C) 1 eV   D) 10^2 eV   E) 10^8 eV.

8 In ionic crystals the Frenkel defect concentration is always greater than the Schottky defect concentration.

A) true   B) false.

9 Vacancies play an important role in

A) deformation twinning   B) self-diffusion
C) climb   D) cross-slip.

10 This is a micrograph of

A) grain boundaries   B) annealing twins
C) climb of edge dislocations   D) slip lines
E) a plastically deformed crystal   F) an undeformed crystal.

25 μm

11 The slip plane and the slip direction in the f.c.c. structure are the $\{111\}$ and $\langle 110 \rangle$ respectively. Consequently the number of slip systems in the f.c.c. structure is

A) 12   B) 6   C) 24   D) 3.

12 The Burgers vector of an edge dislocation is perpendicular to the dislocation.

A) true   B) false.

13 To obtain the Burgers vector of a dislocation, Burgers circuits are drawn

A) first around the dislocation, and then, with the same number of jumps, in the perfect crystal
B) first around the dislocation and then, with one fewer jump, in the perfect crystal
C) first in the perfect crystal, and then with the same number of jumps, around the dislocation
D) first in the perfect crystal, and then with one fewer jump, around the dislocation
E) in an anticlockwise direction
F) in a clockwise direction.

14 Both edge and screw dislocations can climb.

A) true   B) false.

15 The total energy of a dislocation is proportional to

A) G B) G^2 C) b D) b^2 E) b^{-1}

where G and b are the shear modulus and Burgers vector respectively.

16 The core energy of a dislocation is only about one-tenth of the elastic strain energy.

A) true   B) false.

17 The energy of an edge dislocation is greater than that of a screw dislocation in the same material.

A) true   B) false.

18 The Burgers vector of an imperfect dislocation is

A) always less than one lattice spacing
B) always more than one lattice spacing
C) an integral number of lattice spacings
D) not an integral number of lattice spacings.

19 A dislocation of Burgers vector b_1 will be split into two dislocations of Burgers vectors b_2 and b_3 when

A) it is geometrically possible   B) $b_1 > b_2 + b_3$
C) $b_1^2 > b_2^2 + b_3^2$ D) $b_1^2 = b_2 + b_3$
E) b_1 is parallel to b_2 and b_3.

20 Dislocation motion in f.c.c. materials is represented in the diagram on the next page (i) b_1 is the Burgers vector of a . . . (ii) b_2 is the Burgers vector of a . . .

A) perfect dislocation   B) imperfect dislocation
C) screw dislocation   D) partial dislocation
E) unit dislocation.

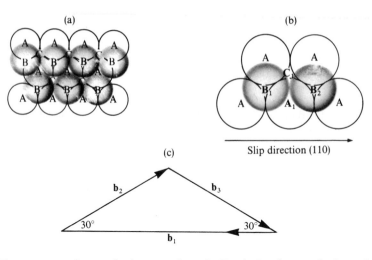

(a)   (b)

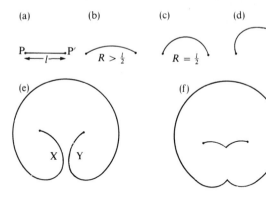

(c)

21 The sequence of events in the operation of a Frank–Read source is shown in (a), (b), (c), (d), (e), and (f). The maximum stress to operate the source

A) corresponds to event (b)   B) corresponds to event (c)
C) corresponds to event (d)   D) corresponds to event (e)
E) is proportional to l F) is proportional to l^{-1}
G) is much greater for screw dislocation than an edge dislocation.

(a)   (b)   (c)   (d)

(e)   (f)

22 The diagram below illustrates

A) grain growth   B) deformation twinning
C) spheroidization   D) vacancy formation
E) a process that reduces the total grain boundary energy
F) clustering of self-interstitials.

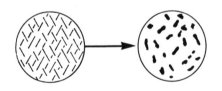

23 Annealing twins are formed in many f.c.c. metals and alloys during

    A) casting   B) plastic deformation
    C) rapid quenching from a high temperature   D) recrystallization.

24 Deformation twins are only formed in b.c.c. metals.

    A) true   B) false.

Each of the sentences in questions 25–34 consists of an assertion followed by a reason. Answer:

    A) If both assertion and reason are true statements and the reason is a correct explanation of the assertion

    B) If both assertion and reason are true statements but the reason is *not* a correct explanation of the assertion

    C) If the assertion is true but the reason contains a false statement

    D) If the assertion is false but the reason contains a true statement

    E) If both the assertion and reason are false statements.

25 Nitrogen and oxygen generally occupy interstitial sites in metals *because* they are gases.

26 When carbon is alloyed with a metal it normally substitutes for the parent metal atoms, i.e. forms substitutional atoms, *because* it is a small atom.

27 The slip planes in the b.c.c. structure are the {100} planes *because* they are some of the more densely packed planes in the b.c.c. structure.

28 Plastic deformation does not take place by the shearing of complete blocks of a crystal (block shear mechanism) *because* the stress for this process is of the order of $G/30$ (where G is the shear modulus), which is much higher than the stresses required to produce plastic deformation.

29 Dislocations can move at stresses as low as $G/10^3$ (where G is the shear modulus) *because* they are non-equilibrium defects.

30 An edge dislocation cannot cross-slip *because* it has an uniquely defined slip plane.

31 The glide motion of a screw dislocation is not so restricted as that of an edge dislocation *because* a screw dislocation can climb.

32 Dislocations are non-equilibrium defects *because* the increase in enthalpy associated with them outweighs the entropy contribution to the free energy.

33 When a crystal is plastically deformed the dislocation density decreases *because* some of the dislocations reach the surface of the crystal.

34 Grain boundaries are line defects *because* they are formed during plastic deformation.

Answers

1 A	**2** B	**3** B, D, F	**4** A, F
5 B, D	**6** C	**7** C	**8** B
9 B, C	**10** D, E	**11** A	**12** A
13 A, F	**14** B	**15** A, D	**16** A

17 A   **18** D   **19** A, C   **20** (i) A, E
                                                                  (ii) B, D

21 B, F   **22** C, E   **23** D   **24** B
25 B   **26** D   **27** D   **28** A
29 B   **30** A   **31** C   **32** A
33 D   **34** E

9

MECHANICAL PROPERTIES

9·1 Introduction

The mechanical properties of a material are concerned with the effects of stress on the material. From our own experience we know that materials can react in a number of ways to an applied stress. For example, if we apply a stress to a stainless steel dish by dropping it on to a hard floor, the dish will not break but it will probably be dented. On the other hand, if we are unfortunate enough to repeat this 'experiment' with, say, a wine-glass, the glass will invariably break. From this we conclude that stresses can produce a shape change (e.g., the dent), but may also cause a material to break or fracture. Both fracture and the different phenomena responsible for shape changes (not only permanent shape changes) will be discussed in this chapter.

9·1·1 The stress–strain curve

The simplest mechanical test to visualize is the tensile test in which an increasing tensile stress is applied to a specimen as in Fig. 9·1(a), and the resulting changes in length are monitored. The *stress*, σ, is defined as the force per unit cross-sectional area of specimen. Thus if the force is P newtons and the cross-sectional area is A square metres, then the stress is given by $\sigma = P/A$ N/m².

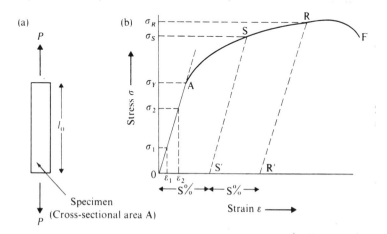

Fig. 9·1 The tensile test: (a) an increasing stress $\sigma = P/A$ is applied to the specimen; (b) the resulting stress–strain curve

In practice the N/m^2 is such a small quantity that MN/m^2 (meganewtons per square metre) are often used. The *engineering strain*, ϵ, is defined as the fractional change in length of the specimen, i.e., $\Delta l/l_0$, where l_0 is the initial length and Δl the change in length. A typical stress–strain curve obtained from a tensile test is shown in Fig. 9·1(b).

At the beginning of the test the material extends elastically, the strain being directly proportional to the stress and the specimen returns to its original length on removal of the stress. (We neglect here any complications due to anelastic and viscous deformation, which will be discussed in Sections 9·3 and 9·4.) Beyond the elastic limit the applied stress produces plastic deformation so that a permanent extension remains after removal of the applied load. The ratio applied load/original cross-sectional area is termed the engineering *stress* and this continues to increase with elongation due to work-hardening (or strain-hardening) until the *ultimate tensile stress* or the *tensile strength* (maximum load/original cross-sectional area) is reached. At this point a neck begins to develop somewhere along the length of the specimen and further plastic deformation is localized within the neck. After necking has begun the nominal stress decreases until the material fractures at the point of minimum cross-sectional area within the neck and very little useful information can be obtained from that part of the test. It should be noted that a curve of engineering stress versus engineering strain will have the same shape as the original load–extension curve obtained directly from the tensile testing machine. Some typical engineering stress–strain curves for different materials are shown in Fig. 9·20, and will be discussed in Section 9.6.

9·2   Elastic deformation

9·2·1   Characteristics of elastic deformation

We begin our detailed study with the elastic region of Fig. 9·1(b), namely, the region O–A, where stress and strain are proportional to one another. Thus elastic deformation is described by an equation of the form:

$$\sigma = E\epsilon \tag{9·1}$$

where E is a constant called *Young's modulus*. The equation is known as *Hooke's law*, although Hooke originally only stated that $\Delta l \propto P$. This linear relationship between stress and strain when a material is deforming elastically in tension is also found under other conditions of stressing. Thus, for the simple shear stress situation we have

$$\tau = G\gamma \tag{9·2}$$

where τ and γ are the shear stress and strain respectively and G is the shear modulus. While if a hydrostatic pressure P_H is applied to a specimen so that the volume V changes elastically by an amount ΔV, a *bulk modulus*, K, may be defined by the equation

$$P_H = K\Delta V/V \tag{9·3}$$

Elastic deformation is *instantaneous*, that is, if we suddenly increase the stress from σ_1 to σ_2 the strain immediately changes from ϵ_1 to ϵ_2. Elastic deformation is also completely *reversible* and if the stress is reduced to its former value of σ_1 the strain falls back to ϵ_1. In fact, if the specimen is unloaded ($\sigma = 0$) the specimen immediately reverts to its initial size and shape.

9·2·2 Atomic mechanism of elastic deformation

The important characteristics of elastic deformation are that it is instantaneous and reversible, and that stress and strain are linearly proportional to one another. In order to understand the basis of these characteristics we must look at what is happening on the atomic scale during elastic deformation.

When an external tensile force is applied to a crystal of length l_0, the atoms are immediately pulled further apart and this is manifested as an elastic elongation of the crystal equal to Δl. We have already used a plot of force against interatomic distance [Fig. 7·4(b)] and the earlier diagram is reproduced in Fig. 9·2. The atoms are pulled by the external force from their equilibrium positions, say from a to b in Fig. 9·2, until the applied force is balanced by the increase, ΔF, in the attractive force between the atoms. The elastic strain of the crystal is $\Delta l/l_0$ and this is equal to $\Delta r/r_0$, where Δr is the displacement of the atoms from their equilibrium positions and r_0 is the equilibrium interatomic distance. On the removal of the external force the atoms return to their equilibrium positions under the action of the interatomic forces; this is why elastic deformation is reversible.

In practice, elastic deformation only produces small strains, rarely in excess of $\frac{1}{2}\%$, hence the part of the force–displacement curve that we are concerned with is around the equilibrium spacing r_0. Figure 9·2 shows that the force–displacement curve can be regarded as approximately linear near to r_0 for small changes of r; it follows that under these circumstances the external

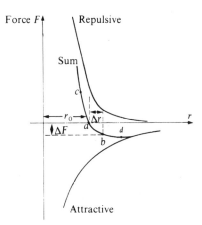

Fig. 9·2 Interatomic force plotted against interatomic distance

force required to elastically deform a material will be linearly proportional to the displacement, and that stress will be linearly proportional to strain (Hooke's law).

An approximate estimate of the Young's modulus can be obtained from the interatomic forces by observing that, for a single atom in a crystal during elastic deformation,

$$\sigma \approx \Delta F/r^2$$

$$\epsilon \approx \Delta r/r_0$$

Now Young's modulus E is equal to σ/ϵ, and as the displacements are small during elastic deformation $r \approx r_0$, therefore

$$E \approx \frac{\Delta F}{r_0 \Delta r} = \frac{1}{r_0}\left(\frac{\partial F}{\partial r}\right)_{r=r_0} \tag{9.4}$$

In Chapter 7 the relationship between F and r was described mathematically by Eqn (7·14):

$$F = -\frac{nC_1}{r^{n+1}} + \frac{mC_2}{r^{m+1}}$$

The value of r_0 was found by putting $F = 0$ so that

$$-\frac{nC_1}{r_0^{n+1}} + \frac{mC_2}{r_0^{m+1}} = 0$$

The differential $(\partial F/\partial r)_{r=r_0}$ is obtained from these two equations. Thus

$$\left(\frac{\partial F}{\partial r}\right)_{r=r_0} = \frac{n(n+1)C_1}{r_0^{n+2}} - \frac{m(m+1)C_2}{r_0^{m+2}} = \frac{nC_1(n-m)}{r_0^{n+2}}$$

and substituted into Eqn (9·4) it gives for Young's modulus

$$E \approx \frac{nC_1(n-m)}{r_0^{n+3}} \tag{9.5}$$

This equation is only approximate because it relates to two atoms only, ignoring the influence of the other neighbouring atoms, but it illustrates roughly the relationship between Young's modulus and the atomic parameters, particularly the interatomic spacing. As the interatomic spacing, and in some cases the bonding, varies with direction in a single crystal, the Young's modulus is dependent on the direction of the stress in relation to the crystal axes, i.e., single crystals are elastically *anisotropic*. A typical degree of variation of E with direction is shown in Fig. 9·3.

Equation (9·5) has been applied to a wide variety of materials. Accurate calculations have been made on a range of ionically bonded materials, such as lithium fluoride and sodium chloride, and the relation shown to be approximately correct, with $n = 1$ (i.e., the value used to determine the Coulomb forces of electrostatic attraction between unlike charges).

This low value of n also applies to the low melting-point metals such as lithium, sodium, and potassium, which have metallic bonding. Higher

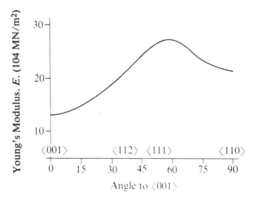

Fig. 9·3 The anisotropy of Young's modulus for Fe – 3%Si single crystals

melting-point metals have a higher value of Young's modulus, consistent with the atomic bond strength increasing with a decrease in r_0. For the metals tungsten, molybdenum, nickel, iron, and aluminium $n \approx 4$, showing that the simple Coulombic attraction is no longer applicable. For covalent bonded materials such as carbon, silicon, silicon carbide, and germanium $n \approx 2$.

Although in this section we have concentrated on the Young's modulus, the conclusions are applicable to the other elastic moduli, e.g., the shear modulus varies with the orientation of a single crystal, and if the atomic parameters of a material are such that the material has a high Young's modulus it will also generally have a high shear modulus, and vice versa. This latter point may be quickly confirmed by reference to Appendix 3.

The above reasoning cannot be applied to polymers because the elastic deformation in these materials is caused primarily by the bending of carbon bonds and not by bond stretching as in crystals. The elastic constants of a polymer are normally 10–300 times less than for metal.

9·2·3 Elastic deformation of an isotropic material

Although single crystals are elastically anisotropic, a polycrystalline material in which the grains, or crystals, are randomly orientated behaves *isotropically*, i.e., properties are independent of direction. Amorphous materials, except when produced in such a manner as to cause some alignment of the molecular structure (a preferred orientation), are also isotropic.

For isotropic materials there exist simple relationships between the elastic moduli E, G, and K. These relationships include another elastic constant, called *Poisson's ratio*, which we must first define. A specimen of any material, be it single crystal, polycrystal, or amorphous, generally contracts in the direction transverse to that in which it is being extended. If the transverse dimension of the specimen is d, then the transverse strain would be $\epsilon_t = -\Delta d/d$ in the elastic region (negative since Δd would be negative), and Poisson's ratio is defined as

$$\nu = -\epsilon_t/\epsilon \tag{9·6}$$

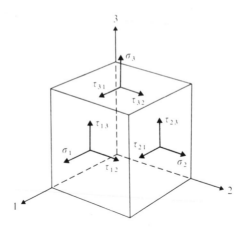

Fig. 9·4 The tensile (σ) and shear (τ) stresses on a cubical element

where ϵ is the tensile strain. The relationships between the elastic constants for an isotropic material are

$$E = 2G(1 + \nu) = 3K(1 - 2\nu)$$

If a stress is applied which is not purely tensile, or purely a shear stress, the behaviour of a material is more complicated than described above. To begin to understand the macroscopic elastic behaviour of isotropic materials it is helpful to have some idea of the elementary concepts of stress analysis. For this purpose let us consider simply a cubic element of the material as depicted in Fig. 9·4. We suppose that the faces of the cube are subjected to normal tensile stresses σ_1, σ_2, and σ_3 acting in the directions 1, 2, and 3 respectively (these are equivalent to the x, y, and z directions of a conventional set of axes). Due to the stress σ_1 the cube will be extended in the 1 direction by a strain $\epsilon_1 = \sigma_1/E$ [cf. Eqn (9·1)] and the cube will contract transversely in the 2 and 3 directions due to the effect described by Poisson's ratio [see Eqn (9·6)]. Thus, due to the stress σ_1 there will be tensile strains of $-\nu\sigma_1/E$ in both the 2 and 3 directions.

Similarly, due to the σ_2 tensile stress there will be tensile strains of σ_2/E in the 2 direction and $-\nu\sigma_2/E$ in the 1 and 3 directions.

It is permissible to add these small elastic strains together, so that three equations are sufficient to describe the deformation of the cube element, namely

$$\epsilon_1 = \frac{1}{E} \ (\sigma_1 - \nu\sigma_2 - \nu\sigma_3)$$

$$\epsilon_2 = \frac{1}{E} \ (\sigma_2 - \nu\sigma_3 - \nu\sigma_1) \qquad\qquad (9\cdot7)$$

$$\epsilon_3 = \frac{1}{E} \ (\sigma_3 - \nu\sigma_1 - \nu\sigma_2)$$

For the most general case, of course, it is necessary to include the possibility of shear stresses being present on the faces of our elemental cube. To describe such complex stress states it is necessary to apply two shear stresses to each face of the cube. For example, on the face of the cube normal to the 1 direction, there would be the shear stresses τ_{12} and τ_{13}, the shear stress τ_{12} acting in the 2 direction and the shear stress τ_{13} acting in the 3 direction. Since τ_{12} and τ_{13} are mutually perpendicular, the resultant shear stress on this plane would be $\tau = \sqrt{(\tau_{12}^2 + \tau_{13}^2)}$. Using Eqn (9·2), we recognize that there will be shear strains of

$$\gamma_{12} = \tau_{12}/G$$

$$\gamma_{13} = \tau_{13}/G$$

associated with these shear stresses.

On the other faces of the cube there will exist shear stresses such as τ_{21} and τ_{31}, etc. Because the element is in equilibrium, there can be no net couple on it, so that $\tau_{21} = \tau_{12}$ and $\tau_{31} = \tau_{13}$, etc. The mathematical notation becomes neater if we denote σ_1 by the symbol σ_{11} and σ_2 by the symbol σ_{22}. Similarly, ϵ_1 is denoted by ϵ_{11}, τ_{12} by σ_{12}, etc. The complete set of equations describing elastic behaviour can then be written as

$$\epsilon_{11} = \frac{1}{E} \; (\sigma_{11} - \nu\sigma_{22} - \nu\sigma_{33})$$

$$\epsilon_{22} = \frac{1}{E} \; (-\nu\sigma_{11} + \sigma_{22} - \nu\sigma_{33})$$

$$\epsilon_{33} = \frac{1}{E} \; (-\nu\sigma_{11} - \nu\sigma_{22} + \sigma_{33}) \qquad (9·8)$$

$$\gamma_{12} = \sigma_{12}/G$$

$$\gamma_{23} = \sigma_{23}/G$$

$$\gamma_{31} = \sigma_{31}/G$$

The quantities σ_{11}, σ_{12}, σ_{23}, etc. make up the components of the stress σ which is called a *tensor* quantity; quite often, such equations as (9·8) are written more compactly in either tensor notation or matrix form. Tensor notation allows the development of the mathematical theories of the plastic flow of solids under complex stresses, whilst matrix algebra is convenient for solving problems of stress analysis on a computer. For example, Eqns (9·8) can be written more compactly by defining stress and strain vectors

$$[\sigma] = [\sigma_{11} \; \sigma_{22} \; \sigma_{33} \; \sigma_{12} \; \sigma_{23} \; \sigma_{31}]^T$$
$$[\epsilon] = [\epsilon_{11} \; \epsilon_{22} \; \epsilon_{33} \; \gamma_{12} \; \gamma_{23} \; \gamma_{31}]^T \qquad (9·9)$$

and a matrix

$$[\Phi_E] = \begin{bmatrix} 1/E & -\nu/E & -\nu/E & 0 & 0 & 0 \\ -\nu/E & 1/E & -\nu/E & 0 & 0 & 0 \\ -\nu/E & -\nu/E & 1/E & 0 & 0 & 0 \\ 0 & 0 & 0 & 1/G & 0 & 0 \\ 0 & 0 & 0 & 0 & 1/G & 0 \\ 0 & 0 & 0 & 0 & 0 & 1/G \end{bmatrix} \tag{9.10}$$

which represents the 'stiffness' or elastic behaviour of the material in response to the total stress system $[\sigma]$ defined above. Equation (9.8) then can be written much more compactly as

$$[\epsilon] = [\Phi_E][\sigma] \tag{9.11}$$

During elastic deformation under the complex stress system σ_{11}, σ_{12}, σ_{31}, etc. in which all six independent stress components may have values, the resultant strain of the element of an isotropic material can be determined from Eqn (9.11) [or (9.8)].

9.2.4 Significance of the elastic moduli

The elastic moduli are important in that they enable us to calculate the elastic strain resulting from the application of a given stress. There are many situations in which a knowledge of the elastic strains is essential, for example, the elastic deformation in the moving parts of an engine needs to be known if close clearances are to be maintained. Furthermore, when more than one material is used in a structure we can make sure that the strains in the different materials are compatible and that the materials share the load in the way that we want them to.

An important property of a material is its ability to absorb the energy of a blow without failing, i.e., without fracture or plastic flow. A material can absorb energy by straining elastically and this stored energy per unit volume is often referred to as the *elastic resilience*.

The word 'resilience' implies correctly that this stored energy is recoverable as mechanical work done by the body as the applied forces are reduced to zero. If we consider a cube of sides l_0 strained in tension by a force which increases from zero to a value F so that the sides move apart by a distance δl, the total work done will be equal to the mean force × distance through which the force acts, viz., $\frac{1}{2}F\delta l$. As the volume of the cube is l_0^3, the elastic resilience or strain energy density is given by $\frac{1}{2}F\delta l/l_0^3$. Now, by definition the tensile stress is $\sigma = F/l_0^2$ and the tensile strain is $\epsilon = \delta l/l_0$, therefore the elastic resilience in terms of stress and strain is

$$\tfrac{1}{2}F\delta l/l_0^3 = \tfrac{1}{2}\sigma l_0^2 \delta l/l_0^3 = \tfrac{1}{2}\sigma\epsilon$$

As $\sigma = E\epsilon$ [Eqn (9.1)] we can introduce the Young's modulus into the expression, giving

$$\tfrac{1}{2}\sigma\epsilon = \tfrac{1}{2}\sigma^2/E = \tfrac{1}{2}\epsilon^2 E \qquad\qquad (9\cdot12)$$

Similarly in an elastic shear deformation, from Eqn (9·2),

$$\text{elastic resilience} = \tfrac{1}{2}\tau\gamma = \tfrac{1}{2}\tau^2/G \quad\text{or}\quad \tfrac{1}{2}\gamma^2 G \qquad\qquad (9\cdot13)$$

For hydrostatic deformation, from Eqn (9·3),

$$\text{elastic resilience} = \tfrac{1}{2}P_H\theta = \tfrac{1}{2}P_H^2/K \quad\text{or}\quad \tfrac{1}{2}\theta^2 K \qquad\qquad (9\cdot14)$$

where $\theta = \Delta V/V$ in Eqn (9·3).

The strain energy density stored when a body is on the point of failure is sometimes called the *modulus of elastic resilience*. This is about $2\cdot5 \times 10^6$ J/m³ for a strong steel, which works out at about 2×10^{-4} eV per atomic bond. Compared with bond energies which are of the order of two to four electron volts it is seen that the strain energy stored is extremely small on an atomic scale.

9·3 Anelastic deformation

Unlike elastic deformation, which is instantaneous on the application of a stress, *anelastic deformation* is time dependent, i.e., the strain lags behind the stress (Fig. 9·5). Anelastic deformation is the result of *time dependent* processes, such as the movement of point defects in response to the stress and the straightening out of kinked or coiled molecular chains, occurring within the material. On removal of the applied stress, these processes will reverse and

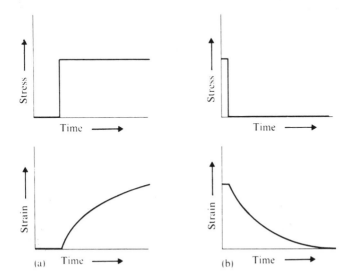

Fig. 9·5 Anelastic deformation – the strain lags behind the stress: (a) applying a stress; (b) removing a stress

the material will gradually revert to its original size and shape, i.e., anelastic deformation is reversible.

Although anelastic deformation occurs in all materials the strain associated with it may not be significant. For example, in metals the anelastic strain is small and its contribution to the total strain negligible compared with that from elastic and plastic deformation.

A material's capacity to damp out vibrations, called *damping capacity* or *internal friction*, is a function of the lag between the applied stress and strain and hence is related to the anelastic deformation. If the anelastic strain is small, the lag between stress and strain is also small and the damping capacity is low. This is why most metals have low damping capacities and do not damp out vibrations.

9·4 Viscous deformation

A liquid cannot support a shear stress and flows irreversibly and continuously when a stress is applied. This is referred to as *viscous flow* and it also occurs in some solids, especially at elevated temperatures. For example, it is common in glasses and polymers, although generally in the latter it is also accompanied by elastic and anelastic deformation. (Viscous flow in glasses and polymers are discussed further in Chapters 11 and 12 respectively.) In contrast, viscous flow in metals is rare and is only found under specific conditions of high temperatures and low stresses.

When a material flows viscously the resulting shear strain γ is a function of both the shear stress τ and time, t. Ideal viscous behaviour is represented by *Newton's Law* of viscous flow:

$$\tau = \eta \; \frac{\mathrm{d}\gamma}{\mathrm{d}t} \tag{9·15}$$

where η is the viscosity. For ideal or Newtonian behaviour η is independent of the rate of shear and the curve of τ against $\mathrm{d}\gamma/\mathrm{d}t$ is as shown in Fig. 9·6(a). A glass above the glass transition temperature behaves as a Newtonian material. There are two deviations from ideal behaviour which are often encountered and these are illustrated in Figs. 9·6(b) and (c). The non-linear relationship between τ and $\mathrm{d}\gamma/\mathrm{d}t$ of Fig. 9·6(b) is termed *shear thinning* and is due to a reversible decrease in viscosity with increasing shear rate. Shear thinning is a desirable characteristic for paints. During brushing, which imposes high strain rates, it is preferable for the viscosity to be not too high. However, after application, the paint should not run extensively under the action of gravitational forces and therefore the viscosity must not be too low. There is an opposite effect to shear thinning, called *shear thickening*, where the viscosity increase with increasing shear rate, but this is rarely observed. The second deviation from Newtonian flow is a *yield value*, that is, a critical stress below which apparently no flow occurs. Above the yield value flow may either be Newtonian, as shown in Fig. 9·6(c), or non-ideal.

Viscous flow in solids normally involves the thermally activated movement of atoms or molecules within the material. The rate of a thermally activated

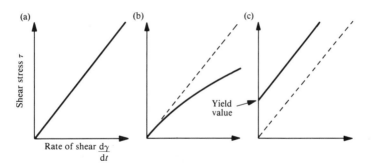

Fig. 9·6 Viscous flow: (a) ideal Newtonian behaviour; (b) non-ideal behaviour, shear thinning; (c) non-ideal behaviour, yield value

process such as this is controlled by the Boltzmann exponential factor, as discussed in Chapter 7. Consequently, over limited ranges of temperatures the viscosity varies with temperature according to

$$\frac{1}{\eta} = A \exp\left(-Q/kT\right) \tag{9.16}$$

where A is a constant and Q is the activation energy for viscous flow.

9·5 Plastic deformation

9·5·1 Dislocations and stress–strain curves

In the previous chapter we have seen how plastic deformation in crystalline materials, whether in single or polycrystalline form, occurs by the movement of dislocations. In this section we will concentrate on the stress–strain curves obtained from tensile tests and will correlate these curves with the movement and interactions of the dislocations responsible for plastic flow.

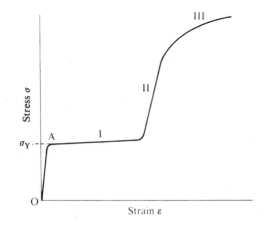

Fig. 9·7 Generalized stress–strain curve for a single crystal

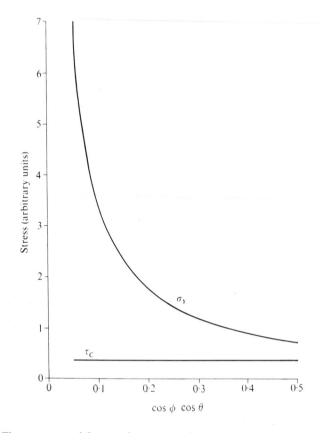

Fig. 9·8 The variation of the tensile stress σ_y of a single crystal with orientation. In contrast, the critical resolved shear stress τ_c is constant

Figure 9·7 is a generalized stress–strain curve for a single crystal. The details of the stress–strain curve, for example the length of stage I, depend on many factors such as the crystal structure, orientation, and purity of the specimen; however, we will not concern ourselves with these finer points but will concentrate on the general features of the curve.

The initial linear region, O–A, of the curve is the elastic region and was discussed in Section 9·2. We are interested in the plastic deformation that takes place in stages I, II, and III of the curve.

First let us consider the tensile stress at the beginning of stage I, i.e., the tensile stress σ_y required for the onset of plastic deformation. This stress σ_y, which is called the *yield stress*, depends on the orientation of the crystal with respect to the tensile axis, in particular, as shown in Fig. 9·8, it is a function of the angles the tensile axis makes with the normal to the slip plane and the slip direction (angles ϕ and θ respectively in Fig. 9·9). This is not surprising as it is the *shear* stress on the slip plane and in the slip direction that causes dislocations to glide, hence this is the stress that has to be determined. Referring to

Fig. 9·9, the cross-sectional area of the slip plane is $A/\cos\phi$, where A is the cross-sectional area of the crystal, and the tensile force, P, resolved in the slip direction is $P\cos\theta$. Therefore, the resolved shear stress, τ, on the slip plane and in the slip direction is given by

$$\tau = (P\cos\theta) \div (A/\cos\phi) = \frac{P}{A}\cos\theta\cos\phi$$

$$= \sigma\cos\theta\cos\phi \qquad (9\cdot17)$$

This equation is known as *Schmid's law* and the term $\cos\theta\cos\phi$ as the *Schmid factor*. If the different tensile yield stresses from tests on single crystals of varying orientations are inserted into Eqn (9·17), a constant value, called the *critical resolved shear stress*, τ_c, is obtained for τ (Fig. 9·8).

We know that crystals have many slip systems, e.g., c.p.h. and f.c.c. have three and twelve systems respectively, and for a given stress direction each system will have different values for ϕ and θ and therefore different values for the Schmid factor. The system that operates at the onset of stage I, referred to as the *primary* slip system, is that on which the critical resolved shear stress is first reached. Reference to Eqn (9·17) and Fig. 9·8 indicates that this is the system with the highest Schmid factor.

As shown in Fig. 9·7, stage I, which is termed the *easy glide* region, is approximately linear. In this region the rate at which the material is becoming harder to deform is low. A measure of this rate is the slope $d\sigma/d\epsilon$ which is called the *work-hardening rate*. This is low because only the primary slip system is active, i.e., dislocations are only moving on one set of parallel planes. This means that the dislocations do not interfere with each others motion, consequently they move large distances unhindered and many reach

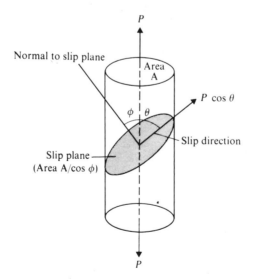

Fig. 9·9 Determination of the resolved shear stress on the slip plane and in the slip direction for a single crystal

the surface of the crystal. In stage II, the *linear hardening* region, dislocations are generated and move on intersecting slip planes (more than one slip system is operating). The dislocations become entangled within the crystal thereby forming obstacles to further dislocation motion. The number of such obstacles increases with increasing strain, thus making plastic deformation more and more difficult–this is reflected in the high work-hardening rate in stage II. When the stress is sufficiently high, cross-slip commences which enables screw dislocations to bypass obstacles to their motion. This leads to the *parabolic* curve of stage III with its reduced work-hardening rate.

Figure 9·1(b) shows a typical stress–strain curve for a polycrystalline material. The plastic deformation of a polycrystal is complex as, if the material is not going to fall apart at the grain boundaries, the strain in each grain must be accommodated by all the neighbouring grains. Because of this constraint, many slip systems have to operate in each grain before plastic deformation can commence. Not all of the required slip systems can be favourably orientated for slip, i.e., some must have low Schmid factors, hence the tensile yield stress of a polycrystalline material is always high. In addition, the grain boundaries act as barriers to dislocations and this further hinders the onset of plastic deformation.

A consequence of the many slip systems operating, and the high stresses involved, is that the stress–strain curves of polycrystals do not exhibit stages I and II shown by single crystals. The work-hardening rate of a polycrystal up to about 10% strain ($\epsilon = 0\cdot1$) is usually greater than that found in stage III for a single crystal; at higher strains however, the two work-hardening rates are very similar.

9·5·2   Strengthening mechanisms

As plastic deformation occurs by the movement of dislocations, anything that hinders their movement increases the stress required for the onset of plastic flow. It has already been mentioned that dislocations interfere with each others' motion and that grain boundaries act as barriers to dislocations, so both of these may be considered as strengthening mechanisms and will be discussed in more detail in this section. In addition, the strengthening achieved by alloying (solution and dispersion hardening) will be described.

(a) *Work-hardening*. In the previous section we saw that as plastic deformation proceeds dislocations interact with one another making further deformation more and more difficult. This phenomenon is called *work- or strain-hardening* and is best explained by reference to Fig. 9·1(b). Consider plastically deforming a material S%, i.e., deform to the point S on the stress–strain curve, and then unload. The material will have been work-hardened and on retesting will not yield until the stress σ_S is reached.

Work-hardening is a commonly employed strengthening mechanism but it has certain disadvantages. For example, by deforming S% we achieved a considerable increase in the yield stress from σ_y to σ_S. However, if we deform a further S% [to point R on the curve of Fig. 9·1(b)], because the work-hardening rate decreases with strain, the increment in the yield stress is smaller, i.e., $\sigma_S - \sigma_y > \sigma_R - \sigma_S$. Moreover, in the latter case the material is

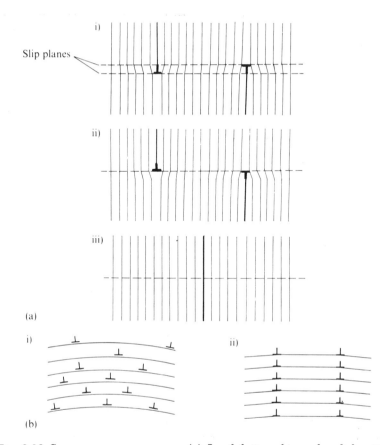

Fig. 9·10 Some recovery processes: (a) Annihilation of two edge dislocations (i) two edge dislocations on parallel slip planes, (ii) climb of dislocations onto the same slip plane, (iii) dislocations come together and annihilate. (b) Polygonization (i) 'random' distribution of edge dislocations, (ii) the same dislocations in a low-energy configuration of low angle grain boundaries

approaching the point of fracture (R is near to F), giving little safety margin if the material were used for structural purposes in this condition.

Materials often have to withstand elevated temperatures during service, and work-hardening is not a good strengthening mechanism in these circumstances. The reason for this is that dislocations are non-equilibrium defects and therefore, when solid state diffusion is sufficiently rapid, dislocation rearrangement and annihilation will take place in order to reduce the free energy of the material. At temperatures of $0·3$–$0·4\ T_m$, where T_m is the melting point of the material in degrees kelvin, *recovery* takes place. Recovery results in a slight decrease in the yield stress due to a reduction in the number of dislocations and the rearrangement of the dislocations, but the grain structure remains unaffected. Two recovery processes, both of which involve the thermally activated climb of edge dislocations, are illustrated in Fig. 9·10.

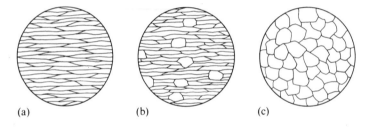

Fig. 9·11 Recrystallization: (a) deformed grains of high dislocation density; (b) new grains of low dislocation density are nucleated and begin to grow; (c) specimen consists solely of the new grains, i.e., it is fully recrystallized

These are the mutual annihilation of two edge dislocations, thereby reducing the dislocation density [Fig. 9·10(a)], and the rearrangement of dislocations, known as *polygonization* into configurations with lower energy [Fig. 9·10(b)].

At higher temperatures *recrystallization* takes place and the yield stress falls to its original value of σ_y. Recrystallization is the nucleation and growth of *new grains of low dislocation density* (Fig. 9·11). The free energy of the material is lowered by the significant reduction in the dislocation density that occurs during recrystallization; in other words, the energy of the dislocations is the 'driving force' for recrystallization. It follows that a heavily deformed material with a large dislocation density will recrystallize more rapidly than a material that has been only slightly deformed (Fig. 9·12).

It is interesting to note that some materials recrystallize at room temperature, e.g., lead for which room temperature is approximately $0.5\ T_m$. Consequently, it is impossible to work-harden lead at room temperature and this is why lead is so soft and malleable. If a material is deformed, or *worked*, at a temperature that is sufficient for concurrent recrystallization it is said to be *hot-worked*. Thus lead is hot-worked at room temperature, whilst a temperature in excess of 900°C is required to hot-work mild steel. *Cold-working* is the deforming at temperatures at which recrystallization does not occur.

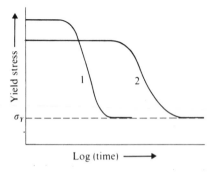

Fig. 9·12 Effect of the amount of prior deformation on the rate of recrystallization. Specimen 1 was deformed more than specimen 2

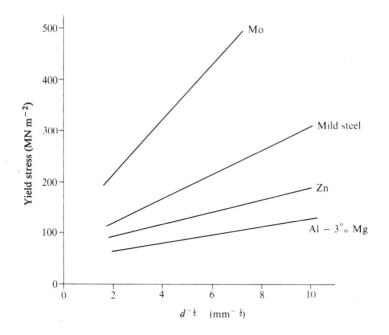

Fig. 9·13 Effect of grain size on yield stress

(b) *Grain boundary hardening.* Grain boundaries act as physical barriers to dislocations causing them to 'pile-up' and hindering their motion. Obviously, the greater the grain boundary area per unit volume the more difficult is dislocation movement and the higher the yield stress. The effect of grain boundaries on the yield stress is given by the *Petch* equation

$$\sigma_y = \sigma_i + k_y d^{-1/2} \tag{9·18}$$

where d is the grain diameter, and σ_i and k_y are constants at a given temperature: σ_i is a measure of the intrinsic resistance of the material to dislocation motion, while k_y is associated with the ease of operation of dislocation sources. Figure 9·13 shows that a linear relationship between yield stress and $d^{-1/2}$ is obtained for a variety of metals in accordance with this expression.

(c) *Solution hardening.* A common way of increasing the hardness and yield stress of a material is to introduce solute atoms into the solvent lattice, so forming a *solid solution* or alloy. The solute atoms, whether they be substitutional or interstitial, will cause *local elastic strains* in the material. These local strains hinder the motion of dislocations and hence increase the strength of the material.

Substitutional atoms produce *symmetrical* strain fields and only give a gradual increase in yield stress with concentration, typically $G/20$–$G/10$ per atom fraction (G is the shear modulus). Theoretical models for the hardening due to substitutional atoms predict that the yield stress (or the critical resolved shear stress in the case of single crystals) is directly proportional to

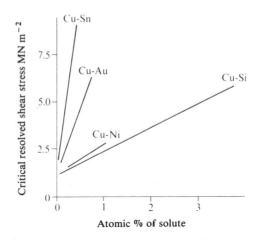

Fig. 9·14 Effect of substitutional alloying additions on the critical resolved shear stress of Cu single crystals (J.O. Linde, and S. Edwards, *Arkiv Fysik*, 1954, **8**, 511)

the concentration for concentrations less than about 10%; this is confirmed by experiment, as illustrated by Fig. 9·14. The reason that the different solutes in Fig. 9·14 vary in their effectiveness as strengtheners is due to their atomic size relative to that of the solvent. If the size difference is large, as for Sn in Cu (see Table 10·1), the strengthening is more marked than when the solute and solvent atoms are of similar size, for example Si in Cu.

Some interstitial atoms e.g., carbon in b.c.c. iron, are associated with strain fields which are high in one direction compared to the transverse directions. These cause rapid strengthening of the order of $2G$–$3G$ per atom fraction. The greater the difference between the longitudinal and transverse strains caused by the interstitial, the more considerable is the strengthening. It has been suggested that the concentration dependence of the yield stress for interstitial additions differs from that for substitutional atoms and is $\sigma_y \propto c^{1/2}$.

In the solid solutions discussed so far there has been no preference for like, or unlike, atoms to be nearest neighbours, that is, the solutions have been random. This is not always the case and the situation can arise where solute atoms prefer solvent atoms to be their neighbours. This leads to an ordered arrangement of the atoms. If the ordering is slight and on a local scale, it is called *short-range order*. However, in the extreme, the solute and solvent atoms arrange themselves on two distinct lattices throughout the crystal and the material is said to be exhibiting *long-range order* [Fig. 9·15(a)]. The plastic deformation of ordered materials, especially those with long-range order, is complex but a simplified view of dislocation motion in these materials will be adequate for our purpose. When a unit dislocation moves through an ordered material it breaks up the special distribution of solute atoms and produces a fault in the ordered stacking sequence, known as an *anti-phase domain boundary*, in its wake [Fig. 9·15(b)]. This requires a great deal of energy and if the deformation of ordered crystals took place by the motion of single unit dislocations then these materials would have much

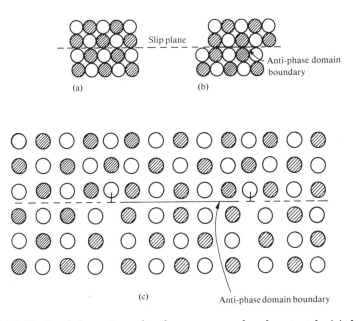

Fig. 9·15 Plastic deformation of a long-range ordered material: (a) before deformation, (b) after the passage of a unit dislocation which has created anti-phase domain boundary, and (c) a superlattice dislocation

higher flow stresses than random solid solutions. In reality the unit dislocations in ordered materials move in pairs, the first dislocation creates the anti-phase domain boundary and the second destroys it [Fig. 9·15(c)]. Consequently the stress required for the plastic deformation of an ordered material does not differ greatly from that for the same material in the disordered condition. This arrangement of two unit dislocations is called a *superlattice* dislocation and the similarity between a superlattice dislocation and an extended dislocation in the f.c.c. lattice (see Section 8·3·5) is apparent. A superlattice dislocation is two unit dislocations separated by a ribbon of anti-phase domain boundary and an extended dislocation is two partial dislocations separated by a ribbon of stacking fault.

(d) *Dispersion hardening*. Many materials consist of two or more phases and often one of the phases is in the form of small particles distributed throughout the material. The presence of a dispersion of small particles increases the strength of the material by an amount that depends on the volume fraction and size of the particles, and hence on the interparticle spacing S_p. Figure 9·16(a) shows a dislocation approaching an array of particles. If the dislocation cannot cut the particles, it has to bow between them and the maximum stress required for this is when the radius of curvature $R = S_p/2$ [Fig. 9·16(b)]. Thus, using Eqn (8·8), the stress required is

$$\tau = \frac{Gb}{R} = \frac{2Gb}{S_p} \qquad (9·19)$$

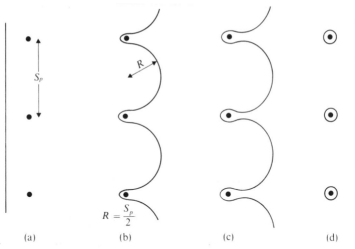

Fig. 9·16 The sequence of events when a dislocation bows past particles by the Orowan mechanism

The sequence of events shown in Figs. 9·16(c) and (d) occurs at stresses less than $2Gb/S_p$ (note the similarity with the operation of the Frank–Read source described in the previous chapter), therefore the dislocation passes the particles leaving loops around them. This process of bowing past particles and leaving a loop is called the *Orowan* mechanism. Figure 9·17 is an electron

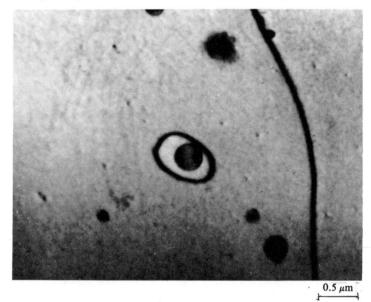

Fig. 9·17 Electron micrograph of deformed copper containing a dispersion of silica particles. A dislocation loop has been left around the particle in the centre of the micrograph. (Courtesy of F.J. Humphreys)

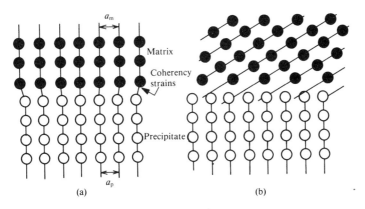

Fig. 9·18 The interface between a precipitate particle and the matrix: (a) coherent interface with coherency strains (the difference between a_p and a_m is exaggerated in order to emphasize the coherency strains); (b) incoherent interface with no matching of planes at the interface

micrograph of deformed copper containing a dispersion of silica particles and the dislocation loop around the silica particle in the centre of the micrographs is clearly visible.

A dispersion of particles may be produced by a number of methods. For example, the silica dispersion in the copper of Fig. 9·17 was produced by *internal oxidation*. A solid solution of Cu–Si was made, then oxygen diffused into the material and allowed to react with the silicon to form silica. Silver with a dispersion of Al_2O_3 is produced this way and is often used for contacts in electrical equipment. Another common method of forming a dispersion is by heat-treating a metastable solid solution so that a second phase precipitates. This is called *precipitation hardening* and certain Al–Cu alloys, which will be discussed in the following chapter, are hardened this way.

In the case of precipitation hardened systems the particles may be metastable and have a *coherent* interface with the matrix. An interface is said to be coherent when there is matching at the interface of the crystal planes in the matrix with those in the precipitate (Fig. 9·18). The metastable precipitates may be small clusters of only about 100 solute atoms or they may have a well developed crystal structure different from that of the matrix. Both types of particle are cut by the dislocations rather than the dislocations bowing past the particles by the Orowan mechanism. Particle cutting can give considerable strengthening and it has been shown that there are several different mechanisms contributing to the hardening. A detailed discussion of all the contributions is not necessary at this stage and we will just mention two mechanisms in order to demonstrate the general principles. Firstly, as seen in Fig. 9·18, a coherent interface can lead to elastic strains in the matrix because the interplanar spacings a_p and a_m are not exactly the same. Dislocations interact with these elastic strains, known as *coherency strains*, in a similar way (but more strongly) to the elastic strains around a solute atom. The second contribution is illustrated in Fig. 9·19, which shows that new

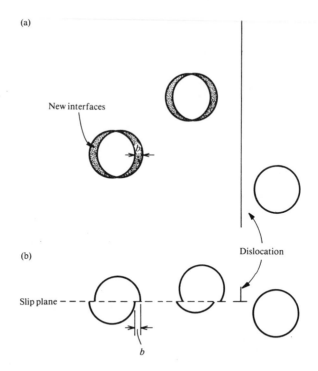

Fig. 9·19 Diagram illustrating the formation of new precipitate-matrix interfaces when particles are cut by a dislocation (a) plan view of slip plane (b) side view of slip plane

precipitate-matrix surfaces are produced when particles are cut. There is an energy per unit area associated with these surfaces and hence additional work, i.e., a higher stress, is required to move the dislocations.

9·5·3   *Plasticity theory for an isotropic material*

The previous sections have dealt in detail with the microscopic aspects of plastic deformation. However, it is often convenient when analysing plastic flow to consider the material to behave as an isotropic continuum. This approach has resulted in the development of a mathematical theory of plasticity by means of which it is possible to establish relationships between stress and strain in an analogous manner to that which exists for elasticity (see Section 9·2·3).

In a book of this type it is not possible to do more than briefly describe the philosophy on which plasticity theory rests. Many of the concepts are reinforced by consideration of the microbehaviour described earlier. For example, since plastic flow proceeds by the movement of dislocations, which results in the slipping of planes of atoms over one another, it is reasonable to assume that no volume change takes place during plastic straining. Thus, the effective value for Poisson's ratio becomes $\frac{1}{2}$ for the following reason. It can be seen from Eqns (9·7) that the volumetric strain (the fractional change in volume) of a cubical element such as that in Fig. 9·4 is, for small strains, the sum of the strains, viz.

$$\epsilon_1 + \epsilon_2 + \epsilon_3 = \frac{1 - 2\nu}{E} (\sigma_1 + \sigma_2 + \sigma_3)$$

This will be zero if $\nu = \frac{1}{2}$, so that the stress–strain relations for plastic flow will be characterized by including a contraction ratio of $\frac{1}{2}$.

Similarly, since the linear relationships between stress and strain are only valid for small strains, it is usual to write the equations of plasticity in terms of small incremental changes of strain, $\delta\epsilon$, and stress, $\delta\sigma$, which are small compared with the components of ϵ and σ. If a small increase, $\delta\sigma$, is made in the stress whilst the material is plastic the resulting incremental strain, $\delta\epsilon$, will be made up of an incremental elastic part, $\delta\epsilon^E$ and an incremental plastic part, $\delta\epsilon^P$. The incremental elastic strain is proportional to the incremental change of stress, according to Eqn (9·7), and has components such as:

$$\delta\epsilon_{11}^E = \frac{1}{E} (\delta\sigma_{11} - \nu\delta\sigma_{22} - \nu\delta\sigma_{33})$$

On the other hand the incremental plastic strain must be assumed to be dependent on the total stress level (again not inconsistent with the ideas of dislocation motion) and can be written in the form

$$\delta\epsilon_{11}^P = \delta\lambda(\sigma_{11} - \tfrac{1}{2}\sigma_{22} - \tfrac{1}{2}\sigma_{33}) \tag{9·20}$$

so that the total strain increment is

$$\delta\epsilon_{11} = \delta\epsilon_{11}^E + \delta\epsilon_{11}^P \tag{9·21}$$

The parameter $\delta\lambda$ introduced in Eqn (9·20) represents the effective 'plastic stiffness' of the material. If Eqn (9·20) is to be useful, it must be possible to relate $\delta\lambda$ to measurable stresses and strains on a whole specimen rather than to those on our hypothetical cube, which cannot be measured directly. To do this it is necessary to introduce the concepts of an effective stress, $\bar{\sigma}$, and effective strain, $\delta\bar{\epsilon}$. These are necessary to provide a way of describing the 'effectiveness' of the complex stress and strain states in the cube in producing plastic flow. The effective stress can be defined by reference to the fact that, in any complex stress system acting on a cubical element, there will be one set of coordinate axes (i.e., one set of directions for the cube edges) for which the shear stresses on each face are zero and the normal (direct) stress is either a maximum or a minimum. These stresses are known as the *principal stresses*, and the planes on which they act (the cube faces) are called the *principal planes*. If we denote the principal stresses by σ_1, σ_2, σ_3 then one possible definition of the effective stress is

$$\bar{\sigma} = \frac{1}{\sqrt{2}} \sqrt{[(\sigma_1 - \sigma_2)^2 + (\sigma_2 - \sigma_3)^2 + (\sigma_3 - \sigma_1)^2]} \tag{9·22}$$

It can be seen from this equation that, when $\sigma_2 = \sigma_3 = 0$ (i.e., the uniaxial state of stress which exists in the tensile test), $\bar{\sigma} = \sigma_1$ and the effective stress is equal to the stress which exists in the tensile test.

Having defined the effective stress, a similar definition can be made for the effective strain increment and we then relate these to the factor $\delta\lambda$ in Eqn (9·20) as follows:

$$\delta\lambda = \delta\bar{\epsilon}/\bar{\sigma}$$

Thus $\delta\lambda$ can be related to measured stresses and strains. Using the above equations, it is now possible to write Eqn (9·21) in full:

$$\delta\epsilon_{11} = \frac{1}{E}(\delta\sigma_{11} - \nu\delta\sigma_{22} - \nu\delta\sigma_{33}) + \frac{\delta\epsilon}{\bar{\sigma}}(\sigma_{11} - \tfrac{1}{2}\sigma_{22} - \tfrac{1}{2}\sigma_{33}) \qquad (9·23)$$

Six equations of this type are needed to describe fully the stress–strain relationship during plastic flow. It is sometimes again more convenient for the computer solution of real problems to write these equations in matrix form, similar to that of Eqn (9·11), and many of the most recent developments in the theory have depended on such techniques.

In practical situations it is often necessary to try and predict the behaviour of materials when subjected to plastic flow. The equations of plasticity theory permit this to be done to a certain extent, although it must be understood that only the stress–strain relations have been briefly discussed here, for the purpose of this text. To solve a real problem it is necessary to treat each cubical element separately and to ensure that the forces beween cubes are in equilibrium. Hence an additional *equation of equilibrium* is required. Similarly, the strains of neighbouring cubes must be matched to one another, for their faces must meet or there would be voids in the solid. This condition is met by an *equation of compatibility*. Another set of equations called boundary conditions ensure that the boundary surfaces of the specimen are the shape given in the problem and that the stresses on them also conform to the actual applied stresses. Such a complicated set of equations can only rarely be solved algebraically to give an analytical expression relating stress and strain. It is therefore normally necessary to use numerical methods. Finite difference or finite element methods using the matrix formulation of the equations suitable for the computer thus provide the main modern tools of analysis for such problems.

It must be emphasized that the mathematical theory just outlined is for an idealized material which is both isotropic and homogeneous, i.e., has same properties in all directions and in all places. Thus, it is implicit in the theory that the stress–strain curves in compression and tension are equal and opposite, but in practice these curves may be very different. Furthermore, a preferred orientation may be induced in a material during manufacture resulting in mechanical anisotropy. Nevertheless this theory represents an important step forward in our ability to predict the plastic behaviour of materials under stress.

9·6 Mechanical testing

9·6·1 *The tensile test*

As described earlier in this chapter, in the tensile test the specimen is gradually elongated under tension, the load and the extension being recorded continuously. It will be remembered that the load–extension curve obtained in this way is identical to a plot of engineering stress against engineering strain.

Some typical engineering stress–strain curves for different materials are shown in Fig. 9·20. For all materials the stress–strain relationship depends on the chemical composition, the heat-treatment and the method of manufacture. In ductile materials like aluminium, copper, and annealed mild steel, which exhibit large strains before failure, the engineering stress at fracture is lower than the tensile strength although the actual (true) stress in the neck will be higher. In brittle materials such as cast iron, high strength steel, tungsten carbide, and high strength aluminium alloys necking does not occur.

Rubber has an exceptional behaviour, exhibiting an extended region in which the stress only slightly increases with strain, followed by a rapid hardening phase. This is related to its molecular structure, since the poly-isoprene $(C_5H_8)_n$ molecules of which it is made are in the form of long chains having a backbone of carbon atoms bonded directly together with side links to CH_3 groups and hydrogen atoms. As will be described later, the chains in such a polymer are not straight but form an irregular tangle which straightens out under applied stress and it is only when they have become almost aligned to the applied tension, after a large extension, that the elastic strength of the interatomic bonds comes into play.

The properties of concrete depend on the mix (i.e., the proportion of sand, cement, aggregate, and water) and the curing time, as described in more detail in Chapter 11.

Returning to the tensile test, it is possible to obtain a better understanding of the plastic flow of materials by introducing the concepts of the true stress and true strain. Thus, if the current (instantaneous) load is P, the instantaneous cross-sectional area is A_i and the true stress σ_T, then

$$\sigma_T = \frac{P}{A_i} \tag{9·24}$$

Similarly, the engineering strain does not give a meaningful measure of the overall strain when the specimen has lengthened considerably. We define the true strain ϵ_T as the integral of the elemental incremental strain dl/l, where l is the instantaneous length, between the initial length l_0 and the current (instantaneous) length l_i. Thus

$$\epsilon_T = \int_{l_0}^{l_i} \frac{dl}{l} = \ln \frac{l_i}{l_0} \tag{9·25}$$

The true stress, σ_T, and the true strain, ϵ_T, can then be regarded as the effective stress, $\bar{\sigma}$, and effective strain, $\bar{\epsilon}$, introduced in the discussion leading to

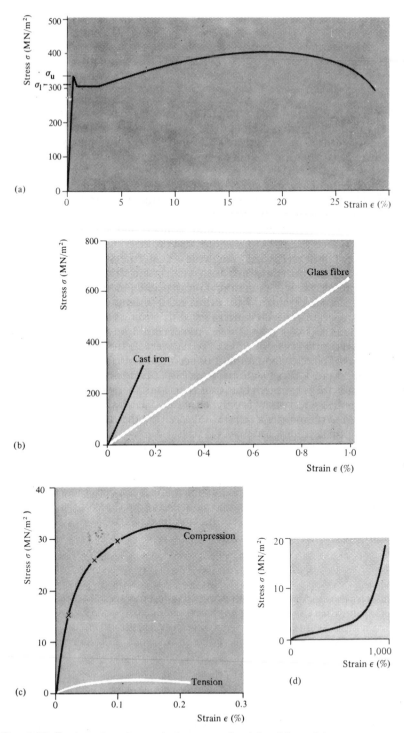

Fig. 9·20 Engineering stress–strain curves for (a) mild steel (σ_u = upper yield point and σ_l = lower yield stress); (b) brittle solids; (c) concrete in tension and compression; (d) vulcanized rubber

Eqn (9·23) for a complex state of stress in a specimen being plastically deformed. Since it is only the plastic part of the total strain which is of interest in that analysis the elastic strain σ_T/E should, strictly speaking, be deducted from ϵ_T; thus

$$\bar{\epsilon} = \ln \frac{l_i}{l_0} - \frac{\sigma_T}{E}$$

The true stress–true strain curve always has a continually rising characteristic as would be expected when consideration is given to the mechanism of work-hardening which is the interaction of mobile dislocations with each other and with dislocation tangles (Fig. 9·21).

The occurrence of necking can be predicted from the true stress–true strain curve by an analysis originally due to Considère. The instantaneous load is

$$P = \sigma_i A_i \tag{9·26}$$

At necking the load reaches a maximum value, so that $dP/P = 0$. Differentiating Eqn (9·26) gives

$$\frac{dP}{P} = \frac{d\sigma_i}{\sigma_i} + \frac{dA_i}{A_i} \tag{9·27}$$

Now, during plastic flow the volume of the specimen remains substantially constant (neglecting the generally small elastic straining of the specimen), therefore $A_i l_i = $ constant. Differentiating this gives

$$\frac{dA_i}{A_i} = -\frac{dl_i}{l_i} = -d\epsilon_T \qquad \text{[from Eqn (9·25)]}$$

Substituting this into Eqn (9·27) and putting $dP/P = 0$ we find that, at maximum load,

$$\frac{d\sigma_T}{d\epsilon_T} = \sigma_T \quad \text{or} \quad \frac{d\bar{\sigma}}{d\bar{\epsilon}} = \bar{\sigma} \tag{9·28}$$

Hence necking will occur in a tensile test at a strain at which the work-hardening rate equals the effective (true) stress.

An empirical relationship which gives an approximate fit to the behaviour of some materials is

$$\bar{\sigma} = C\bar{\epsilon}^n \tag{9·29}$$

where C and n are constants. The exponent n is often called the strain-hardening exponent and lies between 0·1 and 0·5 for most metals.

Applying Eqn (9·29) to Eqn (9·28), we find that necking occurs when

$$nC\bar{\epsilon}^{n-1} = C\bar{\epsilon}^n \tag{9·30}$$

i.e., when $\bar{\epsilon} = n$, so the physical significance of n is that it is a measure of the effective (true) strain at necking.

The value of the engineering strain at necking, which can be derived from Eqn (9·30) is referred to as the *uniform elongation*.

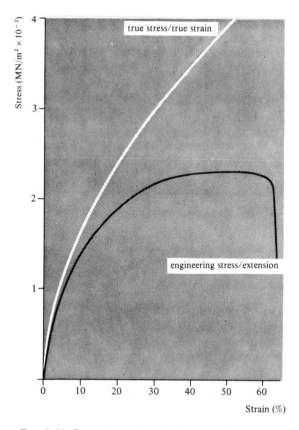

Fig. 9·21 True stress–true strain curve for copper

The relationship between the engineering strain ϵ and the true strain ϵ_T may be determined as follows

$$\epsilon = \frac{l_i - l_0}{l_0} = \frac{l_i}{l_0} - 1$$

Thus

$$\frac{l_i}{l_0} = 1 + \epsilon$$

and

$$\epsilon_T = \ln \frac{l_i}{l_0} = \ln(1 + \epsilon) \tag{9·31}$$

Combining Eqns (9·28) and (9·31) we can easily show that, at the commencement of necking,

$$\frac{d\sigma_T}{d\epsilon} = \frac{\sigma_T}{1 + \epsilon} \tag{9·32}$$

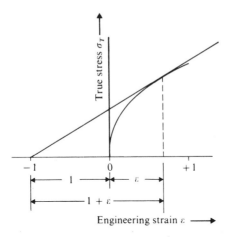

Fig. 9·22 Considère's construction

This leads to Considère's construction for determining the point of necking on the curve of true stress versus engineering strain, as illustrated in Fig. 9·22.

Other terms referred to in the tensile test are the *ductility*, defined as the strain at fracture and usually quoted as a percentage, the *reduction in area* at the neck (another measure of the ductility), and the *proof stress*. The 0·1% proof stress, for example, is the point on the engineering stress–strain curve from which, on removal of the stress and elastic recovery (as illustrated for example by the lines SS′ and RR′ in Fig. 9·1), there would remain 0·1% permanent plastic strain.

Referring to Fig. 9·20 it can be seen that brittle materials such as cast iron show little or no plastic deformation before fracture, i.e., they are not ductile. Materials like this are dangerous because high local tensile stresses occur in structures at points of *stress concentration*, such as holes and corners and can lead to brittle failure. Ductile materials such as copper exhibit considerable plastic flow before fracture in tension, so that stress concentrations are 'relieved' by plastic flow. It will be noticed that the diagram for mild steel is different from other materials in that it exhibits a sharply defined *upper yield point* followed by a period of deformation at a *lower yield stress*. During this phase the plastic deformation is non-uniform along the specimen. It begins at points of stress concentration and propagates along the specimen in the form of bands of deformation (bands of moving dislocations) known as *Lüder's bands*. When the Lüder's bands have covered the whole specimen, the deformation becomes uniform and the material work-hardens in the normal way.

This phenomenon can be explained in terms of an interaction between the interstitial carbon and nitrogen atoms, which are always present in steel, and dislocations. The carbon and nitrogen atoms preferentially occupy interstitial sites near dislocations and in so doing hinder the movement of the dislocations. The dislocations are said to be *pinned* by interstitial *atmospheres*. When the upper yield point is reached, at points of stress concentration the dislocations either break away from their interstitial atmospheres or multiplication occurs producing mobile (unpinned) dislocations. The stress required to move the dislocations (the lower yield stress) is less than the unpinning or multiplication stress (the upper yield point).

9·6·2   The compression test

Brittle materials such as concrete are generally tested in compression since this is the mode of stressing in which they are most frequently used. Necking does not occur but problems may arise due to friction between the ends of the specimen and the loading platens. Friction leads to non-uniform deformation and the specimen tends to form a barrel shape, as shown in the photograph of Fig. 9·23, especially when ductile materials (which undergo large deformation) are tested.

A big advantage of the compression test is that the true stress–true strain curve can be estimated from the results, up to much larger strains that can be obtained in the tension test. This is often of value when it is required to predict the forces involved in the large-strain deformation of materials when they are being formed by processes such as forging, rolling, and extrusion. Various methods have been devised to reduce or mitigate the effects of friction and non-uniform deformation at the loading platens, and hence reduce barrelling of the specimen.

Fig. 9·23 Photograph of barrel-shaped distortion produced in a compression test

9·6·3 The hardness test

The hardness test is used as a quick, inexpensive method of assessing the mechanical properties of a material. The *hardness* of a material is determined by pressing an indenter into its surface and measuring the size of the impression. The bigger the impression the softer the material. Although the earliest form of this test employed a spherical ball of 10 mm diameter, made from hard steel or tungsten carbide (known as the Brinell hardness test), later forms of the test employ pyramidal or conical indenters which have the virtue that the impression remains geometrically similar irrespective of its size. In the Vickers test a diamond cut in the form of a square pyramid having an apex angle of 136° is used and the Vickers hardness number (VHN) is defined as the load/surface area of the impression, which can be shown to be

$$\text{VHN} = 1 \cdot 854\, P/d^2 \; \text{kgf/mm}^2 \tag{9·33}$$

where P is the applied load (kgf) and d is the distance across the diagonal of the impression (mm), as shown in Fig. 9·24.

The units given in Eqn (9·33) have been used for hardness for a long time and are still those commonly quoted for metals. However, in recent years the hardness test has been increasingly used for ceramics and generally these hardnesses are given in the more acceptable SI units of MPa (MN/m^2).

In the Brinell test, the Brinell hardness number (BHN) may be found from the expression

$$\text{BHN} = \frac{2P}{\pi D[D - \sqrt{(D^2 - d^2)}]} \; \text{kgf/mm}^2 \tag{9·34}$$

where D is the diameter of the ball (mm) and d the diameter of the impression (mm). For consistent results the diameter of the indentation d should be between $0 \cdot 3D$ and $0 \cdot 6D$. Under these conditions fairly soft materials have the same Brinell and Vickers hardness numbers, since the geometries of the impressions are not too dissimilar.

The hardness test involves indenting a material by a process of plastic deformation which has been shown to correspond with approximately 8% strain of the material. Thus the hardness is a function of the yield stress σ_y and the work-hardening rate of the material. As a rough guide the VHN is

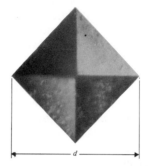

Fig. 9·24 Indentation from a Vickers hardness test

between 0·2 and 0·3 times σ_y, where σ_y is in MN/m², for hard materials and 0·3 σ_y for metals (see Table 9·1).

Table 9·1
Typical values of VHN and yield stress

Material	VHN (kgf/mm²)	Yield stress (MN/m²)
Diamond	8,400	54,100
Alumina	2,000	11,300
Boron	2,500	13,400
Tungsten carbide	2,100	7,000
Beryllia	1,300	7,000
Steel	210	700
Annealed copper	47	150
Annealed aluminium	22	60
Lead	6	16

A diamond conical indenter of 120° included angle is used in the Rockwell test and the depth of the indentation measured.

9·6·4   The impact test

We know that materials can fail in a brittle or ductile manner, the latter occurring only after appreciable plastic flow. Plastic flow, as has already been explained, occurs by the movement of dislocations under the action of a shear stress. Brittle fracture involves tensile separation with little or no plastic flow, so that stress systems which involve high ratios of tensile to shear stress are likely to favour brittle failure. Similarly any material composition or heat-treatment giving a low ratio of tensile strength to shear strength will tend to promote brittle fracture.

Materials such as cast iron or marble are normally brittle but can be made to exhibit ductile behaviour, by superimposing very high all-round hydrostatic pressure. Apart from the stress system, other important variables affecting the propensity for a material to fail in a brittle manner are the rate of straining and the temperature. Pitch flows slowly even at room temperature but will break into pieces if hit with a hammer (showing the effect of strain rate). Glass will flow and can be moulded into almost any shape at elevated temperatures but it is certainly brittle at room temperature (showing the effect of temperature).

Materials which are normally ductile may behave in a brittle fashion if sharp notches or cracks are present in them, particularly at high rates of strain. This is because high local tensile stresses are induced in the neighbourhood of the crack. Furthermore, if the stress situation is examined in more detail it is found that triaxial tensile stresses are set up which reduce the maximum shear stress. Shear stresses are required for dislocation motion and therefore if the maximum shear stress is reduced plastic deformation is less likely and brittle behaviour is favoured.

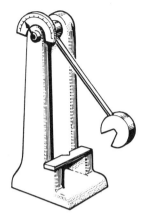

Fig. 9·25 Schematic diagram of an impact testing machine

A commonly used test for brittleness is the Charpy test, illustrated in Fig. 9·25. In this test a standard notch is cut in a standard test specimen which is struck under impact conditions by a heavy weight forming the end of a pendulum. The notch serves to introduce triaxial tensile stresses into the specimen, encouraging brittle failure to occur. The bar of the material is 55 mm long, with cross section 10 mm × 10 mm, and has a V-notch 2 mm deep of 45° included angle and a root radius of 0·25 mm. The bar is clamped in the machine as shown in Fig. 9·25 and the weight released from a known height so as to strike the specimen on the side opposite the notch and induce tensile stresses in it. After breaking the specimen the pendulum swings on and the height to which it rises on the other side is measured. Thus the energy absorbed in breaking the bar may be determined and if this is low the specimen is brittle. The notch brittleness of the specimen can also be assessed from the appearance of the fracture, the proportion of the bright crystalline area being a rough measure of brittleness.

Typically, a tough steel absorbs in the region of 130 joules of energy in a Charpy test at room temperature. A typical result from notch tests on mild steel over a range of temperatures is shown in Fig. 9·26. It will be seen that there is a rapid rise in energy absorbed at 280K as the temperature is raised. This is the *ductile–brittle transition*: for V-notched specimens the *transition temperature* is taken to be that at which 13–15 joules of energy are absorbed. Above this temperature much more plastic flow occurs in the notch so that more energy is absorbed. Any metallurgical treatment which will move the test curve shown in Fig. 9·26 over to the left of the diagram will improve the usefulness of the steel. The unfortunate characteristics of having a ductile–brittle transition is not typical of all metals as shown by the curve for aluminium in Fig. 9·26. Aluminium is ductile at all temperatures although at ambient and elevated temperatures less energy is required to fracture aluminium than mild steel–we say that the mild steel is tougher at these temperatures.

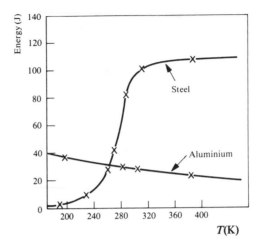

Fig. 9·26 Notch test curves for mild steel and aluminium

9·6·5 Fatigue and creep testing

It is estimated that 90% of all mechanical failures are caused by fatigue, i.e., the failure of a structure under the repeated application of a load far smaller than that required to cause failure in one application. Fatigue failures often occur in a catastrophic manner with no gross distortion preceding collapse. The size and location of the cracks formed in a structure by the fatigue process often make their detection during routine inspection almost impossible.

Laboratory fatigue tests can be carried out on four types of testing machine: (1) push-pull (axial loading); (2) rotating bending (rotating a cantilever beam with a stationary end load); (3) reversed bending (applying an alternating bending moment to a flat plate specimen); (4) torsion (applying torque). The stress cycle is often of sinusoidal form owing to the nature of the testing machines and the *stress amplitude* σ_a is usually kept constant [Fig. 9·27(a) and (b)]. Most tests are performed with the *mean stress* $\sigma_m = 0$, as

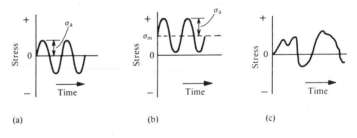

Fig. 9·27 Stress cycle for fatigue testing: (a) $\sigma_m = 0$; (b) σ_m is positive; (c) random σ_m

shown in Fig. 9·27(a) but a non-zero value of σ_m may also be used [Fig. 9·27(b)]. In practice the fluctuating loads imposed on a structure during service may be very different from the regular load cycling so far described. This can be overcome to a certain extent in the laboratory by using servo-hydraulic testing machines which are programmed to give a good approximation to random loading such as that given in Fig. 9·27(c). In extreme cases, such as in the aircraft industry where a fatigue failure could lead to a major disaster, large-scale models or even full-size structures are tested under simulated service conditions; for example, a complete Concorde airframe has been tested under fluctuations of load and temperature designed to simulate service conditions.

To investigate the fatigue behaviour of a material a series of specimens of the material is tested to failure at different values of the stress amplitude σ_a and each test gives one point on a diagram of σ_a versus the logarithm of the number of cycles to failure (known as an S–N curve). Most materials have S–N curves of the type shown by copper in Fig. 9·28, and the fatigue strength or *endurance limit* must be quoted relative to a specified number of cycles, the most common being the stress amplitude which will cause failure in 10^7 cycles. On the other hand, some materials, e.g., most steels and titanium alloys, have a fatigue curve of the type shown by mild steel in Fig. 9·28. For such materials there is a definite value of stress amplitude below which fatigue failure will never occur and this value, the *fatigue limit*, is quoted in some material specifications.

We will discuss fatigue further in Section 9·8, but before finishing with mechanical testing we should briefly look at creep testing. Under certain combinations of stress and temperature, all materials when subjected to a

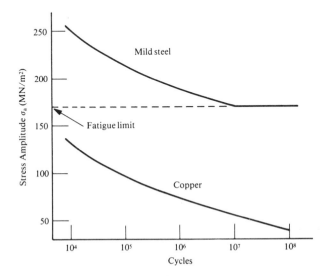

Fig. 9·28 Results of fatigue tests for steel and copper presented as S–N curves

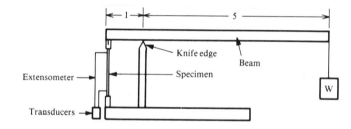

Fig. 9·29 Schematic diagram of a creep machine

constant stress will exhibit an increase of strain with time. This phenomenon is called *creep* and most materials creep to a certain extent at all temperatures, although the engineering metals such as steel, aluminium, and copper creep very little at room temperature. High temperatures lead to rapid creep, which is often accompanied by microstructural changes.

Creep tests are normally carried out in tension for ductile materials such as metals and in compression, or three point bending, for brittle materials like ceramics. The force on the specimen is usually applied via a simple beam system as illustrated in Fig. 9·29. For this machine the force on the specimen is five times the load W because the ratio of the lengths of the beam on either side of the knife-edge is 5:1. Such a simple machine gives a constant load throughout the test, but more sophisticated beams and attachments are available which keep the stress constant on the specimen as its cross-sectional area changes due to creep. The extensometry and the transducers have to be sensitive to small displacements as often the strains in a creep test are not large. The results from a creep test are generally presented as a graph of creep strain against time as shown for lead in Fig. 9·30. The strains experienced during creep are often much less than those shown by lead but the shape of the creep curve of Fig. 9·30 is typical for most materials. The deformation mechanisms occurring in the regions AB, BC and CD of the creep curve are discussed in Section 9·9.

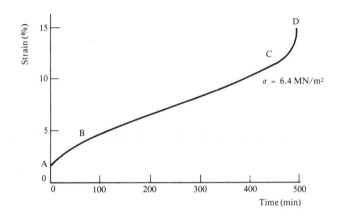

Fig. 9·30 Creep curve of strain against time for lead at room temperature

9·7 Fracture

In general five basic types of fracture are recognized, namely ductile, brittle, fatigue, creep, and environment-assisted fractures, although sometimes a fracture may be difficult to classify as it has characteristics common to more than one of these, e.g., a corrosion–fatigue fracture. Creep, fatigue, and environment-assisted fractures will be discussed later in the chapter. In this section we will concentrate on ductile and brittle fractures and introduce the relatively new concept of fracture toughness. But first let us estimate the strength of an ideal, defect-free material.

9·7·1 Ideal fracture stress

Consider applying an increasing tensile stress to an ideal material containing no flaws or defects. When the stress reaches some critical value, the ideal fracture stress, the atomic bonds will fail across an atomic plane that is perpendicular to the tensile stress. We can calculate the ideal fracture stress using similar reasoning to that employed to estimate the Young's modulus in Section 9·2·2. Referring again to Fig. 9·2, if the tensile force at point d is F_{max} and the stress σ_{max}, then

$$F_{max} \approx \sigma_{max} r_0^2$$

In order to make the mathematics easier, we approximate the force-displacement curve by a half sine wave; therefore, in terms of stress, we have

$$\sigma = \sigma_{max} \sin \frac{2\pi r_0 \epsilon}{\lambda}$$

where λ is the wavelength of the sine wave. Now σ_{max} is a constant and the simplest way to evaluate it is to consider the situation at small strains, but this does not mean the answer is only applicable at small strains. At small strains $\sin \theta \approx \theta$ and Hooke's law is obeyed, hence,

$$\sigma \approx \sigma_{max} \frac{2\pi r_0 \epsilon}{\lambda}$$

then substituting $\sigma = E\epsilon$ (Hooke's law) and rearranging we get

$$\sigma_{max} \approx \frac{E\lambda}{2\pi r_0} \tag{9·35}$$

For most materials this result predicts that σ_{max} lies approximately within the range $E/5$ to $E/30$. In practice materials fracture at stresses much lower than σ_{max}, for example, ordinary glass has a fracture stress of about 50 MN/m² which is $E/1,000$ while alloys or iron and nitrogen can even be heat-treated to give a fracture stress as low as 5 MN/m² ($E/40,000$). Materials do not achieve the theoretical ideal fracture stress because of the presence of flaws and defects. Flaws, such as surface cracks, lower the stress for brittle failure whilst line defects (dislocations) are responsible for initiating ductile fractures.

9·7·2 Brittle fracture–Griffith's theory–fracture toughness

Brittle fracture occurs suddenly, at a stress well below the ideal fracture stress, and is preceded by very little, if any, plastic deformation. Griffith was the first to offer an explanation for the low fracture strength of brittle materials. He postulated that in a brittle material there are small cracks which act to concentrate the stress at their tips. We now deal with the theory which he developed from this idea and which now bears his name.

It can be shown that, for a crack of elliptical section (Fig. 9·31) with a stress, σ, applied perpendicular to its long axis of length $2a$, the stress σ_{tip} at its tip is given by

$$\sigma_{tip} = 2\sigma \left(\frac{a}{\rho} \right)^{1/2} \tag{9·36}$$

where ρ is the radius of curvature at the tip. This stress exists within a distance of approximately ρ of the tip and if it exceeds the ideal fracture stress (that is, the strength of the bonds) it may be assumed that the crack will propagate through the material.

However it is better to consider a thermodynamic criterion for the growth of a crack. This was the approach taken by Griffith who argued that there were two energies to be taken into account when a crack propagated–a release of elastic strain energy and an increase in surface energy. The reader will recall from the previous chapter that solid–gas interfaces have an energy associated with them. Thus, as the material separates along a crack, new surfaces are being created and therefore a certain amount of energy must be provided to create them (i.e., some work must be done). Now before the crack propagates, elastic strain energy is stored in the material. This is released when the material relaxes as the crack spreads. Griffith supposed that the crack propagates when the released strain energy is just sufficient to provide the surface energy necessary for the creation of the new surfaces.

The elastic strain energy for unit volume has been given in Eqn (9·12) as $\frac{1}{2}\sigma^2/E$. For purposes of calculation we take the crack to have unit width perpendicular to the plane of the paper in Fig. 9·31. Very near to the crack faces the stress falls to zero and very far from the crack it is unchanged, so we assume that roughly a region of radius a around the crack is relieved

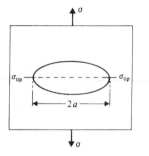

Fig. 9·31 Model for the Griffith crack theory

of its elastic energy. This would, for unit width, give a total elastic energy of

$$\frac{\sigma^2}{2E} \times \text{area} \times \text{width} = \frac{\sigma^2}{2E} \cdot \pi a^2$$

Properly, the strain field should be integrated from infinity to the surface of the crack, which makes the elastic energy U_E available per unit width twice the above value, i.e.

$$U_E = \frac{\sigma^2 \pi a^2}{E} \tag{9·37}$$

If the surface energy per unit area is γ joules per square metre then the surface energy for a crack of length $2a$ and unit width will be

$$U_s = 4\gamma a \tag{9·38}$$

where we multiply by two because there are two faces. The total energy change U is therefore given by

$$U = 4\gamma a - \frac{\sigma^2 \pi a^2}{E}$$

and these terms are plotted as a function of crack length a in Fig. 9·32. The total energy is a maximum at length a_{critical} and consequently a crack of that length is unstable and reduces the total energy by growing. The maximum in U occurs when

$$\frac{\mathrm{d}U}{\mathrm{d}a} = \frac{\mathrm{d}U_s}{\mathrm{d}a} - \frac{\mathrm{d}U_E}{\mathrm{d}a} = 0 \tag{9·39}$$

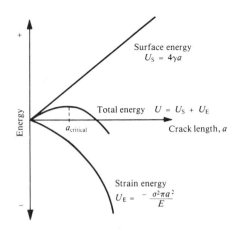

Fig. 9·32 Variation in the surface energy, strain energy and total energy with crack length

From this equation we can see that the crack will grow when the strain energy release rate (dU_E/da) is equal to, or greater than, the rate of change of the surface energy with crack length (dU_s/da). Substituting for U_s and U_E we obtain

$$\frac{dU}{da} = \frac{d}{da}(4\gamma a) - \frac{d}{da}\left(\frac{\sigma^2 \pi a^2}{E}\right) = 0$$

therefore

$$4\gamma - \frac{2\sigma^2 \pi a}{E} = 0 \tag{9.40}$$

and so

$$\sigma = \left(\frac{2\gamma E}{\pi a}\right)^{1/2} \tag{9.41}$$

This is the Griffith equation and shows that the stress necessary to cause a crack to propagate varies inversely as the square root of the crack length. Hence the fracture stress of a brittle material is determined by the length of the largest crack existing before loading. Furthermore, once a crack begins to spread, the stress required for propagation falls as a is increasing, and the crack accelerates rapidly.

The above analysis applies to a crack in a thin plate under stress (known as *plane stress* conditions) but its general features are retained in more complicated analyses, e.g., for a thick plate, which gives conditions of *plane strain*, Eqn (9.41) becomes:

$$\sigma = \left(\frac{2\gamma E}{\pi a (1 - \nu^2)}\right)^{1/2} \tag{9.42}$$

where ν is Poisson's ratio. The relationship between fracture stress and crack length was first confirmed by measuring the strengths of glass containing sharp cracks of known lengths.

During crack propagation it is possible for a small amount of plastic deformation to occur and this requires extra energy. In such cases, e.g., the brittle failure of steel at low temperatures, the Griffith equation [Eqn (9.41)] is modified to

$$\sigma = \left(\frac{E(2\gamma + \gamma_p)}{\pi a}\right)^{1/2} \tag{9.43}$$

where γ_p is the energy expended in plastic deformation per unit area of crack. γ_p may be tens of times greater than γ.

Let us continue to discuss the idea of the energy for crack propagation arising from the release of elastic strain energy in order to develop the concept of *fracture toughness*. Equation (9.39) demonstrates the importance of the rates, with respect to crack length, of the strain energy and surface energy. These rates are plotted against crack length in Fig. 9.33; this figure emphasizes that dU_s/da has a constant value of 4γ, whereas $-dU_E/da$ increases with crack length according to $2\sigma^2 \pi a/E$ [refer to Eqn (9.40)].

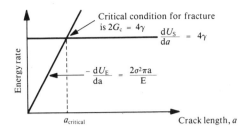

Fig. 9·33 Energy rates (with respect to crack length) as a function of crack length showing that the critical strain energy release rate $G_c = 2\gamma$

We now introduce G, which is the strain energy release rate per crack tip, and as the crack under discussion has two crack tips we have $2G = -dU_E/da$. When G reaches a critical value, known as the *critical strain energy release rate* G_c, a crack will propagate. It can be seen from Fig. 9·33 and Eqn (9·40) that $2G_c = 4\gamma$, hence

$$G_c = 2\gamma \tag{9·44}$$

For a material exhibiting brittle behaviour but with some local plastic deformation

$$G_c = 2\gamma + \gamma_p \tag{9·45}$$

It follows that the original [Eqn (9·41)] and modified [Eqn (9·43)] Griffith equations for plane stress conditions can be written as

$$\sigma = \left(\frac{EG_c}{\pi a}\right)^{1/2} \tag{9·46}$$

and for conditions of plane strain

$$\sigma = \left(\frac{EG_c}{\pi a (1 - \nu^2)}\right)^{1/2} \tag{9·47}$$

G_c is a measure of the fracture toughness of a material and as such it is a material property just like the Young's modulus, yield stress, etc., G_c is usually given in units of kJ/m² and the smaller G_c the less tough, or more brittle, is the material as is illustrated by the values given in Table 9·2.

Table 9·2
Fracture toughness of a variety of materials

Material	G_c (kJ/m²)	K_c (MN/m³/²)
High strength steel	40	92
Co–Cr–Mo–Si alloy	2·5	25
Polymethyl methacrylate	0·6	1·5
Epoxy resin	0·2	0·8
Alumina	0·06	4·5
Glass	0·003	0·4

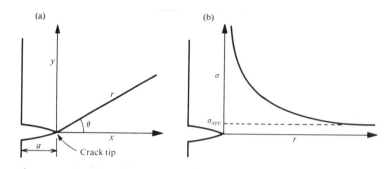

Fig. 9·34 (a) The coordinate system used and (b) the stress as a function of r according to the Eqn (9·47) (σ_{app} is the applied stress)

There is another parameter which is frequently encountered for quantifying the fracture toughness of a material, namely the *critical stress intensity factor K_c*. We have already seen that the introduction of a crack into a material causes a redistribution of stress with the greatest stress being at the crack tip. It is possible to use elasticity theory to calculate the stresses at any point (r, θ) in the vicinity of the crack tip (Fig. 9·34), and the results of such an analysis are

$$\sigma_{yy} = \frac{K}{\sqrt{(2\pi r)}} \cos{\tfrac{1}{2}\theta} \left(1 + \sin{\tfrac{1}{2}\theta} \sin{\tfrac{3}{2}\theta}\right)$$

$$\sigma_{xx} = \frac{K}{\sqrt{(2\pi r)}} \cos{\tfrac{1}{2}\theta} \left(1 - \sin{\tfrac{1}{2}\theta} \sin{\tfrac{3}{2}\theta}\right) \qquad (9·48)$$

$$\tau_{xy} = \frac{K}{\sqrt{(2\pi r)}} \cdot \left(\cos{\tfrac{1}{2}\theta} \sin{\tfrac{1}{2}\theta} \cos{\tfrac{3}{2}\theta}\right)$$

where K is the stress intensity factor. It should be noted from these equations that all crack tip stresses are directly proportional to K. In turn K depends on

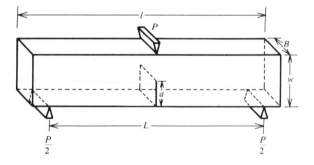

Fig. 9·35 A single edge notched bend specimen for the determination of K_c

the applied load, P and specimen geometry, e.g., for the single-edge notch bend (SENB) specimen illustrated in Fig. 9·35 K is given by

$$K = \frac{3PL\,Ya^{1/2}}{2BW^2} \tag{9·49}$$

where L, B, W, and a are defined in Fig. 9·35 and Y is a function of a/W.

As we increase the load on an SENB specimen during testing K will increase in accordance with Eqn (9·49) and the crack will propagate when K reaches the critical value K_c. K_c is a materials parameter and has dimensions of $MN/m^{3/2}$ and, as for G_c, the lower the value the less tough the material (Table 9·2). K_c and G_c must be related as they are both measures of the fracture toughness of a material, and the following equations show the simple relationships that exist between these parameters:

$$G_c = \frac{K_c^2}{E} \qquad \text{(plane stress)}$$

$$G_c = \frac{K_c^2}{E}(1 - \nu^2) \qquad \text{(plane strain)} \tag{9·50}$$

9·7·3   Ductile failure

Ductile failure occurs after extensive plastic deformation. In fact the energy for a ductile failure is needed mainly for the plastic deformation associated with the crack process and in comparison the amount required for the creation of new surfaces is negligible. Generally, there are three successive events

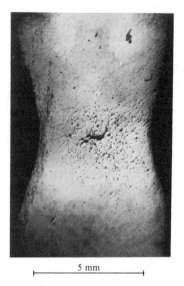

5 mm

Fig. 9·36 Cavities linking up perpendicular to the applied tensile stress in the necked region of a copper specimen [K.E. Puttick, *Phil. Mag.* **4**, 964 (1959)]

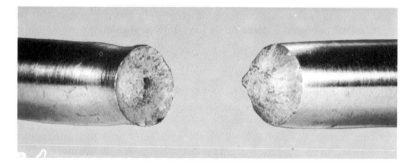

Fig. 9·37 A typical cup and cone ductile fracture resulting from a tensile test

involved in a ductile failure. First small cavities or voids are formed, and in the case of a tensile test this takes place in the neck where the deformation is concentrated. This is followed by the linking of the cavities to form a crack that spreads in a plane approximately perpendicular to the applied tensile stress (Fig. 9·36). Finally the crack propagates to the surface of the sample by shearing in a direction approximately 45° to the tensile axis to give a '*cup and cone*' type fracture (Fig. 9·37).

A ductile fracture surface appears dull when viewed with the naked eye, whereas the fracture surface of a metal that has failed in a brittle manner is shiny. Closer examination of the surfaces with a scanning electron microscope reveals marked topographical differences. The ductile failure has a characteristic 'dimpled' appearance which is a manifestation of the cavities accompanying the crack process. It is now accepted that the nucleation of the cavities during plastic deformation is associated with second phase particles and inclusions and this relationship between cavity and particle is clearly shown in Fig. 9·38(a). This explains why the ductility of a material increases as the purity is raised, i.e., the purer the material the fewer second phase particles and inclusions. A simple model for cavity nucleation at particles due to interfacial decohesion has been proposed by Gurland and Plateau. Their approach was similar to that of Griffith for brittle fracture in that they assumed that the release of elastic strain energy must be sufficient to create the new surfaces. Indeed the following expression which they derived for the tensile stress required for particle–matrix decohesion has similarities with the Griffith equation [see Eqn (9·41)]

$$\sigma = \frac{1}{q} \left(\frac{E\gamma}{d_p} \right)^{1/2} \tag{9·51}$$

where q is the stress concentration factor at the particle of diameter d_p, γ is the surface energy and E is the weighted average of the elastic moduli of the particle and the matrix. If this simple model is extended to take account of the energy of plastic deformation local to the particle, Eqn (9·51) is modified to

$$\sigma = \frac{1}{q} \left(\frac{E\gamma}{d_p} \right)^{1/2} + \frac{\sigma_y}{q} \left(\frac{\Delta V}{V} \right)^{1/2} \tag{9·52}$$

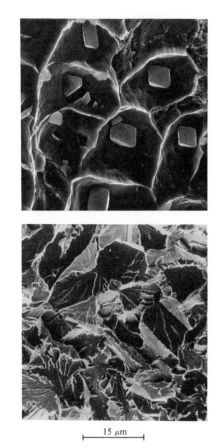

(a)

(b)

15 μm

Fig. 9·38 Scanning electron micrographs of fracture surfaces of a mild steel: (a) ductile failure, (b) brittle cleavage failure (Courtesy of T.J. Baker)

where σ_y is the yield stress of the matrix with particles present, V is the volume of a particle and ΔV is the volume plastically deformed around the particle. Both of these equations indicate the importance of particle size on the initiation stress.

In contrast to the 'dimpled' topography of the ductile fracture, a metal which has failed in a brittle manner has a surface consisting of flat, shiny facets [Fig. 9·38(b)]. This is because the fracture has taken place by the separation of grains along specific crystallographic planes, called the *cleavage planes*, e.g., in b.c.c. iron the cleavage planes are the {100}.

9·8 Fatigue

Earlier in Section 9·6·5 we saw that one way of presenting fatigue data was by the use of S–N curves. An S–N curve, such as that given for copper and steel in Fig. 9·28, is only applicable for a given value of the mean stress σ_m. Experi-

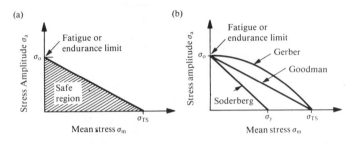

Fig. 9·39 Effect of mean stress on fatigue. (a) Goodman relationship; it is safe to operate within the shaded area. (b) Comparison of the Gerber, Goodman and Soderberg relationships

ments have demonstrated that the greater σ_m the shorter is the fatigue life and a number of empirical relationships have been proposed for this effect. These relationships are known by the names of their originators.

Let us first study the relationship due to *Goodman* by reference to Fig. 9·39(a), which is a plot of stress amplitude σ_a against mean stress. On this plot σ_0 is the fatigue limit, or the endurance limit for a specified number of cycles if the material does not exhibit a fatigue limit, for $\sigma_m = 0$. σ_0 is joined by a straight line to the tensile strength σ_{TS} which is taken to be the maximum possible value for σ_m. Thus the equation for this line is

$$\sigma_a = \sigma_0 \left[1 - \left(\frac{\sigma_m}{\sigma_{TS}} \right) \right] \tag{9·53}$$

According to Goodman if the point representing σ_a and σ_m for given service conditions lies below the straight line in the area shown shaded in Fig. 9·39(a), then the component will not fail by fatigue.

The alternative relationships are shown, together with that proposed by Goodman, in Fig. 9·39(b). The *Soderberg* plot is similar to that of Goodman but with σ_0 joined to the yield stress σ_y. It follows that the Soderberg equation is simply Eqn (9·53) with σ_{TS} replaced by σ_y. The *Gerber* relationship is parabolic and is given by

$$\sigma_a = \sigma_0 \left[1 - \left(\frac{\sigma_m}{\sigma_{TS}} \right)^2 \right] \tag{9·54}$$

Clearly the safest criterion is that due to Soderberg.

The reader will be aware that the stress amplitude varies for many components subject to cyclic loading during service. It is important to be able to predict the life of a component under such conditions and several empirical relationships are available for this purpose. The simplest and best known is *Miner's Law* which is an attempt to quantify the damage that occurs at the different stress amplitudes and hence to assess the cumulative damage. Suppose a component undergoes n_1 cycles under cyclic loading conditions that would lead to failure after N_1 cycles, then n_1/N_1 is taken as a measure of the damage that has occurred. Similarly, if the stress amplitude is changed for n_2 cycles and at this stress amplitude failure would take place after N_2 cycles, then n_2/N_2 is a measure of the damage for this period. According to Miner's Law the component will fail when the sum of the n/N ratios for the various loading conditions is equal to unity, i.e.

$$\frac{n_1}{N_1} + \frac{n_2}{N_2} + \frac{n_3}{N_3} \ldots = 1 \tag{9·55}$$

Let us now turn our attention to what is happening on the microscopic scale during cyclic loading. We can consider a fatigue failure to consist of three stages–stage 1 is crack initiation, stage 2 is fatigue crack growth, and stage 3 is the final catastropic fracture. For large structures such as bridges and oil rigs there are many nucleation sites for a fatigue crack, e.g., surface imperfections and defects associated with welds. In such circumstances, the growth stage of a fatigue crack is the most important. In other cases, where the surface finish is good, the initiation of the fatigue crack may occupy much of the fatigue life. It has been estimated that in well finished laboratory specimens, which are used for S–N curve determination, the initiation stage may be as much as 90% of the lifetime at low stress amplitudes reducing to 10% at higher stress levels. Stage 3 of a fatigue failure is, of course, always rapid compared with the preceding stages.

9·8·1 Stage 1

In specimens with smooth surfaces the initiation of the fatigue crack involves the localized movement of dislocations on slip planes at approximately 45° to the tensile axis. This localized deformation gives rise to slip bands which are similar in appearance to the slip lines described in Section 8·3·1. There is, however, a major difference between the fatigue slip bands and slip lines; on repolishing the surface of a specimen slip lines are removed as they are surface steps, whereas the slip bands remain. For this reason the fatigue slip bands are known as *persistent slip bands*. The persistent slip bands are associated with intrusions and extrusions at the surface from which a fatigue crack propagates for a few grain diameters (Fig. 9·40).

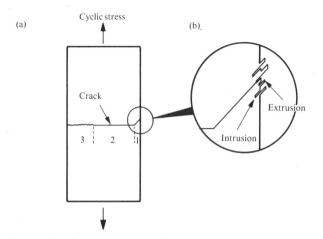

Fig. 9·40 Fatigue failure. (a) The three stages; (b) stage 1

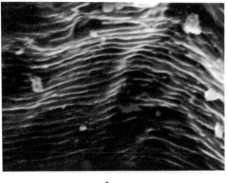

2 μm

Fig. 9·41 Fatigue striations in stainless steel (Courtesy Matcon)

9·8·2   Stage 2

As shown in Fig. 9·40 in this stage the crack changes direction and becomes roughly normal to the stress axis. The fracture has a series of fine, non-crystallographic ridges known as *fatigue striations* (Fig. 9·41). Each striation is produced by a single stress cycle, and there are several models to account for their formation. One such model is illustrated in Fig. 9·42 and shows that a striation is a consequence of an increment of crack growth and concomitant blunting and resharpening of the crack tip by plastic deformation. Figure 9·42(a) corresponds to the zero load condition after a number of cycles; striations 1, 2, and 3 were formed during previous cycles. When a small tensile stress is applied the crack opens up and shear stresses are set up at the crack tip [Fig. 9·42(b)]. As the applied tensile stress increases plastic deformation occurs due to the shear stress and this leads to crack extension and blunting [Fig. 9·42(c)]. Once the compression period of the cycle is

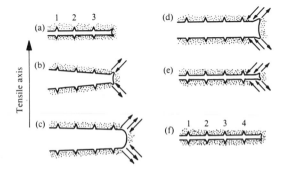

Fig. 9·42 Formation of a fatigue striation during one cycle: (a) zero load, (b) small tensile stress, (c) large tensile stress, (d) small compressive stress, (e) maximum compressive stress, (f) zero load

entered the crack begins to close and the shear stresses are reversed [Fig. 9·42(d)]. Finally, at the point of maximum compressive stress the crack is almost closed and the reverse plastic flow that has taken place during the compression half-cycle has resulted in the formation of a new striation and resharpening of the crack tip.

Clearly the propagation of a fatigue crack is closely related to the stress situation at the crack tip and we know from the previous section in this chapter that all crack tip stresses are directly proportional to the stress intensity factor, K. It is not surprising, therefore, that a relationship between K and stage 2 fatigue crack growth has been found. In cyclic loading K will vary throughout the cycle and the stress intensity factor range ΔK is defined as the difference between K at the maximum load (K_{max}) and K at the minimum load (K_{min}), i.e.

$$\Delta K = K_{max} - K_{min}$$

If the minimum stress is compressive K_{min} is normally taken to be zero. A logarithm–logarithm plot of the crack growth rate da/dN against ΔK is given in Fig. 9·43. The lower limit to the curve ΔK_T is called the threshold stress intensity factor range for crack growth. Whether there is a true threshold or not is under debate, but certainly at low ΔK values the rate of growth becomes vanishingly small. At high ΔK values, where K_{max} is approaching the critical stress intensity factor K_c, crack growth is very rapid and the failure modes are characteristic of normal static failures, e.g., brittle cleavage failure. The normal fatigue stage 2 range is for intermediate values of ΔK where the plot is linear. We therefore have for stage 2

$$\frac{da}{dN} = C(\Delta K)^m \tag{9·56}$$

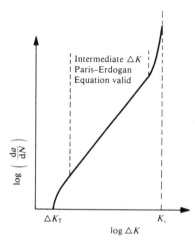

Fig. 9·43 Graph showing the relationship between fatigue crack growth and ΔK

which is known as the *Paris–Erdogan equation*. C and m are constants for a particular material; m is usually between 2 and 4 for most ductile metals.

9.8.3   Stage 3

Stage 3 occurs when the fatigue crack has grown to such an extent that the component finally fails in a catastrophic manner. The final fracture mode varies with material and service conditions; it may be a simple ductile overload failure due to the cross-sectional area of the component having been reduced so much by the fatigue crack, or it may be brittle cleavage failure as a result of K_{max} reaching K_c.

A consequence of the different stages is that a fatigue failure can often be recognized from the macroscopic appearance of the fracture. Two distinct

(a)

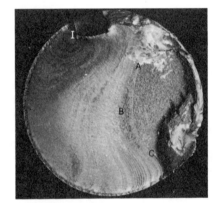

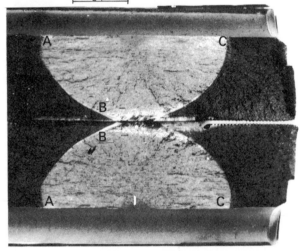

(b)

Fig. 9·44 Fatigue failures. (a) Crankshaft–the failure was initiated in the corner of the keyway (Courtesy of T.J. Baker). (b) Wall of a pressure vessel (both fracture faces shown). I is the point of initiation, ABC is the boundary between stage 2 fatigue crack growth and stage 3 final failure

zones can frequently be distinguished as shown by the examples in Fig. 9·44. There is a relatively smooth zone which corresponds to stage 2 fatigue crack growth. This zone has been smoothed by the continual rubbing together of the cracked surfaces during the cyclic loading but it also sometimes contains concentric rings, called *clam shell markings*, which indicate successive positions of the crack. The clam shell markings are usually made visible by corrosion and the significance of the markings is that they focus back to the point of initiation of the fatigue crack. The second zone is produced by the final failure and has a rougher surface which may be shiny or dull depending on the fracture mode.

9·9 Creep

A typical creep curve is shown in Fig. 9·30. It can be conveniently divided into sections:
(a)   0A, the initial 'instantaneous' strain, which is usually elastic.
(b)   AB, a period of continuously decreasing creep rate, known as *primary creep*.
(c)   BC, a period of dynamic steady state where the creep rate is constant and at a minimum–*secondary* or *steady state creep*.
(d)   CD, a period of accelerating creep culminating in fracture, known as *tertiary creep*.

All materials, under specific test conditions, can exhibit a creep curve of the form shown in Fig. 9·30, and we will now discuss some of the equations used to describe the curve. At the same time, a physical explanation of the three stages of creep will be given for crystalline materials.

9·9·1   Primary creep

Creep in this period is solely due to dislocation movement in crystalline materials. Work-hardening occurs more rapidly than any concomitant recovery processes so the creep rate decelerates with time.

Two alternative equations are used to describe primary creep. The first is

$$\epsilon = \alpha \log t \tag{9.57}$$

where α is a constant and t is time. This equation usually only applies at very low temperatures, but it does hold for a variety of materials, e.g. aluminium, rubber, and glass.

A more widely obeyed equation, which is applicable at higher temperature and stress, is

$$\epsilon = \beta t^{m} \tag{9.58}$$

where β is a constant and m ranges from 0·03 to 1·0 depending on the material, stress, and temperature.

9·9·2   Secondary creep

During this phase the rate of recovery is sufficiently fast to balance the rate of work-hardening, so that the material creeps at a steady rate. Structural observations show that polygonization is an important recovery process during secondary creep. Grain boundary sliding is also a feature of polycrystalline materials deforming by steady-state creep. However, the contribution to the total strain from grain boundary sliding is generally small, less than 10%.

The equation for secondary creep is, of course, simply the equation for a straight line, viz.

$$\epsilon = Kt \qquad (9.59)$$

where K is a constant, which is temperature- and stress-dependent. This is because recovery processes are temperature- and stress-dependent; the temperature dependence, for example, arises from the change in the rate of diffusion with temperature (discussed in Chapter 7). In fact the creep rate, which is equal to the constant K, can often be represented by an expression of the form

$$d\epsilon/dt = A\sigma^n \exp(-Q/RT) \qquad (9.60)$$

where A and n are constants; n is usually in the range 3 to 7 for metals and 1 to 2 for polymers. Q is called the *activation energy for creep* and is approximately equal to the activation energy for diffusion.

At very high temperatures in metals and alloys creep can take place by vacancy migration and no dislocation motion is involved. Stress-directed diffusion of vacancies from grain boundaries which are under tension to compression stressed boundaries occurs as shown in Fig. 9.45(a) and leads to a creep strain. This mechanism is known as *Herring–Nabarro creep* and the creep rate is given by

$$d\epsilon/dt = \frac{\sigma a^3 D}{2d^2 kT} \qquad (9.61)$$

where σ, k and T have their usual meanings, a^3 is the atomic volume, D is the diffusion coefficient, and d is the grain diameter. The significant feature of

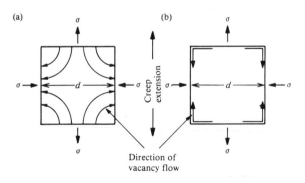

Fig. 9.45 Schematic diagram of diffusion creep in a grain of size d. (a) Herring–Nabarro creep; vacancy flow through the grain. (b) Coble creep; vacancy flow along the grain boundaries

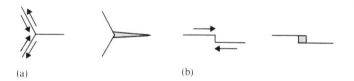

(a)   (b)

Fig. 9·46 Formation of micro-cracks by grain boundary sliding

this equation is the d^{-2} creep rate dependence. Instead of the stress-directed vacancy migration taking place through the grain it can proceed along the grain boundaries [Fig. 9·45(b)]; this is called *Coble creep* and leads to a d^{-3} creep rate dependence. It follows that if diffusion creep, either Herring–Nabarro or Coble, is likely during high-temperature service then a small grain size is undesirable.

9·9·3 Tertiary creep

The accelerating creep rate is associated with the formation of voids or micro-cracks at the grain boundaries. The voids are formed either by vacancy coalescence at the boundaries, which produces rounded voids, or by grain boundary sliding. Two models for the initiation of micro-cracks by sliding are illustrated in Fig. 9·46 and as can be seen from this figure, and also from the micrograph of Fig. 9·47, grain boundary sliding results in angular or wedge-shaped voids. These tend to be formed at lower creep temperatures or at higher creep stresses than the vacancy voids. The voids grow and link up and the material eventually fails in an *intercrystalline* manner, i.e., fails at the grain boundaries. The elongation to fracture in creep is often much less than in the conventional tensile test, e.g., nickel and aluminium alloys may fail after 1 to 3% strain under creep conditions, whereas they exhibit ductilities of 30% in tensile tests at room temperature.

200 μm

Fig. 9·47 Micrograph of creep voids in stainless steel (Courtesy Matcon)

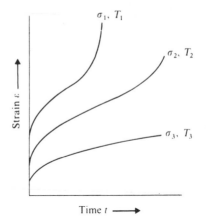

Fig. 9·48 Effect of temperature and stress on the creep curves. If $\sigma_1 = \sigma_2 = \sigma_3$ then $T_1 > T_2 > T_3$; alternatively, if $T_1 = T_2 = T_3$ then $\sigma_1 > \sigma_2 > \sigma_3$

The creep curve of Fig. 9·30 is a generalized creep curve and whether or not a material exhibits the three stages of creep depends on the stress and temperature of testing. The effect of temperature and stress is illustrated in Fig. 9·48. If this figure is taken to represent a series of tests at constant stress, i.e., $\sigma_1 = \sigma_2 = \sigma_3$, then the temperatures are in the order $T_1 > T_2 > T_3$; alternatively, for a series of constant temperature tests, i.e., $T_1 = T_2 = T_3$, the relative magnitude of the stresses is $\sigma_1 > \sigma_2 > \sigma_3$.

High-temperature creep is usually of most concern to engineers, particularly in the design of aircraft, gas turbine components, pressure vessels for high temperature chemical processes, etc. For most practical purposes, creep is usually unimportant in steels below 300°C, or high-temperature nickel-base alloys below 500°C, but must be considered in aluminium alloys at 100°C and in polymers at room temperature. Design criteria usually specify that a particular strain (say 0·1%), or fracture, must not occur in less than the anticipated lifetime of the part.

If the lifetime of a component is specified in terms of strain, then equations such as (9·60) and (9·61) would be used to obtain the approximate combination of material, stress and temperature to satisfy the specification. However if fracture, which is often referred to as *creep rupture*, is specified then the selection of the material would involve the use of creep rupture time (t_r)–temperature–stress extrapolation procedures of which the *Larsen–Miller parameter* is an example. The Larsen–Miller parameter (LM) is

$$LM = T(C + t_r)$$

and it has a constant value at a given stress for a particular material (C is a constant). Hence from a master curve for a particular material of stress versus LM it is possible to predict the creep rupture time for any combination of stress and temperature. For example, Fig. 9·49 gives the stress dependence of the Larsen–Miller parameter for two high-temperature nickel alloys.

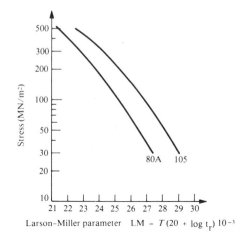

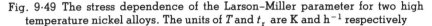

Fig. 9·49 The stress dependence of the Larson–Miller parameter for two high temperature nickel alloys. The units of T and t_r are K and h^{-1} respectively

Let us estimate the creep rupture time of Nimonic 80A at a stress of 300 MN/m² and a temperature of 800°C. From the graph LM = 22·8 × 10³ at a stress of 300 MN/m²; but also LM = $T(C + \log t_r)$ = 1073 (20 + log t_r) for Nimonic 80A at 800°C, therefore 22·8 × 10³ = 1073 (20 + log t_r) which gives t_r = 14·3 h.

We have seen that many different deformation mechanisms may occur when a material is stressed, e.g., dislocation glide, dislocation creep, diffusion creep, and that equations which relate strain rate, temperature, and stress are available for these mechanisms. From these equations, together with experimental data, it is possible to determine which is the dominant mechanism, i.e., which mechanism permits the fastest strain rate, under

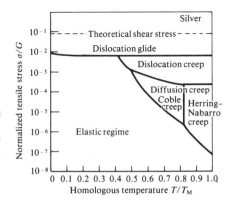

Fig. 9·50 Deformation map for pure silver of grain size 32 μm [M.F. Ashby, *Acta Met.* **20**, 887 (1972)]

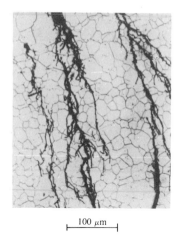

100 μm

Fig. 9·51 Transgranular stress-corrosion cracking of stainless steel due to chlorides (*ASM Metals Handbook*, 8th edn, Vol. 7, p. 135)

specific conditions of stress and temperature for a given material. It is convenient to be able to summarize the information from this type of calculation in a diagrammatic form and this is achieved by diagrams called *deformation maps*. A deformation map is a diagram, with axes of normalized stress (stress/shear modulus) and homologous temperature (temperature/melting point), divided in areas in each of which a particular mechanism is dominant. An example of a deformation map is given in Fig. 9·50, in which the elastic regime is where the strain rate predicted for the various flow mechanisms would be too small to be measured. Often, superimposed on the maps are contours of constant strain rate so the maps can also be used to give an approximate value for the strain rate under given conditions. Therefore the diagrams present the relationships between stress, temperature, and strain rate–if any two of these are specified the map may be used to determine the third and to identify the dominant deformation mechanism.

9·10   Environment-assisted cracking

Even though an environment may not cause general corrosion it can markedly reduce the resistance to crack propagation in the presence of a stress, whether the stress be externally applied or a residual stress due to cold working. This is termed *stress corrosion* and most structural metals are subject to this form of cracking in some environment. Examples of relatively mild environments which lead to stress corrosion in common metals are chloride solutions with stainless steel and solutions containing ammonia with brass. A stress corrosion failure may be transgranular as in the case of chloride solutions and stainless steel (Fig. 9·51) or intergranular as exemplified by the fracture of brass in an ammonia solution.

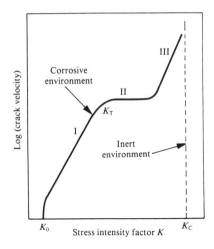

Fig. 9·52 The variation in crack velocity with stress intensity factor for corrosive and inert environments

The concept of the stress intensity factor K, which was introduced earlier (Section 9·7·2), has been applied to the analysis of stress corrosion failures. We have learnt that under the action of a static stress in an inert environment a crack will not propagate until the critical stress intensity factor K_c is reached. In contrast to this behaviour, in a corrosive environment crack growth commences at values of the stress intensity factor which are less than K_c. In these circumstances the crack velocity–K relationship is as depicted in Fig. 9·52 with three distinct regions of crack growth. K_0 is the stress corrosion limit and between K_0 and K_T, i.e., region I, crack propagation is controlled by the rate of chemical reaction at the crack tip. At intermediate values of K, in region II, diffusion of the corrosive species to the crack tip is rate controlling. Finally, in region III the crack propagates in a mixed mode with both stress corrosion and normal fracture contributions.

Hydrogen cracking or *hydrogen embrittlement* is another form of environment assisted fracture. As the name suggests this type of fracture is the result of hydrogen from a corrosion reaction being absorbed by the metal and causing embrittlement. Many steels are prone to hydrogen embrittlement, e.g., oil or gas contaminated with H_2S has been responsible for innumerable failures in high strength steels used for well casings and risers, but it is not confined to ferrous alloys and can occur in refractory metals, titanium alloys and aluminium alloys. In many ferrous and aluminium alloys the absorbed hydrogen segregates to the grain boundaries and reduces G_c [see Eqns (9·44)–(9·47)] so leading to intergranular failure.

Other forms of environment-assisted cracking are also encountered, such as liquid metal embrittlement and corrosion fatigue, but these will not be discussed as sufficient information has been given in this section to demonstrate the significant effect that the environment can have on the mechanical behaviour of materials.

Problems

9·1 A 20,000 N load is applied to a steel bar of cross-sectional area 6 cm². When the same load is applied to an aluminium bar it is found to give the same elastic strain as the steel. Calculate the cross-sectional area of the aluminium bar. (Young's modulus for steel $= 2 \cdot 1 \times 10^5$ MN/m² and for aluminium $= 0 \cdot 703 \times 10^5$ MN/m².)

9·2 For copper Young's modulus $= 1 \cdot 26 \times 10^5$ MN/m² and the shear modulus $= 0 \cdot 35 \times 10^5$ MN/m². Calculate the ratio of the tensile strain to the shear strain if the same energy density is produced by a tensile stress alone and a shear stress alone.

9·3 Using the table of physical properties in Appendix 3 together with the Periodic Table of the Elements, plot (a) boiling point against Young's modulus; (b) Young's modulus against valency; (c) density against valency for all the elements which you would regard as metals. What general conclusions can be drawn regarding the relationship between these various factors and the strength of the metallic binding forces?

9·4 In the forging of steel the ingots are heated to high temperatures at intervals during the working process. Explain why and discuss the basic mechanisms which are involved in the process.

9·5 How would you distinguish between a ductile and a brittle fracture? Why is the fracture strength of real materials lower than the ideal breaking strength?

9·6 Explain why the strength of glass plate may be increased by etching off its surface with hydrofluoric acid. A glass plate has a sharp crack of length 1 μm in its surface. At what stress will it fracture when a tensile force is applied perpendicular to the length of the crack? (Young's modulus $= 0 \cdot 7 \times 10^5$ MN/m² and surface energy $= 0 \cdot 3$ J/m². Assume plane stress conditions.)

9·7 Explain why in a tensile test the level of the true stress–strain curve is higher than that of an engineering stress–strain curve. What would be the relative positions of the two curves in a compression test in which barrel-shaped distortion occurs, and why?

9·8 For what sorts of material is a hardness test most useful? How would a knowledge of the hardness of a material be of help to the machinist, the design engineer, the testing engineer, and the mineralogist?

9·9 A round bar of metal is 9 mm diameter and it is observed that a length of 250 mm extends by an amount of 0·225 mm under a load of 11·8 kN. At the same time its diameter contracts by 0·00227 mm. Determine Young's modulus and the shear modulus for the metal.

9·10 An empirical equation which represents the strain-hardening behaviour of many engineering metals is

$$\bar{\sigma} = A(B + \bar{\epsilon})^n$$

Show that the 'uniform elongation' for such a material would be

$$\bar{\epsilon} = n - B$$

9·11 In a state of pure shear it can be shown that, if the maximum shear stress is τ_s, the principal stresses are

$$\sigma_1 = -\sigma_2 = \tau_s, \sigma_3 = 0$$

In pure tension the principal stresses are

$$\sigma_1 = 2\tau_t, \sigma_2 = \sigma_3 = 0$$

where τ_t is the maximum shear stress in the tensile specimen.

Using Eqn (9·22) show that yielding occurs in a tension specimen when the maximum shear stress in it is $(\sqrt{3})/2$ times the maximum shear stress for yielding in pure shear.

9·12 A rotating component is constructed from a material with an endurance limit (10^7 cycles) of 750 MN/m² at zero mean stress and a tensile strength of 1600 MN/m². Estimate whether it would be safe to run the component at a stress amplitude of 400 MN/m² and a mean stress of 450 MN/m².

Self-Assessment questions

1 If a force of 6 N is applied to a flat tensile specimen of length 6 cm, width 1 cm and thickness 0·3 cm, the stress on the specimen is

A) 1·8 N cm²
B) $1·8 \times 10^{-12}$ MN m²
C) 20 N/m²
D) 0.2 MN/m²
E) 1 MN/m².

2 The elastic strain obtained on applying a stress to a material is

A) time-dependent
B) instantaneous
C) partially permanent
D) reversible
E) directly proportional to the stress
F) inversely proportional to the stress.

3 Poisson's ratio is the ratio of transverse strain to tensile strain during elastic deformation.

A) true   B) false.

4 The stresses acting on a cubical element of a material are depicted in the diagram.
   (i) The stresses σ_1, σ_2, and σ_3 are . . .
   (ii) The stresses τ_{12}, τ_{13}, τ_{21}, etc., are . . .

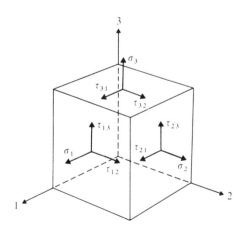

A) tensile stresses
B) shear stresses
C) often denoted as σ_{11}, σ_{22}, etc.
D) often denoted as σ_{12}, σ_{13}, etc.
E) often denoted as τ_{11}, τ_{22}, etc.
F) related such that $\sigma_1 = \sigma_2$, $\sigma_2 = \sigma_3$, etc.
G) related such that $\tau_{12} = \tau_{13}$, $\tau_{21} = \tau_{23}$, etc.
H) related such that $\tau_{12} = \tau_{21}$, $\tau_{13} = \tau_{31}$, etc.

5 The elastic resilience of a material is

A) the stored energy per unit volume during elastic deformation
B) the stored energy per unit volume associated with dislocations
C) given by $1/2\ \epsilon^2 E$
D) given by $1/2\ \sigma\epsilon$
E) given by $E\ \partial\epsilon/\partial t$, where σ, ϵ and E are stress, strain and Young's modulus respectively.

6 Anelastic deformation like elastic deformation, is instantaneous.

A) true   B) false.

7 The critical resolved shear stress is independent of orientation.

A) true   B) false.

8 This is a generalized stress–strain curve for a single crystal. During stage II

A) only the primary slip system operates
B) the primary slip system operates
C) slip occurs on more than one slip system
D) the work-hardening rate is less than in stage III
E) the work-hardening rate is greater than in stage I.

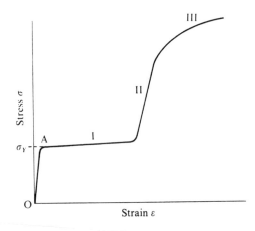

9 The stress–strain curves of polycrystalline materials do not exhibit stages I and II shown by single crystals.

A) true   B) false.

10 Work-hardening is a useful strengthening mechanism but it has the following disadvantages:

A) only useful for two-phase materials
B) decreases the ductility of the material
C) not suitable if the material is to be used at an elevated temperature
D) only applicable to single crystals.

11 The presence of a dispersion of small particles increases the strength of a material by an amount that depends on the

A) volume fraction of the particles
B) valency of the solute atoms
C) the size of the particles
D) the viscoelastic deformation rate of the particles.

12 In the plasticity theory for an isotropic material the effective value of Poisson's ratio is

A) $\frac{1}{3}$ B) $\frac{1}{2}$ C) 0   D) 1   E) 2.

13 The effective stress $\bar{\sigma}$ is a function of the principal stresses σ_1, σ_2, and σ_3.

A) true   B) false.

14 According to Griffith, a crack propagates when the released elastic strain energy is just sufficient to provide the surface energy necessary for the creation of the new surfaces.

A) true   B) false.

15 According to Griffith's theory for brittle fracture, the stress required to propagate a crack is

A) less than the ideal fracture stress
B) more than the ideal fracture stress
C) proportional to $a^{1/2}$, where $2a$ is the crack length
D) proportional to $a^{-1/2}$
E) proportional to γ the surface energy
F) proportional to $\gamma^{1/2}$.

16 This fracture resulted from a tensile test. The fracture

A) is a brittle fracture
B) is a ductile fracture
C) was preceded by considerable plastic deformation
D) probably started at a surface crack
E) is called a 'cup and cone' fracture.

17 Necking occurs in a tensile test at a strain at which the work-hardening rate equals the effective (true) stress.

   A) true   B) false.

18 The true strain at necking of a material with a strain-hardening exponent of 0·2 is

   A) 0·1   B) 0·2   C) 0·4   D) 0·8.

19 The true strain is given by

   A) $\ln \dfrac{l_1}{l_0}$ B) $\dfrac{\Delta l}{l_0}$ C) $\dfrac{\epsilon}{l_0}$ D) $\ln(1 + \epsilon)$ E) $\displaystyle\int_{l_0}^{l_1} \dfrac{dl}{l}$

   where ϵ, l_0, and l_1 are the engineering strain, the initial length, and the current length respectively.

20 This stress–strain curve

   A) is for pure copper   B) is for mild steel
   C) shows an upper yield point   D) shows a lower yield stress
   E) is for a brittle material.

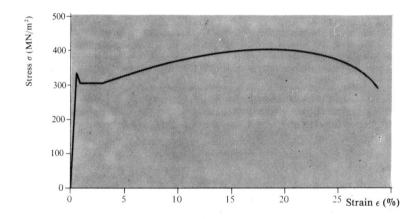

21 The fatigue resistance of a material is reduced by

   A) permanent residual compressive stresses
   B) a mean positive (tensile) stress
   C) chemically or mechanically hardening the surface
   D) poor surface finish.

22 In primary creep the work-hardening occurs more rapidly than any concomitant recovery processes.

   A) true   B) false.

23 In secondary creep the

   A) recovery rate is greater than the work-hardening rate
   B) recovery rate is equal to the work-hardening rate
   C) creep strain ϵ is given by $\epsilon = Kt$ where K is a constant and t is the time
   D) creep strain is given by $\epsilon = Kt^{1/3}$
   E) creep rate is independent of temperature.

24 The Paris–Erdogan equation

A) relates creep rate and temperature
B) relates fatigue crack growth and ΔK
C) is applicable to tertiary creep
D) is applicable to stage 2 of a fatigue failure
E) gives the stress for void formation during a ductile failure.

25 For a material exhibiting brittle behaviour with some local plastic deformation the critical strain energy release rate is given by $G_c = 2\gamma + \gamma_p$.

A) true   B) false.

26 In fatigue the mean stress σ_m

A) is equal to twice the stress amplitude
B) is equal to one-quarter of the stress amplitude
C) does not effect fatigue life
D) has an effect on fatigue life which can be analysed by means of the Goodman relationship
E) is determined by the ease of dislocation climb.

27 This diagram shows three strain–time plots for the same material at different stresses (σ_1, σ_2, and σ_3) and different temperatures (T_1, T_2, and T_3). These plots

A) are creep curves
B) are fatigue curves
C) are conventional tensile test curves for a single crystal
D) indicate that $\sigma_1 > \sigma_2 > \sigma_3$ (assuming $T_1 = T_2 = T_3$)
E) indicate that $T_3 > T_2 > T_1$ (assuming $\sigma_1 = \sigma_2 = \sigma_3$).

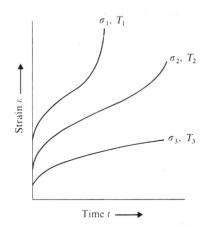

Each of the sentences in questions 28–39 consists of an assertion followed by a reason. Answer:

A) If both assertion and reason are true statements and the reason is a correct explanation of the assertion
B) If both assertion and reason are true statements but the reason is *not* a correct explanation of the assertion
C) If the assertion is true but the reason contains a false statement
D) If the assertion is false but the reason contains a true statement
E) If both the assertion and reason are false statements.

28 Elastic strain is directly proportional to the applied stress *because* the force displacement curve is parabolic near to the equilibrium spacing.

29 Single crystals are elastically anisotropic *because* the interatomic spacing, and in some cases the bonding, varies with direction.

30 Over a limited temperature range the viscosity of a solid increases exponentially with increasing temperature *because* viscous flow normally involves the thermally activated movement of atoms or molecules within the material.

31 In a tensile test on a single crystal, slip commences in the slip system with the highest Schmid factor *because* this is the system with the highest resolved shear stress for a given tensile stress.

32 Interstitial atoms cause a rapid increase in yield stress with concentration (typically $2G$–$3G$ per atom fraction, where G is the shear modulus) *because* their strain fields are symmetrical.

33 In a tensile test the true stress is always greater than the engineering stress *because* the cross-sectional area decreases with strain.

34 Brittle materials are generally tested in compression *because* barrelling of the specimen in compression enhances necking.

35 The hardness test is a slow, expensive method of assessing the mechanical properties of a material *because* the hardness is a function of the yield stress and the work-hardening rate of the material.

36 Specimens for impact testing are never notched *because* a notch introduces triaxial tensile stresses, which encourage brittle fracture.

37 All materials have a fatigue limit *because* fatigue cracks are never initiated at the surface.

38 Creep failure in polycrystalline metals is normally intercrystalline *because* the secondary creep rate is stress dependent.

39 All crack tip stresses are directly proportional to the stress intensity factor *because* the critical stress intensity factor and the critical strain energy release rate are a measure of the toughness.

Answers

1 D	2 B, D, E	3 A	4 (i) A, C
			(ii) B, D, H
5 A, C, D	6 B	7 A	8 B, C, E
9 A	10 B, C	11 A, C	12 B
13 A	14 A	15 A, D, F	16 B, C, E
17 A	18 B	19 A, D, E	20 B, C, D
21 B, D	22 A	23 B, C	24 B, D
25 A	26 D	27 A, D	28 C
29 A	30 D	31 A	32 C
33 A	34 C	35 D	36 D
37 E	38 B	39 B	

10

MICROSTRUCTURE AND PROPERTIES

10·1 Introduction

Most materials of industrial significance consist of more than one atomic or molecular species, that is, they are not single component systems. For example, the basic components of steel are iron and carbon, although there are other components (elements) present. The components may be distributed throughout the material in a variety of ways and, even under equilibrium conditions, a material may be single- or multi-phase depending on the temperature and composition. In practice, the situation is further complicated by the fact that materials may exist in metastable conditions for long periods of time without changing to the equilibrium condition. A knowledge of the microstructure of a material, which is the number of phases present, their distribution, volume fraction, shape, and size, is essential since many properties are structure-sensitive, e.g., plastic and magnetic properties. This chapter is concerned with the understanding of the formation of phases, how the stability of phases may be represented in diagrams called *equilibrium phase diagrams*, and how we can use these diagrams to interpret microstructures. Finally, at the end of the chapter we will look at the structure and properties of some commercial materials. We begin by considering what happens when two components are mixed together.

10·2 Solid solutions and intermediate phases

When two liquids A and B (which may also be liquid metals) are mixed together in varying proportions, they do so in one of the following ways:
(a) The liquids are *completely miscible* in one another over the whole composition range from pure A to pure B, i.e., at any composition a homogeneous single-phase solution is formed with the atoms or molecules of one of the components randomly dispersed in the other. The component (liquid) in excess is called the solvent and the other the solute. An example of two completely miscible liquids is water and ethyl alcohol (whisky can be diluted as much as one likes!).
(b) The liquids are only partially soluble, or miscible, in each other. Thus, if there is only a small amount of solute and a large amount of solvent, a single-phase, homogeneous solution is formed. If more solute is added the limit of solubility is reached and two solutions form, which, on standing, separate into two layers. We now have a two-phase system. Each phase (layer) will have one component as the solvent with a limited

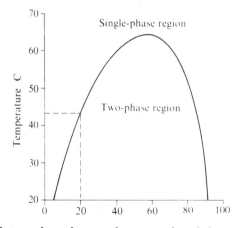

Fig. 10·1 Equilibrium phase diagram for water-phenol showing that a two-phase mixture can be transformed into a single-phase homogeneous solution by raising the temperature

quantity of the other dissolved in it. Phenol and water behave in this way at room temperature. However, if we heat-up the two-phase mixture, it will change into a single-phase, homogeneous solution at a specific temperature which depends on the overall composition. We can represent the range of temperature and composition over which the two-phase mixture is stable on a diagram of temperature against composition. Such diagrams are called *equilibrium phase diagrams* and the phenol–water diagram is given in Fig. 10·1. We can see from this diagram that if we had, say, a 20% phenol mixture, it would be two-phase at room temperature, but would change to a single-phase solution above about 43°C.

Generally, the more dissimilar are the components, both chemically and in atomic or molecular size, the more restricted is the partial solubility. Consequently, very dissimilar liquids are nearly, but not quite, completely insoluble in each other, e.g., oil and water.

The discussion so far has been concerned with liquids but the concepts introduced, and indeed the terminology, are equally applicable to solids. Thus, two components may be completely or partially soluble in each other in the solid state.

In a *solid solution* the solute atoms are distributed throughout the solvent crystal, the crystal structure of the solvent being maintained. As described in Chapter 8, the solute atoms can be accommodated in two different ways. If they occupy interstitial positions, as shown in Fig. 10·2(a), we have an *interstitial solid solution*. Alternatively, if they replace the solvent atoms as shown in Fig. 10·2(b), the resulting arrangement is called a *substitutional solid solution*. Which of these is formed depends mainly upon the relative sizes of the solvent and solute atoms. The situation is similar to the choice of structures in

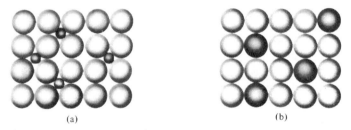

(a)   (b)

Fig. 10·2 (a) Interstitial and (b) substitutional solid solutions

ionic solids discussed in Chapter 6. In general, interstitial solid solutions can only form when the solute atom diameter is 0·6 or less of the atomic diameter of the solvent.

The distances between atoms in metals, as between ions in ionic crystals, approximately obey an additive law, each atom or ion being packed in a structure as if it were a sphere of definite size as discussed in Chapter 6. Actually, the radius of the sphere for any atom or ion is not a constant size but varies according to the number of neighbours it has, that is, it depends on the coordination number, defined in Chapter 6. There is, for example, a 3% contraction in the radius when passing from 12-fold (close-packed) to 8-fold coordination. However, for purposes of comparison, atomic radii given in books of tables are based on the size for 12-fold coordination. In Table 10·1 these radii are given; in this table the figures marked with an asterisk are elements which form crystals of very low coordination. In these cases the quoted figure is half the smallest interatomic distance. It is emphasized that Table 10·1 gives the radii of complete (neutral) atoms; where an atom has lost one or more electrons it becomes an ion and, naturally, has a smaller radius. It is ionic radii which are given in the table in Chapter 6.

Since the commercially important metals range from cobalt (1·25 Å) to lead (1·75 Å), it will be seen that the atoms which can go into interstitial solutions in these metals must have radii less than 0·75 to 1·05 Å. This effectively limits the possibilities to the first six elements, hydrogen to sulphur. It should be noted that this includes carbon and the interstitial solid solution of carbon in iron is the basis of steel.

When atoms of the two components are more nearly the same size a substitutional solid solution is formed. For complete solid miscibility the two components must have the same crystal structure, but even when this is so, if the atoms differ in size by more than 14%, the solubility will be restricted. A simple view of the thermodynamics of the situation will explain this behaviour. As solute is added to the solvent the degree of disorder increases, therefore the entropy of the material increases (see Section 7·16). This increase in entropy is independent of the sizes of the atoms and is determined solely by the concentration. At the same time, the solute atoms introduce elastic strains into the solvent lattice, i.e., some of the solvent atoms are slightly displaced from their equilibrium positions by an amount which depends on the relative sizes of the solute and solvent atoms. This gives an increase in the enthalpy which is dependent on atomic size–the greater the size difference the larger the enthalpy increase. Now, increases in enthalpy and entropy act in opposite sense on the free energy as $\Delta G = \Delta H - T\Delta S$ [Eqn (7·23)]. For low solute concentrations the entropy term outweighs the enthalpy leading to a decrease in free energy and the formation of a stable, homogeneous solid solution. However, at higher concentrations the entropy does not increase so rapidly with solute content and the enthalpy predominates. This results in the single-phase solid solution becoming energetically unfavourable compared with a two-phase mixture as the concentration is increased.

Table 10·1
Atomic radii

Element	Radius (Å)	Element	Radius (Å)	Element	Radius (Å)
H	0·46	Ir	1·35	Mg	1·60
O	0·60	V	1·36	Ne	1·60
N	0·71	I	1·36†	Sc	1·60
C	0·77	Zn	1·37	Zr	1·60
B	0·97	Pd	1·37	Sb	1·61
S	1·04†	Re	1·38	Tl	1·71
Cl	1·07†	Pt	1·38	Pb	1·75
P	1·09†	Mo	1·40	He	1·79
Mn	1·12†	W	1·41	Y	1·81
Be	1·13	Al	1·43	Bi	1·82
Se	1·16†	Te	1·43†	Na	1·92
Si	1·17†	Ag	1·44	A	1·92
Br	1·19†	Au	1·44	Ca	1·97
Co	1·25	Ti	1·47	Kr	1·97
Ni	1·25	Nb	1·47	Sr	2·15
As	1·25†	Ta	1·47	Xe	2·18
Cr	1·28	Cd	1·52	Ba	2·24
Fe	1·28	Hg	1·55	K	2·38
Cu	1·28	Li	1·57	Rb	2·51
Ru	1·34	In	1·57	Cs	2·70
Rh	1·35	Sn	1·58	Rare ⎱	⎰ 1·73 to
Os	1·35	Hf	1·59	Earths ⎰	⎱ 2·04

†Estimated from half the interatomic distance in the pure material.

We not only have to consider crystal structure and atomic size when examining the extent of solid solubility, but also valency. When the valencies of the components differ markedly there will be a tendency, as mentioned in Chapter 6, to form *intermediate or intermetallic compounds*. These compounds are normally formed at, or near, compositions corresponding to a simple ratio of components, such as AB, AB_2, and may exist over a composition range around the simple ratio. They often have a crystal structure which differs from that of either of the components.

The effects of atomic size and valency are well illustrated by the equilibrium phase diagrams for silver (f.c.c.) alloyed with the f.c.c. elements gold, copper, and aluminium. Silver and gold are completely miscible in each other as they have the same atomic sizes and are both monovalent. Copper is also monovalent, but the Cu and Ag atoms differ in size by about 11% with the result that they are only partially soluble in each other. Aluminium has atoms of similar size to silver but is trivalent; thus intermediate compounds are formed in the Ag–Al system.

10·3 Equilibrium phase diagrams

In the previous section we have seen that the number of phases present in a material may be affected by the temperature and the composition, and that the regions of phase stability can be represented on an equilibrium phase diagram. In this section the more important features found in the equilibrium phase diagrams for solids will be described. Furthermore, we will show how these diagrams may be used (1) to determine the relative amounts and compositions of the phases and, (2) sometimes to deduce the morphology of the phases present.

In referring to compositions in this chapter, we shall normally quote the percentage by weight of each component, sometimes abbreviated to wt %. Occasionally, however, it will be useful to refer to the percentage by number of atoms, or atomic percent (at %).

In the following it is assumed that there is complete miscibility in the liquid phase, although there are some notable exceptions to this, e.g., Al–Pb, Zn–Pb.

10·3·1 Complete solid miscibility

It has already been mentioned that it is possible to form a continuous range of solid solutions between two components, e.g., Ag–Au, Cu–Ni, NiO–MgO and now we want to study the equilibrium phase diagram for this situation, termed complete solid miscibility. Generally, information from a number of experimental techniques is brought together in order to construct a phase diagram. However, *thermal analysis* is the most widely used technique in the determination of phase diagrams. Thermal analysis is the monitoring of the heat that is evolved during an *exothermic* transformation or absorbed during an *endothermic* transformation. The simplest form of thermal analysis is the recording of heating and cooling curves; these curves exhibit a change in slope or a plateau when a transformation is occurring. More sensitive to small thermal changes are the techniques of *differential thermal analysis* (DTA) and *differential scanning calorimetry* (DSC). In DTA the temperature of the specimen under investigation is compared with that of an inert standard as the temperature is raised or lowered at a controlled rate. DSC also uses an inert standard but instead of recording temperature differences the heat evolved or absorbed during a transformation is quantified. We will employ the simple cooling curve technique in our study of phase diagrams, although DTA plots will be presented for comparison in the discussion on the eutectic system. First, let us look at the cooling curves for some alloys in the Cu–Ni system and from the data obtained construct the equilibrium phase diagram.

Referring to Fig. 10·3(a), the cooling curve for pure copper contains a flat portion, AB, during which complete solidification takes place at constant temperature with liberation of latent heat. When 20 wt % nickel is added the flat portion is no longer present but there is a transition region of temperature, A_1B_1, with completely liquid mixture at A_1 going to complete solid at B_1. At temperatures in between there is a mixture of solid and liquid which will remain in equilibrium so long as the temperature is held constant. With 60 wt % nickel similar behaviour occurs over the transition region A_2B_2, but when 100% nickel is reached the transition from liquid to solid, A_3B_3, is again flat, occurring at a fixed temperature. From these results the points A, A_1, A_2, and A_3, which correspond to the lowest temperatures at which the solution is entirely liquid, can be plotted as in Fig. 10·3(b). This line is called the *liquidus* curve. Similarly, a curve through the points B defines the highest temperature at which the solution is a solid and is called the *solidus* curve. The points B are not well defined on cooling curves and the solidus is usually determined from heating curves. The resulting graph is the equilibrium phase diagram for Cu–Ni and is typical for a system exhibiting complete solid miscibility.

Not only does the equilibrium phase diagram tell us how many phases are present in a given material at a particular temperature, but also the compositions and relative proportions of the phases. For example, at temperature T_2 an alloy of X % Ni consists of liquid and solid in equilibrium. To obtain the compositions of these phases a horizontal, isothermal line is drawn, called a *tie-line*, and the intercepts of the tie-line with the phase boundaries, i.e., the solidus and liquidus, give the relevant compositions, namely liquid C' % Ni and solid D' % Ni. The proportions by weight of the two phases are given by the *lever rule* and are in the ratio of the lengths CE and ED. In fact:

weight of solid × ED = weight of liquid × CE

The rule can be stated in the following way. Let the point representing the composition and temperature be the fulcrum of a horizontal lever. The lengths of the lever arms from the fulcrum to the boundaries of the two-phase field multiplied by the weights of the phases present must balance. We can

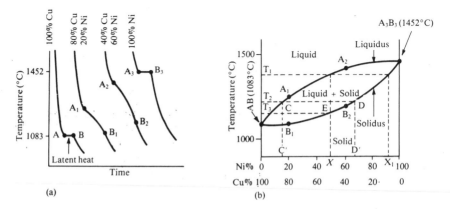

Fig. 10·3 (a) Cooling curves and (b) equilibrium phase diagram for the copper–nickel system

take our alloy of composition $X\%$ Ni and derive the lever rule from first principles. Let the weight of the alloy be W_A and the weights of the liquid of composition C' % Ni and the solid of D' % Ni at temperature T_2 be W_L and W_S respectively. Clearly the weight of the alloy W_A is just the sum of the weights of the liquid and solid phases, i.e.

$$W_A = W_L + W_S$$

Similarly, the weight of nickel in the alloy is the sum of the weight of nickel in the two phases, therefore,

$$\frac{X}{100} W_A = \frac{C'}{100} W_L + \frac{D'}{100} W_S$$

Combining these two equations gives

$$W_S(D' - X) = W_L(X - C')$$

which is

$$W_S \times ED = W_L \times CE$$

Now let us look in detail at the solidification of the alloy containing X % Ni, which will be typical of the solidification of any alloy in the system. Solidification begins at temperature T_1 and the appropriate tie-line shows that the first solid is of composition X_1 % Ni. If a series of tie-lines is drawn at lower temperatures in the two-phase region, it can be seen that the proportion of solid increases with decreasing temperature, and that the composition of the solid forming at any given temperature changes with temperature along the solidus. Correspondingly, the composition of the remaining liquid follows the liquidus. When the cooling-rate is slow enough to maintain equilibrium, the solid formed earlier changes composition by diffusion so that at any temperature all the solid is of the same composition. Thus, when the temperature has fallen to T_2 all the solid is of composition D' % Ni and the remaining liquid is of composition C' % Ni. Solidification continues in this manner and is completed at temperature T_3. The microstructure of the alloy so formed will consist of grains, or crystallites, of homogeneous solid solution of composition X % Ni.

It is common for the cooling rate to be too rapid for the composition of the solid to change significantly by diffusion during solidification. This results in a non-equilibrium structure in which the composition varies throughout the grains, i.e., the grains are heterogeneous. The grains are said to be *cored*. Figure 10·4 is a micrograph of a cored solid solution; the variations in composition are shown by the different tones in the micrograph within each grain.

Cored structures are common in practice as the rate of cooling in chill casting, and even with sand casting, is generally much too rapid for equilibrium to be maintained. A cored structure will generally persist indefinitely at room temperature; it is therefore a *metastable* structure. However, it will revert to the equilibrium condition of homogeneous grains when annealed at an elevated temperature.

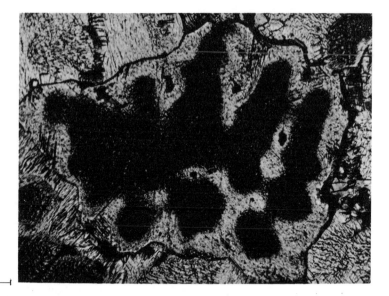

75 μm

Fig. 10·4 Cored grains in a chill cast Cu – 5% Sn solid solution

10·3·2   Partial solid miscibility

Although in many systems the components are completely soluble in the liquid state, often they are only partially miscible in each other in the solid state, e.g., Cu–Al and MgO–CaO. Partial solid miscibility results in two distinct types of equilibrium phase diagram, namely the eutectic and the peritectic.

(a) *Eutectic.* To study the eutectic phase diagram, we will use the same approach as was used for complete solid miscibility, namely, we will examine the cooling curves [Fig. 10·5(a)] for a number of alloys in a particular system, lead–tin, and from these plot the liquidus and solidus. For comparison DTA plots, that is, graphs of the temperature difference between the sample under investigation and an inert standard as a function of temperature, are also given for two of the alloys [Fig. 10·5(c)]. The DTA plots show distinct minima corresponding to the endothermic melting processes that occur on heating.

The distinctive feature of the cooling curves of Fig. 10·5(a) is the *arrest* shown by a number of the alloys at about 180°C. The equilibrium phase diagram constructed from the thermal analysis data is given in Fig. 10·5(b). In this diagram α and β are solid solutions of tin in lead and lead in tin respectively. It can be seen that over a considerable composition range the solidus is horizontal, corresponding to the arrests at 180°C. The temperature of the horizontal portion of the solidus is called the *eutectic temperature* and the composition at which the liquidus meets this section of the solidus is the *eutectic composition*, C_e. It is important to note that the eutectic composition is very unlikely to be the equiatomic (50%–50%) composition, but varies from system to system, being about 62 wt % Sn ($\sim$ 73 at % Sn) in this case.

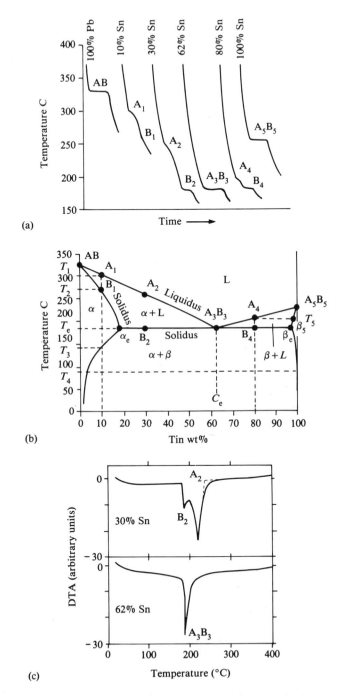

Fig. 10·5 Use of thermal analysis for the determination of the phase diagram for the lead–tin system: (a) cooling curves; (b) phase diagram; (c) DTA (heating) plots

Let us follow the solidification of some alloys in the Pb–Sn system. The first alloy we will consider is Pb–10 wt % Sn, which is typical of any alloy containing less tin than the composition shown as α_e in Fig. 10·5(b). Under equilibrium conditions this alloy will solidify in the manner already described for complete solid miscibility, i.e., solidification will start at temperature T_1 and finish at T_2, giving homogeneous grains of α solid solution. The interesting point about this alloy is that as it cools in the solid state, the *solubility limit* of tin in lead is reached at temperature T_3. Therefore, at temperatures below T_3 the β phase begins to be *precipitated* out of the solid solution. At T_4 the compositions of the α and β phases which will exist in equilibrium are those at the *boundaries* of the α phase region and the β phase region, namely Pb–4 wt % Sn and virtually 100% Sn respectively. The proportions will be given by the lever rule, viz.

$$\text{(weight of } \alpha)(10 - 4) = \text{(weight of } \beta)(100 - 10)$$

i.e.   (weight of α)(6)   = (weight of β)(90)

Thus, the amount of β precipitate is $[6/(90 + 6)]100\% \approx 6\%$.

Any alloy with more tin than the composition β_e will behave in a similar manner to the Pb–10 wt % Sn alloy except that β solid solution forms first and α precipitates out if the solid solubility limit is exceeded.

As mentioned in the previous chapter, the significance of precipitation is that in some alloy systems it can lead to strengthening. This will be discussed further in the section on commercial alloys.

The phase diagram and the cooling curve show that an alloy of the eutectic composition completely solidifies at the eutectic temperature T_e. At the eutectic temperature all the liquid, which is of course of the eutectic composition C_e, solidifies into an intimate mixture of the α and β phases. The compositions of the α and β phases in the eutectic mixture are α_e and β_e respectively, so the eutectic reaction may be written as

$$L_{C_e} \rightleftharpoons (\alpha_e + \beta_e) \tag{10·1}$$

where L_{C_e} represents the liquid.

As the temperature falls below T_e the compositions of the phases will, under equilibrium conditions, change by solid state diffusion. When temperature T_4 is reached the compositions of the phases in equilibrium will be the same as for the Pb–10 wt % Sn alloy but the proportions will be different. There will be far more β in the eutectic alloy. According to the lever rule we have, at T_4,

$$\text{(weight of } \alpha)(62 - 4) = \text{(weight of } \beta)(100 - 62)$$

i.e.   (weight of α)(58)   = (weight of β)(38)

so the amount of β present in $[58/(38 + 58)]100\% \approx 60\%$.

It will be noted that the eutectic alloy has the lowest melting point, much lower than the melting points of the pure components. For this reason the eutectic alloy is the basis of soft solder, which is used in the electrical industry for joining wires in circuits.

Now consider an alloy lying between the compositions C_e and β_e, say one containing 80 wt % Sn. Solidification will start at temperature T_s with the formation of the solid, having composition β_s [Fig. 10·5(b)]. Solidification down to the eutectic temperature is similar to that described for complete solid miscibility, i.e., as the temperature falls the amount of solid β increases and the compositions of the solid and liquid follow the solidus and liquidus respectively. Therefore at T_e we have *primary* β solid of composition β_e and the remaining liquid is of composition C_e. This liquid then undergoes the eutectic reaction, hence

$$\beta_e + L_{C_e} \rightleftharpoons \beta_e + (\alpha_e + \beta_e) \tag{10·2}$$

The microstructure of the alloy consists of crystallites of primary β within a background, called the *matrix*, of the eutectic mixture. This characteristic microstructure of a tin-rich alloy is shown in Fig. 10·6(a).

Alloys in the composition range α_e to C_e solidify in an analogous manner but with α as the primary phase. Figure 10·6(b) is a micrograph of an alloy containing 40 wt % Sn, showing the primary α phase (black) in a eutectic mixture matrix. The exact form of the primary phase, and of the eutectic mixture, varies from system to system. For example the primary phase in antimony-rich alloys of the lead–antimony system has cubic shape (Fig. 10·7) while the eutectic in the copper–phosphorus system is more lamellar (plate-like) than the eutectics shown in Figs. 10·6 and 10·7 (see the micrograph of Fig. 10·8).

(b) *Peritectic.* As we are now more familiar with the construction and use of equilibrium phase diagrams, we need not spend so much time on the peritectic diagram. A generalized peritectic diagram is shown in Fig. 10·9 and, as for the eutectic, it can be seen that a section of the solidus is horizontal. The temperature corresponding to this section is the *peritectic temperature*, T_p. An example of a simple peritectic phase diagram occurs in the Co–Cu system. The solidification and precipitation behaviour of an alloy

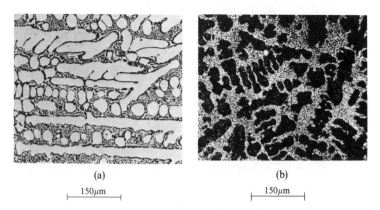

(a)   (b)

⊢ 150μm ⊣   ⊢ 150μm ⊣

Fig. 10·6 Microstructures of lead–tin alloys: (a) tin-rich (Pb – 70% Sn); (b) lead-rich (Pb – 40% Sn). (From J. Nutting, and R.G. Baker, *The microstructure of metals*, Institute of Metals, London, 1965)

50μm

Fig. 10·7 Antimony-rich lead–antimony alloy, showing a cubic primary phase in a eutectic mixture matrix

40μm

Fig. 10·8 The eutectic mixture in the copper–phosphorus system

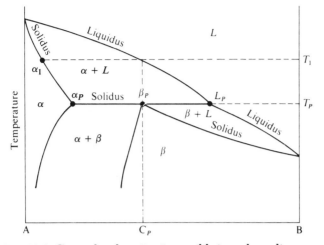

Fig. 10·9 Generalized peritectic equilibrium phase diagram

lying between pure A and α_p is the same as that for alloys between O and α_e in the eutectic diagram of Fig. 10·5(b). Furthermore, alloys in the range L_p to pure B solidify in an identical manner to an alloy in a system that exhibits complete solid miscibility. Therefore, only between the compositions α_p and L_p do we encounter new solidification behaviour.

First let us consider an alloy of the *peritectic composition* C_p. Solidification commences at T_1 with the formation of primary α of composition α_1. Solidification of α continues until T_p is reached when the solid will be of composition α_p and the remaining liquid L_p. At the peritectic temperature, all the liquid reacts with all the α_p to form solid β of composition $\beta_p = C_p$, that is

$$L_p + \alpha_p \rightleftharpoons \beta_p \qquad (10\cdot3)$$

This is the *peritectic reaction*.

The solidification of an alloy of composition between α_p and β_p is similar; at T_p we again have L_p and α_p but in different proportions (more α_p present). This time all the liquid reacts with some of the α_p to form solid β_p

$$L_p + \alpha_p \rightleftharpoons \alpha_p + \beta_p \qquad (10\cdot4)$$

Thus the microstructure consists of primary α surrounded by a partial, or complete, network of the β phase.

Finally, for an alloy which lies in the composition range β_p to L_p, the reaction that occurs at the peritectic temperature is

$$L_p + \alpha_p \rightleftharpoons L_p + \beta_p \qquad (10\cdot5)$$

In other words, some of the liquid reacts with all of the α_p to give solid β_p. After this reaction the remaining liquid solidifies as β phase and, under equilibrium conditions, the final structure will consist of homogeneous grains of β solid solution.

It must be emphasized that in this section on partial solid miscibility we have been dealing with solidification under ideal equilibrium conditions. Fast rates of cooling, which are often found in practice, can result in (1) coring, as described in Section 10·3·1, (2) modifications of the form of the phases, and (3) changes in the relative proportions of the phases.

10·3·3 Eutectoid reaction

So far we have concentrated on transformations involving a liquid phase, the only solid state transformation we have discussed being the precipitation that takes place when the solid solubility limit is exceeded. There is, however, a number of structural changes that occur in the solid state, e.g., those called eutectoid, peritectoid, and miscibility gap. We will study the eutectoid reaction as a knowledge of the reaction is essential if the heat-treatment and microstructures of steels are to be understood.

The copper–aluminium system exhibits a eutectoid reaction and the appropriate section of the phase diagram is given in Fig. 10·10, where T_E and C_E are the *eutectoid temperature* and the *eutectoid composition* respectively. The similarity between the eutectoid and eutectic transformations is obvious when Fig. 10·10 is compared with 10·5(b). The important difference is that all the phases involved in the eutectoid reaction are solid, either solid solutions or intermediate phases (in the case of Cu–Al, β and γ are intermediate phases and α a solid solution).

Consider cooling the eutectoid alloy, Cu–11·8 wt % Al, from the β phase field. The alloy will exist as single phase β until the eutectoid temperature T_E is reached. At T_E the following reaction occurs by solid state diffusion

$$\beta_E \rightleftharpoons \alpha_E + \gamma_E \tag{10·6}$$

That is, the β phase decomposes into an intimate mixture, called the *eutectoid mixture*, of the α and γ phases. The eutectoid mixture has a characteristic lamellar morphology as shown in Fig. 10·11(a). The proportion of α_E and γ_E are

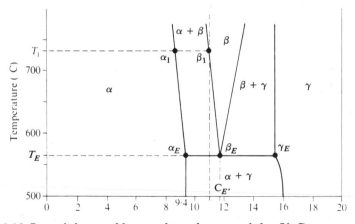

Fig. 10·10 Part of the equilibrium phase diagram of the Al–Cu system which displays a eutectoid reaction

(a)

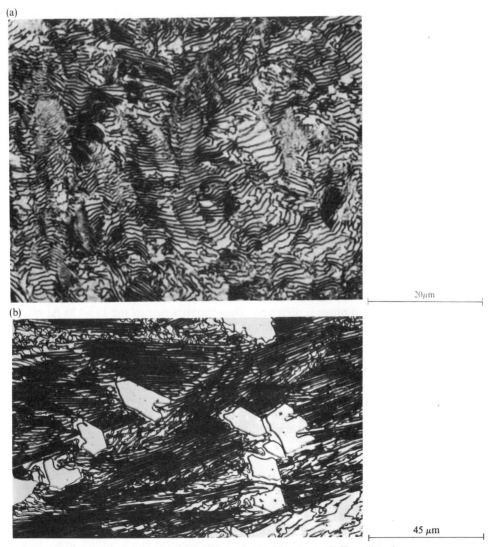

20 μm

(b)

45 μm

Fig. 10·11 The microstructure of: (a) a Cu – 11·8% Al eutectoid alloy; (b) a hyper-
eutectoid alloy

$$\text{(weight of } \alpha)(11\cdot8 \ - \ 9\cdot4) = \text{(weight of } \gamma)(15\cdot6 \ - \ 11\cdot8)$$

i.e.,   (weight of α)(2·4)   = (weight of γ)(3·8)

therefore the percentage of α present is $3\cdot8/(3\cdot8 + 2\cdot4)100 \approx 61\%$.

Let us now turn our attention to the structural changes that occur during the cooling of an alloy removed from the eutectoid composition, e.g., a Cu–11 wt % Al alloy. The boundary of the β phase field is cut at temperature T_1 and α, of composition α_1, begins to be precipitated. As the temperature falls more and more α is precipitated, the α and β phases changing composition by diffusion down the curves $\alpha_1 - \alpha_E$ and $\beta_1 - \beta_E$ respectively. Hence, when the eutectoid temperature is reached we have α_E and β_E. The β_E then decomposes by the eutectoid reaction

$$\alpha_E + \beta_E \rightleftharpoons \alpha_E + (\alpha_E + \gamma_E) \tag{10.7}$$

and the final structure consists of pro-eutectoid α (that which existed before the eutectoid reaction) in a matrix of the eutectoid mixture. The amount of α present, counting both pro-eutectoid and eutectoid α, is clearly greater than for the eutectoid alloy. In fact

$$(\text{weight of } \alpha)(11.0 - 9.4) = (\text{weight of } \gamma)(15.6 - 11.0)$$

i.e.,   (weight of α)(1.6)   = (weight of γ)(4.6)

giving approximately 74%, by weight, of α.

Alloys, such as the 11 wt % Al alloy, which contain less solute than the eutectoid composition are termed *hypo-eutectoid*. If an alloy contains more solute than the eutectoid composition, it is said to be a *hyper-eutectoid*. (The prefixes 'hypo-' and 'hyper-' may also be used in connection with the eutectic transformation.) The structural changes in a hyper-eutectoid alloy, e.g., Cu–13 wt % Al, are similar to those in a hypo-eutectoid alloy except that the pro-eutectoid phase is different. Thus the structure of the 13 wt % Al will be pro-eutectoid γ in a eutectoid matrix [Fig. 10.11(b)].

We have now considered the more important features of phase diagrams in some detail. Some systems, for example copper and zinc (i.e., brass), show several different phases with variations of temperature and composition and the equilibrium phase diagram appears very complex indeed (Fig. 10.37). However, the principles discussed above can be applied to all diagrams, and the compositions and proportions of the various phases present under equilibrium conditions for any alloy composition can be determined at any temperature.

10.4   Free energy and equilibrium phase diagrams

In Chapter 7 we learned that the equilibrium condition of a material was that it should have the lowest possible free energy. We have also seen that equilibrium phase diagrams are a means of representing the equilibrium condition of a material as a function of temperature and composition. It follows that there must be a close connection between the temperature and composition dependences of the free energy and equilibrium phase diagrams. This relationship will be demonstrated for the complete solid miscibility and eutectic systems. Before we do this, however, we must define a new thermodynamic parameter called the *chemical potential* μ.

Consider the case of two components A and B which form a solid solution α. When A and B are placed in contact, they spontaneously dissolve in one another, because the free energy of the solution is lower than the sum of the free energies of the components. It is possible to define a quantity called *chemical potential* which measures the tendency of component A for chemical change when in the phase α formed by A and B. A similar chemical potential can be defined for component B. In fact the chemical potential is the free energy of a component (e.g., A or B) in a phase (e.g., α or β) and is also known as the partial molar free energy. Now consider a case in which A and B form two phases, α and β, and let these be placed in contact with one another. If α and β are in equilibrium then the free energy of the mixture must be a minimum with respect to changes in the compositions of α and β, and the chemical potentials of A and B in the phase α are equal to the respective chemical potentials in the phase β. If, on the other hand, the chemical potentials differ, the mixture is not in equilibrium, and its free energy can be reduced by a rearrangement of the components A and B between the phases. Thus component A is redistributed until its chemical potential is the same in all the phases present, i.e., equilibrium is reached. Similarly for component B. This may result in a reduction in the number of phases (e.g., $\alpha + \beta \rightarrow \alpha$) or further phases may become involved (e.g., $\alpha + \beta \rightarrow \alpha + \gamma$).

Figure 10·12 is a curve of free energy against composition for two components, A and B, which are completely soluble in each other. It can be shown that the intercepts on the free energy axis of the tangent at any composition C give the chemical potentials, μ_A and μ_B, of the components A and B in the solution of composition C. This concept is true for all free energy curves and is basic to the following discussion.

The equilibrium phase diagram for a system exhibiting complete solid miscibility is given in Fig. 10·13(a). A number of temperatures are marked on the phase diagram and the corresponding free energy curves for the liquid and solid phases at these temperatures are given in Fig. 10·13(b), (c), (d), etc. At the temperature T_1, the free energy curve for the liquid is lower than for

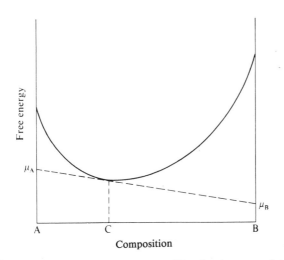

Fig. 10·12 Free energy curve versus composition for two completely miscible components. The tangent at C gives the chemical potentials of the components A (μ_A) and B (μ_B) in a solution of composition C

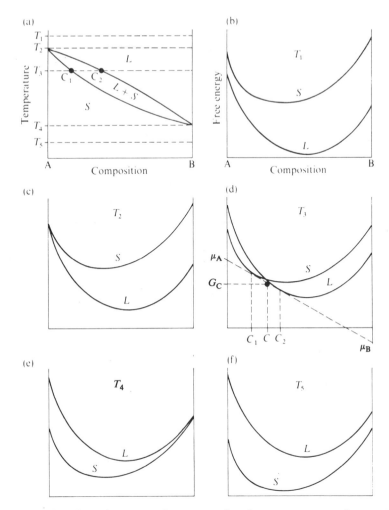

Fig. 10·13 The phase diagram and corresponding free energy curves for a system exhibiting complete solid miscibility

the solid, therefore the equilibrium state is liquid over the complete composition range A to B. Temperature T_2 is the melting point of component A, and we see that at this temperature solid A and liquid A have the same free energy. The free energy curves cross at temperature T_3 and a common tangent has been drawn to the two curves at compositions C_1 and C_2. This means that the chemical potentials of components A and B in solid of composition C_1 are equal to those in liquid of composition C_2, i.e., these two phases are in equilibrium. Thus any alloy of composition between C_1 and C_2, such as alloy C, is two-phase and its free energy, G_c, lies on the tangent and is lower than the free energy of liquid, or solid, of the same composition. At T_4, which is the melting point of component B, the free energies of solid B and liquid B are the same. For all other compositions the solid has the lower free energy and so is the equilibrium phase. Finally, at T_5 the free energy curve for the solid is below that for the liquid at all compositions, hence only solid can exist under equilibrium conditions.

If two components have the same crystal structure and yet are only partially miscible in the solid state, it is a consequence of the free energy curve for the solid having two minima as shown in Fig. 10·14(a). Alternatively, when two components have different crystal structures, and so are only partially miscible, we have two superimposed free energy curves, which effectively give a curve with two minima [Fig. 10·14(b)]. The analysis that follows, which is based on a free energy curve for the solid with two minima, is therefore applicable to either type of system.

The phase diagram and free energy curves at various temperatures for a eutectic are given in Fig. 10·15. Liquid has the lower free energy for all compositions at temperature T_1 and is therefore, as shown by the phase diagram, the equilibrium phase over the complete composition range A to B. At temperature T_2 there is a common tangent to the solid and liquid curves at C_1 and C_2 respectively. Consequently, the equilibrium state of an alloy between compositions A and C_1 is solid α, between C_1 and C_2 is two-phase solid plus liquid, and between C_2 and B is liquid. At lower temperatures, e.g., T_3, the free energy curve for the solid cuts the curve for the liquid twice, giving two common tangents corresponding to the solid α + liquid and solid β + liquid phase fields. There is only one tangent at T_e showing that, at the eutectic temperature, liquid of composition C_e is in equilibrium with solid of composition C_7 and C_8. Finally, at T_4 no liquid is present under equilibrium conditions and the common tangent touches the curve for the solid at C_9 and C_{10}. Thus at all compositions between C_9 and C_{10} a two-phase structure exists consisting of solid of compositions C_9 and C_{10}.

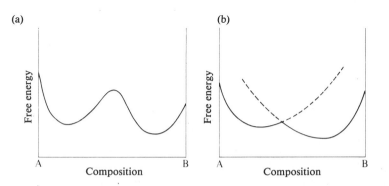

Fig. 10·14 The free energy curves that lead to partial solid miscibility; (a) components A and B have the same crystal structure; (b) components A and B have different crystal structures

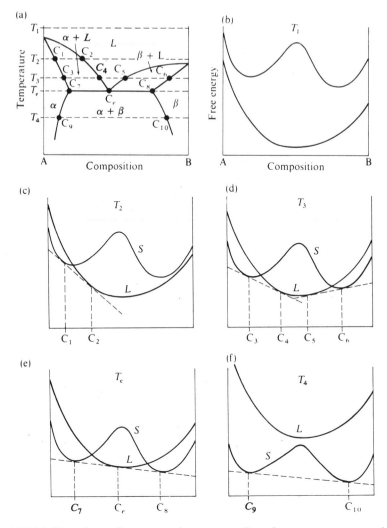

Fig. 10·15 The phase diagram and corresponding free energy curves for a eutectic system

10·5 Nucleation and growth

We have seen that when the temperature of a material is altered a phase transformation may occur. The transformation may involve only a change in structure, examples being solidification and allotropic changes in a pure element such as iron. In other cases, however, the transformation may result in a change in composition as well as structure, e.g., the precipitation that takes place when the solid solubility limit is exceeded. Both types of transformation commonly proceed by the *nucleation and thermally activated growth* mechanism.

As the name suggests, the nucleation and thermally activated growth mechanism can be conveniently divided into two stages, namely a nucleation stage and a growth stage. The first stage is *nucleation*, which is the formation of small grains, or *nuclei*, of the new phase, each just a few atoms in size, in the old phase. *Growth* of these nuclei then occurs by material being transferred, generally by diffusion, from the old phase and into the new phase.

In practice a phase transformation does not always take place exactly at the 'transition temperature', for example, liquids may undercool before solidification commences, and precipitation may not occur immediately the solid solubility limit is exceeded. Let us look at the latter in more detail; the relevant part of the phase diagram showing the solid solubility limit is given in Fig. 10·16. When an alloy of composition C is rapidly quenched to room temperature from the α phase field, e.g., from temperature T_S, the precipitation of β is suppressed and a metastable state is attained which is called a *supersaturated* solid solution. This procedure is called solution treatment. If the metastable solid solution is then heat-treated at an elevated temperature T_A in the two-phase region, precipitation of the β phase occurs and the alloy reverts to the equilibrium two-phase condition. This latter treatment, termed *aging*, can be carried out at various temperatures below the 'transition temperature' T_T and this affects the final microstructure. A quantitative thermodynamical analysis for the microstructural dependence on the aging temperature is possible, but complex, and a simple qualitative approach will suffice for our purposes.

Over a normal range of aging temperatures, the rate of nucleation increases with decreasing temperature. In contrast, the growth rate, because it is generally diffusion controlled, is slower the lower the aging temperature. As a result, the temperature dependence of the *overall transformation rate*, which is a function of the nucleation and growth rates, is as shown in Fig. 10·17. Near to T_T, say at T_1, the overall transformation rate is slow, even though the growth rate is fast, because the nucleation rate is low. Aging at this temperature will produce a coarse dispersion of β particles as only a few

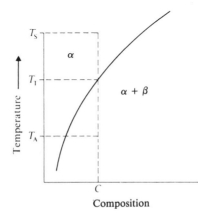

Fig. 10·16 Part of a phase diagram showing the solid solubility limit and the solution treatment (T_S) and aging (T_A) temperatures

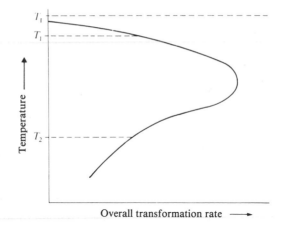

Fig. 10·17 Overall transformation rate as a function of temperature for a transformation occurring by the nucleation and growth mechanism

nuclei are formed and these grow rapidly. At much lower temperatures, such as T_2, the nucleation rate is high but the growth rate very low, again giving a slow overall transformation rate. This time, however, due to the many nuclei present and the low growth rate, aging produces a fine dispersion of the β phase.

The overall transformation rate curve of Fig. 10·17, and our deductions from it, are applicable to all nucleation and thermally activated growth processes, thus for solidification the temperature T_T in Fig. 10·17 is the melting point of the material. We can now understand why rapidly cooled liquids generally produce a finer microstructure than slowly cooled liquids. Rapid cooling causes considerable undercooling resulting in a large number of solid nuclei, small in size and densely packed.

If, instead of plotting rate against temperature, we plot the start and finish times for a transformation at various temperatures on a temperature-logarithm (time) graph we obtain a curve which is a mirror image of the overall transformation curve (Fig. 10·18). Such a curve is known as a Time–Temperature–Transformation (TTT) diagram and is commonly employed to present the temperature dependence of a transformation; see, for example, the discussion later in this chapter on steels.

Up to now we have not discussed where the nuclei appear in a material undergoing a transformation. When the nucleation is completely at random throughout the material, it is said to be *homogeneous*. Homogeneous nucleation rarely occurs in practice as it is energetically easier for nucleation to commence at structural imperfections. For example, during solidification nucleation takes place at the mould walls and on solid impurities in the melt, whilst dislocations, non-metallic inclusions, and grain boundaries are preferred sites for nucleation of precipitate particles in a solid. Nucleation on structural imperfections is called *heterogeneous* nucleation and is the rule rather than the exception.

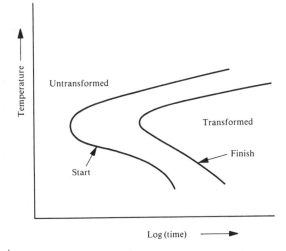

Fig. 10·18 Time–temperature–transformation diagram

10·6 Martensitic transformation

Phase transformations that occur by the nucleation and thermally activated growth mechanism proceed relatively slowly, as shown by the fact that we can prevent the decomposition of a solid solution by rapidly quenching through the solid solubility limit. Furthermore, these transformations usually, but not always as we will see in the following section, result in the formation of the equilibrium phase or phases. There is, however, another type of transformation, called the *martensitic transformation*, which is very rapid and often produces a metastable phase.

In contrast to the nucleation and growth transformation, no diffusion is involved in the martensitic transformation. Instead, it occurs by the systematic coordinated shearing of the lattice of the old phase in such a way that the distance moved by any atom is less than one atomic spacing. This means that an atom retains the same neighbours; hence a martensitic transformation can only lead to changes in crystal structure, and not in composition, of the phases.

The electron micrograph of Fig. 10·19 shows the typical plate-like microstructure formed by a martensitic transformation. Note also the large number of dislocations (the thin black lines) within the plates. The time taken for an individual martensite plate to grow to its final size may be less than 10^{-14} s, even at low temperatures, which confirms that diffusion cannot be involved. The martensitic transformation is therefore extremely rapid and it is impossible to prevent it by rapid quenching. We will come back to the martensitic transformation when we discuss the heat-treatment of steels in the following section.

$2\mu m$

Fig. 10·19 Electron micrograph showing martensite plates with high dislocation
density in a Ti – 1% Si alloy

10·7   Some commercial alloy systems

10·7·1   Iron–carbon system (cast iron and plain carbon steels)

Because of its considerable importance in engineering, the iron–carbon
system has received much detailed study. The iron–carbon, or more properly
the iron–iron carbide phase diagram, is given in Fig. 10·20.

The carbon atom is smaller than the iron atom (see Table 10·1) and dis-
solves interstitially in all three phases, (α, γ, and δ), of iron. The solubility in
f.c.c. γ-iron is at a maximum near 2·0 wt % at 1,130°C, the solid solution
being known as *austenite*. The solubilities in the b.c.c. phases are much
smaller, the maxima being 0·1 wt % at 1,492°C in δ-iron and 0·03 wt % at
723°C in α-iron, the latter being called *ferrite*.

Iron and carbon form an intermediate compound, iron carbide, which
contains 6·67 wt % carbon. This has the formula Fe_3C, is called *cementite*,
and is extremely hard and brittle. Cementite is involved in both of the important transformations that occur in the iron–iron carbide system. First, as
shown by the phase diagram (Fig. 10·20), there is a eutectic reaction at
1,130°C where liquid of composition 4·3 wt % C solidifies into a eutectic
mixture of austenite and cementite. The eutectic mixture is known as
ledeburite. The other important transformation is the eutectoid transformation which occurs at 723°C; austenite of composition 0·8 wt % C decomposes
into a eutectoid mixture of ferrite and cementite, called *pearlite*.

Let us examine the microstructure and properties of some ferrous alloys.
In this discussion we shall first consider *plain carbon steels and irons* then

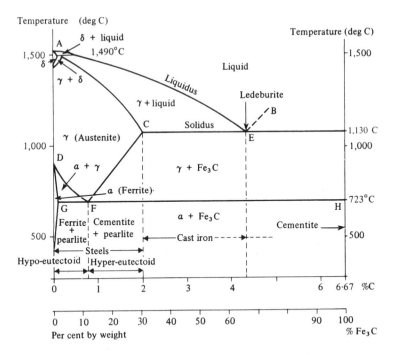

Fig. 10·20 Phase diagram for the iron–iron carbide system

later steels containing alloying additions such as nickel, chromium, tungsten, and molybdenum.

Pig-iron is the product of the blast furnace and has a carbon content in the range 2–4% but generally about 3·5 wt %. *Wrought iron* is made by refining pig-iron and it contains as little as 0·05 wt % C and small amount of refining slag, principally iron silicate. When pig-iron is just remelted, but not refined, and cast in its final form it is known as *cast iron*. The microstructure, and hence properties, of cast iron vary with the rate of cooling and the presence of elements such as sulphur, manganese, and silicon.

When the rate of cooling is fast, nearly all the carbon in a cast iron exists as cementite. This makes the material hard and brittle and it is only used in applications where hardness and wear resistance are important, e.g., grinding and crushing machinery. The fracture surface of such iron has a white, or silvery, appearance and consequently it is called *white cast iron*.

Most white cast irons are hypo-eutectic and a typical microstructure is given in Fig. 10·21. This microstructure is best understood by following the phase transformations that occur as the iron is cooled. Primary austenite is formed in increasing amounts until the eutectic temperature is reached, at which point the remaining liquid solidifies as the eutectic mixture, ledeburite. Using the lever rule in conjunction with the phase diagram, it can be seen that the proportion of cementite increases as the temperature falls from the eutectic temperature to the eutectoid temperature, i.e., precipitation of cementite occurs between 1,130° and 723°C. On cooling to below the

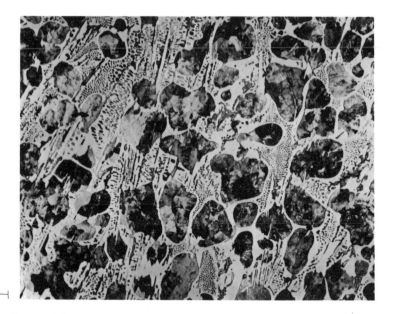

75μm

Fig. 10·21 Microstructure of a hypo-eutectic white cast iron

eutectoid temperature all the austenite transforms to the eutectoid mixture, pearlite. The lamellar structure of the pearlite is not resolved in the micrograph and it appears grey; the larger grey regions correspond to the pearlite formed from the primary austenite, and the finer regions to that formed from the austenite in the ledeburite. The cementite remaining from the ledeburite is white. The transformations that take place in a hyper-eutectic white cast iron are very similar, except that primary cementite is formed instead of primary austenite. The microstructure, which is shown in Fig. 10·22, consists of primary cementite platelets in a matrix of cementite and pearlite from the ledeburite, which is not resolved by the microscope.

When pig-iron is cooled slowly, or if it contains for example more than 3 wt % Si, much of the carbon occurs in the free form as flakes of graphite. The fracture surface is dark, as fracture takes place along the soft graphite flakes, and so the iron is referred to as *grey cast iron*. Grey cast iron is relatively soft and weak, but it is easily machined. The microstructure of a grey cast iron is shown in Fig. 10·23. The large, black graphite flakes are the result of the eutectic reaction $L \rightleftharpoons \gamma + $ graphite, which is the true equilibrium reaction and is therefore favoured by slow cooling. The matrix in Fig. 10·23 is mainly pearlite.

Now let us turn our attention to lower carbon contents, i.e., the steels. The solidification of steel is generally less important than the changes that occur on cooling from the austenitic region and so we will concentrate on the latter. If a steel of the eutectoid composition is air-cooled from the austenite phase field, it is said to be *normalized*, and it will have a 100% pearlitic structure similar to the eutectoid structure shown in Fig. 10·11(a). Slower cooling

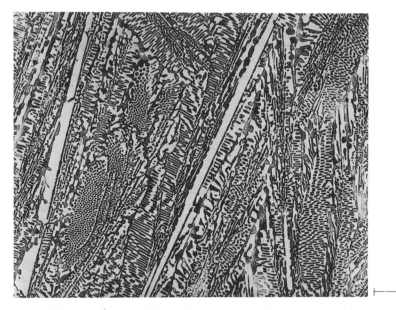

75μm

Fig. 10·22 Microstructure of a hyper-eutectic white cast iron

75μm

Fig. 10·23 Micrograph of a grey cast iron showing the black graphite flakes

rates, such as experienced in furnace cooling, give coarser structures and there may even be some spheroidization. Subsequent prolonged heating at temperatures near to, but below, the eutectoid temperature produce complete spheroidization and coarsening of the cementite. In this condition, for a given carbon content, the steel has the minimum hardness and maximum ductility (see Table 10·2 on page 276).

The pearlitic transformation, as it involves solid state diffusion, can occur at temperatures well below the eutectoid temperature of 723°C. This is true for all eutectoid reactions and represents a major difference between eutectoid and eutectic transformations–eutectic transformations always take place at temperatures very close to the eutectic temperature. Thus it is possible to quench a eutectoid steel from the austenite phase field to, say, 650°C and then isothermally transform to pearlite at that temperature. In fact pearlite in plain carbon steels may be obtained by quenching to and then transforming at temperatures in the range 450° to 723°C. As we will see later in this section, a TTT curve can be constructed for the isothermal transformation of austenite to pearlite.

The preceding discussion has concentrated on steels of the eutectoid composition. Steels of other compositions behave in a similar manner except that on cooling from the austenite condition a pro-eutectoid phase is formed before the pearlitic transformation starts. For hypo-eutectoid steels, i.e., less than 0·8 wt % C, the pro-eutectoid phase is ferrite which is soft and ductile (Fig. 10·24), whereas for high carbon, hyper-eutectoid steels the pro-eutectoid phase is hard, brittle cementite. Because of this change in the type and proportion of phases with carbon content, the mechanical properties of pearlitic steels with a given thermal history vary considerably with composition. As shown by the stress–strain curves for steels in the normalized condition (Fig. 10·25), as the carbon content increases the steels have higher yield stresses and tensile strengths but lower ductilities.

We have seen that austenite in plain carbon steels, that is steels containing no major alloying additions, isothermally transforms to pearlite in the temperature range 450°C to 723°C. If austenite is quenched to and then transformed at lower temperatures in the range 250°C to 550°C, a different reaction occurs, namely the *bainitic transformation*. Note that there is an overlap of the temperature ranges for the pearlitic and bainitic transformations and both reactions occur when isothermally transforming in the range 450°C to 550°C.

The product of the bainitic transformation is termed *bainite* and essentially it consists of lath-shaped ferrite and cementite, although the exact details of the microstructure vary with transformation temperature. The bainitic reaction is unusual in that it has some characteristics of a martensitic transformation and other characteristics typical of a nucleation and thermally activated growth process. For example, as observed in martensite, bainite has a relatively high dislocation density, although generally not as high as found in martensite. On the other hand, in common with nucleation and growth transformations, a TTT curve can be constructed for the bainitic reaction.

When a steel is rapidly quenched from the austenitic condition, the pearlitic and bainitic transformations are suppressed and the *martensitic*

(a) |——— 100 μm ———|

(b) |—— 10 μm ——|

Fig. 10·24 Microstructure of hypo-eutectoid steel in the normalized condition. (a) Light micrograph. The pro-eutectoid phase is ferrite and is the lighter colour. The dark regions are pearlite which is unresolved at this magnification. (b) Scanning electron micrograph with the pearlite resolved into lamellar ferrite and cementite (Courtesy H.M. Flower)

transformation occurs resulting in a typical martensitic structure as shown in Fig. 10·26. As no long-range diffusion is involved in the martensitic transformation all the carbon that was in solution in the austenite remains in solution in the martensite. Martensite is, therefore, supersaturated with respect to carbon and consequently the normal b.c.c iron structure is distorted to a body-centred tetragonal (b.c.t.) structure [Fig. 10·27(a)]. The greater the carbon content of the martensite the greater the distortion, as can be seen by the increase in the c/a ratio for the b.c.t. structure with increasing carbon [Fig. 10·27(b)].

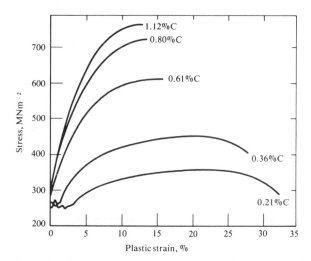

Fig. 10·25 Effect of carbon content on the stress–strain curves of pearlitic steels

50μm

Fig. 10·26 Martensitic structure in a 1·5% carbon steel quenched from the austenite phase field. There is some retained austenite (white in this structure) which would not be present for a hypo-eutectoid steel

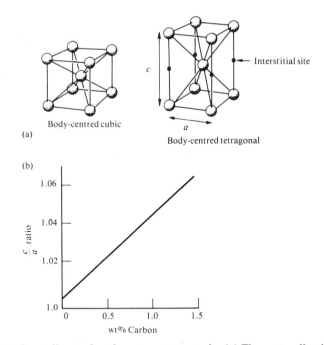

Fig. 10·27 Crystallography of martensite in steels. (a) The unit cells of the b.c.c. and b.c.t. structures (not all the interstitial sites shown would be occupied). (b) The increase in the c/a ratio of the b.c.t. martensite with increasing carbon content

On rapidly cooling a steel the martensitic transformation commences at the martensite start temperature, M_s. The transformation continues as the temperature falls and is completed at the martensite finish temperature, M_f. The extent of the transformation to martensite depends on the temperature, but is independent of time. In other words, if we quench a steel to a temperature of between M_s and M_f a certain amount of martensite will rapidly be formed, but if we continue to hold the steel at that temperature the amount of martensite will not increase. To increase the proportion of martensite we have to lower the temperature. The M_s and M_f temperatures are composition dependent, and in the case of carbon they decrease with increasing carbon content (Fig. 10·28). An important feature to note about the M_f curve is that it is below room temperature for plain carbon steels containing more than 0·7 wt % C. This means that if we quench a eutectoid or hyper-eutectoid steel to room temperature the martensitic transformation will not go to completion and some austenite will be retained as shown in the micrograph of Fig. 10·26. The retention of austenite affects the hardness of quenched steels. The hardness of quenched steels containing up to 0·7 wt % C increases markedly with carbon content due to the strong solid solution hardening effect of the interstitial carbon atoms in the martensite. However, above 0·7 wt % C the increasing proportion of retained austenite, which is relatively soft, balances the increasing hardness of the martensite and as a result the hardness of the steel remains essentially constant (Fig. 10·28).

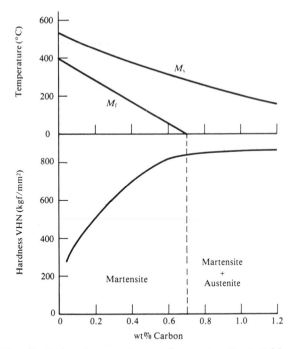

Fig. 10·28 The effect of carbon content on the martensite start M_s and finish M_f temperatures and on the hardness of plain carbon steels

Steel in the martensitic condition is too hard and brittle for normal service (Table 10·2). However, the properties are improved by reheating, that is *tempering*, to a temperature below the eutectoid temperature. During tempering carbon is removed from solution by the precipitation of iron carbides, mainly cementite, and the matrix reverts to the b.c.c. structure. At the higher tempering temperatures some spheroidization and coarsening of the carbides occurs and also the dislocation substructure associated with the martensitic transformation is modified by processes analogous to recovery

Table 10·2
Effect of heat treatment on the hardness and ductility of eutectoid (0·8% C) steel

Heat treatment	Microstructure	Hardness VHN (kgf /mm²)	Ductility (%)
Air-cooled from γ-region (normalized)	Pearlite	280	15
Furnace-cooled from γ-region	Coarse pearlite and some spheroidization	210	20
Heated just below eutectoid temperature	Completely spheroidized	175	
Quenched from γ-region	Martensite	850	Negligible
Quenched then tempered for 1 h at 500°C	Precipitates of iron carbide in ferrite	400	13

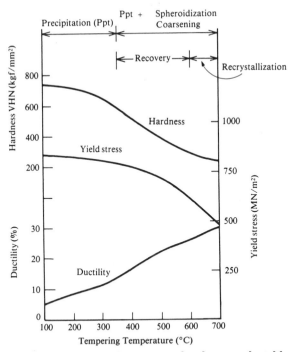

Fig. 10·29 Graph showing the decrease in hardness and yield stress and the increase in ductility on tempering a 0·55% C steel

and recrystallization. The temperature ranges of these structural changes are marked on Fig. 10·29, which gives the mechanical properties as a function of tempering temperature. From this figure we can see that tempering is accompanied by an increase in ductility at the expense of strength as shown by the decrease in yield stress and hardness.

Mild steel, which contains less than 0·25 wt % C (Fig. 10·24), is used in the ferrite–pearlite condition and is easily cold-formed and machined. It is a general-purpose steel for applications where strength is not particularly important, such as nails and car bodies. Higher carbon steels may be used in the pearlitic state or in the quenched and tempered condition depending on the properties required. *Medium carbon steel* (0·25–0·50 wt % C) is employed for components needing a higher strength than that of mild steel coupled with good toughness, e.g., hand tools, crank shafts. *High carbon steel* (>0·50 wt % C) is hard but lacks the ductility and toughness of the lower carbon content steels. It is, therefore, often found in applications where wear is the prime consideration as in cutting tools.

10·7·2   Alloy steels

So far our discussion of steels has concentrated on plain carbon steels, that is iron-carbon alloys containing only small amounts of impurities, e.g., less than 0·5% Mn and 0·5% Si. However the microstructure and properties of steels can be altered significantly by the introduction of alloying elements. Alloying elements have various different effects but the most important are associated with:

(a) stabilization of ferrite or austenite,
(b) changes in the rates of transformations, and
(c) carbide formation.

Let us look at each of these in turn.

Elements which are more soluble in ferrite than in austenite, such as silicon, chromium, vanadium, and molybdenum, tend to favour the formation of ferrite and hence are known as *α-stabilizers*. In contrast elements which are more soluble in austenite tend to stabilize that phase and are called *γ-stabilizers*, e.g., manganese, nickel, cobalt. The well known *18/8 austenitic stainless steel*, which contains 18% Cr and 8% Ni is a good example of the effect of a γ-stabilizer. Chromium is present because it reduces corrosion and oxidation by forming a thin, chromium-rich film on the surface of the steel. A binary Fe–18% Cr alloy would be ferritic at room temperature, but on adding 8% Ni the γ-phase field is expanded to such an extent that an 18/8 stainless steel is austenitic at room temperature. Austentic stainless steels are non-magnetic and exhibit good ductility.

The effect of alloying elements on the rates of the pro-eutectoid, pearlitic and bainitic transformations is best studied by means of TTT diagrams (see Fig. 10·18). First let us examine the diagram for a plain carbon steel of the eutectoid composition. In this case there is no pro-eutectoid phase but the diagram must take into account the pearlitic and bainitic reactions, and the individual TTT curves for these reactions are shown in Fig. 10·30(a). The two curves overlap so much that the resulting combined TTT diagram for a eutectoid steel consists, by chance, of one single continuous curve [Fig.

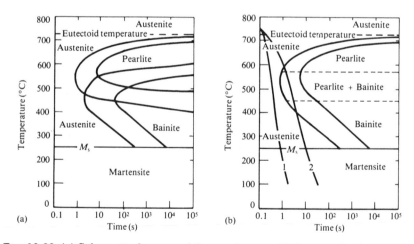

Fig. 10·30 (a) Schematic diagram of the overlapping TTT curves for the pearlitic and bainitic transformations in plain carbon steel of the eutectoid composition (0·8 wt % C). (b) TTT diagrams for the same steel showing a single continuous curve with two superimposed cooling rates designated 1 and 2

10·30(b)]. Figure 10·30(b) shows that the decomposition of austenite is most rapid at about 550°C where the transformation to a mixture of pearlite and bainite starts in just under 1 s. Thus it is possible by rapid quenching to suppress the pearlitic and bainitic transformations and produce a martensitic structure, with a little retained austenite, throughout a small (say less than 5 mm diameter) sample of a eutectoid steel; see cooling rate 1 on Fig. 10·30(b). The cooling rate varies with the size of the sample and the quenching medium, e.g., water gives a more rapid quench than oil. It follows that the resultant microstructure will depend on these two factors. When we quench a larger sample (>5 mm diameter) in the same medium as before, the cooling rate will decrease from the surface to the centre of the sample and for this larger sample the rate in the centre is represented by cooling rate 2 on the TTT diagram. This will be slow enough for some pearlite to be produced as well as the martensite.

A typical TTT diagram for a plain carbon hypo-eutectoid steel is shown in Fig. 10·31(a). The essential differences between this diagram and that for the eutectoid steel are (a) the rates of the pearlitic and bainitic transformations are faster, i.e., the curve is displaced to shorter times, and (b) the addition of a curve for pro-eutectoid ferrite. The TTT diagram for a hyper-eutectoid steel is very similar, except, of course, the pro-eutectoid phase is cementite. As the reactions are faster for hypo- and hyper-eutectoid steels it is even more difficult to obtain predominantly martensitic structures in samples of reasonable size. Furthermore, problems can arise if very fast quenches, such as a brine quench, are employed due to thermal stress induced cracking, which is termed *quench cracking*.

The ability of a steel to form martensite on quenching is termed the *hardenability*–the greater the hardenability the easier it is to form martensite. In order to increase the hardenability the rates of the pro-eutectoid, pearlitic

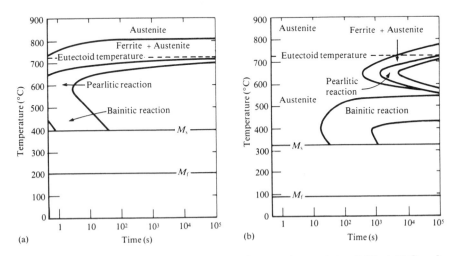

Fig. 10·31 TTT diagrams for (a) a plain carbon steel containing 0·35 wt % C and (b) a 1·5 % Ni – 1·1 % Cr – 0·3 % Mo alloy steel of the same carbon content

and bainitic reactions must be slowed down and this may be achieved by alloying. All the common alloying elements in steel, with the exception of cobalt, decrease the reaction rates and displace the TTT curve to longer times, as illustrated by the diagram for a Ni–Cr-Mo steel [Fig. 10·31(b)]. It follows that martensite can be produced in this alloy steel with less severe quenching, and hence throughout components of larger sections, than in the case of the plain carbon steels. Note also that the alloying additions have separated the curves for the pearlitic and bainitic reactions.

Alloying additions enable us to obtain steels in the martensitic condition more easily but they also affect the tempering characteristics. Firstly, the precipitation and coarsening of the iron carbides may be altered by the presence of an alloying addition. For example, silicon slows down the precipitation and coarsening of cementite and this can slightly reduce the fall-off in hardness observed for plain carbon steels. More important is that some alloying elements called the strong *carbide formers*, e.g., Cr, Mo, Ti, V, and W, form alloy carbides such as V_4C_3 and Mo_2C which are more stable than the iron carbides. Due to the slow rate of diffusion of these substitutional alloying elements, the alloy carbides do not form until tempering temperatures in the range 450–600°C are reached. The precipitation of the alloy carbides results in an increase in hardness, termed *secondary hardening*, at high tempering temperatures (Fig. 10·32). Secondary hardening enables us to produce steels

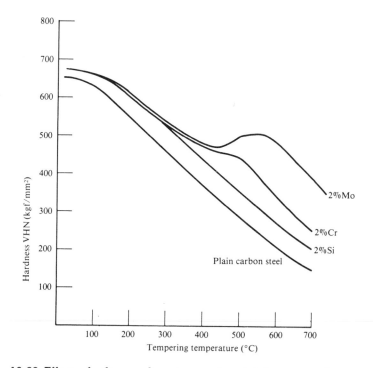

Fig. 10·32 Effect of silicon, chromium and molybdenum on the tempering characteristics of a 0·35 % C steel

with a high hardness and strength but with a better ductility than that exhibited by a plain carbon steel tempered to the same hardness by lower temperature tempering. Also, because of the slow diffusion of the carbide forming elements their carbides do not readily coarsen, consequently alloy steels so hardened may be used for high temperature applications.

10·7·3 Aluminium–copper system

If resistance to dislocation motion were all that was required of a strong material we could use intrinsically hard solids like silicon carbide for engineering applications. Such materials are, however, brittle and it is more often a combination of ductility and strength that is required. The ductility not only makes it possible to work the material to a required shape but gives a safety factor during service, i.e., if it is accidentally stressed the material will deform plastically rather than fracture in a catastrophic manner. We have already seen (Section 9·5) how a dispersion of small particles can strengthen an intrinsically soft material, but what is also important is that, with careful control of microstructure, this strengthening can be achieved without a marked loss in ductility. Aluminium alloys containing small amounts of copper (2–4·5 wt % Cu) are capable of being strengthened this way, the dispersion being produced by *precipitation* from a supersaturated solid solution. These include the well-known alloy 'duralumin' which is widely used where strength coupled with light weight is required.

We discussed precipitation earlier in this chapter and saw that it was possible to control the precipitate dispersion by solution treatment and subsequent aging in the two-phase region (refer to Fig. 10·16). The appropriate part of the Al–Cu phase diagram is similar to that shown in Fig. 10·16 and is given in Fig. 10·33. Thus, we can produce a supersaturated solid solution of, say, an Al–4% Cu alloy, by rapidly quenching from the temperature T_s. The behaviour of this alloy differs from the general precipitation behaviour described earlier in the structural changes that occur during aging. On aging, the Al–Cu alloy does not always transform directly to the equilibrium two-phase condition of $\alpha + \theta$, where θ is $CuAl_2$. Instead, the precipitation process may take place via intermediate, metastable phases; the complete sequence of events is

$$\alpha_{ss} \rightarrow \alpha + GP \rightarrow \alpha + \theta'' \rightarrow \alpha + \theta'' \rightarrow \alpha + \theta$$

where α_{ss} is the supersaturated solid solution, while GP, θ' and θ' are metastable phases. For interest, GP stands for Guinier–Preston zones, after the investigators who discovered them by using X-ray diffraction techniques. Further details, such as the compositions and structures of these intermediate phases, are not essential to this discussion.

An explanation for the precipitation behaviour of Al–Cu can be given in terms of an activation energy which has to be provided for a phase change to proceed. The concept of activation energy as an energy barrier was fully developed in Chapter 7 (Section 7·17) and the various barriers in the Al–Cu system are represented in Fig. 10·34. It can be seen that a large activation energy, Q_θ, is associated with the change from the supersaturated solid solution to the equilibrium two-phase condition of $\alpha + \theta$. Much smaller

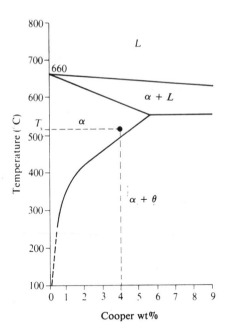

Fig. 10·33 The aluminium-rich section of the aluminium–copper equilibrium phase diagram

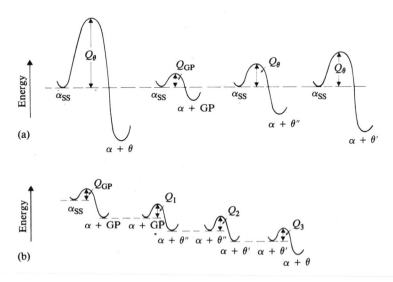

Fig. 10·34 The activation barriers for transformations in aluminium-rich Al–Cu alloys: (a) the barriers when transforming from the supersaturated solid solution directly to another phase; (b) the barriers corresponding to the sequence of events
$$\alpha_{ss} \rightarrow \alpha + GP \rightarrow \alpha + \theta'' \rightarrow \alpha + \theta' \rightarrow \alpha + \theta$$

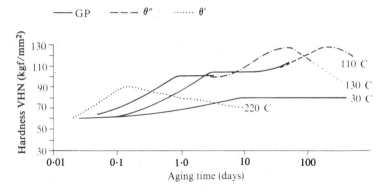

Fig. 10·35 Ageing curves for an Al – 4% Cu alloy

activation barriers have to be surmounted for changes from the super-saturated solid solution to intermediate conditions (namely, Q_{GP}, $Q_{\theta''}$, and $Q_{\theta'}$), and from one intermediate phase to another (Q_1, Q_2, and Q_3). Thus, when the alloy transforms via the metastable, intermediate phases it is following the path of lowest activation energy at each stage.

Whether or not the complete sequence of events is followed, in a given alloy, depends on the aging temperature. At high ageing temperatures, considerable thermal energy is available to aid transformations and hence the changes with high activation energy, namely $\alpha_{ss} \rightarrow \alpha + \theta$ and $\alpha_{ss} \rightarrow \alpha + \theta'$, occur readily. At low aging temperatures only changes with low activation energy can take place. Figure 10·35 illustrates this point as well as showing the marked effect that microstructure has on the hardness of an Al-4 wt % Cu alloy. An $\alpha + \theta'$ structure is shown in Fig. 10·36.

Maximum hardness in this system is associated with the precipiation of particles of the metastable θ'' which are cut by dislocations during deformation. The Orowan mechanism (see Section 9·5) is operative when θ' or the equilibrium phase θ are present and a lower hardness results.

10·7·4 Copper–zinc system

The phase diagram of the copper–zinc system is complex (Fig. 10·37), but we need only concern ourselves with the composition range 0 to about 50 wt % zinc, which represents the series of alloys known as *brasses*, which are used widely where ductility or ease of machining are of importance.

Copper is capable of dissolving up to 38 wt % zinc in substitutional solid solution, so forming the α brasses. The hardness of α brass as a function of zinc content is shown in Fig. 10·38. Up to about 10 wt % zinc the relationship between hardness and concentration is approximately linear in accordance with the discussion on solution hardening in the previous chapter. At greater zinc concentrations individual zinc atoms are less effective as hardeners due to the overlapping of their individual elastic strain fields. All α brasses are ductile and can be severely cold-worked, consequently they are widely used for pressings and drawn tubing.

1 μm

Fig. 10·36 Electron micrograph of the $\alpha + \theta'$ structure in an Al – 4% Cu alloy
(Courtesy H.M. Flower)

Zinc in excess of 38 wt % leads to the formation of the β phase and most of the cast brasses have a two-phase ($\alpha + \beta$) structure. A cast and annealed ($\alpha + \beta$) structure is shown in Fig. 10·39. The β phase is hard at room temperature but soft at elevated temperatures, therefore ($\alpha + \beta$) brasses are commonly hot-worked, at a temperature at which they are predominantly β phase. Because of the presence of the β phase, ($\alpha + \beta$) brasses are harder than α brasses (see Fig. 10·38).

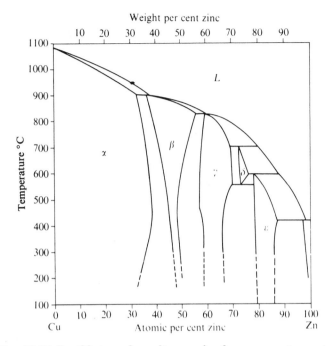

Fig. 10·37 Equilibrium phase diagram for the copper–zinc system

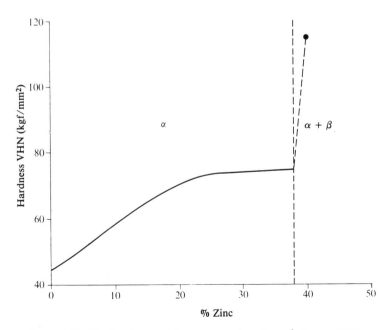

Fig. 10·38 The hardness of brasses as a function of zinc content

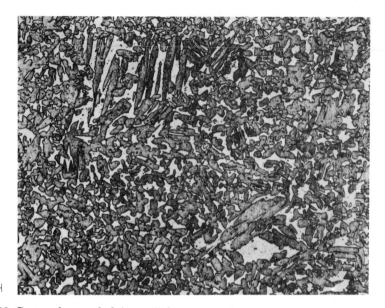

Fig. 10·39 Cast and annealed (α + β) brass. α is dark (etched in ammonium persulphate)

Problems

10·1 Using the Sn–Pb equilibrium phase diagram of Fig. 10·5(b) and applying the lever rule, answer the following questions.
  (a) What is the eutectic composition?
  (b) What are the compositions of the solid phases in equilibrium at the eutectic temperature?
  (c) For an alloy containing 90 wt % Sn, what fraction exists as the α phase at 220°C?
  (d) What fraction of this alloy is liquid just above the eutectic temperature?
  (e) Determine the fractions of α and β phases just below the eutectic temperature in the 90 wt % Sn alloy and the eutectic alloy.
  (f) What fraction of the total weight of the 90 wt % Sn alloy will have the eutectic structure, just below the eutectic temperature?

10·2 Draw free energy curves versus composition for the solid and liquid phases in a peritectic system at:
  (a) a temperature above the peritectic temperature T_p, where a solid phase is present.
  (b) T_p.
  (c) a temperature below T_p where a liquid phase is present.

10·3 Using the iron–carbon diagram of Fig. 10·20 answer the following questions:
  (a) A 0·4 wt % C steel is cooled from 1,600°C to room temperature under equilibrium conditions. State the sequence of phases and the temperatures at which they occur.
  (b) Determine the relative weights and compositions of the phases present at 800°C.
  (c) What are the weights of ferrite and cementite present in pearlite?
  (d) What is ledeburite?

10·4 Sketch and label TTT diagrams for a hypo-eutectoid plain carbon steel and a eutectoid plain carbon steel. By reference to the diagrams explain how you would produce samples with the following microstructures:

(a) martensite with some retained austenite,
(b) ferrite and martensite (a small amount of pearlite may be present),
(c) ferrite and pearlite, and
(d) pearlite and martensite.

Self-Assessment questions

1 For complete solid miscibility the two components must have

A) similar melting points
B) the same crystal structure
C) similar size atoms
D) identical free energies at room temperature
E) similar hardness
F) the same valency.

2 An intermediate or intermetallic compound, such as AB_2, always has the crystal structure of one of the components A or B.

A) true   B) false.

3 The proportions by weight of two phases in equilibrium can be determined from the phase diagram using a simple law of mixtures.

A) true   B) false.

4 An alloy of the eutectic composition does not solidify over a temperature range but completely solidifies at the eutectic temperature.

A) true   B) false.

5 The eutectic reaction is

A) $S_1 \rightleftharpoons S_2 + S_3$ B) $L \rightleftharpoons S_1 + S_2$
C) $L_1 + S_1 \rightleftharpoons L_2 + S_2$ D) $L_1 + S_1 \rightleftharpoons S_2 + S_3$
E) a martensitic transformation
F) found in systems that exhibit complete solid miscibility.

(S = Solid; L = Liquid.)

6 With reference to the lead–tin phase diagram on p. 288, which of the following statements apply to a Pb–80 wt % Sn alloy?

A) the first solid to solidify is of composition α_e
B) the first solid to solidify is of composition β_e
C) the first solid to solidify is of composition A_4
D) the first solid to solidify is of composition β_5
E) when solidification is complete the microstructure of the alloy consists of an $(\alpha + \beta)$ eutectic mixture
F) when solidification is complete the microstructure of the alloy consists of primary β crystals in an $(\alpha + \beta)$ eutectic mixture matrix
G) when solidification is complete the microstructure of the alloy consists of primary α crystals in an $(\alpha + \beta)$ eutectic mixture matrix
H) the cooling curve shows one discontinuity
I) the cooling curve shows two discontinuities
J) the cooling curve shows three discontinuities.

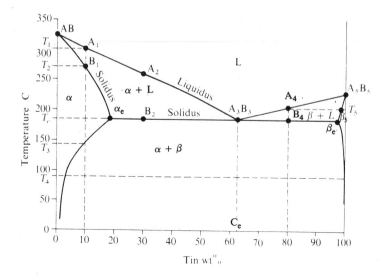

7 An alloy of the peritectic composition does not solidify over a temperature range but completely solidifies at the peritectic temperature.

A) true   B) false.

8 A solid is produced from two components A and B. The equilibrium condition of this solid is two phase, $\alpha + \beta$, when

A) the α phase and the β phase have the same free energy
B) the α phase and the β phase have the same crystal structures
C) the chemical potentials of A and B in the α phase are equal to the chemical potentials in the β phase
D) A and B have similar melting points
E) the two phase $\alpha + \beta$ structure corresponds to the lowest free energy state.

9 The free energy curves at temperature T_3 for the solid and liquid phases in a system that exhibits complete solid miscibility are given in the diagram. The diagram shows that

A) liquid is the equilibrium state for all compositions

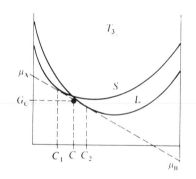

B) solid is the equilibrium state for all compositions
C) only alloys of composition between C_1 and C_2 are completely liquid
D) only alloys of composition between C_1 and C_2 are completely solid
E) alloys of composition between C_1 and C_2 are two-phase (S + L)
F) μ_A is the chemical potential of component A in solid of any composition
G) μ_A is the chemical potential of component A in solid of composition C_1
H) μ_A is the chemical potential of component A in liquid of composition C_2.

10 The overall transformation rate as a function of temperature for a nucleation and growth process is shown below. At high temperatures, such as temperature T_1,

A) the nucleation rate is high   B) the nucleation rate is low
C) the growth rate is high   D) the growth rate is low
E) the resulting microstructure will be coarse
F) the resulting microstructure will be fine.

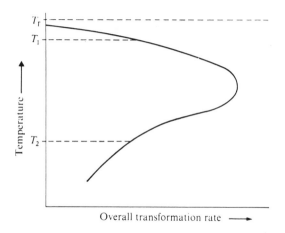

Overall transformation rate ⟶

11 A martensitic transformation

A) involves the precipitation of an equilibrium precipitate
B) is extremely rapid
C) can only lead to a change in the crystal structure of the phases
D) can lead to a change in composition of the phases
E) does not involve diffusion.

12 The micrograph on the next page

A) is a light (optical) micrograph   B) is an electron micrograph
C) shows a two-phase structure   D) shows a martensitic structure
E) is of an annealed α-brass.

2μm

13 The eutectoid mixture in steel is

   A) a mixture of ferrite and cementite
   B) a mixture of ferrite and austenite
   C) a mixture of austenite and cementite
   D) called pearlite
   E) called ledeburite.

14 A simple cast iron with no major alloying additions

   A) is remelted and cast pig iron
   B) is the product of the blast furnace
   C) has a carbon content in the range 0·2 to 2·0 wt % C
   D) has a carbon content in the range 2·0 to 5·0 wt % C
   E) always contains pearlite
   F) always contains martensite.

15 The micrograph on the next page

   A) is a light (optical micrograph)   B) is an electron micrograph
   C) is of a white cast iron   D) is of a grey cast iron
   E) shows large black flakes of cementite
   F) shows large black flakes of graphite.

75μm

16 An Al–4 wt % Cu alloy can be hardened by solution treatment and subsequent aging.

    A) true   B) false.

17 This is the aluminium-rich end of the Al–Cu phase diagram. Rapidly quenching an Al–4 wt % Cu alloy from temperature T_s to room temperature produces

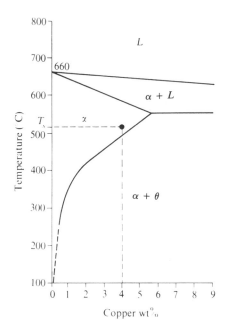

A) a single-phase material   B) a two-phase material
C) a supersaturated solid solution   D) a metastable material
E) the room temperature equilibrium state of the alloy
F) a martensitic structure.

18 Martensite in steel is

A) hard   B) ductile   C) b.c.c.   D) b.c.t.
E) an equilibrium phase.

19 Tempering of martensite in steels is accompanied by an increase in ductility and a decrease in hardness.

A) true   B) false.

20 Precipitation hardening in the Al–Cu system

A) is due to the presence of metastable phases
B) is due to the presence of the equilibrium phase, θ
C) is accompanied by a marked loss in ductility
D) involves a solution treatment and an ageing treatment.

21 The ductility of a normalized plain carbon steel increases with carbon content.

A) true   B) false.

22 Bainite

A) is commonly found in brasses
B) consists of lath-shaped ferrite and cementite
C) is a product of the decomposition of austenite
D) is only stable at temperatures above 800°C.

23 Brasses are copper–zinc alloys.

A) true   B) false.

24 The ability of a steel to form martensite on quenching is termed the hardenability.

A) true   B) false.

25 Alloying additions to steel

A) generally decrease the pearlite and bainite reaction rates
B) are important because of their effect on the peritectic reaction
C) always decrease ductility and hardness
D) may produce secondary hardening
E) promote quench cracking
F) may be α or γ stabilizers.

Each of the sentences in questions 26–35 consists of an assertion followed by a reason. Answer:

A) If both assertion and reason are true statements and the reason is a correct explanation of the assertion
B) If both assertion and reason are true statements but the reason is *not* a correct explanation of the assertion
C) If the assertion is true but the reason contains a false statement
D) If the assertion is false but the reason contains a true statement
E) If both the assertion and reason are false statements

26 Cored solid solutions are common in practice *because* the rate of cooling in chill casting, and even with sand casting, is generally too rapid for equilibrium to be maintained.

27 Only solid phases are involved in a eutectoid reaction *because* it occurs in the Fe–C system.

28 Homogeneous nucleation rarely occurs *because* it is energetically easier for nucleation to take place at structural imperfections.

29 The overall transformation rate in a nucleation and thermally activated growth process is temperature dependent *because* the nuclei of the new phase are just a few atoms in size.

30 The fracture surface of a grey cast iron is dark *because* failure takes place along the weak cementite plates.

31 A supersaturated Al–Cu solid solution does not always transform directly to the equilibrium two-phase condition of $\alpha + \theta$ *because* the transformation follows the path of maximum enthalpy.

32 Quenched plain carbon hyper-eutectoid steels have some retained austenite *because* their M_f temperatures are below room temperature.

33 A brass containing 40% zinc cannot be hot-worked *because* it has a two-phase $\alpha + \beta$ structure.

34 Secondary hardening is observed on tempering certain alloy steels *because* of the precipitation of alloy carbides which are more stable than iron carbides.

35 Martensite in steel is hard and brittle *because* of the large number of substitutional atoms present.

Answers

1 B, C, F	**2** B	**3** B	**4** A
5 B	**6** D, F, I	**7** B	**8** C, E
9 E, G, H	**10** B, C, E	**11** B, C, E	**12** B, D
13 A, D	**14** A, D, E	**15** A, D, F	**16** A
17 A, C, D	**18** A, D	**19** A	**20** A, D
21 B	**22** B, C	**23** A	**24** A
25 A, D, F	**26** A	**27** B	**28** A
29 B	**30** C	**31** C	**32** A
33 D	**34** A	**35** C	

11

CERAMICS AND COMPOSITES

11·1 Introduction

The word ceramic derives from the Greek 'Keramos' which means 'burnt-stuff' or pottery. The advent of ceramics was in even earlier times than that of the ancient Greeks and as such they represent the oldest man-made materials. House bricks, earthenware pots and porcelain cups are everyday examples of the use of ceramic materials. However, nowadays the word ceramic has come to be applied to a much wider range of materials than those used to make these commonplace items. We will classify as ceramics all non-metallic and inorganic solids and the following materials fall under this classification: the traditional clay ceramics, the new 'purer' ceramics, e.g., alumina and silicon nitride, glasses, glass-ceramics, and cement.

Composites consist of a mixture of two distinct materials; one of the materials is known as the reinforcing phase and the other as the matrix. The reinforcing phase, which is normally in the form of fibres although particulates are also used, is embedded in the matrix. The reason for employing such a mixture is to optimize the mechanical properties, as far as possible, by utilizing the best features of the reinforcing and matrix materials. Even though there are some examples of the use of composites in ancient civilizations, e.g., straw-reinforced mud bricks, composites have only been developed to the point where they are used in both technological and everyday applications in the last few decades. The most commonplace composites consist of ceramic fibres, which are strong, in a low-density polymer matrix, although other combinations of materials are also encountered.

Table 11·1

Bonding in ceramics (where a single bond type is given it comprises over 70% of the bonding)

Material	Bonding
Si	Covalent
SiC	Covalent
Si_3N_4	Covalent
NaCl	Ionic
MgO	Ionic
Mica	Ionic
Al_2O_3	Covalent–ionic
SiO_2 (quartz)	Covalent–ionic
Soda-lime glass	Covalent–ionic

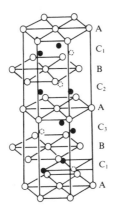

Fig. 11·1 Crystal structure of α-Al_2O_3 (hexagonal close packed structure). The aluminium sites (layers C_1, C_2 and C_3) which are only two-thirds full, are sited between the hexagonal layers (A, B) of oxygen atoms

11·2 Structure of ceramics

Ceramics are usually of a hard, brittle nature and the bonding is ionic, covalent or has mixed ionic-covalent characteristics (Table 11·1).

Most ceramic materials are crystalline, although their crystal structures are generally more complex than the simple metallic structures (see for example Figs. 6·9 and 11·1). Familiar naturally occurring ceramic materials which are

pore

Fig. 11·2 Microstructure of a complex ceramic, a chrome-magnetic brick. PS is primary spinel, S is secondary spinel, P is periclase and Si is silicate phase. (Courtesy R. Webster of the Dyson Group)

500 μm

obviously crystalline to the naked eye are the large calcite ($CaCO_3$) crystals found in limestone rock, and the gemstone ruby, which is only Al_2O_3 with some chromium impurity. However, ceramics can also exist in an amorphous or glassy state and often the microstructure of a ceramic is complex with both crystalline and glassy phases present (Fig. 11·2). Silica (SiO_2) is an example of a ceramic that may be produced in either the crystalline or the glassy state, and the following discussion of this material will show the relationships between these states.

If silica is melted and cooled very slowly it will crystallize at a particular temperature T_m, called the freezing or melting point, in an identical manner to that of a metal. The specific volume as a function of temperature exhibits a discontinuity at the melting point as shown by the full curve in Fig. 11·3(a). Silica can crystallize in a number of forms all of which can be regarded as a network of oxygen ions, following the cubic or hexagonal type of lattice, with silicon ions in the tetrahedral spaces between them. This is shown schematically in two dimensions in Fig. 11·4(a). If the silica is cooled more rapidly from the molten state, it is unable to attain the long range order of the crystalline state and the temperature dependence of the specific volume is given by the dashed curve of Fig. 11·3(a). The temperature T_g on this curve is called the *glass transition temperature*, which is not a well-defined temperature and depends on the cooling rate Fig. 11·3(b). The slope of the curve between T_g and T_m is the same as that above T_m, indicating that there is no change in structure at T_m. This means that between T_g and T_m the material is a *supercooled liquid*. There is a change in slope at the glass transition temperature but no marked discontinuity as shown by the slowly cooled material at the melting point. Furthermore, again unlike the slowly cooled material at T_m, there is no evolution of latent heat at T_g. These observations suggest that the

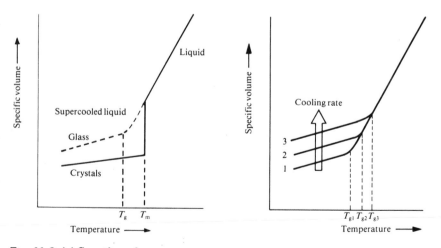

Fig. 11·3 (a) Specific volume versus temperature curves showing the relationship between the liquid, crystalline and glassy states. (b) Effect of cooling rate on the T_g (cooling rate 3 > 2 > 1)

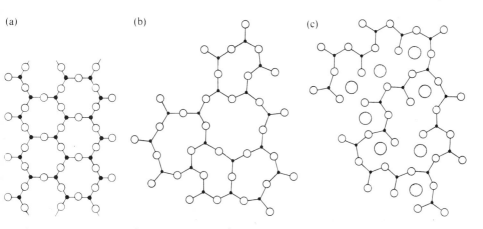

Fig. 11·4 (a) Crystalline structure of silica, (b) network structure of glassy silica, (c) soda-silica glass. Open circles indicate oxygen atoms, black dots silicon atoms, and large circles sodium atoms

state below T_g, termed the glassy state, is very closely akin to the liquid state. This is indeed the case, the glassy state consists of a short range ordered network as shown in Fig. 11·4(b). This is a metastable structure and will very slowly tend to change to the lower free energy crystalline form. A glass is called a *vitreous* solid and if the material transforms to the crystalline state it is said to have been *devitrified*. At room temperature the rate at which devitrification to the crystalline state occurs is infinitely slow. Consequently glass articles manufactured thousands of years ago are not significantly, if at all, more crystalline than they were when they were first produced. However, if molecular mobility is increased by raising the temperature devitrification can proceed at a faster rate. Devitrification is also enhanced by the presence of foreign particles in the glass which act as nucleation sites for crystallization. Unplanned-for partial devitrification, due to heterogeneous nucleation during glass production, is undesirable as the glass may become opaque and its strength reduced.

One of the important features of the structure of glass is that it is a very open network and can easily accommodate atoms of different species, such as sodium, potassium, calcium and boron atoms. We will illustrate the effect of such atoms by concentrating on two specific examples, namely soda-silica glass and Pyrex glass.

First let us look at soda-silica glass, which contains monovalent sodium ions. The addition of sodium to silica decreases the silicon/oxygen ratio as, in order to maintain electrical neutrality, one Si^{4+} ion must be removed for the addition of every four Na^+ ions. It follows that, whereas in pure silica every oxygen atom is bonded to two silicon atoms [Fig. 11·4(b)], when sodium is present some of the oxygen atoms are only bonded to one silicon atom and are said to be *non-bridging*. Thus, the continuity of the network is disrupted, and as a consequence sodium is called a *network modifier*. The non-bridging oxygen atoms and the disrupted network are illustrated in Fig. 11·4(c). As sodium breaks up the network structure it produces significant changes in the properties of the glass. For example, at high temperatures the viscosity of a soda-silica glass is much less than that of pure silica and so it is easier to fabricate into shape. The change in viscosity is very marked–at 1,400°C the viscosities of silica and silica with a 20% Na_2O addition are 10^{11} N s/m² and 10 Ns/m² respectively. Ordinary window, plate and container glass all have about 15% Na_2O in them as well as other additions. It is not normally possible to make a soda-silica glass with more than 50% Na_2O as above this amount there are so many non-bridging oxygen ions that a network structure cannot form. The material crystallizes instead of forming a glass.

In contrast to soda-silica glass, Pyrex glass contains only small amounts of network modifiers. Instead it has about 14–15% of *glass formers*, mostly B_2O_3 in addition to silica. Glass formers, as their name suggests, contribute to the network formation. The characteristics of Pyrex glass, e.g. high viscosity, resistance to chemical attack and low coefficient of expansion, are due to the network being undisrupted.

We have seen that defective glass can result from unplanned devitrification but if crystallization is controlled new types of material, called *glass-ceramics*, can be produced. Most 'oven-to-table' ware and ceramic hobs found in the modern kitchen are manufactured from a particular range of glass-ceramics which have low coefficients of thermal expansion. Glass-ceramics are polycrystalline materials formed by the controlled devitrification of glasses of carefully chosen compositions which give a large number of nuclei for crystallization. After shaping the component whilst the material is still in the glassy state, a two-stage heat treatment is followed to transform the

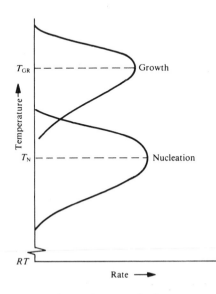

Fig. 11·5 Nucleation rate and growth rate at the heat treatment temperatures of T_N and T_{GR} for the production of a glass-ceramic

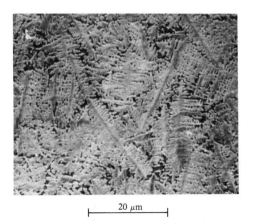

20 μm

Fig. 11·6 Micrograph of 'Silceram' glass-ceramic showing a large volume fraction of fern-like crystals with a small amount of residual glass

glass to a glass-ceramic. First the glass is held at a low temperature, T_N, to produce a large number of well dispersed nuclei (Fig. 11·5). The glass is then heated to a higher temperature, T_{GR}, at which the crystal growth rate is a maximum. On holding at this temperature the crystalline phase, which may be of different structure and composition to the nuclei, grows upon the nuclei until crystallization is almost complete with only a small amount of residual glass remaining (Fig. 11·6). A glass-ceramic, like a glass, has negligible porosity and its mechanical properties are intermediate between that of a glass and the new 'purer' ceramics such as Al_2O_3 and Si_3N_4 (Table 11·2, p. 309).

11·3   Production of ceramics other than glass and cement

11·3·1   Raw materials

The most widely used raw materials are *clays*, and they are generally used without any major chemical purification. The term clay covers a huge range of natural substances differing greatly in appearance, texture, chemical composition, and properties. Clays contain many phases but the important phases are the clay minerals such as kaolinite ($Al_2O_3 \cdot 2SiO_2 \cdot 2H_2O$). The amount and type of clay minerals in a clay varies considerably; English china clay may contain up to 90% of kaolinite, whereas some brick clays have only 30% of this mineral with considerable quantities of other minerals such as montmorillonite and illite. The common characteristics of all clay minerals are a sheet-like structure and the ability to absorb water on the surface and between these sheets. The water molecules act in a similar way to a plasticizer in a polymer, consequently wet clays are malleable and are said to be *hydroplastic*.

Many of the newer ceramics are produced from raw materials which have to be first processed. We will look in detail at the processing of *alumina*

(Al_2O_3) as being typical of this type of material because of the large quantities produced throughout the world. Alumina is produced chiefly from bauxite, which is a mixture of oxides of aluminium, silicon, and iron, by the Bayer method. First the ore is ground then treated with hydroxide solution under pressure at about 165°C to give sodium aluminate

$$Al_2O_3 + 2NaOH = 2NaAlO_2 + H_2O$$

The ferric oxide impurity is removed at this stage by filtering but the silica forms sodium silicate and remains in solution with the sodium aluminate.

The sodium aluminate is unstable and precipitation of aluminium hydroxide takes place on passing carbon dioxide gas through the solution

$$2NaAlO_2 + CO_2 + 3H_2O = Na_2CO_3 + 2Al(OH)_3$$

The $Al(OH)_3$ is separated from the sodium silicate, which remains in solution, by filtration. Finally the $Al(OH)_3$ is calcined at around 1100°C to form alumina

$$2Al(OH)_3 = Al_2O_3 + 3H_2O$$

11·3·2   Forming processes

Having acquired the raw material, it must be fabricated into the required shape. Several shaping techniques are available, and the one chosen depends largely on the characteristics of the raw material and the size, shape, and desired properties of the finished component. We will discuss the three most widely employed techniques: namely hydroplastic forming, slip casting, and powder pressing.

We have seen how water makes clays malleable and it follows that in this state they are easy to shape. Forming of wet clays by applying an appropriate stress is termed *hydroplastic forming*. The most common method of hydroplastic forming is to *extrude* the malleable clay through a die orifice by means of a rotating screw as shown in Fig. 11·7. (Extrusion is also extensively used for thermoplastics and metals.) After extrusion into shape, the clay has enough strength to maintain that shape and to be carefully handled. Pipes and bricks are produced by extrusion.

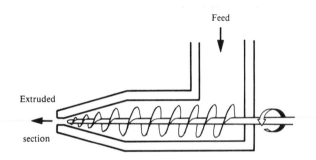

Fig. 11·7 Hydroplastic extrusion of clay. The feed is clay which has been shredded and the air removed

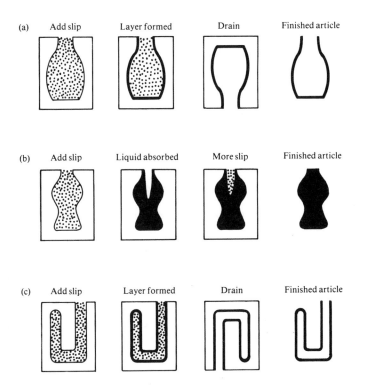

Fig. 11·8 Slip casting (a) basic process, (b) production of a solid article, (c) double wall casting

In *slip casting* the ceramic material has to be available as a suspension, or *slip*, of particles in a liquid. Slip casting depends on the ability of a mould, made of a porous material such as plaster of Paris, to absorb the liquid, which is commonly water, from the slip so leaving an even layer of particles of the ceramic on the mould walls. When this layer reaches the required thickness the remaining slip is poured off [Fig. 11·8(a)]. The article is then allowed to dry until strong enough to be removed from the mould; this may take only a few minutes for a thin-walled article but can extend into several hours for components with thick sections.

Slip casting is particularly suitable for complex and irregular shapes, for example, household articles such as teapots and jugs (the famous Delft pottery is made this way). There are variations of the basic slip casting process which enable us to produce solid articles and articles with both internal and external surfaces of good quality. Solid articles can be made by the repeated building-up of the ceramic layer [Fig. 11·8(b)] and a double wall casting method is employed to achieve the smooth and well-defined external and internal surfaces that are required for sanitary ware [Fig. 11·8(c)].

Now let us turn our attention to the third of the forming processes, *powder pressing*. Powder pressing consists of compacting dry or slightly damp

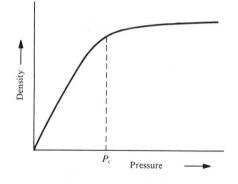

Fig. 11·9 Effect of pressure on the density of a powder compact

powder at a sufficiently high pressure so that a relatively dense and strong article, which can be handled, is formed. The density of the pressed compact increases with applied pressure up to a certain pressure, P_c, but raising the pressure further has little effect (Fig. 11·9). In general the density, and hence properties, will vary throughout the article because of pressure differences due to friction at the die walls. This is illustrated for unidirectional pressing in Fig. 11·10(a). The density variation can be reduced by pressing simulta-

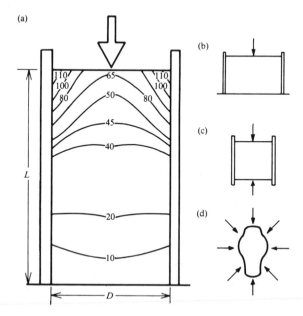

Fig. 11·10 Powder pressing. (a) Pressure distribution, and hence density varia-tion, in a unidirectional pressing [after P. Duwez and L. Zwell, *Met. Trans.*, **185**, 137 (1949)]. Methods of reducing the density variation are (b) reduce L/D ratio, (c) press from both ends and (d) isostatic pressing

neously from both ends and/or by reducing the length, L, to diameter, D ratio [Fig. 11·10(b)(c)]. However, the greatest uniformity of density is obtained by the application of pressure from all directions, which is known as *isostatic pressing*. The powder to be compacted is encased in a rubber mould and the pressure applied via a fluid [Fig. 11·10(d)].

There are limitations on the size and shape of the articles that can be formed by powder pressing and it tends to be mainly used for the production of high-density components from the newer ceramics, e.g., alumina spark plugs.

11·3·3   Post-forming processes

Ceramics formed by slip casting, hydroplastic forming, and in some cases powder pressing have to be dried before final firing at an elevated temperature. After shaping, the particles of the ceramic are not in intimate contact being separated by a thin continuous layer of water. During the *drying process* the water evaporates. Initially the rate of drying is rapid and a considerable amount of shrinkage takes place. This rapid drying stage has to be carefully controlled by using environments of constant humidity and temperature if warping and cracking of the formed component are to be avoided. Once enough water has evaporated so that the ceramic particles are in contact the movement of the residual water is restricted, and therefore the rate of drying decreases and further shrinkage is minimal.

We now come to *firing*, which is the final stage in the production of a ceramic. Firing is the densification of a dried component by heating to an elevated temperature. Densification during firing occurs by the processes of sintering and vitrification.

Sintering is the consolidation of a powder by means of prolonged use of elevated temperatures which are, however, below the melting point of any major phase of the ceramic. Figure 11·11 shows that sintering involves the replacement of high energy solid–gas interfaces by lower energy solid–solid interfaces (grain boundaries). It is this reduction in total interface energy that

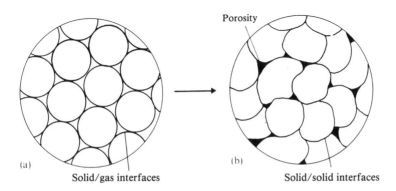

Fig. 11·11 The effect of sintering: (a) powder with high energy solid/gas interfaces; (b) solid with lower energy solid/solid interfaces. Note also the porosity

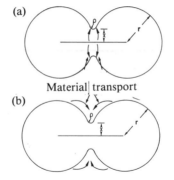

Fig. 11·12 Sintering mechanisms: (a) transport by lattice diffusion and (b) transport by evaporation–condensation

is the driving force for the sintering process. Clearly sintering requires the movement of atoms or molecules through the component and it has been found that the mechanism of mass transport, e.g., lattice diffusion, surface diffusion, and evaporation–condensation (Fig. 11·12), varies from ceramic to ceramic.

Equations, which tell us how the contact area between the particles changes as a function of important variables such as time t and temperature T of firing, have been derived for each of these transport mechanisms. One such equation, which is for the densification via lattice diffusion, is

$$\frac{x}{r} = \left(\frac{40\gamma a^3 D}{kT}\right)^{1/5} r^{-3/5} t^{1/5} \tag{11·1}$$

where x is the interparticle neck radius and r the particle radius [see Fig. 11·12(a)], D is the self-diffusion coefficient, γ is the surface energy, a^3 is the atomic volume and k is Boltzmann's constant. This equation shows that the rate of sintering decreases with time but increases on raising the temperature through the exponential temperature dependence of D. Indeed, this is true no matter which mechanism is operative and Eqn (11·1) may be written in a general form which is applicable to all mechanisms, i.e.,

$$x = Ct^m \tag{11·2}$$

The coefficient of proportionality C depends on various factors such as diffusion coefficient, vapour pressure, etc., which are specific to a particular mechanism, but it always increases on raising the temperature. The exponent m is less than unity and its value is characteristic of the transport mechanism, e.g., 1/3 for evaporation–condensation, 1/5 for lattice diffusion and 1/7 for surface diffusion. It follows from Eqn (11·2) that in most practical situations the amount of sintering that has taken place, and hence the final micro-structure (percent porosity, etc.), is more sensitive to the temperature than the time of sintering.

The sintering mechanisms discussed so far have not involved a liquid. There is an industrially important mechanism, known as *reactive liquid-phase sintering*, which occurs in the presence of a small amount of liquid. For this mechanism to be successful, it is essential for the liquid to wet the solid, i.e., low contact angle between the liquid and solid, and for the solid to be reasonably soluble in the liquid. Material transfer takes place by solution, transport through the liquid and precipitation. The transport through the liquid by diffusion is responsible for the temperature dependence of this sintering process and therefore, as the diffusion coefficient in a liquid usually changes less markedly with temperature than in a solid, reactive liquid-phase sintering is less sensitive to temperature variations than solid-state sintering. Another advantage of this mechanism is that it is generally more rapid than its solid-state counterparts.

As sintering proceeds the density of the ceramic increases, but the final product will always contain some porosity although with a good quality modern ceramic such as alumina it may only be a few percent. Unfortunately, grain growth, which has a detrimental effect on the mechanical properties, may also occur during sintering. Therefore, a compromise has to be reached between maximizing the density and minimizing grain growth. A modified procedure sometimes employed is *hot pressing* which is simultaneous forming and firing. This has the advantage over normal sintering in that higher densities and finer grain sizes may be achieved at lower temperatures.

Vitrification is the densification in the presence of a viscous liquid and is the process that takes place in the majority of ceramics produced on a large scale, e.g., all clay-based ceramics. It differs from reactive phase sintering in many ways but particularly in the amount of liquid present during firing. More liquid is present in vitrification, although the proportion has to be kept to less than about 40% to prevent distortion (*slumping* and *warping*) of the component under the forces of gravity. Initially there is a rapid increase in density as the individual solid particles are drawn together by the liquid phase, but the rate of densification falls off once a solid skeleton has been established. Vitrification is increased by reducing the particle size, lowering the viscosity and increasing the surface tension. In most cases surface tension is not changed significantly by composition or temperature and therefore the important variables are the particle size and the viscosity of the liquid. Clearly we have some control over the particle size but the viscosity of the liquid phase is the variable which can be altered the most. Not only is viscosity changed by composition (Section 11·2) but, as you will recall from Chapter 9, it is very temperature dependent (Fig. 11·13). This means that the firing temperature has to be carefully controlled. On cooling from the firing temperature the liquid phase solidifies as a glass thereby further increasing the bonding between particles and so forming a solid material, albeit with some porosity.

A solid compact of a few specialized ceramics is produced not by either sintering or vitrification but by a chemical reaction, the process being termed *reaction bonding*. This technique is used particularly for silicon-based ceramics such as silicon nitride and silicon carbide, and we will discuss briefly the former. The raw material is silicon which is mixed with a polymer binder and formed by pressing or extrusion. After forming, the binder is burnt out at a relatively low temperature to give porous silicon. The component is then fired in a nitrogen atmosphere at just below the melting point of silicon. As the silicon is porous the nitrogen penetrates to the interior of the component and the following reaction occurs

$$3Si + 2N_2 = Si_3N_4$$

Once a solid skeleton of silicon nitride has been formed the temperature is raised to increase the reaction rate, but even so the process may take a few days. The porosity content of the finished product is high (10–30%).

11·4   Production of glass

11·4·1   Melting of glass

To produce molten glass we heat the raw materials in either a number of ceramic pots, each holding up to 750 kg, placed in a furnace or in a large, refractory lined, tank furnace with a capacity up to 10^6 kg of glass. The major constituent of glass is silica, which is obtained from sand, and to this is added other raw materials in the appropriate proportions. For example, in the production of soda–lime–silica glass the additions are the carbonates of calcium and sodium which decompose on heating to give the oxides and the evolution of CO_2. It is usual to add to the raw materials up to 30% of scrap glass of the same composition as the glass being produced. The addition of the scrap glass, known as *cullet*, gives not only an economic benefit but also improves the rate of fusion as the cullet conducts heat better and has a lower melting point than the powdered constituents.

We can view the melting of glass as consisting of three distinct stages. The first involves the evaporation of water and various chemical reactions, such as the decomposition of carbonates, sulphates, and nitrates, and results in a highly viscous liquid full of bubbles. The second stage, known as *refining*, is concerned with the removal of the bubbles by raising the temperature and the

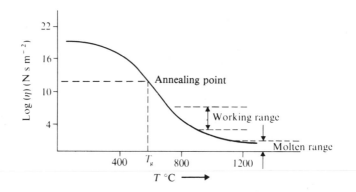

Fig. 11·13 The temperature dependence of the viscosity of a soda-silica glass. Low temperature data are generally not available due to experimental difficulties associated with measuring viscosity values greater than about 10^{14} N s/m^2

addition of chemical refining agents which liberate large bubbles that sweep the smaller bubbles to the surface. At this stage the glass is fluid ($\eta = 5$ to 10 Ns/m²) and the final stage in the melting process is to cool until the glass has a higher viscosity ($\eta = 10^3$ to 10^7 Ns/m²) suitable for working (Fig. 11·13). In order to put these viscosity values into perspective the reader might like to refer to the following values for some fluids near room temperature: air, $0·18 \times 10^{-4}$; water, 10^{-3}; treacle, 10^2; pitch, 10^9 Ns/m².

11·4·2 Glass forming and annealing

Everyone is familiar with the use of glass for bottles and windows and so we will take these as examples for forming processes.

Bottles are produced by *blow-moulding*, a technique which is also used for polymeric materials. In its simplest form blow-moulding consists of blowing by means of compressed air a blob of molten glass within a mould so that the glass takes up the shape of the mould. The problem with this one-stage process is that the walls of the article, in this case a bottle, vary in thickness [Fig. 11·14(a)]. Fortunately we can overcome this problem by a two-stage operation. In the first stage a partially blown bottle, called a parison, is produced which has a large, but controlled, variation in wall thickness. The parison is so designed that when blown in a second (finishing) mould a bottle is produced with a uniform wall thickness [Fig. 11·14(b)].

As in all glass-forming processes the viscosity of the glass is important at all stages in the operation. The initial temperature of the glass and the speed of the operation must be controlled so that the viscosity is low enough to permit

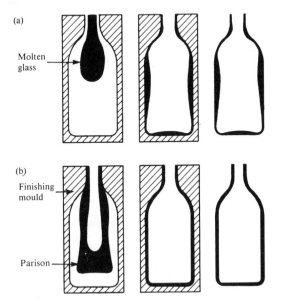

Fig. 11·14 Blow-moulding of a glass bottle. (a) Single-stage operation giving a variation in wall thickness. (b) Two-stage operation giving an even wall thickness

easy blowing, i.e., the viscosity has to be in the working range (Fig. 11·13). However, once blowing is completed and the bottle is ready to be removed from the finishing mould the viscosity must be high enough for the shape to be maintained, in fact the outer surface must be below T_g.

Glass for windows has to be of constant thickness and free from surface blemishes if we are to avoid distortion of the view. For a long time the only way to achieve these qualities was by an expensive procedure of grinding and polishing of sheet glass, but in 1959 high quality flat glass was produced for the first time by the then revolutionary *float process* which does not involve mechanical finishing of the glass. In the float process the molten glass leaves the furnace in a continuous ribbon up to 4 m wide and then floats on the surface of a bath of molten tin. The bath of molten tin is surrounded by an atmosphere of nitrogen to prevent oxidation of the tin as tin oxide imparts a bloom to glass. With good temperature control it is possible to produce parallel-sided flat glass with an excellent surface finish as the top surface of the glass is protected by the nitrogen atmosphere and the bottom surface by the liquid tin.

Glass is a poor conductor of heat and therefore during forming it is inevitable that there will be marked temperature gradients and that the cooling rate will differ considerably throughout the component. Obviously the outer surface experiences the most rapid cooling rate and will cool to below T_g and become relatively rigid while the interior is still at higher temperature and in a less viscous state. If you refer to Fig. 11·3(b) you will see that slow cooling rates are associated with low T_g values and large volume changes during the liquid-to-glass transition. Thus on cooling through T_g the interior wants to contract more than the surface layer. The rigid outer surface restricts, to a certain extent, the rearrangement of the molecules in the interior with the result that thermal stresses are set up which may augment any applied stress and lead to premature failure during the final stages of production or in service. It is, therefore, necessary to eliminate these stresses by *annealing* at a temperature where there is a small amount of molecular motion, namely near to T_g. In the manufacture of both bottles and plate glass the annealing is carried out immediately after forming by passing the glass component slowly through a long furnace, called a *lehr*, which has an entry temperature near to T_g and an exit temperature of about room temperature.

11·5 Mechanical properties of ceramics

The reader will recall from the section on fracture in Chapter 9 that when a stress is applied to a material with a crack or flaw the stress is concentrated at the crack tip. The stress concentration may be relieved by flow (permanent deformation) of the material, by propagation of the crack or by a combination of these. If the material flows easily crack propagation requires a great deal of energy, and is therefore difficult, and the material behaves in a ductile manner. On the other hand, if flow is difficult crack propagation is favoured and the material is brittle. The most noticeable characteristic of ceramics is that they are hard and brittle at room temperature (see Table 11·2). Therefore

Table 11·2

Room temperature mechanical properties of some ceramics

Material	Crystal	Slip system	No. of independent systems	Hardness† (GN/m²)	K_{1C}† (MN/m³/²)
Glass-ceramic‡	—	—	—	7·0	2·0
Soda-lime glass	—	—	—	5·5	0·7
Al_2O_3	Hexagonal	{0001} ⟨11$\bar{2}$0⟩	2	14·0	4·9
Si_3N_4	Hexagonal	{10$\bar{1}$0} ⟨0001⟩	2	18·5	4·0
MgO	NaCl	{110} ⟨1$\bar{1}$0⟩	2	9·3	1·2
SiC	Cubic (ZnS)	{111} ⟨1$\bar{1}$0⟩	5	20·0	4·0
Porcelain	Multiphase	—	—	—	1·0

†The hardness and K_{1C} are structure sensitive and will vary around the values quoted according to the composition, grain size and porosity content of the ceramic
‡Silceram

the question that we must answer is 'why is flow in ceramics so restricted?'

Let us first consider crystalline ceramics where any flow would take place by dislocation motion. For a dislocation to move it must overcome the intrinsic lattice friction stress (the Peierls–Nabarro stress) τ_{PN} which is given by Eqn (8·2)

$$\tau_{PN} \sim G \exp\left(-2\pi w/b\right)$$

where G and b have their usual meaning and w is the width of a dislocation. As w appears in an exponential term it markedly affects the value of τ_{PN}; the greater w the lower is τ_{PN} and the easier it is for dislocation motion. The width of a dislocation is determined by the shape of the energy versus atomic displacement curve which in turn depends on the type of bonding. Metals are bonded primarily by a free-electron gas and the energy of an atom is relatively insensitive to displacement. This leads to wide (5 to 10 atomic spacings) dislocations, a low Peierls–Nabarro stress and ductile materials. In contrast covalent bonds are directional and the bond energy is sensitive to the angle between the atoms and therefore displacement. This gives narrow (1 to 2 atomic spacings) dislocations, a high Peierls–Nabarro stress, limited dislocation mobility and brittle behaviour. Some crystalline ceramics are ionically bonded, which is non-directional, and results in relatively wide dislocations and a moderate Peierls–Nabarro stress. This is why ceramics which exhibit primarily ionic bonding, such as MgO, show some ductility in single crystal form.

All ceramics are brittle at room temperature in the polycrystalline condition. This is a consequence of the magnitude of τ_{PN} and the lack of slip systems. For a polycrystalline material to plastically deform each grain must be able to accommodate the deformation of the neighbouring grains. This means that each grain must be capable of undergoing an arbitrary change in shape otherwise voids and cracks will form at the grain boundaries leading to failure. *Von Mises* has shown that five independent slip systems are needed to satisfy this requirement and so permit a polycrystalline material to deform plastically. A slip system was defined in Section 8·3·1 and an independent slip system is one whose operation produces a change in shape which cannot be produced by a combination of slip on the other independent slip systems.

In general for metals the slip direction is the closest packed direction and the slip plane the closest packed plane. The same is also applicable to ceramics with the following provisos:

(a) slip must replace the ions on the appropriate sub-lattice, and
(b) the slip plane should be such that like ions are not brought into juxtaposition during gliding.

These provisos can reduce the number of slip systems in a ceramic. For example, for an f.c.c. metal there are twelve slip systems of the $\{111\} \langle 110 \rangle$ type, of which five are independent, giving ductile behaviour in the polycrystalline state. However the NaCl structure, which is taken up by NaCl and MgO, may be viewed as two interpenetrating f.c.c. lattices (see Section 6·5) but the slip systems are not the $\{111\} \langle 110 \rangle$ but $\{110\} \langle 110 \rangle$, and only two of these are independent. It follows that MgO, although exhibiting some ductility when a single crystal, is brittle in the polycrystalline form due to insufficient independent slip systems (Table 11·2).

If we now turn our attention to the non-crystalline ceramics, that is glass, the possibility of flow by dislocation motion does not exist as there is no periodic lattice. Flow in glass has to take place by the thermally assisted motion of molecules past one another. However by definition a glass is below T_g at ambient temperature and hence thermally activated molecular motion cannot occur to a significant extent. Thus on the application of a stress molecules are unable to move any distance and consequently glass is brittle.

As the temperature is raised thermal activation causes the molecular vibrations to become more and more violent. At temperatures above T_g the thermal energy plus the applied stress are sufficient to move molecules past one another and dramatic changes occur in the mechanical properties of a glass as illustrated by the viscosity versus temperature curve of Fig. 11·13. Such large changes in mechanical behaviour with increasing temperature do not occur for most crystalline ceramics. For some crystalline ceramics additional slip systems operate at elevated temperatures and these, combined with the primary slip systems can give five independent systems. For example in MgO the $\{001\} \langle 1\bar{1}0 \rangle$ slip systems become active, which gives an additional three independent slip systems making five independent systems in all. Thus polycrystalline MgO, which is ionically bonded, is reasonably ductile at temperatures in excess of 1700°C; however, this is unusual and most polycrystalline ceramics exhibit very little ductility even at elevated temperatures.

Let us return to the brittle behaviour of ceramics. The literature provides ample evidence that if identical mechanical tests are carried out on a number of specimens of a given ceramic a large variation will be obtained in the measured fracture stresses. In contrast similar experiments performed on a ductile metal would give only a small scatter in the tensile strength. This difference is due to fracture in the ceramic occurring from flaws, whereas the ductile failure of a metal is a more homogeneous process involving dislocation motion (the reader is referred to Section 9·7·3 on ductile failure). In most cases the flaws which lead to the catastrophic failure of ceramics are surface defects that propagate under the action of a tensile stress. In a ceramic specimen or component there will be a distribution of flaw sizes and flaw orientation and location with respect to the applied stress. The variation in strength is a direct consequence of the flaw distribution. In order to characterize fully the fracture strength of a ceramic it is clearly necessary to define in some way the variation in strength. This is done by describing the strength variation by an appropriate statistical distribution function. The most widely applicable, and hence the most commonly used, distribution function is the empirical *cumulative distribution function* due to Weibull. According to this function the survival probability, P, of a specimen of volume V is given by

$$P = \exp\left[\left(\frac{-V}{V_0}\right)\left(\frac{\sigma}{\sigma_0}\right)^m\right]$$

where σ is the fracture stress, V_0 is unit volume, and σ_0 and m are constants. The *Weibull modulus m* is the important constant as it characterizes the width of the fracture stress distribution; the higher m, the narrower the distribution, i.e., the smaller the scatter in the measured fracture stresses. For modern high-quality engineering ceramics m is usually in the range 5 to 10 and is never significantly more than 20. Metals on the other hand have much higher values of typically 50 to 100.

The strength may be improved by increasing the fracture toughness K_{1C}, by reducing the number and size of the flaws, or by having a compressive residual stress acting against the applied tensile stress. The strength of glass, for example, can be increased by improving the surface finish or by putting the surface into compression; the latter is the basis for *toughened glass*. If the surface is in compression the applied stress has to first overcome the residual compressive stress before the surface flaws are subject to a tensile stress. Earlier, we discussed how thermal stresses, which are set up during the cooling of glass, have a detrimental effect on strength and that, as a result, glass components are generally annealed in order to relieve these stresses. Surprisingly, toughened glass is produced by the controlled rapid cooling which puts the surface in compression and the interior in tension. Not only can toughened glass withstand high stresses, but when it does eventually fail it disintegrates into many small cubes which are relatively harmless compared to the large jagged sections obtained with normal glass. This desirable failure behaviour is due to residual tensile stress in the interior augmenting the applied tensile stress and will be familiar to any reader who has experienced a broken automobile windscreen which was manufactured from toughened glass.

Another novel way of improving the surface properties of glass is by altering the composition of the surface layer. For example, a tin compound is diffused into the outer surface of milk bottles to the depth of 1 μm. This not only increases the abrasion resistance of the glass, and therefore causes less cracks to be produced, but also increases the stress required for crack propagation, presumably by changing the surface energy and Young's modulus and hence K_{1C}.

11·6   Wear and erosion resistance

Wear and erosion are terms which are in common usage and everyone has encountered a solid component which, even to the layman, shows obvious signs of wear or erosion. It is somewhat surprising therefore to find that the mechanisms of wear and erosion are not well understood and that there is not even consensus as far as terminology is concerned. Wear is sometimes categorized into four main mechanisms, namely, abrasive, adhesive, fatigue and corrosive. The first of these, *abrasive wear*, which is the loss of material from the surface due to rubbing against the asperities on the surface of another solid or due to the grinding action of small particles, is the most relevant to ceramics. Particles are also responsible for *erosion* which is defined as the loss of material as a result of the impingement of particles onto a surface.

In general ceramics have good wear and erosion resistance and consequently are used in applications such as valve faces and for lining chutes and pipework. The good resistance of ceramics is often attributed to their high hardness, and certainly there is a general correlation between wear resistance and hardness when materials exhibiting a wide range of properties are compared. However the wear and erosion resistance also depends on other material properties as well as a number of external parameters such as surface roughness, presence of a liquid, size of abrading particles, angle of impact of impinging particles, etc. In order to illustrate some of the factors involved let us examine erosion resistance in more detail.

Erosion, which is usually quantified by the mass or volume removed from a surface by unit mass of impinging particles, varies markedly with the angle of impingement or impact. This is shown for a ductile metal and a ceramic in Fig. 11·15. We can see from this figure that the erosion of the two classes of

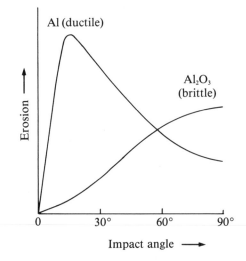

Fig. 11·15 Comparison of the effect of impact angle on the erosion of a ductile metal (aluminium) and a brittle ceramic (alumina)

material have very different impact angle dependencies. Brittle materials, such as ceramics, characteristically exhibit maximum erosion under conditions of normal incidence, whereas for a metal the erosion reaches a maximum at an oblique angle of around 20° to 40°. A simple explanation is that the maximum erosion occurs at oblique angles for metals because material is removed by the gouging action of the particles. In contrast, material is removed by microfracture in the case of ceramics; the extent of the microcracking associated with a single impact increases with the depth of penetration of the erodant particle, which in turn is a maximum at normal incidence. The depth of penetration, and hence the erosion, also increases as the velocity of the impacting particles is raised. A number of equations relating the erosion, V_E, at a given impact angle to the characteristics of the impinging particles (e.g., velocity v, density ρ and radius R) and the mechanical properties of the ceramic have been proposed. Although these equations differ in detail they are essentially of the following form

$$V_E \propto v^a R^b \rho^c K_{1C}^d H^e$$

where the exponents a, b, c, d and e are constants. The exponent d is negative, hence the higher the fracture toughness K_{1C} the better the erosion resistance. There is some uncertainty about the effect of hardness on the erosion of ceramics and both negative and positive values of e may be found in the literature. Finally, turning our attention to the eroding particles, as the exponents a, b and c are all positive, the greater the velocity and density, and the larger the size of the particles, the more significant is the erosion.

11·7 Thermal shock

Earlier in the chapter we discussed how changes in temperature result in thermal stresses in glasses as a consequence of the glass transition. Rapid fluctuations in temperature can induce stresses not only in glasses but in all ceramics. In these circumstances ceramics, with a few notable exceptions such as silicon nitride and certain glass-ceramics, are prone to cracking and a loss in strength, that is, ceramics have poor *thermal shock resistance*. When one bears in mind that most ceramic production involves a high temperature stage and that ceramics are often used in high temperature applications, it is clear that during production and service there is the possibility of a component being subjected to thermal shock.

Thermal shock resistance is usually assessed by measuring the retained strength after quenching into water from elevated temperatures. A typical set of results, in this case for two aluminas, is shown in Fig. 11·16. It can be seen that there is no degradation in strength until the temperature difference between the elevated and quench-medium temperatures exceeds a critical value ΔT_c. Above ΔT_c the thermal stresses are sufficient for the formation of microcracks which reduce the strength. The reduction in strength can be drastic and the reader will appreciate that it is therefore important to have an understanding of the factors which control the thermal shock characteristics.

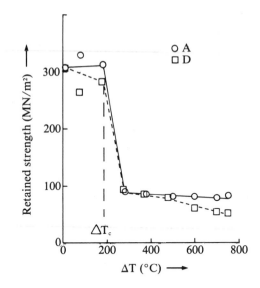

Fig. 11·16 Retained strength of two aluminas (A 99.5% and D 87% purity) as a function of ΔT (Courtesy I. Thompson)

The maximum surface stress, σ_m, developed when a material experiences an infinitely fast quench over a temperature range ΔT is given by

$$\sigma_m = \frac{E\alpha\Delta T}{1 - \nu} \qquad (11\cdot3)$$

where E is Young's modulus, α is the coefficient of thermal expansion and ν is Poisson's ratio. If σ_m is less than the fracture stress, σ_f, of the ceramic then no microcracking will occur; this is the situation for temperature changes of less than ΔT_c. Conversely, if $\sigma_m \geqslant \sigma_f$, which applies for temperature changes equal to or greater than ΔT_c, then microcracking will be initiated. Thus we can write

$$\sigma_f = \sigma_m = \frac{E\alpha\Delta T_c}{1 - \nu}$$

which on rearranging gives for the critical temperature change

$$\Delta T_c = \frac{\sigma_f(1 - \nu)}{E\alpha} \qquad (11\cdot4)$$

From this equation we can deduce that the requirements for good thermal shock resistance, i.e., a high ΔT_c, are a high fracture stress and low Young's modulus and coefficient of thermal expansion. Practical experience has established that silicon nitrides are able to tolerate more drastic temperature changes without damage than aluminas and this is consistent with the values of ΔT_c calculated from Eqn (11·4) for these materials (Table 11·3).

The reader may have noticed that the calculated ΔT_c *for alumina is less*

Table 11·3
Comparison of the materials properties relevant to shock resistance for a silicon nitride and an alumina

	Reaction-bonded silicon nitride	99·5% purity alumina
Young's modulus E (GN/m^2)	166	382
Coefficient of thermal expansion α (K^{-1})	2×10^{-6}	$8 \cdot 5 \times 10^{-6}$
Fracture stress σ_f (MN/m^2)	210	332
Poisson's ratio ν	0·27	0·27
Thermal conductivity at 50°C K (W/m K)	15	9
Critical temperature change† ΔT_c (K)	462	75

†Calculated from Eqn (11·4)

than that obtained from the experimental data presented in Fig. 11·16. This is because Eqns (11·3) and (11·4) are only applicable for infinitely fast quenches which are not achieved in practice. In order to take this into account two further parameters have to be considered, namely, the heat transfer coefficient, h, between the ceramic and the quench medium and the thermal conductivity, K, of the ceramic. These are introduced into our calculation for thermal stress and ΔT_c via a dimensionless parameter, β, known as *Biot's modulus*

$$\beta = ah/K$$

where a is the heat transfer length and is usually taken as the ratio of specimen volume to surface area. The 'infinitely fast quench' model is modified by the addition of some function of β into Eqns (11·3) and (11·4), for example,

$$\Delta T_c = \sigma_f \frac{(1 - \nu)}{E\alpha}\left(A + \frac{B}{\beta}\right)$$

which is applicable at small values of β; A and B are constants. The function of β is such that a high thermal conductivity is beneficial as far as shock resistance is concerned. Reference to the values for K in Table 11·3 shows that this is another factor that contributes to the better performance of silicon nitride compared to alumina.

11·8   A commercial ceramic system: the silica–alumina system

Equilibrium phase diagrams are also useful for studying non-metallic materials and the diagram for the silica–alumina system is given in Fig. 11·17.

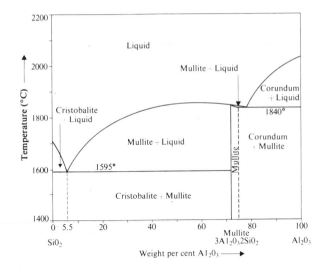

Fig. 11·17 Equilibrium phase diagram for the SiO₂-Al₂O₃ system

Hydrated silica–alumina compounds occur widely in natural form in the earth's crust, and the various phases have therefore been identified by the names given to the natural minerals.

Materials based on the SiO₂-Al₂O₃ system are refractory, that is, they are capable of withstanding high temperatures, and are used in high temperature environments such as kilns and furnaces. The performance of a refractory material under load at high temperatures depends mainly on the amount of liquid phase present at the working temperature, and on its viscosity. Bearing this in mind let us examine some materials in the silica–alumina system.

Silica refractories contain under 1 wt % Al₂O₃ and can be used under load at temperatures up to 1,650°C, decreasing with increasing Al₂O₃ content. Silica refractories are resistant to attack by iron oxide and acid slags and therefore have been of special importance to the steel industry.

Refractories containing up to about 50 wt % Al₂O₃ are produced from blends of naturally occurring clays called *fireclays*. From the phase diagram it can be seen that alumina contents near to the eutectic composition of 5·5 wt % Al₂O₃ should be avoided, as the material would be useless under load as soon as the eutectic temperature was reached. As the Al₂O₃ content is increased so the proportion of liquid decreases and the performance of the material in service improves. For example, a superduty fireclay brick contains typically 42 wt % Al₂O₃ and will deform only 5% in 10 minutes at 1,600°C under a compressive stress of 0.2 MN/m².

To improve the refractory properties still further we have to increase the Al₂O₃ content. This is achieved by adding a non-clay ingredient, rich in alumina, to the fireclay. Such an ingredient is calcined bauxite and the final product is termed a *bauxite* or *high alumina refractory*. A bauxite refractory may contain from 50 to 80 wt % Al₂O₃, and the best quality bricks can

10 μm

Fig. 11·18 Fracture surface of alumina showing the fine grain structure

operate at 1,800°C, although if they are under load the service temperature
may have to be reduced to 1,650°C.

Refractories containing more than 80 wt % Al_2O_3 are made from fused
pure alumina bonded by a little fireclay. They have better refractory
properties than bauxite refractories and good resistance to both oxidation
and reduction. Their resistance to *spalling* (breaking of corners and flaking)
is however poor.

Finally we come to *alumina*, a term which is applied to materials containing more than about 90% Al_2O_3 and some of the best quality, high purity
commercial aluminas consist of greater than 99·5% Al_2O_3. Alumina is
produced in a fine polycrystalline form by sintering. The fine polycrystalline
nature of alumina is clearly revealed on examining a fracture surface as the
fracture is intercrystalline (Fig. 11·18). Alumina is very hard and exhibits
good wear and erosion resistance. In addition it maintains its strength to high
temperatures and can stand for short periods working temperatures in the
region of 1,900°C. Not only is it a useful material on account of its
mechanical properties but also because it is a good electrical and thermal
insulator. Al_2O_3 is not exceptional in having good insulating properties and
one of the chief uses of ceramics is as insulators. Their good insulating
properties arise because all the valency electrons are occupied in the
covalent–ionic bonds and none is left free to conduct electricity or to transfer
thermal energy. Alumina is the white ceramic used in spark plugs for internal
combustion engines.

11·9 Two technical ceramics – zirconias and Sialons

Although ceramics have been used by man for thousands of years there have
been considerable advances made in ceramic technology in recent years. A
new class of ceramics materials has been developed which has come to be

known as *engineering, special* or *technical ceramics*. Zirconias and Sialons are examples of this new class of ceramics.

Zirconia can exist in three crystalline forms, the cubic (c), the tetragonal (t) and the monoclinic (m). At normal cooling rates the high temperature c-phase transforms at 2,680°C by a nucleation and thermally activated growth process (see Section 10·5) to the t-phase. In contrast the lower temperature tetragonal-to-monoclinic transformation is martensitic and therefore has many features in common with the well-known martensitic transformation in metals, and particularly in steel–the reader is referred to Sections 10·6 and 10·7. The transformation to the m-phase is accompanied by a 3% increase in volume and in pure zirconia the transformation commences at about 1,150°C, i.e., the M_s temperature is 1,150°C. In pure zirconia the transformation proceeds unabated and the volume expansion causes severe cracking; consequently it is impossible to produce components from pure zirconia. In order to produce crack-free components from zirconia ceramics it is necessary to control the martensitic transformation by means of alloying. The alloying additions used are other oxides, such as magnesia (MgO), calcia (CaO) and yttria (Y_2O_3), which are termed *stabilizing oxides*. Let us discuss the effect of one of these stabilizing oxides, Y_2O_3, in more detail.

The yttria-rich section of the ZrO_2–Y_2O_3 phase diagram of Fig. 11·19 shows that yttria additions in excess of about 9 mol % completely stabilize the c-phase to room temperature. A zirconia with the cubic structure stabilized by a large yttria addition is called a *fully stabilized zirconia* (FSZ). The main uses of FSZ's are in electrical applications, such as oxygen sensors, fuel cells, etc.

An intermediate amount of yttria, e.g., 6 mol %, gives a *partially stabilized zirconia* (PSZ), which has a microstructure containing the three main crystal forms c, t and m. A PSZ is normally fired at an elevated temperature, which depends on the composition but is of the order of 1,750°C and in the cubic phase field; it is then heat treated at a lower temperature (1,300–1,400°C) in the cubic plus tetragonal phase field to produce a fine dispersion of particles of the t-phase. On cooling to room temperature some, but not all, of the t-particles transform martensitically and become monoclinic. Those particles that remain in the tetragonal state do so because the elastic constraint of the cubic matrix opposes the volume change associated with the martensitic transformation. The t-particles are therefore in a metastable state. It has been found that the inhibition of the martensitic transformation to the monoclinic depends on the size of the particles; the smaller the particle the more likely it will remain in the metastable tetragonal state. PSZ's have a low thermal conductivity, which is effectively invariant with temperature, and a high coefficient of thermal expansion. These characteristics make them suitable for thermal barrier coatings on metal substrates. The low conductivity results in a large temperature difference across the PSZ coating, thus protecting the metal substrate in high temperature environments, and the high thermal expansion coefficient reduces the thermal expansion mismatch between the ceramic coating and the substrate.

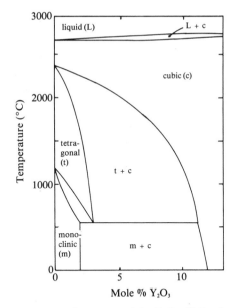

Fig. 11·19 Zirconia-rich section of the ZrO_2–Y_2O_3 phase diagram

Finally we come to low yttria additions as typified by ZrO_2–3 mol % Y_2O_3. These zirconias have a fine-grained microstructure (less than 1 μm) and are fully tetragonal; the metastable tetragonal state being a consequence of the fine grain size. These zirconias are known as *tetragonal zirconia polycrystals*, TZP's.

Both TZP's and PSZ's have good mechanical properties, especially the TZP's which have K_{IC} values in the range 10–15 MN/m$^{3/2}$. The good toughness of these two types of zirconia is associated with the martensitic t→m transformation which increases the toughness by two distinct mechanisms. Firstly, if a restricted number of particles undergo the transformation during cooling from the fabrication temperature, a fine distribution of microcracks is produced. The microcracks increase the toughness by interacting with a propagating crack, causing deflection and blunting of the crack [Fig. 11·20(a)]. However, this toughening mechanism, known as *microcrack toughening*, occurs at the expense of strength, which is reduced by the microcracks acting as flaws. The stress field at a crack tip can induce a metastable t-particle to transform to the monoclinic. This is the basis of the second toughening mechanism, *transformation toughening*, where the propagation of a crack is hindered by both the transforming particles at the crack tip and by the compressive back-stress due to the transformed particles in the crack wake [Fig. 11·20(b)]. Transformation toughening, unlike microcrack toughening, does not have a detrimental effect on strength.

The second example of a technical ceramic is the Sialons, which are closely related to silicon nitride, Si_3N_4. The reader will recall that Si_3N_4 produced by reaction bonding has a high porosity content, which is obviously undesirable.

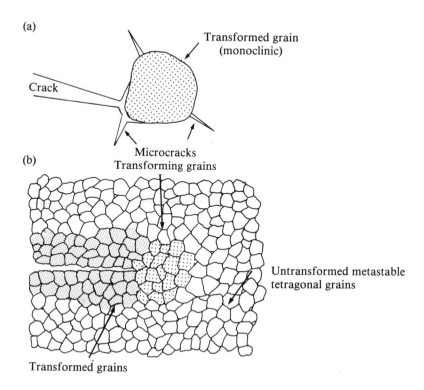

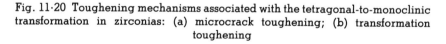

Fig. 11·20 Toughening mechanisms associated with the tetragonal-to-monoclinic transformation in zirconias: (a) microcrack toughening; (b) transformation toughening

Denser material can be made by hot-pressing but the process is limited to simple shapes. Thus, although Si_3N_4 is a successful technical ceramic in its own right (its particularly good thermal shock resistance was discussed earlier in the chapter), there are considerable problems with production. The attraction of the Sialons is that they have similar properties to Si_3N_4 but are easier to manufacture.

The crystal structure of β-Si_3N_4 is built up of SiN_4 tetrahedra as shown in Fig. 11·21. This structure is capable of accommodating the simultaneous substitution of aluminium atoms for silicon and oxygen for nitrogen to give a range of solid solutions termed β-Sialons. The chemical formula of the β-Sialons is

$$Si_{6-z}Al_zO_zN_{8-z}$$

where z denotes the level of substitution of aluminium for silicon and oxygen for nitrogen and has a maximum value of 4·5. By appropriate control of composition, including the addition of oxides such as magnesia, a reasonable

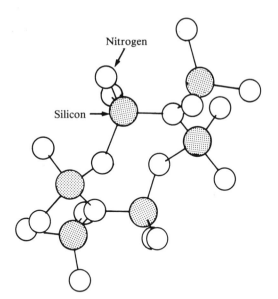

Fig. 11·21 Crystal structure of β-Si_3N_4. Sialons are obtained by substituting silicon atoms with aluminium and nitrogen with oxygen

amount of liquid is formed at relatively low temperatures which facilitates densification by vitrification so that hot-pressing is not necessary.

A Sialon produced in this manner would have a good combination of properties at room and intermediate temperatures (Table 11·4) but they cannot be used in applications where the operating temperature continuously exceeds about 1,000°C. This operational limit is due to the presence of a continuous intergranular glassy phase which softens at high temperatures. We may obtain Sialons with improved high temperature performance through the addition of yttria, Y_2O_3. As before, densification takes place via a liquid phase, but in this case there is Y_2O_3 in the liquid, and under normal cooling conditions an intergranular glassy phase is again formed. However by a post-firing heat treatment at about 1,400°C, or by controlled cooling

Table 11·4
Properties of Sialons and reaction-bonded (RB) and hot-pressed (HP) silicon nitride

	RB Si_3N_4	HP Si_3N_4	Sialon + glass	Sialon + YAG
Strength, 20°C (MN/m²)	210	800	945	
Strength, 1,200°C (MN/m²)	175	490		600
Weibull modulus, 20°C	15	20	11	
Toughness (MN/m³/²)	4·0		7·7	
Thermal conductivity (W/m K)	10–15	15–20	22	
Thermal expansion coeff. (K^{-1})	$2–3 \times 10^{-6}$	$2–3 \times 10^{-6}$	3×10^{-6}	

from the firing temperature, the glass reacts to give crystalline yttrium-aluminium–garnet (YAG, $Y_3Al_5O_{12}$) at the grain boundaries and a lightly changed Sialon composition

$$Si_5AlON_7 \quad + \quad Y\text{–}Si\text{–}Al\text{–}O\text{–}N \quad = \quad Si_{6-z}Al_zO_zN_{8-z} \quad + \quad Y_3Al_5O_{12}$$

Sialon with   glass   Sialon   YAG
  $z = 1$

The YAG results in the Sialons retaining strength up to higher temperatures of the order of 1,400°C.

11·10   Cement and concrete

In terms of quantity the most extensively used man-made structural material is concrete, which is produced by mixing mineral lumps (stones), called *aggregate*, with water and cement. The aggregate, which consists of coarse and fine lumps, may comprise about three-quarters of the volume of the concrete. The rest of the volume is occupied by the *cement paste*, which is formed by the reaction of water with the cement, and air voids. In many ways concrete may be viewed as a composite material with the cement paste as the matrix and the aggregate as the filler material.

To manufacture cement calcareous deposits, such as limestone, and clays are ground together and fed into a rotary kiln maintained at temperatures of 1,400–1,600°C. During the heating a broad sequence of reactions occurs between the raw materials, the most important being the formation of a liquid in the temperature range 900–1,250°C and compound formation at temperatures above 1,280°C. This heating process is known as *clinkering* and the product of mainly calcium silicates as clinker. Finally the clinker is ground with a few percent of gypsum to give the fine powder that is so familiar to everyone.

Table 11·5
The constitution of Portland cement

Constituent	Symbol	Weight %
Dicalcium silicate ($2CaO \cdot SiO_2$)	C_2S	28
Tricalcium silicate ($3CaO \cdot SiO_2$)	C_3S	46
Tricalcium aluminate ($3CaO \cdot Al_2O_3$)	C_3A	11
Tetracalcium alumino ferrite ($4CaO \cdot Al_2O_3 \cdot Fe_2O_3$)	C_4AF	8
Gypsum ($CaSO_4$)	—	3
Magnesia (MgO)	M	3
Calcium oxide (CaO)	C	0·5
Sodium oxide (Na_2O)	N	0·5
Potassium oxide (K_2O)	K	

The composition of cement varies but the most commonly used formulation is that of *Portland* cement, the constitution of which is given together with a short-hand notation for the constituents in Table 11·5. In this notation, which is universally used by cement scientists, each oxide is designated by a single letter corresponding to the first letter of the chemical symbol for the cation, e.g., $CaO = C$, $SiO_2 = S$, $Al_2O_3 = A$, $H_2O = H$, etc.

Mixing cement with water produces a plastic workable paste. For some time the characteristics remain unchanged, and this period is known as the *dormant* or *induction* period. At a certain stage the paste begins to stiffen to such a degree that though still soft it becomes unworkable. This is known as the *initial set*. The *setting* period follows in which the paste continues to stiffen until it can be regarded as a rigid solid, *i.e., final set.*

Setting and hardening are brought about by the hydration of the cement constituents. On adding water there is an initial fast reaction with the tricalcium aluminate

$$C_3A + 6H \rightarrow C_3AH_6$$

which evolves considerable heat of hydration. If allowed to proceed unhindered this reaction would cause rapid setting and a rise in temperature without any significant contribution to the strength of the concrete (Fig. 11·22). In Portland cement we control this reaction by the addition of gypsum. In the presence of gypsum C_3A hydrates to form a high-sulphate calcium sulphoaluminate known as *ettringite*. This coats the C_3A grains thus retarding further hydration and so regulating the rapid set that would otherwise occur.

The principal subsequent hydration reactions involve the calcium silicates C_2S and C_3S, which make up about 75% of the cement. The hydration of these compounds contribute significantly to the strength of the concrete. The calcium silicates hydrolyse to form calcium hydroxide and less basic calcium silicate hydrate, which is assumed to have the composition $C_3S_2H_3$ on complete hydration. The reactions are approximately represented by

$$2C_3S + 6H \rightarrow C_3S_2H_3 + 3CH$$

and

$$2C_2S + 4H \rightarrow C_3S_2H_3 + CH$$

where CH is calcium hydroxide. The C_3S takes about 30 days to reach 70% of its ultimate strength, while the C_2S only reaches about two-thirds of its final strength in six months at normal temperatures (Fig. 11·22). Thus concrete goes on hardening for years.

The resulting hydrate is poorly crystallized and produces a porous solid defined as a rigid gel. This gel is sometimes called *tobermorite* gel, but its similarity to this naturally occurring structure is weak and so it is preferable to refer to the hydrates simply as calcium silicate hydrate (CSH).

On the microstructural level the hydration products consist of a heterogeneous mixture of hexagonal platelets of $Ca(OH)_2$, needles of

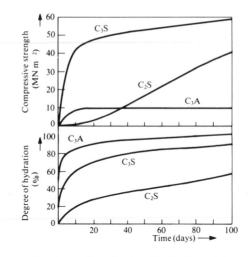

Fig. 11·22 Compressive strength and rate of hydration of the pure constituents of cement

ettringite and thin foils with colloidal dimensions of CSH [Figs. 11·23(a), (b)]. The volume of the hydration products is more than twice that of the anhydrous cement, and so as hydration proceeds the products gradually grow and fill up the spaces between the cement grains. The formation of points of contact causes stiffening of the paste and eventually their number so increases that the mobility of the cement grains is restricted and the paste becomes rigid, i.e., sets. With this setting mechanism in mind we can see that the choice of the term concrete, which is derived from the Latin word 'concretus' meaning to grow together, is particularly apt.

The composition of Portland cement can be varied to give the appropriate properties for a specific application. There are four main types of Portland cement commonly available: ordinary (the composition of which is given in Table 11·5), rapid hardening, low heat and sulphate resisting. The *rapid hardening cement* reaches about twice the strength of the ordinary cement over the first 24 h of setting and this is useful when manufacturing pre-cast components or when working in a low-temperature environment. Rapid hardening is achieved by increasing the proportion of C_3S and grinding the cement more finely to increase the surface area and so enhance the hydration reactions. If, however, we are building a large structure from concrete, such as a dam, care must be taken to minimize the temperature rises associated with the evolution of the heat of hydration as these may set up thermal stresses and damage the structure. It follows that a rapid rate of hydration is undesirable and that hydration reactions that evolve little heat are preferable. A *low-heat cement* therefore has reduced amounts of C_3S and C_3A as these hydrate rapidly (see Fig. 11·22) and have high heats of hydration compared to C_2S. Finally, it has been found that concrete structures deteriorate when in contact with water- or soil-containing sulphates. The hydration products of C_3A are responsible for the deterioration and hence the problem is overcome in *sulphate-resisting cements* by reducing the C_3A content to below 5% and

(a)

(b)

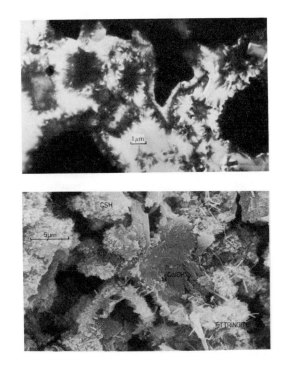

Fig. 11·23 Microstructure of cement (Courtesy K. Scrivener): (a) Transmission electron micrograph of 12 h old cement showing fine CSH gel surrounding dark grains of unhydrated cement; (b) scanning electron micrograph of the fracture surface of a two-day old cement, showing CSH together with crystals of Ca(OH)$_2$ and needles of ettringite (examples labelled)

increasing the proportion of C$_4$AF. A summary of the main constituents of cement and their characteristics and roles in concrete technology is given in Table 11·6.

In order to obtain high strength it is important to use the correct ratio of water to cement. If there is too little water, incomplete hydration and entrapped air gives a porous, and therefore weak structure. Full hydration requires a water/cement (w/c) ratio by weight of about 0·3 but the strength falls if too much water is added as evaporation of the free water leaves pores. The dependence of the compressive strength, σ_c, on w/c is illustrated in the graph of Fig. 11·24 and may be mathematically expressed by an equation of the form

$$\sigma_c = A/B^{1 \cdot 5(w/c)} \tag{11·5}$$

where A is an empirical constant usually taken to be about 100 MN/m^2 and B is a constant that depends on the type of cement. If you have ever made concrete you will be well aware that hand mixing and pouring is easier at higher w/c ratios but unfortunately, as just described, this leads to reduced

Table 11·6

Characteristics and significance of the main constituents of cement

Approximate chemical composition	$3CaO \cdot SiO_2$ (C_3S)	$2CaO \cdot SiO_2$ (C_2S)	$3CaO \cdot Al_2O_3$ (C_3A)	$4CaO \cdot Al_2O_3 \cdot Fe_2O_3$ (C_4AF)
Rate of hydration	Rapid	Slow	Very Rapid	Rapid
Rate of strength development	Rapid	Slow	Rapid	Medium
Ultimate strength	High	High	Low	Low
Heat of hydration	Medium	Low	Very high	Medium
General comments	Proportion increased in rapid hardening cement, reduced in low-heat cement	Responsible for the continued hardening of cement over long periods of time	Proportion reduced in low-heat and sulphate-resisting cements	Gives grey colour to cement. Proportion increased in sulphate-resisting cement

strengths. Nowadays mechanical mixing and consolidation by applying vibrations are common practice; this allows stiffer mixes to be used and produces concrete with less voids and entrapped air which results in about a 15% improvement in compressive strength over concrete made in the traditional manner.

Concrete has many favourable features, e.g., castable into any shape, economical, fire-resistant, but has the disadvantage of a low tensile strength and low ductility. Although concrete is poor in tension, its performance is improved if it is allowed to set around steel wire in the form of mesh or rods welded together. This is called *reinforced concrete* and the steel wire hinders the propagation of cracks through the concrete. The reader may recall the stress–strain curves given in Chapter 9, which showed that concrete was about eight times as strong in compression as in tension [Fig. 9·20(c)]. Thus, even greater strengthening may be obtained if the concrete is put permanently into a state of compression. This is achieved by elastically deforming the steel reinforcing wires in tension while allowing the concrete to set around them. When the concrete has set the tensile stress on the wires is removed and the wires contract elastically. This elastic contraction produces compressive stresses in the concrete, which is known as *prestressed concrete*. The production of prestressed concrete is illustrated in Fig. 11·25. Prestressed concrete is a very attractive structural material as concrete is inexpensive and since it also protects the steel from rust, maintenance is eliminated.

11·11 Composites

The breaking strength of strong solids is, as was seen in Chapter 9, sensitive to the presence of cracks. If the material is in the form of a bundle of fibres then the only cracks that matter must necessarily be short–across the width of the

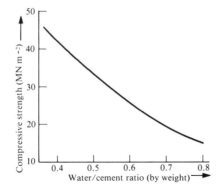

Fig. 11·24 Compressive strength of concrete at 28 days as a function of the water/cement ratio

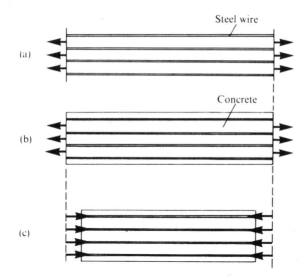

Fig. 11·25 Production of prestressed concrete: (a) tensile stress applied to steel wires; (b) concrete added and allowed to set whilst tensile stress is maintained; (c) tensile stress removed – the concrete is now in compression

fibre – and are limited to the fibres in which they exist. A solid made up from such fibres may therefore be expected to have greater fracture strength than would a solid lump of the strong material of the same total volume.

The fibres must, somehow, be joined together: hemp fibres can be twisted together to form long strands which are then coiled together to form a rope. Strong brittle fibres can be produced (Table 11·7) but their strength is very dependent upon surface damage and they cannot be twisted into a rope form.

Table 11·7
Properties of various fibres used in composites

	Specific gravity	Young's modulus (GN/m^2)	Tensile strength (GN/m^2)
Glasses			
E-Glass	2·5	73	3·5
S-Glass	2·5	86	4·6
D-Glass	2·2	52	2·4
SiO$_2$	2·2	74	5·9
Polycrystalline ceramics and multiphase			
Alumina	3·2	173	2·1
Carbon	1·8	544	2·6
Boron	2·6	414	2·8
Silicon carbide	4·1	511	2·1
Whiskers			
Alumina	3·9	1,550	20·8
Boron carbide	2·5	448	6·9
Graphite	2·2	704	20·7
Silicon nitride	3·2	379	7·0
Metals (cold worked)			
Tungsten	19·3	345	2·9
Molybdenum	10·2	335	2·2
Austenitic stainless steel	7·9	200	2·4
Eutectoid steel	7·8	240	4·0
Organic			
Kevlar (aromatic nylon)	1·45	130	2·7

However, if a suitable embedding material is used it can serve to hold the
fibres together and at the same time protect the fibre surfaces from damage
due to abrasion. The embedding material, called the *matrix*, has two other
functions; it separates the fibres so that cracks cannot run from one fibre
to another and it binds to the fibre surface so that the load can be transferred to the fibres. Both metals and polymers have been found to be

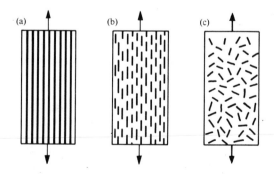

Fig. 11·26 Examples of composites: (a) continuous unidirectional fibres, (b)
aligned discontinuous fibres, (c) randomly orientated discontinuous fibres

satisfactory matrix materials, and currently there is much interest in the development of ceramic matrix composites.

The reinforcing phase in a composite may be in the form of particles or fibres. Usually the greatest improvement in mechanical performance is obtained by the use of fibres and hence most composites are fibre-reinforced composites. In this chapter we will therefore concentrate on the mechanical properties of fibre-reinforced composites. The fibres may be *continuous* or *discontinuous*, i.e., in short lengths, and may be aligned or randomly orientated (Fig. 11·26). Analysis of the mechanical properties is simplest for a unidirectionally reinforced composite with continuous fibres [Fig. 11·26(a)] and we will therefore study this first and then modify the results for an aligned discontinuous fibre composite.

11·12 Mechanical properties of continuous fibre composites

11·12·1   Strength

Let us consider the situation where the continuous fibres are aligned in the direction of the applied force. Some of the force P_c applied to the composite is taken by the fibres (P_f) and the remainder by the matrix (P_m)

$$P_c = P_f + P_m \tag{11·6}$$

Writing this equation in terms of stress σ and cross-sectional area A gives

$$\sigma_c A = \sigma_f A_f + \sigma_m A_m \tag{11·7a}$$

where the subscripts c, f and m refer to composite, fibres, and matrix respectively. It is usually more convenient to use the volume fraction of fibres V_f (V_f = volume of fibres/volume of composite = A_f/A) and the volume fraction of matrix V_m ($V_m = 1 - V_f$ = volume of matrix/volume of composite = A_m/A), hence we rearrange Eqn (11·7a) into the following

$$\sigma_c = \sigma_f V_f + \sigma_m V_m \tag{11·7b}$$

It should be emphasized that σ_c is the tensile stress on the composite at any specified strain and is not the stress at failure of the composite, i.e., it is not the fracture stress or the tensile strength, σ_c^{TS}, of the composite. To determine expressions for the tensile strength we have to consider the relative mechanical properties of the fibres and the matrix. In particular it is the relative values for the strains to failure which are paramount. Figure 11·27 shows two sets of stress–strain curves for fibres and matrix; it can be seen that in (a) the tensile failure strain of the fibre, ε_f^*, is less than that of the matrix ε_m^*, whereas in (b) the fibre has the higher tensile failure strain.

Let us first study a fibre–matrix combination where $\varepsilon_f^* < \varepsilon_m^*$ as shown in Fig. 11·27(a). The failure sequence will depend on the volume fraction of the fibres. Provided the volume fraction of fibres is above a certain minimum value V_{min} the composite will fail when the fibres fail, thus the tensile strength σ_c^{TS} of the composite is given by

$$\sigma_c^{TS} = \sigma_f^{TS} V_f + \sigma_m' V_m \tag{11·8}$$

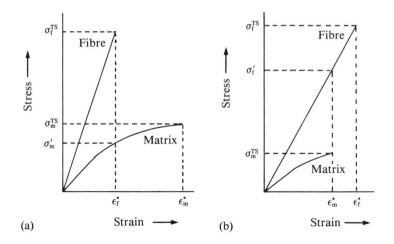

Fig. 11·27 Stress–strain curves for two different fibre–matrix combinations: (a) ductile matrix, i.e., $\varepsilon_f^* < \varepsilon_m^*$; (b) brittle matrix, i.e., $\varepsilon_f^* < \varepsilon_m^*$

where σ_f^{TS} is the tensile strength of the fibres and σ_m' is the tensile stress borne by the matrix at the failure strain of the fibres and the composite. This type of failure is illustrated schematically in Fig. 11·28(a) and a specimen that has failed in this manner is shown in Fig. 11·29(a).

For small volume fractions of fibres below V_{min} there is sufficient matrix material present to carry the load as the fibres break. The composite, therefore, only fails when the stress on the matrix reaches the tensile strength of the matrix σ_m^{TS} and multiple fracture of the fibres can occur [Fig. 11·28(b)]. In these circumstances the fibres do not contribute to the strength and the tensile strength of the composite is

$$\sigma_c^{TS} = \sigma_m^{TS} V_m \tag{11·9}$$

Equations (11·8) and (11·9) are plotted in Fig. 11·30, and from the graph we can see that V_{min} may be evaluated by equating these two equations, namely

$$\sigma_{c(min)}^{TS} = \sigma_f^{TS} V_{min} + \sigma_m' V_m = \sigma_m^{TS} V_m$$

Now $V_m = 1 - V_f$, therefore

$$V_{min} = \frac{\sigma_m^{TS} - \sigma_m'}{\sigma_f^{TS} + \sigma_m^{TS} - \sigma_m'} \tag{11·10}$$

In fact the fibres do not lead to strengthening until a *critical volume* V_{crit}, which is slightly larger than V_{min}, has been exceeded (Fig. 11·30). To determine V_{crit} we put $\sigma_c^{TS} = \sigma_m^{TS}$ and $V_m = 1 - V_{crit}$ into Eqn (11·8)

$$\sigma_m^{TS} = \sigma_c^{TS} = \sigma_f^{TS} V_{crit} + \sigma_m'(1 - V_{crit})$$

$$\therefore V_{crit} = \frac{\sigma_m^{TS} - \sigma_m'}{\sigma_f^{TS} - \sigma_m'} \tag{11·11}$$

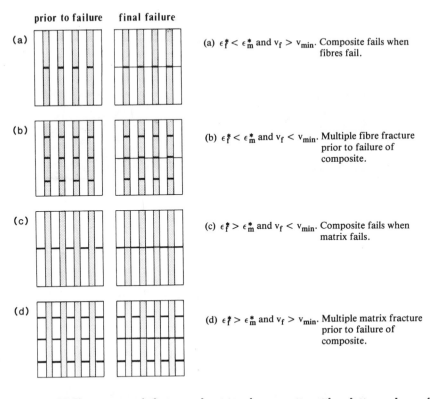

prior to failure   final failure

(a) $\epsilon_f^* < \epsilon_m^*$ and $v_f > v_{min}$. Composite fails when fibres fail.

(b) $\epsilon_f^* < \epsilon_m^*$ and $v_f < v_{min}$. Multiple fibre fracture prior to failure of composite.

(c) $\epsilon_f^* > \epsilon_m^*$ and $v_f < v_{min}$. Composite fails when matrix fails.

(d) $\epsilon_f^* > \epsilon_m^*$ and $v_f > v_{min}$. Multiple matrix fracture prior to failure of composite.

Fig. 11·28 Variation in failure mechanism of composite with relative values of strains to failure of the fibres (ε_f^*) and the matrix (ε_m^*) and with volume fraction of the fibres (V_f).

As σ'_m is generally much smaller than both σ_m^{TS} and σ_f^{TS} this equation approximates to

$$V_{crit} \sim \sigma_m^{TS}/\sigma_f^{TS} \tag{11·12}$$

Now we will consider the situation where $\varepsilon_f^* > \varepsilon_m^*$. Again two different fracture sequences may be envisaged depending on V_f. The matrix, being the most brittle component, fractures before the fibres and the load is therefore transferred to the fibres. At low volume fractions of fibres, the fibres are unable to support the load and hence the composite fails [Fig. 11·28(c)]; the tensile strength of the composite is given by

$$\sigma_c^{TS} = \sigma'_f V_f + \sigma_m^{TS} V_m \tag{11·13}$$

where σ'_f is the stress on the fibres at the failure strain of the matrix [see Fig. 11·27(b)]. In contrast if there is a high volume fraction of fibres the large

(a)

(b)

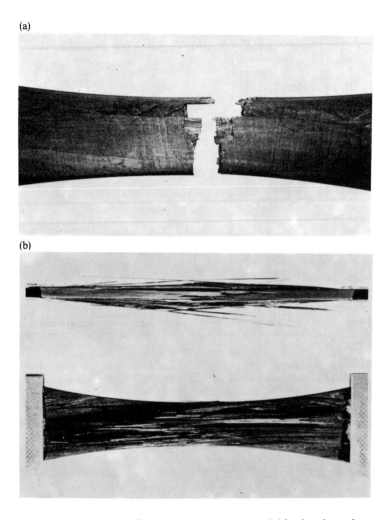

Fig. 11·29 Fracture of two polymer matrix composites: (a) high volume fraction of fibres and ductile matrix, i.e., $\varepsilon_f^* < \varepsilon_m^*$ (b) as (a) but with a brittle matrix such that $\varepsilon_f^* > \varepsilon_m^*$

number of fibres are capable of sustaining the transferred load when the matrix first fractures. Multiple fracture of the matrix occurs [Figs. 11·28(d) and 11·29(b)] and failure of the composite does not take place until the stress on the fibres reaches the fibre fracture stress. Thus at large V_f the tensile strength of the composite is given by

$$\sigma_c^{TS} = \sigma_f^{TS} V_f \tag{11·14}$$

The fracture strength as a function of V_f for a brittle matrix is illustrated in Fig. 11·31. The change in fracture mechanism is at V_{min} which is obtained by equating the two previous equations

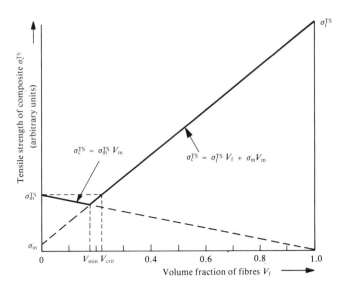

Fig. 11·30 Variation in the tensile strength of a composite where $\varepsilon_f^* < \varepsilon_m^*$ as a function of the volume fraction of continuous unidirectional fibres according to Eqns (11·8) and (11·9)

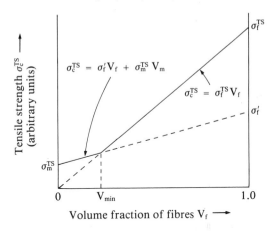

Fig. 11·31 Variation in the tensile strength of a composite where $\varepsilon_f^* > \varepsilon_m^*$ as a function of the volume fraction of continuous unidirectional fibres according to Eqns (11·13) and (11·14)

$$\sigma_c^{TS} = \sigma_f' V_{min} + \sigma_m^{TS} (1 - V_{min}) = \sigma_f^{TS} V_{min}$$

Therefore

$$V_{min} = \frac{\sigma_m^{TS}}{\sigma_f^{TS} - \sigma_f' + \sigma_m^{TS}}$$

So far in the analysis we have assumed that the fracture strength of all the fibres is the same, namely σ_f^{TS}. However, brittle fibres exhibit a considerable scatter in strength due to a variation in flaw size in the same manner as previously discussed in Section 11·5 for monolithic ceramics. The simplest way to attempt to account for the variation in strength is to employ the mean fibre strength σ_{mean} in the equations instead of σ_f^{TS}. A further improvement would be to take into account the distribution in fibre strength by incorporating the Weibull modulus m into the analysis. There are a number of models available that include m but we will only consider the *cumulative weakening model*. In this model it is assumed that the weak fibres will fail and lead to lengths, l_i, of fibres which will not support the full load; l_i is known as the *ineffective length* because fibres of that length are ineffective as far as load bearing is concerned. According to this model the appropriate fibre strength σ_{cum}^* to insert in the equations for the tensile strength of the composite is

$$\sigma_{cum}^* = \sigma_{mean} \left(\frac{l}{l_i me} \right)^{1/m} \frac{1}{\Gamma(1 + 1/m)} \tag{11·15}$$

where Γ is a tabulated gamma function, e is the base of natural logarithms, and l is the gauge length of the composite sample. It is difficult to quantify l_i accurately, but inserting typical values for l_i and the other parameters into Eqn (11·15) shows that σ_{cum}^* exceeds σ_{mean}.

11·12·2   Young's modulus

Initially, on applying a force the composite deforms elastically and the strains experienced by the composite (ε_c) and fibres (ε_f) and the matrix (ε_m) are the same

$$\varepsilon_c = \varepsilon_f = \varepsilon_m \tag{11·16}$$

Remembering that Young's modulus is equal to stress divided by strain and that the stress on the composite is given by Eqn (11·7b) we have for the Young's modulus of the composite

$$E_c = \frac{\sigma_c}{\varepsilon_c} = \frac{\sigma_f V_f + \sigma_m V_m}{\varepsilon_c}$$

But all the strains are equal [Eqn (11·16)], therefore:

$$E_c = \frac{\sigma_f}{\varepsilon_f} V_f + \frac{\sigma_m}{\varepsilon_m} V_m = E_f V_f + E_m V_m \tag{11·17}$$

where E_f and E_m are the Young's modulus of the fibres and matrix respectively. This equation is an example of the *law of mixtures* in that the contribution of the fibres and matrix is additive and in proportion to the volume occupied by each. The law of mixtures would also apply to density.

11·13 Mechanical properties of discontinuous fibre composites

Many composites contain discontinuous fibres which have either been chopped to length or were initially produced in short lengths, e.g., whiskers. It has been found experimentally that the strength of an aligned discontinuous fibre composite is always less than that of an aligned continuous fibre composite with the same volume fraction of fibres. However the longer the discontinuous fibres the stronger the composite, and for very long fibres the strength of a continuous unidirectional composite is approached (Fig. 11·32). To understand this behaviour of discontinuous fibre composites we must study the stress situation around and in the fibres.

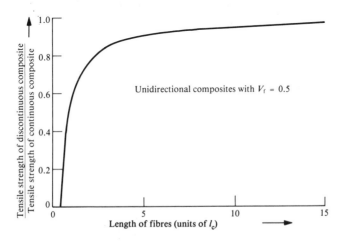

Fig. 11·32 The ratio of the tensile strengths of composites containing discontinuous and continuous fibres as a function of fibre length

Equation (11·16) is not obeyed for a discontinuous fibre composite, and a consequence of the difference in strain in the fibres and in the matrix is that a shear stress is induced in the matrix at the matrix – fibre interface parallel to the fibres. This shear stress has a maximum at the fibre ends and a minimum at the middle of the fibre [Fig. 11·33(a)]. In turn, the shear stress sets up

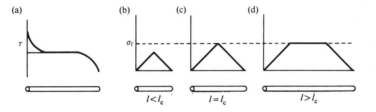

Fig. 11·33 Shear stress in the matrix (a) and the tensile stress in the fibre as a function of fibre length l (b, c and d)

a tensile stress in the fibres which is always zero at the ends and increases towards the mid-length of the fibres [Fig. 11·33(b)]. The maximum tensile stress attainable in the fibres is, of course, the tensile strength σ_f^{TS} and, as shown in Figs. 11·33(c) and (d), this is only achieved when the fibres are equal or greater in length than l_c, which is known as the *critical length*. It can be shown that l_c is related to the shear strength of the matrix τ_m, the tensile strength of the fibres and the diameter d of the fibres by

$$\frac{l_c}{d} = \frac{\sigma_f^{TS}}{2\tau_m} \tag{11·18}$$

The ratio l_c/d is called the *critical aspect ratio* and this ratio and l_c vary considerably from composite to composite (Table 11·8).

Table 11·8
Typical values of the critical length (l_c) and the critical aspect ratio (l_c/d)

Matrix	τ_m (MN/m²)	Fibre	σ_f^{TS} (MN/m²)	d (μm)	l_c/d	l_c (mm)
Ag	55	Al₂O₃ (Whisker)	20,800	2	189	0·38
Cu	76	Tungsten	2,900	2,000	19	38
Al	80	Boron	2,800	100	18	1·75
Epoxy	40	Boron	2,800	100	35	3·5
Polyester	30	Glass	2,400	13	40	0·52
Epoxy	40	Carbon	2,600	7	33	0·23

We can see from Fig. 11·33 that the average tensile stress $\bar{\sigma}_f$ on a fibre of length l, even when l is greater than l_c, must be less than σ_f^{TS}. It follows that the equations derived in the previous section for the strength of aligned continuous fibre composites have to be modified to take into account the reduced load-bearing contribution of the fibres in discontinuous fibre composites. As for continuous fibre composites, the behaviour of discontinuous fibre composites depends on V_f and on the relative values of ε_f^* and ε_m^*. We will only consider the case of an aligned discontinuous fibre composite with a high V_f and $\varepsilon_f^* < \varepsilon_m^*$.

For such a composite the strength may be determined by the initiation of failure by the fracture of the fibres or by the fracture of the matrix. First, let us concentrate on the former failure mode, i.e., the composite fails when the tensile stress on the fibres reaches σ_f^{TS}. Clearly for this to occur l must be greater than l_c; the relationship between l and the average tensile stress on a fibre when the tensile strength of the fibre is reached can be shown to be

$$\bar{\sigma}_f = \sigma_f^{TS}\left(1 - \frac{l_c}{2l}\right) \tag{11·19}$$

The equation for the strength of a continuous fibre composite [Eqn (11·8)] is modified by the substitution of $\bar{\sigma}_f$ for σ_f^{TS} to give

$$\sigma_c^{TS} = \sigma_f^{TS}\left(1 - \frac{l_c}{2l}\right)V_f + \sigma_m'V_m \tag{11·20}$$

It is this equation which is plotted in Fig. 11·32; note that when $l = l_c$ the tensile strength of a discontinuous fibre composite is approximately half that of the corresponding continuous fibre composite as the last term in Eqn (11·20), $\sigma'_m V_m$, is generally small.

The other failure mode occurs when the tensile stress on the fibres is insufficient to cause fracture of the fibres but when the stress in the matrix reaches σ_m^{TS}. The average stress $\bar{\sigma}_f$ when the matrix stress is σ_m^{TS} is

$$\bar{\sigma}_f = \left(\frac{l\tau_m}{d}\right)$$

when $l < l_c$. In this situation the strength of the aligned discontinuous composite is given by the following modification of Eqn (11·8):

$$\sigma_c^{TS} = \left(\frac{l\tau_m}{d}\right) V_f + \sigma_m^{TS} V_m$$

It is apparent from the preceding discussion on strength that discontinuous fibres are less efficient than continuous fibres. It comes as no surprise therefore that this reduced reinforcing efficiency also applies to the elastic properties and the law of mixtures equation [Eqn (11·17)] becomes

$$E_c = \eta E_f V_f + E_m V_m$$

η is a constant which depends on the dimensions of the fibres and on the properties of the fibres and the matrix. It has a value of less than unity, although it approaches unity for long fibres, say greater than about 1 mm, provided there is a strong fibre–matrix bond.

11·14 Anisotropy

So far we have only considered the behaviour of composites with aligned fibres and when the tensile stress is applied along the fibre axis. Of course, in practice the situation can be more complicated as the applied stress system may be complex and the fibres need not be aligned. Still concentrating for the moment on aligned fibre composites, let us see what happens when the tensile stress is applied normal to the fibre axis. In this case the composite is extremely weak, i.e., the *transverse strength* is very low. In fact the transverse strength is usually less than the strength of the matrix, in other words, the fibres rather than reinforcing have a detrimental effect on the transverse strength. When properties vary with orientation the material is said to be *anisotropic*.

Shear failure of the composite may occur with other combinations of stress. In these circumstances the strength is normally dominated by the matrix properties since failure can take place by shear of the matrix without fracturing the fibres.

If we took an aligned fibre composite and measured the tensile strength as a function of the angle ϕ between the stress and fibre axes, we would obtain results similar to those presented in Fig. 11·34. This figure shows that a ϕ

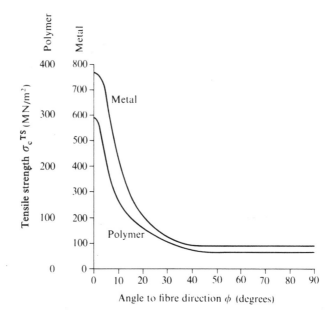

Fig. 11·34 Anisotropy of the tensile strength (σ_c^{TS}) in composites with aligned fibres. The metal matrix composite is aluminium with silica fibres, and the polymer matrix composite is epoxy resin with glass fibres

value of only a few degrees is sufficient to reduce the tensile strength dramatically for both polymer and metal matrix composites. Analysis of the failure modes indicates that tensile fracture occurs only at low angles of less than about 5°. Over an intermediate ϕ range, which depends on the composite under consideration but is typically 5–25°, shear failure dominates. Finally, the failure mode changes yet again at high ϕ values to transverse fracture.

We have discussed previously the law of mixtures equation [Eqn (11·17)] for the Young's modulus of an aligned composite when the tensile stress is applied parallel to the fibre axis. In order to see whether the elastic properties of a composite are also orientation dependent, we will derive an analogous equation for the *transverse modulus* E_T. When a stress σ_T is applied normal to the fibre axis it is assumed that it acts equally on the fibres and on the matrix so that

$$\sigma_m = \sigma_f = \sigma_T$$

This contrasts with the assumption made when the stress is parallel to the fibre axis, namely $\varepsilon_c = \varepsilon_f = \varepsilon_m$ [Eqn (11·16)]. The transverse strains of the fibre, $\varepsilon_{f(T)}$, and the matrix, $\varepsilon_{m(T)}$, are

$$\varepsilon_{f(T)} = \frac{\sigma_T}{E_f} \quad \text{and} \quad \varepsilon_{m(T)} = \frac{\sigma_T}{E_m}$$

respectively. The transverse strain of the composite, $\varepsilon_{c(T)}$, is

$$\varepsilon_{c(T)} = \varepsilon_{f(T)}V_f + \varepsilon_{m(T)}V_m = \frac{\sigma_T V_f}{E_f} + \frac{\sigma_T V_m}{E_m} \tag{11·21}$$

But

$$E_{c(T)} = \frac{\sigma_T}{\varepsilon_{c(T)}}$$

and substituting for $\varepsilon_{c(T)}$ from this equation into Eqn (11·21) gives for the transverse modulus

$$E_T = \frac{E_f E_m}{E_f V_m + E_m V_f} \tag{11·22}$$

To illustrate the difference in the Young's modulus with orientation let us substitute typical values for a polyester resin–glass fibre composite into Eqn (11·17) for the longitudinal modulus and Eqn (11·22) for the transverse modulus. The values used are $V_f = 0.5$, $E_m = 3\ \text{GN/m}^2$ and $E_f = 73\ \text{GN/m}^2$, and these give over a factor of 6 difference in the calculated moduli with E being 38 GN/m² whereas E_T is only 6 GN/m².

It is clear from the preceding discussion that aligned fibre composites are very anisotropic and this is an undesirable characteristic. It might be thought that the solution to this problem is to randomize the orientation of the fibres, and indeed this is often done to varying extents and the properties do become more isotropic. Unfortunately, the isotropic properties of a randomly orientated fibre composite are inferior to the properties parallel to the fibre axis in an aligned composite with the same volume fraction of fibres.

11·15 Toughness

The various toughening mechanisms which may be operative in a composite are illustrated in Figs. 11·20 and 11·35. We have already discussed *transformation toughening and microcrack toughening* (Fig. 11·20) and will not consider them any further except to point out that microcracks may be produced by differences in thermal expansion coefficients of the components as well as by phase transformations.

Figure 11·35(a) shows the *crack bowing* mechanism. Resistance of the reinforcing phase to fracture hinders the progress of the crack and causes it to bow into a non-linear crack front. Bowing reduces the stress intensity, K, on the matrix whilst producing an increase in K on the reinforcing phase. As the extent of the bowing increases so K on the toughening phase rises until fracture of the toughening phase occurs and the crack advances. *Crack deflection* [Fig. 11·35(b)] is similar to crack bowing in that the reinforcing phase perturbs the crack front. Deflection results in a non-planar crack which requires an increase in the applied stress to maintain a sufficient stress intensity at the crack tip for crack propagation. The effectiveness of the crack bowing and crack deflection mechanisms depends on the morphology of the

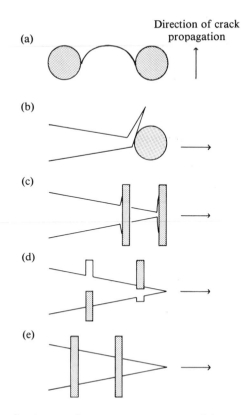

Fig. 11·35 Toughening mechanisms in composites: (a) crack bowing; (b) crack deflection; (c) debonding; (d) pull-out; (e) wake toughening (fibre bridging)

toughening phase; the greatest improvement in toughness is obtained with constituents, such as fibres, with high aspect ratios.

When a crack causes *debonding* [Fig. 11·35(c)] extra energy is required due to the creation of new interfaces. We can measure this energy by pulling a fibre which is partially embedded in the matrix material and recording the stress–strain curve. Debonding takes place at point A on the stress–strain curve and the energy of debonding is given by the area OAB (Fig. 11·36). The energy varies with the length x embedded in the matrix and it can be shown that the maximum energy of debonding, W_D, for a single fibre of diameter d occurs when $x = l_c/2$, or in other words when the tensile stress in the fibre reaches the fracture stress. W_D has been calculated to be

$$W_D = \frac{\pi d^2 (\sigma_f^{TS})^2 l_c}{48 E_f} \tag{11·23}$$

where E_f and σ_f^{TS} are the Young's modulus and fracture stress of the fibre, respectively. If the embedded length is less than $l_c/2$ we can pull the debonded

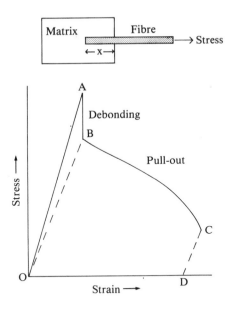

Fig. 11·36 Pull-out test and the resulting stress–strain curve showing the difference in magnitude of the energies of debonding (area OAB) and pull-out (area OBCD)

fibre out of the matrix [Fig. 11·35(d)], but a stress is required due to frictional forces. These frictional forces arise from residual stresses and from the expansion of the fibre diameter as the tensile stress is reduced in accordance with Eqn (9·6). The energy associated with *pull-out* is given by the area OBCD under the stress–strain curve. Comparison of the areas OAB and OBCD indicates that the energy of pull-out is greater than the energy of debonding. This difference may be quantified as the maximum energy of pull-out, W_P, given by

$$W_P = \frac{\pi d^2 \sigma_f^{TS} l_c}{16} \qquad (11·24)$$

Therefore

$$\frac{W_P}{W_D} = \frac{3E_f}{\sigma_f^{TS}}$$

Now E_f/σ_f^{TS} is always greater than unity, sometimes considerably so, e.g., for SiC fibres this ratio can be over 200 (see Table 11·7), and hence pull-out contributes more significantly to toughness than debonding. Reference to Eqns (11·23) and (11·24), and remembering that l_c is inversely proportional to τ_m [Eqn (11·18)], demonstrates that strong, thick fibres and a low matrix shear strength, or interfacial strength, are needed to obtain high energies of debonding and pull-out. Pull-out can clearly be seen in the fracture surface of

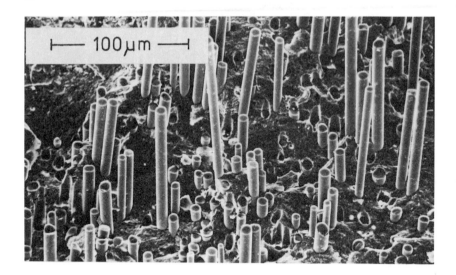

(b)

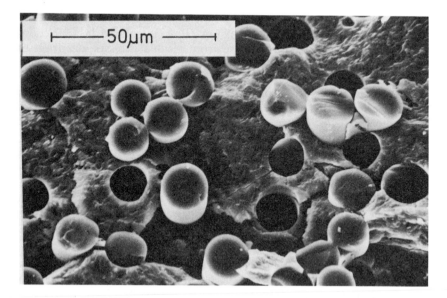

Fig. 11·37 Scanning electron micrographs of the fracture surfaces of two glass-ceramic matrix composites: (a) tough composite exhibiting pull-out; (b) less tough composite with planar fracture and negligible pull-out (Courtesy H.S. Kim, P.S. Rogers and R.D. Rawlings)

the tough composite shown in Fig. 11·37(a). However, if the correct balance between σ_f^{TS} and τ_m is not obtained, fibre fracture occurs with negligible pull-out and consequently there is little toughening [Fig. 11·37(b)].

If some debonding has taken place, but the fibre is strong and has not fractured, then the fibre will bridge the faces in the wake of the propagating crack [Fig. 11·35(e)]. For the crack to continue to grow the faces must open but, in simple terms, crack opening is hindered by the bridging fibre which has to deform elastically. This toughening mechanism, which is known as *wake toughening* or *fibre bridging*, is most effective when strong fibres are used in as high a proportion as possible, but even then the contribution to toughness is thought to be rather limited.

11·16 Comparison of polymer, metal and ceramic matrix composites

Polymer matrix composites, especially glass-reinforced plastics (GRP), are now commonplace. The matrix may be a thermosetting or a thermoplastic polymer: the former, usually a polyester or epoxy, is more common, but the use of fibre-reinforced thermoplastics is increasing much more rapidly. Carbon-fibre-reinforced plastics (CFRP) are considerably more expensive than GRP but can offer much greater strength and stiffness. CFRP is usually based on an epoxy matrix.

Since the properties of composite materials are so heavily influenced by the method of production, let us consider techniques used for forming GRP. The most familiar GRP is glass-reinforced thermosetting polyester, formed by the *hand lay-up* method into components such as car panels or boat hulls. The viscous unsaturated polyester base resins are usually thinned by dilution in styrene; mixing in a small quantity of catalyst starts the slow polymerization reaction during which the constituents crosslink and set. A former is coated with a release agent and a thin initial layer of catalysed polyester applied. This will form a protective *gel coat* to act as a barrier to water, which can otherwise diffuse along and weaken the critical polymer/fibre interface. The core of the composite is then overlaid in successive layers of glass fibre mat and/or tape into which the resin is rolled or brushed. Curing usually takes place slowly at room temperature, but may be accelerated by heating. Although an advantage of this method is that the thickness and direction of reinforcement can be varied at will, many hand lay-up composites use randomly oriented chopped strand mat. A faster alternative for the production of randomly oriented fibre composites is the *spray-up* method. Here the mould/former and its preparation are similar to that in the hand lay-up method but, instead of using mats of fibres, bundled and chopped fibres are blown on through a spray gun. Several more refined techniques, which are capable of achieving better properties and higher production rates, involve compression moulding of pre-impregnated mats (*pre-pregs*) or of pre-mixed resin/fibre *dough-moulding compounds*.

Tubular components such as pressure pipes, in which control of the reinforcement direction is more critical, are usually produced by *filament winding* of resin-impregnated strands or rovings onto a rotating mandril or liner. This simple process can be given great flexibility by numerically controlling the rate of traverse of the winding head along the axis (pressure pipes, for example, may be laminated by winding alternate layers at opposite 45° directions to the axis) and has been extended to produce a variety of complex hollow shapes. Beams of uniform cross-section can be produced continuously by *pultrusion*. The required lamination stack of fibre yarn, mat, veils, etc., converges from separate rolls, passes through a resin bath for impregnation, and is pulled through a heated die which compacts and shapes it in the precise external cross-section required. Curing is completed in a continuous oven.

Whereas these production processes for reinforced thermosets evolved to exploit the directional properties of the material, reinforced themoplastics were originally developed to exploit the speed and flexibility which the processes used for thermoplastics (see Section 12·4) already offered. As a result, glass-filled nylon grades developed for injection moulding, for example, were restricted to very short fibre lengths–little greater than the critical length–by the need for the melt to flow freely. More recently, however, long-fibre injection moulding grades and techniques have been developed. Another current trend is towards the compression moulding of glass-mat thermoplastics, consisting of longer fibres in high-performance thermoplastic matrix materials such as polyetheretherketone (PEEK).

Because most fibres have a greater tensile strength and Young's modulus than polymers, reinforcing a polymer can markedly increase the stiffness and improve the strength. Fibre reinforcement may, or may not, be beneficial as far as toughness is concerned.

Metal matrix composites are probably not so familiar to the reader, but development is continuing in this area. Several methods are used to produce metal matrix composites, but we will only briefly mention three. Two closely related techniques are *powder metallurgy* and diffusion bonding. In powder metallurgy a metal powder is mixed with whiskers or chopped fibres, pressed into the desired shape and fired. During firing the metal powder will sinter into a solid compact as previously described for ceramics (see Section 11·3·3). *Diffusion bonding* employs metal foil instead of powder. Alternate layers of metal foil and fibres are pressed and heated so that sufficient diffusion occurs to bond the foils together in a similar manner to the sintering of a powder. Low melting-point metals can be poured into a mould and allowed to infiltrate around the fibres previously arranged in the mould. This process is known as *liquid metal infiltration* and is particularly suitable for producing small components. Liquid infiltration is usually carried out under pressure applied via a gas or mechanically; when the pressure is applied mechanically the process is known as *squeeze casting*. Most metals have a much higher Young's modulus than polymers, therefore the increase in this constant due to fibre reinforcement is generally not so marked as for polymer matrices. There is usually an improvement in strength, although this is often achieved at the expense of toughness.

Let us consider the mechanical behaviour of polymer and metal matrix composites in more detail by studying the stress–strain curves for aligned fibre composites with $\varepsilon_f^* < \varepsilon_m^*$ and $V_f > V_{min}$. These curves are qualitatively the same for both continuous and discontinuous fibre composites, provided the discontinuous fibres exceed the critical length. In general the deformation proceeds in four stages, although every stage is not present for all composites.

(a) *Stage 1*   Initially, both the matrix and fibres deform elastically and the Young's modulus of the composite, E_c, is given by Eqn (11·17), i.e., a simple law of mixtures. The Young's modulus of a metal matrix is high and so the matrix makes a significant contribution to the modulus of the composite. In contrast, the Young's modulus of a polymer is much lower than that of a metal, and consequently the modulus of the fibres is normally one of two orders of magnitude greater than that of a polymer matrix. It follows that the polymer matrix contributes little to the modulus of the composite, and Eqn (11·17) reduces to

$$E_c \sim E_f V_f \qquad (11·25)$$

In practice the Young's modulus of a polymer matrix composite is even less than $E_f V_f$, even when the fibres are continuous. The reasons put forward for this are (a) the breaking of weak fibres at low strains, and (b) the buckling of fibres during curing. The fact that polymer matrix does not contribute to the modulus of the composite does not matter as long as fibres with a high Young's modulus are used, e.g., carbon fibres. Glass fibres, however, have a low E_f so glass-reinforced polymer (GRP) has a low Young's modulus. The data of Table 11·9 show that an epoxy–carbon fibre composite has a Young's modulus intermediate between that of metal matrix composites and GRP. In many applications the weight of material has to be kept to a minimum; therefore a parameter which is sometimes used to assess elastic characteristics is the *specific modulus* E/ρ, where ρ is the density. The density of polymer matrix composites is low and this means that GRP has a reasonable specific modulus, while a carbon fibre composite has a high specific modulus that is much greater than 0·03 $(GN/m^2)(kg/m^3)$ which is typical of conventional metals such as steel and aluminium alloys (Table 11·9).

Table 11·9

Comparison of the mechanical properties of polymer-matrix composites, metal-matrix composites and metals

Composite	Density (kg/m³)	Young's modulus (GN/m²)	Specific modulus [(GN/m²)/(kg/m³)]	Tensile strength (MN/m²)	Specific strength [(MM/m²)/(kg/m³)]
Aluminium–50% boron	2,700	207	0·077	1,137	0·42
Copper–50% tungsten	14,130	262	0·018	1,207	0·09
Epoxy–58% carbon	1,660	165	0·099	1,517	0·91
Epoxy–72% E glass	2,170	56	0·026	1,640	0·76
Epoxy–72% S glass	2,120	66	0·031	1,896	0·89
Epoxy–63% Kevlar	1,380	83	0·060	1,310	0·95
Aluminium alloy	2,800	71	0·025	600	0·21
Steel (mild)	7,860	210	0·027	460	0·06
Steel (eutectoid)	7,800	210	0·027	3,000	0·38
Brass (Cu–30% Zn)	8,500	100	0·012	550	0·06

(b) *Stage 2*   Stage 2 commences at low strains and is associated with the matrix deforming plastically, whilst the fibres are still elongating elastically. On removal of the load the matrix and the fibres first contract elastically, then the matrix deforms plastically in compression due to the continuing elastic contraction of the fibres. This results in a quasielastic behaviour and the composite almost regains its initial shape and size. The stress–strain curve is not linear in Stage 2 as the matrix is deforming in a non-linear manner. The slope of the stress–strain curve at a given strain is called the *secondary modulus*, E_s, and depends on the elastic properties of the fibres and the work-hardening rate, $d\sigma_m/d\epsilon_m$, of the matrix according to

$$E_s = E_f V_f + (d\sigma_m/d\epsilon_m)V_m \tag{11·26}$$

This stage is particularly significant for metal matrices but is often difficult to detect in the stress–strain curves of polymer-matrix composites. This is because the non-linearity of the polymer matrix is not marked and $d\sigma_m/d\epsilon_m$ is small compared with E_f hence, from Eqn (11·26), E_s is approximately equal to $E_f V_f$.

(c) *Stage 3*   This stage, in which both the matrix and fibres deform plastically, only occurs in certain composites. It is absent in composites containing brittle fibres and is only prominent in composites consisting of strong metal wires in a metal matrix.

(d) *Stage 4*   Finally, the composite fails at a stress σ_c^{TS}, given by Eqn (11·8), if the fibres are continuous and Eqn (11·20) for discontinuous fibres. Just as the specific modulus is sometimes used to characterize elastic behaviour, so the specific strength σ_c^{TS}/ρ is employed for assessing strength. Due to their low densities, and indeed good strengths, polymer matrix composites have high specific strengths which are well in excess of those for conventional metals (Table 11·9).

Finally, let us turn our attention to ceramic matrix composites. Generally ceramics have sufficient stiffness and strength and hence the prime objective

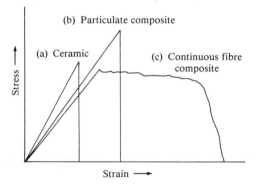

Fig. 11·38 Comparison of the stress–strain behaviour of (a) ceramic, (b) particulate, or short random fibre, composite and (c) a continuous fibre composite

in producing ceramic matrix composites is not to enhance these properties. We have seen in Section 11·5 that the major problem with ceramics is their brittleness and sensitivity to flaws. Thus the main aim of reinforcing a ceramic, whether with fibres or particles, is to improve the fracture toughness. This improvement is illustrated by the stress–strain curves of Fig. 11·38. The failure of a ceramic is catastrophic and the small area under the stress–strain curve is indicative of the low toughness. The failure of a particulate-reinforced, or short random fibre-reinforced, ceramic composite is similar to that of the ceramic matrix in that it is catastrophic; note however that the toughness has increased as demonstrated by the larger area under the stress–strain curve. In contrast the failure of a continuous fibre composite is not catastrophic and the composite maintains a substantial load-carrying capacity well after failure has commenced. Therefore not only has the fracture toughness increased but the failure mode is more desirable. Thus the attraction of both particulate- and fibre-reinforced ceramic composites is the improved toughness which makes them less sensitive to flaws and more reliable during assembly and service.

Ceramic matrix composite production is the least established composite production technology. Due to the high melting point of most crystalline ceramics, techniques involving a melt are unsuitable. Melt infiltration is even difficult with precursor melts for glasses and glass-ceramics as these are more viscous and more reactive than molten metals. The routes which at this stage of development look most promising are: (i) conventional mixing of the components followed by cold pressing and firing or hot pressing, (ii) mixing using a slurry followed by sintering; (iii) mix a ceramic-producing polymer and the toughening phase then pyrolyse the polymer; and (iv) chemical vapour deposition of the matrix, i.e., impregnate a pre-form with a gas which deposits the ceramic.

11·17 Some commercial composites

11·17·1 Carbon–carbon and zirconia-toughened alumina

Two examples of ceramic-matrix composites will be discussed in this section. These have been chosen because of the marked differences in method of production, structure and properties.

Carbon–carbon composites consist of fibrous carbon in a carbonaceous matrix. Although both constituents are the same element their morphology and structure differ greatly from composite to composite. The carbon in these composites may vary from highly crystalline graphite through a wide range of quasicrystalline forms such as *turbostratic* carbon, to amorphous, glassy carbon. By control of the proportions, morphology and crystallography of the constituents, together with the porosity content, it is possible to produce a series of composites ranging from a soft, porous material used for furnace insulation to a hard, high temperature, structural material.

The composite is produced by impregnating the fibrous reinforcement, which is available in many forms from unidirectional lay-ups to multi-directional weaves. Impregnation is achieved by two basic methods: either (i) *carbonization* of an organic solid or liquid, or (ii) *chemical vapour deposition* (CVD) of carbon from a hydrocarbon. Examples of the organic precursor materials for method (i) are pitch and phenolic resin. A typical manufacturing procedure using a phenolic resin would be to cure after impregnation of the fibrous reinforcement and then to pyrolyse at a higher temperature. Shrinkage occurs as the resin decomposes during pyrolysis, and therefore, in order to obtain high densities, a number of impregnation, cure and pyrolysis cycles are required. In the CVD process carbon is deposited by passing a hydrocarbon gas through the fibrous reinforcement which is held at a temperature of about 1,100°C.

The main forte of structural carbon–carbon composites is that they maintain their strength and toughness to high temperatures of the order of 2,000°C. The superior high temperature mechanical properties, especially when assessed in terms of specific strength, of these composites compared to alumina, silicon nitride and to a high temperature nickel alloy (*superalloy*) is illustrated in Fig. 11·39.

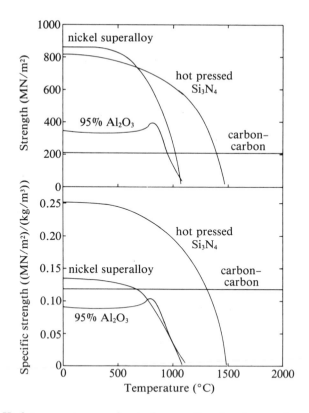

Fig. 11·39 High temperature mechanical properties of carbon–carbon composite and some other high temperature materials

The problem with carbon–carbon composites is that they oxidize and sublime in the presence of oxygen at temperatures in excess of 600°C. Therefore for high temperature applications we must protect the composite by either using an inert atmosphere or a coating. For example, the leading edges and the nose cap of the space shuttle are sealed by a coat of silicon carbide. However, this does not give complete protection as, due to the thermal mismatch between silicon carbide and the underlying composite, microcracks develop in the coating. Consequently additional protection is supplied by impregnating the surface with tetraethylorthosilicate which enhances oxidation resistance by leaving silica in the microcracks.

This chapter has been mainly concerned with fibre-reinforced composites for the simple reason that, in general, fibres produce a greater improvement in mechanical properties than particles. Nevertheless there are some successful particulate-reinforced composites in commercial production, of which *zirconia-toughened alumina* (ZTA) is an example.

ZTA is a composite consisting of a fine-grained polycrystalline alumina matrix reinforced with about 15 vol % ZrO_2 (Fig. 11·40). The ZrO_2 particles, which typically contain 3 mol % of the stabilizing oxide Y_2O_3, are less than 1 μm in size. In contrast to the carbon–carbon composites, ZTA is manufactured by the conventional ceramic technology of pressing and sintering. The Al_2O_3–ZrO_2 mixture for sintering may be obtained by mechanically mixing Al_2O_3 and ZrO_2 but this tends to give inhomogeneous products of variable particle size. Better quality mixtures are prepared by a *sol–gel* route. A ZrO_2–Y_2O_3 *sol*, which is a dispersion of small particles of less than 100 nm, termed colloids, in a liquid, is added to an Al_2O_3 slip and a gel produced. A *gel* is simply a sol that has lost some liquid to give a significant

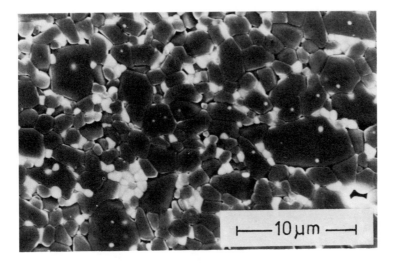

Fig. 11·40 Scanning electron micrograph of zirconia-toughened alumina showing a fine dispersion of small metastable tetragonal zirconia particles (white) in a coarser polycrystalline alumina matrix (dark)

Table 11·10

Room temperature mechanical properties of zirconia-toughened alumina (ZTA)
and the alumina matrix

	ZTA	Alumina
Strength (MN/m^2)	461	287
Weibull modulus m	9·0	10·4
Toughness K_{1C} (MN/m$^{3/2}$)	6·5	3·8
Hardness (GN/m^2)	16·1	12·5

increase in viscocity. The gel is then dried, and after some further treatment,
a fine homogeneous powder suitable for compaction is obtained.

Because of their size, and the stabilizing effect of the Y_2O_3, the zirconia
particles in ZTA are in the metastable tetragonal condition. As previously
explained in our study of zirconia (Section 11·9) the metastable tetragonal
particles undergo a martensitic transformation to the monoclinic structure
under the action of the stress at a crack tip. Thus the same transformation
toughening mechanism (Fig. 11·20) found in some zirconias is also operative
in ZTA. The improvement in the toughness and strength of ZTA over the
corresponding values for the alumina matrix is shown by the mechanical
property data of Table 11·10.

11·17·2   Alumina-reinforced aluminium alloys

Fibre-reinforced aluminium alloys are produced by powder metallurgy or by
squeeze casting. We will only consider those produced by the latter method

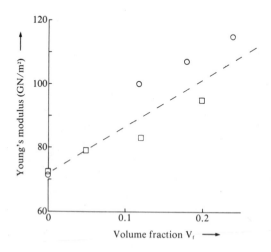

Fig. 11·41 Graph illustrating the increase in the Young's modulus of two
aluminium alloys obtained by reinforcing with α-alumina fibres (circles-Al-9%
Si-3% Cu, J. Dinwoodie *et al.*, 5th Int. Conf. Composite Materials, 1985;
squares-Al-12.5% Si-1% Cu-1% Ni, L. Ackermann *et al.*, 5th Int. Conf.
Composite Materials, 1985)

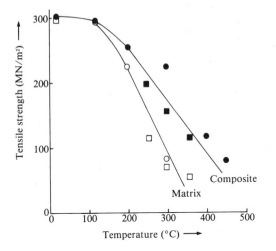

Fig. 11·42 Graph demonstrating the superior strength of aluminium alloy matrix composites at elevated temperatures (circles–Al–9% Si–3% Cu, J. Dinwoodie *et al.*, 5th Int. Conf. Composite Materials, 1985; squares–Al–12.5% Si–1% Cu–1% Ni, L. Ackermann *et al.*, 5th Int. Conf. Composite Materials, 1985)

and, as this is a casting technique, most of the aluminium alloys which have been reinforced are conventional casting alloys, such as Al–Si and Al–Mg base alloys. In squeeze casting of these materials the pre-form, die and ram are preheated before the introduction of the molten alloy at a controlled degree of superheat. Impregnation is then achieved at pressures normally in the range 50–100 MN/m². A variety of fibres have been used for the pre-forms but we will discuss the effects of reinforcing with short, α-alumina fibres which have a tensile strength and Young's modulus of 2,000 MN/m² and 300 GN/m² respectively.

The Young's modulus of aluminium and its alloys is low compared to most structural metals and one of the major benefits of reinforcement with high modulus fibres is the significant increase in stiffness (Fig. 11·41). In contrast there is only a minor improvement in the room temperature tensile strength although, due to the superior properties of the ceramic fibres at elevated temperature, the composites maintain their strength to higher temperatures than the matrix alloys (Fig. 11·42). In common with most fibre-reinforced metals the toughness and ductility of the matrix is degraded; Table 11·11 gives some examples of the low ductilities shown by alumina-reinforced aluminium and aluminium alloys. As far as other mechanical properties are concerned, these composites can show an improvement in both wear and fatigue resistance, e.g., the 10^7 cycles endurance limit of Al–12% Si–1% Cu–1% Ni casting alloy may be increased by 30% by reinforcing with 0.2 volume fraction of alumina.

One of the first commercial applications of these composites has been for the pistons of heavy duty diesel engines. This component is subjected to wear and to thermal and mechanical cyclic stresses at elevated temperatures which,

Table 11·11
Room temperature ductility of composites of aluminium and its alloys containing
0·2 volume fraction of alumina fibres

Matrix	Ductility (%)
Al	4·0
Al–2·5% Mg	3·3
Al–10% Mg	1·3
Al–12% Si–1% Cu–1% Ni	<1

in certain areas, may approach 500°C. Other potential applications for these
lightweight cast composites are brake drums, gears and rotary compressors.

11·17·3   Polymer matrix composites

A wide variety of polymer matrix composites are commercially available;
therefore, rather than discuss one or two of these in detail we will expand on
some of the points made in Section 11·16. The first point to emphasize is that
the polymer matrix may be a thermoset (e.g., epoxy, polyester), or a
thermoplastic (e.g., polycarbonate, acrylic), or even a semicrystalline
material (e.g., polyetheretherketone, nylon). In fact virtually all types of
polymers have, at one time or another, been used as the matrix for some form
of composite.

The reinforcement most widely used with polymers is fibrous. The fibres
are normally carbon, glass or 'Kevlar', although boron fibres find some
applications–mainly in the aerospace industry–and the relatively new polyethylene fibre will no doubt form part of the general market in the future. In
addition, there are many types of particulate reinforcement which differ
greatly in size, shape and composition, and consequently in their effect. An
important feature of many new systems is the combination of different
fibres, sometimes with other reinforcing, or toughening agents, to form
hybrid composites. The aim is to capitalize on the best properties of each
component, and particularly to benefit from any synergistic effect.

There has been a very rapid growth of fibre-reinforced polymer matrix
composites over the last few years, which shows every sign of continuing. The
main reason for this growth is the high specific modulus and specific strength
of these materials (Table 11·9). The high values for these properties means
that in many cases the weight of components can be reduced considerably.
Limiting the weight is an important factor in moving components,
particularly in transport and aerospace, where weight reductions save energy
and give greater efficiency. The wide range of applications for polymer
matrix composites is well illustrated by the examples given in Table 11·12.

There are many manufacturing possibilities for fibre-reinforced plastics,
but it is important to recognize that the particular means chosen, and the
processes involved, can have a profound effect on the final properties of the
composite owing to their effect on microstructure and internal stresses.
Those processes which allow the placement of long fibres in known preferred
directions have a clear advantage in terms of the composite mechanical

Table 11·12
Examples of applications of polymer–matrix composites

Industrial sector	Examples
Aerospace	Wings, fuselages, radomes, antennae, tail-planes, helicopter blades, landing gear, seats, floors, interior panels, fuel tanks, rocket motor cases, nose cones, launch tubes
Automobile	Body panels, cabs, spoilers, consoles, instrument panels, lamp-housings, bumpers, leaf springs, drive shafts, gears, bearings
Boats	Hulls, decks, masts, engine shrouds, interior panels
Chemical	Pipes, tanks, pressure vessels, hoppers, valves, pumps, impellers
Domestic	Interior and exterior panels, chairs, tables, baths, shower units, ladders
Electrical	Panels, housings, switchgear, insulators, connectors
Leisure	Motor homes, caravans, trailers, golf clubs, racquets, protective helmets, skis, archery bows, surfboards, fishing rods, canoes, pools, diving boards, playground equipment

properties over other processes which sacrifice the high potential offered by the fibres for reasonable isotropy of the composite. Processes which allow good control over fibre alignment include vacuum bag, pressure bag and autoclave, filament winding, pultrusion, compression moulding, resin injection, and, to a lesser extent, hand lay-up. Those processes which give little control over fibre alignment include injection moulding, centrifugal casting, reinforced reaction injection moulding (RRIM) and spray-up. These latter techniques are, however, ideally suited for short fibres, and, in general, confer good isotropy to the product. Of the former processes only compression moulding and hand lay-up are routinely used for both long and short fibre composites. Many of the techniques referred to above can cope with both thermoset and thermoplastic resin composites, but care is required in identifying the best method for economical production.

Problems

11·1 Discuss, from the atomic point of view, the factors that determine the mechanical properties of ceramics.

11·2 Using the SiO_2–Al_2O_3 phase diagram of Fig. 11·17 answer the following:
(a) What fraction of the total ceramic is liquid for a 30 wt % Al_2O_3 composition at 1,600°C?
(b) What are the compositions of the phases involved in the eutectic reaction at 1,595°C?
(c) Describe the changes that take place as a 60 wt % Al_2O_3 composition cools from 2,000°C to room temperature under equilibrium conditions.

11·3 What is the cause of thermal stresses in a glass cooled from the melt and how may these stresses be reduced? Describe a process in which thermal stresses are used to an advantage in glass technology.

11·4 Define the following terms: clinker, cement, and concrete. Discuss the main reactions occurring during the hydration of cement making reference to their heats of hydration and their contribution to strength.

11·5 With reference to composites with discontinuous fibres, discuss the significance of the 'critical length'.

From the following date calculate the critical length and hence the tensile strength of the composite:

Fibre			Matrix		
Diameter (μm)	Tensile strength (MN/m^2)	Length (mm)	Shear strength (MN/m^2)	Tensile stress at failure strain (MN/m^2)	Volume fraction (%)
100	3,000	5	100	250	50

11·6 Discuss the elastic and quasi-elastic behaviour of composites.

11·7 During service a ceramic valve which is operating at 250°C is likely to be suddenly quenched to room temperature by a flowing aqueous solution. The valve could be manufactured from either of two ceramics, designated A and B, the properties of which are given below. Discuss which ceramic you would select for this application.

	A	B
Young's modulus E (GN/m^2)	120	350
Coefficient of thermal expansion α (K^{-1})	2×10^{-6}	9×10^{-6}
Fracture stress σ_f (MN/m^2)	230	270
Poisson's ratio ν	0·25	0·27
Thermal conductivity at 50°C K (W/m K)	14	8

Self-Assessment questions

1 The glass transition temperature is

A) a characteristic of all metals

B) associated with the change from a supercooled liquid to a glassy state

C) a well defined temperature for a given material

D) cooling rate dependent

E) identical to the melting point.

2 A glass-ceramic is predominately crystalline with a small amount of residual glass.

A) true   B) false.

3 The addition of a network modifier to silica

A) disrupts the network structure

B) enhances the network structure

C) produces vacancies

D) produces non-bridging oxygen atoms

E) increases the viscosity.

4 Slip casting

A) is the casting of a molten material

B) is the casting of a slip or suspension of particles

C) involves slip by dislocation motion

D) is commonly used for the production of glass articles

E) is particularly useful for complex and irregular shapes

F) involves the use of a porous mould.

5 Unfortunately the density variation obtained by isostatic pressing is greater than that from unidirectional pressing.

A) true   B) false.

6 Vitrification is

A) the same as devitrification

B) the densification in the presence of a viscous liquid phase

C) the transformation from a glassy to a crystalline state

D) important during the firing of clays

E) associated with the hydration to tobermorite gel.

7 The viscosity of glass decreases linearly with increasing temperature.

A) true   B) false.

8 Sintering is the consolidation of a powder at an elevated temperature which is below the melting point of any major phase.

A) true   B) false.

9 SiC is

A) covalently bonded

B) ionically bonded

C) produced by the controlled devitrification of glass

D) produced by reaction bonding

E) soft for a ceramic.

10 The microstructure shown in this micrograph is that of a

A) complex, clay-based, multiphase ceramic

B) glass

C) glass-ceramic

D) alumina

E) sintered material

F) devitrified material.

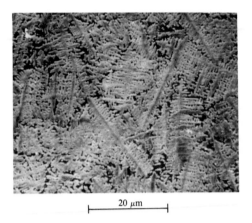

20 μm

11 Refractories in the SiO_2–Al_2O_3 system containing 5 to 50 wt % Al_2O_3 are

A) produced from blends of naturally occurring clays called fireclays
B) produced from fused alumina bonded by a little fireclay
C) produced by melting SiO_2 and casting
D) more refractory the greater the Al_2O_3 content
E) only of commercial importance if of the eutectic composition.

12 Concrete goes on hardening for years.

A) true   B) false.

13 The production of prestressed concrete is shown in the sequence of events (a), (b), and (c).

A) the steel wires are deformed elastically in (a) and (b)
B) the steel wires are deformed plastically in (a) and (b)

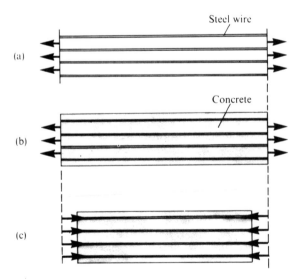

C) in (c) the length of the steel wires is reduced by an externally applied compressive force

D) in (c) the length of the steel wires is reduced due to elastic contraction as the applied tensile force is removed

E) in (c) the concrete is put into tension

F) in (c) the concrete is put into compression.

14 The strength of concrete in compression is about eight times that in tension.

A) true   B) false.

15 C_3A is a constituent of concrete and has the following characteristics:

A) very rapid rate of hydration

B) slow rate of strength development

C) high ultimate strength

D) high heat of hydration

E) in the presence of gypsum hydrates to form ettringite.

16 The proportion of C_3S is increased in rapid hardening cement.

A) true   B) false.

17 The Young's modulus, E_c, of a composite with a polymer matrix is:

A) generally higher than E_c for a metal matrix composite

B) independent of the fibres used for a given polymer matrix

C) approximately proportional to the Young's modulus of the fibres

D) often called the secondary modulus.

18 The strength of an aligned discontinuous fibre composite is always less than that of an aligned continuous fibre composite with the same volume of fibres.

A) true   B) false.

19 The critical volume has to be exceeded before fibres lead to strengthening of a composite with $\varepsilon_f^* < \varepsilon_m^*$.

A) true   B) false.

20 This figure is applicable to composites and shows:

A) the shear stress in a fibre

B) the shear stress in the matrix

C) the tensile stress in a fibre

D) the tensile stress in the matrix

E) the shear stress along the length of the composite

F) the tensile stress along the length of the composite.

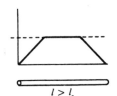

$l > l_c$

21 In a tensile test the deformation proceeds in four distinct stages for all composites.

A) true   B) false.

22 During the initial stage in the deformation of a composite with a metal matrix

A) there is a hyperbolic relationship between stress and strain

B) both the matrix and the fibres deform elastically

C) the Young's modulus of the composite is given by a simple law of mixtures

D) the Young's modulus of the composite is determined primarily by the yield stress of the metal matrix

E) the majority of the fibres break and rotate so that their longitudinal axes are parallel to the stress axis.

23 A composite with aligned fibres is mechanically anisotropic.

A) true   B) false.

24 A high coefficient of thermal expansion is beneficial as far as thermal shock resistance is concerned.

A) true   B) false.

25 The tetragonal-to-monoclinic transformation in zirconia is:

A) a nucleation and thermally activated growth process
B) a martensitic transformation
C) accompanied by an increase in volume
D) unaffected by the addition of oxides such as yttria
E) a high temperature tranformation, i.e., takes place at temperatures in excess of 2,000°C.

26 The transverse strength of an aligned fibre composite is very low, usually less than the strength of the matrix.

A) true   B) false.

27 Carbon–carbon composites:

A) are produced by mixing in a slurry and sintering
B) are produced by carbonization
C) are produced by chemical vapour deposition
D) have good oxidation resistance
E) maintain their strength and toughness up to temperatures of the order of 2,000°C.

Each of the sentences in questions 28–38 consists of an assertion followed by a reason. Answer:

A) If both assertion and reason are true statements and the reason is a correct explanation of the assertion
B) If both assertion and reason are true statements but the reason is *not* a correct explanation of the assertion
C) If the assertion is true but the reason contains a false statement
D) If the assertion is false but the reason contains a true statement
E) If both the assertion and reason are false statements.

28 The dislocations in a covalently bonded ceramic are wide *because* the covalent bond is directional.

29 During sintering the total energy associated with interfaces is reduced *because* high energy solid/gas interfaces are replaced by lower energy solid/solid interfaces.

30 When making concrete it is important to have the correct ratio of water to cement *because* too much water dissolves the aggregate.

31 Ceramics are generally good electrical insulators *because* all the valency electrons are occupied in the covalent-ionic bonds.

32 After manufacture most glass articles are annealed at a temperature near to T_g *because* at T_g there is a change in slope of the specific volume versus temperature curve.

33 During the production of cement care is taken to remove all traces of gypsum from the clinker *because* gypsum increases the rate of hydration of C_3A.

34 The float process produces plate glass with a good surface finish *because* tin oxide from the tin bath diffuses into the glass and hardens the surface.

35 A composite with a polymer matrix has a very low specific modulus *because* the density is low.

36 The critical aspect ratio varies from composite to composite *because* polymers are less dense than metals.

37 The Weibull modulus of a ceramic is lower than that of a metal *because* a ceramic is less tough and more sensitive to flaws.

38 Wake toughening is the most effective and most widely observed toughening mechanism in composites *because* it always occurs before any debonding.

Answers

1 B, D	**2** A	**3** A, D	**4** B, E, F
5 B	**6** B, D	**7** B	**8** A
9 A, D	**10** C, F	**11** A, D	**12** A
13 A, D, F	**14** A	**15** A, D, E	**16** A
17 C	**18** A	**19** A	**20** C
21 B	**22** B, C	**23** A	**24** B
25 B, C	**26** A	**27** B, C, E	**28** D
29 A	**30** C	**31** A	**32** B
33 E	**34** C	**35** D	**36** B
37 A	**38** E		

PLASTICS AND POLYMERS

by P. S. Leevers

12·1 Introduction

Synthetic organic polymers are the most widely used new materials of the 20th century. Their ancestors, naturally occurring macromolecular materials such as natural rubber, wood, cotton, wool and hardened oils, have been known for much longer and valued for the qualities conferred by their basic structure, but the first synthetic polymers were not produced until the 19th century. Synthetic polymers now dominate so many applications–not only as plastics and rubbers but also as fibres, paints and adhesives–that the rate of increase in their consumption is an important barometer of an industrial economy. In this chapter we concentrate mainly on the use of polymers, usually compounded with various additives (e.g., pigments and fillers), to form solid *plastics* from which components and structures can be formed.

Most early plastics were *thermosets,* such as phenol formaldehyde (filled with wood flour to produce 'Bakelite') and melamine formaldehyde. These cure (set) permanently during forming into their final shape, and are subsequently quite heat-resistant. A second major class of polymers, the rubbers or *elastomers,* also owe their spectacular elastic properties to a post-moulding process during which they set and acquire stability. The majority of polymers consumed nowadays are *thermoplastics*, principally the 'commodity' thermoplastics polyvinyl chloride (PVC), polyethylene (PE), polypropylene (PP) and polystyrene (PS). These are repeatedly remouldable under heat and pressure, making their conversion into finished products particularly rapid and economical. The price for this re-formability is their relative sensitivity to temperature, and in particular their tendency to creep under load. Outside the electrical industry, in which their intrinsic high resistivity and dielectric properties were exploited relatively early, these materials originally gained popularity less on their excellence in performance than on their convenience as substitutes for other materials, offering low cost per unit volume, easy processability, and good decorative qualities or transparency.

Since the mid-1970s, however, this situation has changed rapidly. While the cost of the commodity thermoplastics has increased considerably (helping to outdate 'cheap plastic' as a term of abuse), greater understanding of the relationships between their molecular structure and their bulk propeties has directed the steady introduction of new grades, spectacularly improved by modification, blending or filling. The well-established use of a polyamide (PA, 'nylon') in engineering components had already spearheaded a new wave of high-performance 'engineering plastics' such as polycarbonate (PC) and polyoxymethylene (POM, 'acetal'), which have gained respect as reliable load-bearing materials. More recently, it has virtually become possible to *design* a special polymer to a particular portfolio of required properties. Some of these new polymers offer mechanical performance or physical properties which would have seemed unattainable in 1970.

Underlying the enormous diversity of plastics and polymers are common structural features, which distinguish them as a class of material very different from those which we have considered so far. It is the relationships between these structural features and properties which this chapter will introduce. Since the final properties of these materials are heavily influenced by their history and the method by which they have been formed, we will also consider the techniques used to do so.

12·2 Molecular structure

12·2·1 Monomers and polymers

Polymers are defined by their molecular structure: they contain long, covalently bonded chains of repeated sub-units (*mers*). These chains are sometimes branched, and sometimes interconnected at points along their length (by branch chains or by short crosslinks) to form a network.

A *homopolymer* is constructed from a single type of mer. The distinctive characteristics of a polymer can be illustrated by comparing linear polyethylene, whose molecule is a very long chain of i ethylene mers $[-(C_2H_4)-]_i$, with the linear alkane series, $[-C_{2i}H_{4i+2}$ [(Fig. 12·1). Starting from ethane ($i = 1$), the alkanes show steadily changing physical properties as i increases

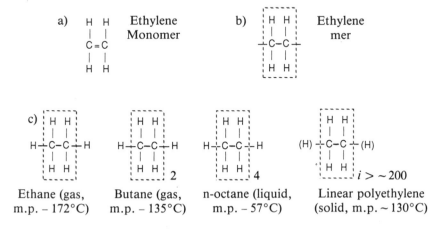

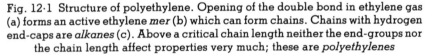

Fig. 12·1 Structure of polyethylene. Opening of the double bond in ethylene gas (a) forms an active ethylene *mer* (b) which can form chains. Chains with hydrogen end-caps are *alkanes* (c). Above a critical chain length neither the end-groups nor the chain length affect properties very much; these are *polyethylenes*

and the chain becomes longer. Ethane itself and butane ($i = 2$) are gases, while *n*-octane ($i = 4$) is a liquid: as we would expect, the melting point and boiling point rise as the size of the molecule increases. From $i = 10$ to 15 (light oils, the paraffins) these substances progress through greases and waxes until, at a critical i of about 200, many physical properties (e.g., melting point and density) settle to values substantially independent of i. The chain ends (in this case just hydrogen atoms) have little influence on properties, and the substance takes on the character of a true *polymer*, in this case high-density polyethylene (HDPE); i is the *degree of polymerization*. Members of the homologous series with $i < 200$ are termed *oligomers*: most commercial plastics are based on polymer chains 10 to 100 times longer. Properties are dominated by the strength of the long, covalently bonded chain backbone (in this case consisting only of carbon atoms), which greatly restricts the mobility of individual mers along it, and by the weaker secondary bonding between mers of adjacent molecules, which restricts inter-chain slippage. Because of this weakness, HDPE is a thermoplastic: its molecules can be separated reversibly by heat or solvent action.

In practice, polyethylene is synthesized from pure ethylene *monomer* by a *chain-growth* mechanism. An *initiator* starts the polymerization reaction by attacking a few monomer units, opening the double bond to provide active seeds from which a reaction proceeds quickly by a domino effect. The reaction may terminate in any of several ways: for example, exchange of a hydrogen atom from one active chain end to another will terminate one by a methyl group (CH_3) and the other by a double bond ($=CH$). The active lifetime of a growing chain, from initiation to termination, might be 10–15 minutes. Other polymers are formed by *step-growth* polymerization reactions. Here all of the monomer units are activated at the two or more sites where they will bond to their neighbours, and they polymerize at a rate which falls as the reactants are consumed. Step-growth polymers include the polyamides (nylons), which polymerize by a reaction between acid and base end-groups on the monomer (Fig. 12·2), and polyesters such as polyethylene terephthalate. Because such reactions often liberate water or other gases, their products are sometimes referred to as *condensation* polymers.

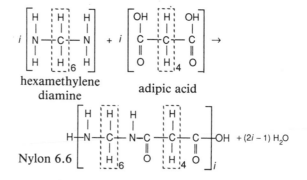

Fig. 12·2 Nylon 6·6, a polyamide, is formed by a *step-growth* reaction in which a molecule of water is condensed out from between reacting monomers

12·2·2 Molecular weight

In considering the properties of ethylene oligomers and polymers we considered *monodisperse* HDPE in which every molecule has the same molecular weight $M = M_r i$, M_r being the molecular weight of the repeat unit. For polyethylene, $M_r = 28$, and the *critical molecular weight* at which the monodisperse oligomer acquires the characteristics of a polymer is $M_c = \sim 4,000$. In practice, polymerization produces a *polydisperse* material containing a wide range of molecular weights, whose population distribution can have a strong influence on mechanical and melt flow properties.

There are several ways of describing this distribution. Given a large enough sample, we could imagine counting the number N_i of chains having each degree of polymerization i and thus molecular weight $M_i = M_r i$. If the distribution of chain lengths (Fig. 12·3) is approximately symmetrical it will peak near the *number-average molecular weight*:

$$\overline{M}_n \equiv \frac{\sum_i N_i M_i}{\sum_i N_i} \tag{12·1}$$

the simple arithmetic mean of the lengths of all the individual chains. Alternatively, we could plot the *weight* of each fraction, $W_i = N_i M_r i$. This distribution would look rather different, emphasizing the fractions of longer and heavier chains, and would peak near the *weight-average molecular weight*:

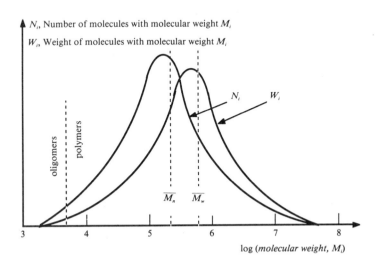

Fig. 12·3 Typical distribution of molecular weights in high-density polyethylene (HDPE), plotted by number and by weight

$$\overline{M}_w \equiv \frac{\sum_i W_i M_i}{\sum_i W_i} \qquad\qquad (12\cdot2)$$

The distributions shown in Fig. 12·3 represent a typical commercial HDPE. The very shortest chains are oligomers having only about 10 repeat units. Few as they are, these contribute to bulk properties, which is why some commercial polyethylenes feel and smell rather 'waxy'. For a monodisperse material $\overline{M}_n = \overline{M}_w$, but for polydisperse materials the ratio $\overline{M}_w/\overline{M}_n > 1$ and its value, typically 5–10 for HDPE, provides a good measure of the breadth of dispersion.

12·2·3   Branching, tacticity and copolymers

A polymer of quite different properties is formed if a *side-group*, in the case of HDPE just −H, is replaced by another complete chain to form a *branched* polymer (Fig. 12·4). In the case of polyethylene, the branched form is *low-density* polyethylene (LDPE), which would typically have a long branch every 500 units or so and a short branch (one or two mers) about every 50 units. Repeated branching in some polymers can lead to a tree-like structure. We will see later that branching prevents molecules from packing closely, affecting not only the density but many other properties as well. Different chain architectures can be produced during polymerization of the same monomer by controlling the reaction pressure and the action of catalysts, providing another means of tailoring polymers for specific properties.

The linear polyethylene structure has been drawn schematically with the backbone C–C bonds lying along a straight line. However, because the valence bonds in a carbon atom are actually directed towards the corners of a regular tetrahedron, as illustrated by the diamond structure (Fig. 6·4), the most extended line that the backbone can actually take is a zig-zag in one plane. The hydrogen atoms in HDPE stick out sideways symmetrically from

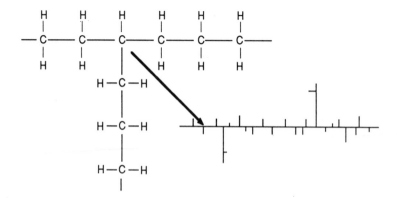

Fig. 12·4 Branching in low-density polyethylene (LDPE)

this plane–one pair upwards then one pair downwards–from the remaining two bonds of each C atom. The C–H bond length, 0·109 nm, is rather less than that of the C–C bond, 0·154 nm. The only other commercially important plastic with a structure as simple as this is polytetrafluoroethylene (PTFE), in which all four hydrogen atoms in the repeat unit are replaced by fluorine.

Table 12.1
Structure, glass transition temperature and crystalline melting temperature of various thermoplastics

Polymer abbreviation, name		Repeat unit	T_g(°C)	T_m(°C)
PE	Polyethylene	$[-CH_2-CH_2-]$	−90	130
PP	Polypropylene	$[-CH_2CH-]$ $\underset{CH_3}{\mid}$	−10	175
PS	Polystyrene	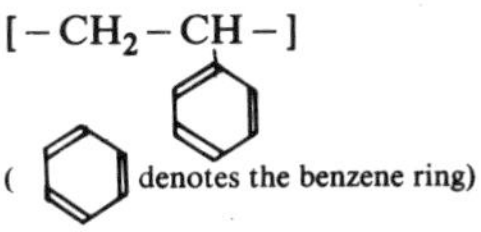 $[-CH_2-CH-]$ with benzene ring ($\bigcirc$ denotes the benzene ring)	95	–
PVC	Polyvinyl chloride	$[-CH_2-CH-]$ $\underset{Cl}{\mid}$	80	200
PMMA	Polymethyl methacrylate	$[-CH_2-\underset{COOCH_3}{\overset{CH_3}{\mid}}C-]$	105	–
PTFE	Polytetra- fluoroethylene	$[-CF_2-CF_2-]$	–	325
PA PA6	Polyamides: Nylon 6	$[-NH(CH_2)_5CO-]$	–	225
PA6·6	Nylon 6·6 etc.	$[-NH(CH_2)_6NH-CO(CH_2)_4CO-]$ (see Fig. 12·2)	–	267
POM	Polyoxy- methylene (acetal)	$[-CH_2-O-]$	−75	175
PET	Polyethylene terephthalate	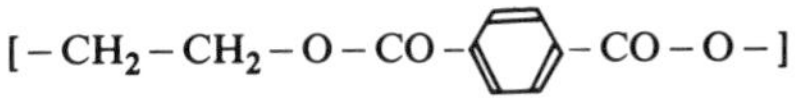 $[-CH_2-CH_2-O-CO-\bigcirc-CO-O-]$	65	255
PC	Polycarbonate	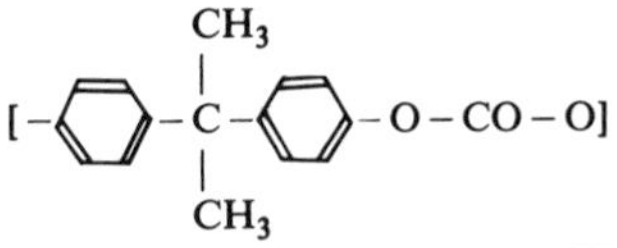 $[-\bigcirc-\underset{CH_3}{\overset{CH_3}{\mid}}C-\bigcirc-O-CO-O]$	150	267

Table 12·1 shows the chemical structure of some other thermoplastics. The most important of the other *polyolefins*, those constructed from simple hydrocarbon monomers, is polypropylene (PP), which is similar in many ways to HDPE. It is formed by polymerization of propylene (propane) gas into linear chains of the mer $-[C(CH_3)H-CH_2]-$, an ethylene group in which one of the hydrogen atoms has been replaced by a methyl group. The substitution of a hydrogen atom by chlorine, on the other hand, produces polyvinyl chloride (PVC), while its replacement by benzene ring yields polystyrene (PS). PE, PP, PVC and PS together make up over 80% of the market in thermoplastics. Polymers whose repeat units, unlike that of PE, are asymmetrical, share the important structural property of *tacticity*, possessing chemically distinct configurations or *stereo-isomers*. In PP, for example, each propylene repeat unit could join the chain end either head (methyl group end) to head, or head to tail; in practice the first form is rare because the side-groups tend to interfere with each other. In the head-to-tail form (Fig. 12·5) the methyl groups can be aligned regularly on the same side of the backbone plane (the *isotactic* form) or can alternate regularly from one side to the other (the *syndiotactic* form). In the *atactic* form there is no significant regularity at all.

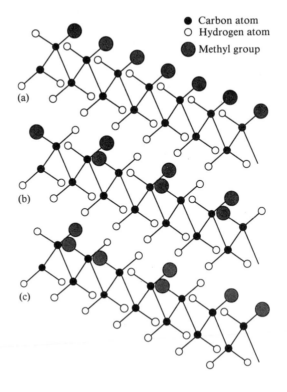

Fig. 12·5 Tacticity of polypropylene: (a) isotactic, (b) syndiotactic and (c) atactic forms

A mixed-composition structure unique to polymers is that of a hetero-geneous chain: a *copolymer*, in contrast to a *homopolymer* in which every mer is the same. Polymerization of a mixture of monomers, say A and B, allows the range of available molecular structures to be expanded even further to produce a material which blends the properties of its constituents. Thus many commercial 'polypropylene' grades, for example, are actually ethylene–propylene copolymers whose exact recipes have been adjusted to optimize properties such as impact resistance. The most basic structural variable in a copolymer is the order of A and B mers along the chain, which can vary from the case of a *random* or *ideal* copolymer in which the chains have no detectable regularity (say ABBAABABAABAAA), through struc-tures in which there are relatively long AAAA or BBBB segments (*block* copolymers) to highly ordered ABABAB chains (*alternating* or *regular* copolymers). Another level of structure is illustrated by *graft* copolymers, which are formed by connection of complete BBB side-branches onto an AAA backbone.

12·3 Mechanics of flexible polymer chains

Having examined the composition of some simple polymer chains, we turn to the way in which they behave as structural elements within a material. The character of a complete chain can first be studied in isolation by forming a very dilute solution of the polymer; in a suitable solvent, each molecule will behave as if it were suspended in free space. This is, in fact, the state in which many properties of a polymer are usually characterized: for example, the viscosity of a dilute polymer solution can be measured to yield a good estimate of its average molecular weight.

12·3·1 Chain conformations

The most striking properties of, for example, a linear polyethylene chain are its *flexibility* and *mobility*. Although a single covalent C–C bond in the backbone is very rigid in both length and direction (giving diamond its extremely high elastic modulus), it can rotate freely. However, if we could somehow pivot two chain sections against each other around a C–C bond, we would find that the bond would tend to settle into one of the three directions in which the hydrogen atoms were furthest apart. The most preferred (lowest energy) conformation of these three is the *trans* state already described for PE, in which successive C–C bonds zig-zag in the same plane. The other two *gauche* states are mirror images of each other at ±120° from this plane (Fig. 12·6). As Fig. 12·7 illustrates, the implications of this are that any possible 'static' equilibrium conformation of a carbon chain backbone can be drawn as a random walk along the network of a diamond lattice–with the restriction, of course, that it cannot cross itself.

A series of *gauche* bonds would wind the chain into a tight coil in which the alternate H atoms would repel each other strongly, so the *trans* state is more stable and zig-zag sequences are common. Nevertheless, a long chain sequence always contains some *gauche* bonds and is therefore much more

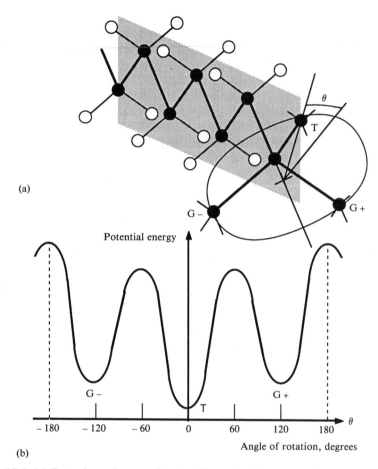

Fig. 12·6 (a) A single carbon–carbon backbone bond can pivot freely, but inter-ference of the attached hydrogen atoms imposes three preferred directions: the *trans* position and two *gauche* positions; (b) the preferred directions correspond to potential energy minima

statically *flexible* than it would be if it were linear, since it can flex and stretch by C–C bond rotation rather than by bond bending or extension. It is also much more compact, because the frequent kinks wind it into a loose random coil with a radius of gyration

$$r_g = CM^{0.5} \tag{12·3}$$

where C is a constant. The diameter of gyration is much smaller than the end-to-end distance of the chain (of the order of 100 nm in HDPE), but an isolated chain still has a very spread-out structure with plenty of space within its coils.

Thermal vibrations can cooperate to rotate the chain at a C–C bond and

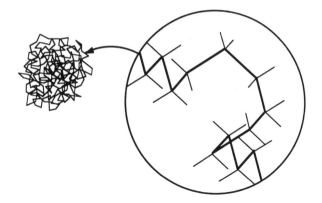

Fig. 12·7 Because three equally spaced rotational positions are preferred at each backbone bond, the preferred *conformations* of a 'linear' polyethylene chain are random walks along the diamond lattice

switch between these statically stable states. Although these three states are more probable because of their lower energy, at any moment in time the bond could be in any position between them. Of course, rotation at just a single bond would require swinging around the entire chain on one side of it–a highly unlikely event! A more probable degree of cooperation between two nearby bonds would allow a short section of chain to rotate between two collinear pivots in a 'crankshaft' motion. As we shall see, it is the cooperative nature of motions needed to cause changes in chain conformation which underly the *time dependence* of solid polymer deformation. For the time being we can visualize random small-scale rotations along a free-floating chain accumulating to compact it into a busily wriggling ball. Bulky side-groups on alternate C atoms, such as the benzene rings in polystyrene, tend to interfere (*steric hindrance*) in the *gauche* state. As a result, an isolated poly-styrene chain is stiffer, less mobile and more extended in space than that of polyethylene, although its sheer length still ensures that it remains crumpled up as a whole.

It is important to realize that *conformational* changes (rotations at backbone bonds) cannot change the tacticity of a chain with asymmetrical repeat units, such as polypropylene. If we were to straighten the chain out as in Fig. 12·5, we would find that the *configuration* remained the same, and that to change it would involve breaking covalent bonds and swapping hydrogen and side-group units.

12·3·2 Entanglements

If we start to concentrate a solution of isolated, coiled-up polymer chains, a point will arrive at which they start to interpenetrate each other's space, tangling together and hindering each other's mobility. This concentrated solution becomes very viscous and begins to acquire cohesion. It is because a minimum length of chain is needed for inter-chain entanglement to serve any structural purpose that there is a transition to a plateau in physical properties, at a molecular weight M_c, marking the transition from oligomer to true polymer. What are usually called 'entanglements' could be visualized as knots or loops between chains, but they are really just statistical entities measuring the degree of connectedness. As the most basic cause of cohesion, they are of fundamental importance to many properties of polymer solutions, melts and solids.

In concentrated solutions, entanglements resist, but do not prevent, flow under stress: the segments along each chain are not tied to their neighbours on adjacent chains. If each chain is sufficiently supple, it can burrow its way through the entangled but loose structure by a snake-like motion known as *reptation*. It is not until the remaining solvent is removed from between chains that van der Waals forces and hydrogen bonding can provide the energy barriers needed to lock chains against slippage. Stronger inter-chain bonding is promoted by regular polar side-groups, for example, by the chlorine atom in polyvinyl chloride (PVC), while in polystyrene the bulky benzene ring on every second C atom blocks slippage. It is important to realize, however, that chains can resist *flow* by mutual slippage but still extend quite freely by bond rotation, while each segment keeps the same neighbours on adjacent chains.

12·4 Thermoplastic melts

'Plastic' means 'mouldable': formable under pressure. Above all other properties it has béen the easy mouldability of plastics, and in particular of thermoplastics, that has sustained their expansion into more and more applications. This picture is slowly changing as new polymers are developed which offer extremely high performance (e.g., the high strength of Kevlar fibres) or unique properties (the low coefficient of friction of poly-tetrafluoroethylene, PTFE) at the cost of much greater difficulty in processing. The *rheology* of thermoplastic melts, the study of their deformation and flow under stress, is an important area of polymer materials science.

Thermoplastic melts behave rather like concentrated polymer solutions. In a solution it is the solvent molecules which separate chains, allowing them to wriggle and slither along against each other and preventing them locking under the action of intermolecular forces; in a melt it is the thermal agitation of the chain segments and side-groups. This gives the bulk fluidity: an applied shear stress will cause a continuous increase in shear strain. Unlike the much smaller molecules in liquids (and even in solids, if the temperature is high enough), which can diffuse relatively easily in any direction, polymer chains have only one, very limited means of fluid mobility: axial reptation. As a result, free diffusion in static polymer melts is extremely slow, and melts surfaces are reluctant to weld together even under pressure. This can be an awkward problem in moulding, creating permanent weaknesses (*weld lines*) where two melt streams must meet and merge.

12·4·1   Viscosity

Because of their chaotically entangled structure, polymer melts are both very viscous (typically having the consistency of toffee) and markedly elastic. Viscous deformation has been discussed in Section 9·4. The *shear viscosity*, η is the ratio of shear stress to shear strain rate in an ideally viscous (Newtonian) material. Just as a shear viscosity η can be defined analogous to the shear modulus G, viscous fluids also have an *extensional viscosity*, η_e, analogous to the elastic modulus E. This is the tensile stress required to maintain unit extensional strain rate. For an ideal *Troutonian* fluid, these constants are related, rather as E and G are for an elastic material, by

$$\eta_e = 3\eta \qquad\qquad (12·4)$$

The shear viscosity of polymer melts is much higher than that of other liquids with similar intermolecular forces because entanglements cause shear across any specific plane to draw the chains which cross this plane in other directions as well [Fig. 12·8(a)]; simple block slip along a single plane as shown in Fig. 8·5 is impossible in polymers. Thus the high viscosity of polymer melts is another illustration of the strengthening effect of chain entanglement as well as of the slow rate of reptation. The principal result is that thermoplastics can seldom be formed by simple casting techniques, since they will not pour quickly enough under gravitational forces.

Fortunately, another intrinsic property of thermoplastic melts greatly assists in their processing. Although at low rates of shear thermoplastic melts are approximately Newtonian [see Eqn (9·15)] with a high *intrinsic* viscosity, at higher rates they *shear-thin*, a property mentioned in Section 9·4. The higher the rate of shear, the lower is their apparent viscosity. This is because two competing processes are at work. Firstly, shear stress tends to extend and disentangle the polymer chains, orienting them along flow lines [Fig. 12·8(b)]. This direct effect of flow, depending only on the extent of shear, *reduces* the viscosity. However, the chains continuously crumple up again and re-entangle by diffusional movements, *increasing* the viscosity, so that the melt must be sheared continuously and rapidly to maintain any degree of disentanglement and reduction in viscosity. The same spontaneous shortening of the chains when stress is removed also underlies the rubbery elasticity of polymer melts.

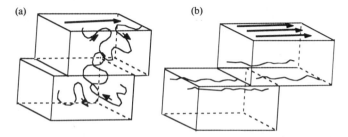

Fig. 12·8 Shear thinning. Polymer melts are very viscous because simple planar slip (a) is made impossible by entanglements, but rapid shear (b) extends and straightens chains faster than they can re-entangle

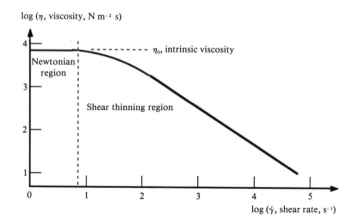

Fig. 12·9 Viscosity as a function of shear strain rate for a typical thermoplastic melt

As Fig. 12·9 illustrates, apparent viscosity falls dramatically with increasing shear strain rate from the Newtonian plateau (the *intrinsic* viscosity) at low rates, at which entanglement dominates. The intrinsic viscosity of a given polymer melt increases with molecular weight as

$$\eta_0 = K\overline{M}_w^{3 \cdot 5} \tag{12·5}$$

for $\overline{M}_w > \overline{M}_c$, where K is a constant. The dependence of viscosity on structural differences between polymers is more complicated, depending on such features as the degree of branching, the inherent flexibility of the chain and the mutual interference between its side-groups. Other factors which influence viscosity are temperature (increasing temperature increases mobility, promotes diffusion and thus decreases viscosity) and pressure (which forces chains closer together and thus increases viscosity).

Although all polymer melts shear-thin, their extensional viscosity η_e generally falls (they *tension-thin*) at a much slower rate with increasing strain rate, so that the ratio η_e/η swiftly rises to much more than the ideal value of 3–sometimes to several orders of magnitude more. Some polymers even *tension-stiffen*. An interesting comparison is that between branched and linear polyethylenes, LDPE and HDPE, which are similar in shear viscosity and elasticity but differ dramatically in extensional viscosity: HDPE tension-thins whilst LDPE tension-stiffens.

12·4·2   *Processing of thermoplastics*

With this picture of an elastic, viscous polymer melt in mind, we can sketch the main processes used to form thermoplastics. The most important of these are *extrusion* and *injection moulding*. Extrusion is used to produce long shapes of constant cross section (such as pipe, rod or narrow plate sections) by pumping the melt through a *die* of the required shape; the principle is similar to that of the *hydroplastic extrusion* of clay, illustrated in Fig. 11·7. The pump is a *plasticating screw* turning in a heated barrel, which continuously draws in granulated solid feedstock and melts and mixes it under increasing pressure. The section emerging from the die (the *extrudate*) is drawn off, perhaps through a sizing die to confer more accurate dimensions, into a cooling water bath. A similar pump unit is used intermittently in an *injection moulding* machine to plasticize a fixed charge of melt, which is then injected through a small orifice into a closed, cold mould. Pressure is maintained for a short time while the moulding cools and freezes. Finally, while cooling is completed and the mould is opened to allow ejection of a finished component, the next charge of melt is prepared. With careful control, injection moulding can be used to produce parts of very high complexity and precision, at relatively high speed. Extrusion and injection moulding both subject the melt to very high shear strain rates as it flows along internal channels (the flights of an extruder screw or the channels in an injection mould).

Other common processes rely partly on the high viscosity of the polymer to maintain shape even when flowing between free surfaces. These include *film blowing*, in which a thin extruded tube is continuously inflated by internal air pressure before it freezes. The greatly extended and thinned-down tube is then either slit to produce sheet or, more commonly, sealed at intervals to form plastic bags. Film blowing is difficult for polymers which tension-thin, since these tend to neck down and rupture at weak points; the material most often film-blown is low-density polyethylene, whose tension-stiffening stabilizes the thickness of the extending bubble. Films of more precisely controlled thickness or finish are produced by *calendering* between rollers. Another process based on extrusion of tube is the *blow moulding* of components such as bottles. This is a discontinuous process, similar to that previously described for a glass (Section 11·4·2) in which a section of tube is inflated in the melt state to fill a cold, closed mould. Continuous extrusion through a die plate perforated by a large number of tiny holes is the first stage of *melt spinning*, used to produce artificial fibres.

The influence of molecular weight distribution on melt viscosity [Eqn (12.5)] and its temperature dependence is reflected in the production of different *grades* of polymer designed for different processing applications. High-density polyethylene again provides a good example. Grades designed for injection moulding will normally have a low average molecular weight with a narrow distribution, to give the low viscosity needed to reduce injection pressures and to allow fast injection without generating too large a temperature rise by shear working. Extrusion, however, involves strain rates 10 to 100 times lower than those during injection moulding and large-sectioned extrusions must partly support their own weight while cooling. High melt viscosity here becomes a real advantage, so extrusion grades tend to have a higher average molecular weight with a wider distribution. Polymethyl methacrylate cast into sheet (e.g., 'Perspex') and polymerized *in situ* has a very high molecular weight and an almost intractably high melt viscosity, while grades designed for inject moulding are of a much lower degree of polymerization.

Two other characteristics of thermoplastic melts significantly affect their processing. *Melt elasticity* is most visible as *die swell* after extrusion: the melt, having been deformed by radial compression to force it through the die, springs back on exit to increase the diameter of the extrudate. Melt elasticity varies widely between polymers, reflecting differences in the persistence of entanglements in the melt. Branched polyethylene is much more elastic in the melt state than linear polyethylene, since individual molecules find it much more difficult to reptate and relieve applied stress.

Polymers melts are relatively brittle. Their tensile strengths, of the order of $1 \, MN/m^2$, are not very different from those of other liquids, but constitute a problem because high viscosity requires large stresses to be applied during forming. A common manifestation of *melt fracture* is the break-up of extrudate emerging too rapidly from an extruder die, under the tensile stresses which accompany die swell.

12·5 Amorphous polymers

Most thermoplastic melts are *amorphous*, their constituent chains randomly coiled and disordered. As the melt cools, the thermal vibration of chain segments, which kept neighbouring chains at bay, diminishes; slippage becomes more and more difficult in the sense that the *coordinated* thermal wriggle which the chain needs in order to slip becomes a rarer and rarer event. The structure now sets to a *rubbery* state in which the elastic modulus is not very much higher than that of the melt, but flow has all but ceased.

12·5·1 The rubbery state

As already mentioned, the enormous extensibility and low modulus characterizing this state have little to do with bond stretching or bending. In the solid state as in the melt, *rubber elasticity* arises from the tendency of long, thermally mobile polymer chains to crumple up to a size much shorter than their end-to-end length. It is best understood by imagining many polymer chains to have been fully extended and used like scaffolding poles to build some sort of loosely jointed structure. As we already know, thermally induced kinking of the chain will quickly crumple the structure back into a dense, collapsed state in which, although individual segments along their length are very mobile, chain conformations are restricted. The tendency of the chains to return to their most probable configuration appears as a restoring force when the network of joints is deformed: they behave as 'entropy springs'.

In a rubbery amorphous polymer these joints between chains are 'entanglements', of which there may be many along each chain. Since these are not permanent knots or bonds, they can still slip under stress, so that rubbery polymers–in contrast to the elastomers which we will consider later–tend to *creep*, or deform permanently. Nevertheless, most of the large extensions possible in rubbery polymers take place by straightening of the chains between a deforming net of *temporary entanglements*.

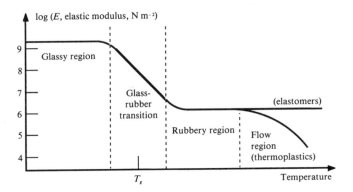

Fig. 12·10 Effect of temperature on elastic modulus for a thermoplastic and an elastomer

12·5·2 The glass transition

As a rubbery polymer is cooled further, the modulus increases as the pivots which allow long-range stretching of the chain freeze up. Eventually a temperature range is reached at which a very large increase in modulus to the order of GN/m² takes place. This is the *glass transition temperature*, T_g, at which rotational flexure of backbone bonds virtually ceases, the structure freezing under the attractive force of secondary bonds into an amorphous rigid glass (Fig. 12·10).

We have already noted the two distinct forms of interatomic bonding present in polymers: primary covalent bonds within molecules, typically of modulus 100 GN/m², and secondary bonds between molecules, whose modulus may be two or three orders of magnitude smaller. If we apply a stress to a block of randomly distributed glassy polymer chains, it is easy to show that most of the extension occurs in the weaker, lower modulus secondary bonds. The modulus of a polymer below T_g is therefore dominated by that of its 'weakest links' and is seldom more than a few GN/m².

Because polymer chains differ widely in their ability to pivot at backbone bonds, T_g also varies, and one of the basic classifications of a polymer is based on whether this is below or above room temperature. Polymethyl methacrylate ($T_g = 105°C$) and isotactic polystyrene ($T_g = 95°C$) are amorphous glassy polymers at room temperature, since backbone rotations in both are severely hindered by bulky side-groups. An amorphous thermoplastic with T_g below room temperature is seldom useful due to its tendency to flow by slippage at entanglements. However, in some polymers, the mobility of chain segments between entanglements can be deliberately increased, reducing T_g, by introducing a *plasticizer* of low molecular weight but low volatility. The small plasticizer molecules effectively lubricate chain motion by swelling the structure and giving segments more room to manoeuvre. The most common example is the use of, for example, dioctyl phthalate to plasticize PVC to a rubbery material suitable for wire and cable covering. For similar reasons, in a polydisperse material the low-M tail of the molecular weight distribution (Fig. 12·3) has a strong influence on T_g, the oligomer fraction itself acting as a plasticizer.

12·5·3   Time dependence

We have already mentioned that conformational changes in polymer chains require cooperation. As a result, *deformational* processes in solid polymers require not only a sufficient temperature but also a sufficient time. Focusing on a chain segment as being the smallest unit of polymer substructure, we can see that its individual mobility is restricted to what can be achieved by bending and stretching its backbone bonds. These glass deformations require hardly any time. All of the larger deformation modes up to and including flow require the segment to move relative to another one, and must wait for coordinated thermal vibration to allow them to sidestep each other.

For a polymer melt in which *all* deformational processes occur, we can sketch the time scales involved. Relative movement by rotation of side-groups or side-chains may occur within 10^{-10} s, whereas even small-scale segmental motion will have to wait for about 10^{-8} s. Deformations requiring large-scale motion (crankshaft or skipping-rope like rotations of short chain sections) will take at least 10^{-6} s. Finally, the deformation process which characterizes melts–viscous flow–is translation of the entire chain (*reptation*) by one segment; this takes at least 10^{-4} s. A typical chain will need about 10^{-4} s to diffuse through its own length, which is why the self-diffusion processes in polymer melts needed for welding to take place are so inconveniently slow. However, remember that chain segments in a sample of polymer at a particular temperature, like the atoms in all substances from monotomic gases to crystalline solids, have a wide spectrum of different vibrational energies (see Section 7·10). Thus, for example, even in a polymer below its glass transition a few segments always have the energy that they need to reptate. In an amorphous polymer, where no well-defined structures with high disruption energies (latent heats) exist, the improbable modes will always happen somewhere if we wait long enough, and an externally applied stress will provide a bias towards their occurrence and against their subsequent reversal. Amorphous thermoplastics under stress will eventually betray significant viscous flow.

12·6   Crystallinity and orientation

As we have just mentioned, amorphous low-density polyethylene (LDPE) would be a rubbery material with poor mechanical properties. In practice, LDPE has much more useful properties because it is not completely amorphous, but *semicrystalline*.

Many polymers can be crystallized, either from an amorphous melt or from a solution, but complete crystallization is virtually impossible (especially from a melt). The sheer length of the chain hinders their extraction from the tangled amorphous state by the higher intermolecular forces within close-packed crystallites, so that for the reasons introduced in the last section, the process is again time and temperature dependent. The rate of crystallization reaches a maximum at some temperature below the crystalline melting point T_m but high enough above T_g for chains to be quite mobile. Quenching from a melt freezes the amorphous state, but slower cooling rates (or annealing) promote the formation of a semicrystalline structure. Because this is usually valuable but processing time is as well, it is common to seed commercial semicrystalline plastics with nucleating additives to hurry the process along.

The tendency of a particular polymer to crystallize depends not only on the cooling rate but also on the flexibility and regularity of its chain structure and the strength of inter-chain bonding. Linear HDPE, for example, has a regular backbone of great flexibility, impeding the disentanglement needed for crystallization, but its regularity allows close packing and HDPE crystallizes readily by up to 90%. Branched LDPE, no less flexible, can only attain around 60% crystallinity because the side-chains interfere with regular packing. Isotactic and, to a lesser extent, syndiotactic polymers are regular in structure and usually crystallize, while polymers with irregularly spaced bulky side-groups such as PMMA are necessarily amorphous. Thus atactic polypropylene (Fig. 12·5) is amorphous, while the isotactic and syndiotactic forms are semicrystalline; most commercial grades are based on the isotactic form, with a crystallinity of about 70%. However polyvinyl chloride, although atactic, can achieve limited crystallinity (about 10%) because hydrogen bonding provides strong intermolecular packing forces. For a polymer of given structure, the degree of crystallinity is also affected by molecular weight, since longer chains are more difficult to draw from amorphous zones into crystallites.

For copolymers, regularity is again the key to successful crystallization. Thus alternating structures (ABABABAB) and copolymers with large block lengths can generally crystallize whilst random, graft and block copolymers lacking long-range regularity cannot.

12·6·1 Lamellae and spherulites

The smallest known crystalline units in polymers are plate-like *lamellae*, a few tens of nanometres thick (that is, just one or two hundred repeat units) but as wide as several micrometres. These precipitate from pure, dilute polymer solutions, extending in width as the isolated and relatively mobile chains fold back and forth repeatedly through the thickness [Fig. 12·11(a)]. Similar structures start to form in cooling polymer melts, but they cannot continue to extend in an orderly and coordinated way into such a dense and chaotically disordered environment. Each lamella rapidly breaks up and branches into an expanding sheaf of tape-like *fibrils*, separated by trapped amorphous material, onto which many individual chains compete to crystallize simultaneously [Fig. 12·11(b)]. Continued branching splays the ends of the sheaf outwards until the structure becomes an expanding sphere–a *spherulite*–drawing chains radially inwards to crystallize tangentially on its surface [Fig. 12·11(c)]. Spherulites can be observed under

(a)

(b)

(c)

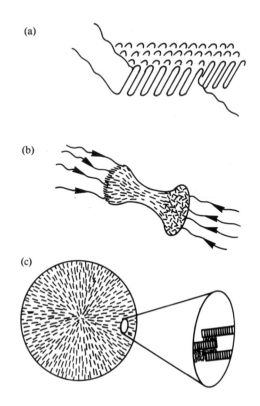

Fig. 12·11 Development and structure of polymer crystals. (a) Lamellae form in dilute solutions by folding of isolated chains. (b) Bunches of lamellar fibrils draw in polymer chains in a melt, eventually forming spherulites (c) containing much trapped amorphous material

an optical microscope and may grow to as much as 100 μm in diameter.

The effects of crystallinity on properties are profound. Within crystallites the closer packing and stronger bonding increase strength and modulus. At temperatures between the glass transition temperature and the melting point of the crystalline phase, overall structural integrity of the amorphous phase is reinforced by chains which bridge and bind adjacent crystallites (*tie molecules*): a polymer such as LDPE, which would otherwise be semi-fluid, becomes a hard rubber. This kind of material has great ductility and it is in this region that semicrystalline plastics are most valuable.

Although a few highly crystalline polymers of low molecular weight do show a sharp transition to a relatively fluid melt, most are processed in a state in which they are more like amorphous viscoelastic solids. Unplasticized PVC, indeed, chemically decomposes at such a low temperature that it is commonly processed *below* the melting point of its sparse crystalline phase. It never truly melts, but is processed as a *gel*, that is, a suspension of solid regions in an amorphous viscous fluid. Structural features can often be seen in moulded or extruded products which were present in the granules from which they were processed. Because most semicrystalline polymers have several levels of scale and organization, they have a broad melting range.

12·6·2 Orientation

The elastic modulus along the axis of perfectly aligned chains in the crystalline phase is extremely high: about 320 GN/m^2 in polyethylene. If all the backbones of adjacent chains could be lined up in the direction of loading in a perfect crystal, we would expect to measure such moduli. In practice the modulus of bulk linear polyethylene is at least two orders of magnitude lower, for two main reasons: the modulus of the amorphous phase is very much lower, and within crystallites there are lamellae oriented in all directions. In each case 'weak links' in the structure reduce the average modulus, which results from a series combination of low and high modulus regions. Much higher moduli can be achieved by lining up the crystalline regions in the required direction. The simplest way in which this can be achieved in flexible chain thermoplastics such as polyethylene and polyamide is by mechanically straightening them out as they cool from the melt; as we have seen, this always occurs to some extent during their processing as a result of shear and extensional flow.

The strengthening and stiffening function of *orientation* is exploited most fully in the process of spinning strong semicrystalline fibres. There are three requirements for high axial modulus and strength: long, high molecular weight, straight molecules having few chain-end defects; good orientation of chains in the axial direction; and good lateral packing to inhibit slip. In melt spinning, extrusion through a fine hole in itself tends to align the chains along each fibre axis, but orientation is vastly improved by drawing the fibres in tension as they solidify. Within short sections of each solidified fibre after drawing, the spherulitic structure is destroyed but chains remain fully extended and have packed together under the action of secondary bonds. However, because the initial melt was highly entangled, it is never possible to unravel it completely except very locally. Just as combing tangled hair tends to concentrate 'knots', so the oriented regions in drawn fibres remain separated by densely entangled amorphous regions (Fig. 12·12). Many polymer chains will start in one oriented section, pass through an amorphous region and end in another oriented one. Additives and impurities in the base polymer also tend to be concentrated in these disordered regions. Nevertheless, moduli of 100 GN/m^2 have been achieved in PE fibres, and even higher moduli have been achieved by cold drawing after extrusion.

As we have already seen in melts, orientation can also occur in the

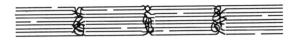

Fig. 12·12 Fibres drawn from semicrystalline polymers contain highly oriented regions separated by disordered amorphous zones

amorphous phase. Deformation at a temperature above T_g will tend to extend chains and rotate them onto the axis of extension, so that although they remain tangled and disordered at a molecular scale there are then more strong covalent bonds along this axis and fewer across it. In geometries other than fibres, of course, orientation can be a very mixed blessing, because high strength and stiffness along the orientation direction is accompanied by low strength and stiffness across it. This can be a particular nuisance in injection-moulded products, where freezing the surface too rapidly can form a layer heavily oriented through its own plane, supported by a lower layer oriented across it. In bad cases this surface layer will easily peel off.

12·7   Thermosets and elastomers

In reviewing the structure and properties of thermoplastics, formed by polymerization of monomers in which a double bond was opened up to form a linear chain, we have seen the importance of reversible secondary bonding between chains in determining properties. Another important class of polymers with very different properties arises if *primary* bonds are formed between chains by chemical crosslinking. These materials, which set (cure) irreversibly, range from the *thermosets*, which are generally glassy in character, through to the rubbery *elastomers*.

12·7·1   Thermosets

Thermosets are usually formed from 'precursor' polymers which, on heating, crosslink at close intervals to form what is effectively a single macro-molecule. Phenol formaldehyde (filled with wood flour to form 'Bakelite'), until recently the main plastic used in electrical components, is usually supplied in powder form. On melting, this lightly crosslinked prepolymer can be moulded into shape, and, while held at high temperature, undergoes the chemical reaction which crosslinks and cures it. Because this reaction is irreversible, the structure is infusible and insoluble: it can only be disrupted irreversibly by destroying covalent chemical bonds by heat or chemical attack. Thus crosslinking confers valuable properties of heat and chemical resistance and dimensional stability, but, by preventing chain slippage, greatly reduces ductility.

Three venerable materials–phenol formaldehyde, urea formaldehyde and melamine formaldehyde–still account for the majority of thermosets used. Engineering thermoplastics capable of operating at relatively high temperatures are displacing these from many of their traditional applications, such as electrical components, mainly because they are much easier to process quickly and because scrap and waste can be reground and reprocessed.

The precursors of thermosetting polyesters are short-chain unsaturated oligomers, less than 10 mers long. These viscous liquids are cured by copolymerization with styrene, a thinner liquid which makes the dissolved polyester considerably easier to form. Heat or the addition of a catalyst initiates a reaction in which the styrene crosslinks at several points along each polyester mer. Unsaturated polyester, urethane and epoxy resins have found expanding new applications as matrices for composite materials such as glass-reinforced plastics (Chapter 11), in which their weaknesses can be compensated for by the properties of the reinforcing fibres. Another rapidly expanding application for thermosetting resins–notably the epoxies–is as high performance adhesives.

12·7·2 Elastomers

This second important group of crosslinked polymers can be formed from *diene* monomers, containing two double bonds. Polymerization opens one double bond but leaves the other intact for use by the crosslinks. The most common of these elastomers are natural rubber (NR, *cis*-polyisoprene), polybutadiene rubber (BR) (see Table 12·2) and copolymers such as styrene–butadiene copolymer rubber (SBR). All are thermoplastic in their unmodified state, but they differ from the isotactic polyolefins by being amorphous even down to temperatures at which they do not appear to be fluid; these are true rubbery thermoplastics, and until stabilized by crosslinking are of little structural use. The raw material for natural rubber (NR) is *latex*, a viscous fluid which contains a linear polymer of polyisoprene. Because of its rotationally rigid double bond, polyisoprene has two stereo-isomers. In *trans*-polyisoprene (gutta percha) the CH_2 groups are on alternate sides of the backbone, and interfere with each other, inhibiting flow. In *cis*-polyisoprene they are on the same side, and this material is a soft, fluid rubber until heating in the presence of sulphur (*vulcanization*) establishes short crosslinks, consisting of short chains of sulphur atoms, between active points on adjacent chains formed by opening up some of the remaining double bonds.

The mechanism of rubber elasticity has already been described. Unlike amorphous rubbery thermoplastics in which the ends of highly-extensible chain sections are marked by entanglements–which readily slip–the chain section 'ends' in elastomers are short, strong primary crosslinks (Fig. 12·13), while the intermolecular forces due to secondary bonds are relatively low. The modulus of the crosslinked elastomer depends primarily on the

Table 12.2
Structure and glass transition temperature of some common elastomers

Elastomer abbreviation, name		Repeat unit	T_g (°C)
NR	Natural rubber (polyisoprene)	$[-CH_2-C=CH-CH_2-]$ $\qquad\quad \mid$ $\qquad\quad CH_3$	−70
BR	Polybutadiene	$[-CH_2-CH=CH-CH_2-]$	−85
CR	Polychloroprene	$[-CH_2-C=CH-CH_2-]$ $\qquad\quad \mid$ $\qquad\quad Cl$	−50

Crosslinks

Extensible rubber chains

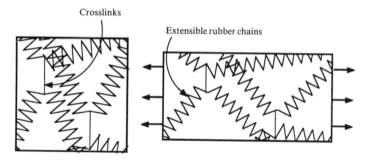

Fig. 12·13 Rubber elasticity in elastomers. Highly extensible rubbery polymer chains extend between points permanently connected by crosslinks

interval between these crosslinks (typically a hundred or so mers): the greater the crosslink density, the higher the modulus.

At extremely large strains (100–200%) a second effect may appear. The chain sections between crosslinks have now straightened out considerably, and are aligned with each other to the extent that secondary bonds start to induce crystallization. The microstructure, as revealed by X-ray diffraction, resembles that of a semicrystalline polymer, link chains in the amorphous regions interpenetrating and connecting the crystalline regions. This, as we have seen, considerably increases the modulus, and also contributes significantly to the great toughness and tearing resistance of rubbers.

12·7·3   Thermoplastic elastomers

Crosslinked rubbers cannot be reprocessed, since, like thermosets, their covalent crosslinks cannot re-form once broken. Rubbery thermoplastics such as plasticized PVC, on the other hand, tend to flow during large extensions. One of the current frontiers of polymer technology is the development of materials which combine the mechanical properties of vulcanized rubbers with the processability of thermoplastics.

Most thermoplastic elastomers currently available are 'triblock' copolymers of styrene and a rubber, in which a centre block of the rubber is terminated by smaller polystyrene end-blocks. Because these two components are immiscible, on cooling from the melt a phase separation takes place, the polystyrene agglomerating into distinct glassy amorphous domains within the rubber matrix. Each blob of polystyrene may entangle together the ends of several hundred polydiene chains, forming a pseudo-crosslink. On reheating, the polystyrene crosslinks melt and the rubber flows. The most common material of this type is styrene–butadiene–styrene (SBS) rubber, although newer variants use polyisoprene or other rubbers to form the matrix. In the *linear* form a rubber chain is terminated by polystyrene segments, whilst in the *radial* form strong primary crosslinks bond up to twenty SBS chains together near the centre.

In another system, a polyether and a polyester are copolymerized. Short sections of the relatively hard polyester crystallize into lamellae, while the intervening polyether sections remain soft and amorphous. Here it is the crystalline lamellae which adopt the crosslinking role, and the structure becomes thermoplastic at the crystalline melting point. Similar mechanisms account for the rubber-like properties both of thermoplastic polyurethanes, which are true block copolymers of chemically distinct sequences, and of plasticized PVC, in which the soft segments are chemically identical to the hard segments, but whose relative motion is effectively lubricated by small plasticizer molecules.

12·8 Mechanical properties

Chapter 9 showed how the mechanical properties of metals and ceramics were measured, and some of the ways in which these properties arose from microstructural characteristics. Having reviewed the structural features of polymers, we are now in a better position to understand how their mechanical properties originate.

12·8·1   Stress–strain behaviour

The most basic properties of an engineering material are its elastic modulus and its ultimate strength. The small-strain moduli of plastics in general are relatively low compared to those of the materials which we have considered so far. More specifically, we have discussed four broad structural classes of polymers: amorphous thermoplastics (in the glassy or rubbery state), semicrystalline thermoplastics, elastomers and thermosets. Most thermoplastics are broadly similar in modulus if they are in the same state. Glassy polymers–glassy thermoplastics, highly crosslinked thermosets and some very highly crosslinked elastomers like ebonite–all have moduli of a few GN/m^2, either above or below the β transition, whilst rubbery plastics and lightly crosslinked elastomers have moduli up to four orders of magnitude lower.

The post-yield strength properties of polymers are more variable. Highly crosslinked thermosets have little or no scope for internal mobility and flow, and like other covalently bonded solids such as SiC they are hard and brittle: a small elastic extension due to covalent bond stretching is followed by fracture. Chains in less densely crosslinked thermosets such as the epoxies, polyesters and polyurethanes can slip against each other a little before the crosslinks break, so that smooth specimens may show a yield point (Fig. 12·14). Again, the glassy thermoplastics behave rather similarly; in tension, unmodified PMMA and PS fracture before yield. Above T_g, however, the degree of crystallinity has a profound effect. We have seen that fibres of semicrystalline polymers become highly oriented when they are extended. A similar phenomenon known as *drawing* is seen in tensile specimens of PA and HDPE: yielding is followed by formation of a neck, which then lengthens, without diminishing in cross section, by drawing in spherulitic material from the shoulders and realigning the already ordered chains along the specimen axis. As for fibres, no volume change occurs during this process, since the

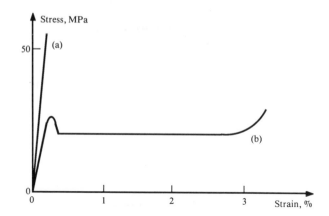

Fig. 12·14 Typical stress–strain curves for (a) thermosets and glassy thermoplastics, and (b) a semicrystalline polymer (e.g., high density polyethylene) showing drawing

proportion of crystalline material remains constant and only its form and
orientation change.

Although it is not unusual to see strength and modulus values quoted for
polymers without any warning for the unwary about their restricted validity,
we are now in a position to see the likely pitfalls of using them in design. The
modulus of a glassy thermoplastic near its T_g will be extremely sensitive to
temperature, and the modulus of a crystalline polymer will depend critically
on its crystallinity (and hence on its thermal history) as well as on the degree
and direction of orientation during moulding or subsequent deformation.
The variable most likely to affect the elastic modulus, however, is the time for
which load is applied.

12·8·2 Creep, recovery and stress relaxation

We have seen in Section 12·5·3 that the origins of large-time effects on
deformation in polymers lie in the diffusion-like motion of chain segments,
which demand cooperation. The effects of time on modulus are best
measured for engineering purposes in two simple experiments.

If a tensile specimen is subjected to a suddenly applied *constant stress*, σ,
for example, by suspending a weight from it, the strain ε will continue to
increase [Fig. 12·15(a)] at a rate which diminishes with time t, demonstrating
creep. The elastic modulus measured in creep is not a constant, as it would be
for a purely elastic material, but continuously increases with time:

$$E_c(t) = \frac{\sigma}{\varepsilon(t)} \qquad\qquad (12\cdot6)$$

However, if the stress is removed, it is not just the initial instantaneous
deflection which is recovered (as it would be for a creeping metal): given
enough time, *all* of the creep deformation will be recovered and the specimen
will have returned to its original length. If a *constant strain* is applied to a

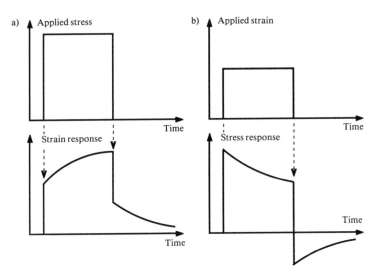

Fig. 12·15 Viscoelasticity in polymers: (a) *creep* strain under constant stress is followed by a recovery on unloading, and (b) stress relaxation at constant strain is reversed when the strain is removed

specimen, say by stretching it and clamping its ends, the stress will decrease at a rate which diminishes with time [Fig. 12·15(b)], demonstrating *relaxation*. This is expressed as a *relaxation modulus*

$$E_r(t) = \frac{\sigma}{\varepsilon(x)} \tag{12·7}$$

which also decreases with time, although the creep and relaxation moduli are not exactly the same. Reversal of the initial strain will induce a compressive stress which will itself eventually relax to zero. These two tests illustrate *viscoelasticity*; the specimen returns 'elastically' to its original state, but, due to its 'viscosity', only after some time.

12·8·3   Time–temperature correspondence

In discussing the time dependence of deformation in thermoplastics under stress (Section 12·5·3) we developed a picture of rubbery and viscous deformations occurring by the thermally activated diffusion of chain segments. The opportunities for a sufficiently mobile segment to move into a nearby 'vacancy' are provided by thermal vibrations whose amplitude changes continuously, so that the opportunity for a segment to move to relieve an applied stress will arrive–given enough time. It is therefore not surprising that when the results from a large series of creep tests are reviewed, a correspondence of sorts can be drawn between the effects of temperature, which increases the average vibration amplitude, and of time under stress.

Figure 12·16 illustrates typical log(modulus, $E_c(t)$) versus log(time, t) curves for an amorphous polymer from a series of creep experiments at fixed

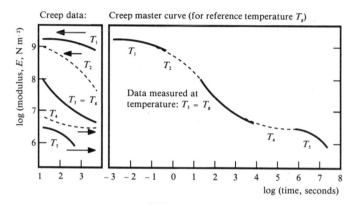

Fig. 12·16 Time–temperature correspondence. Modulus–time data from creep tests at different temperatures can be reduced to a single master curve, by shifting along the time axis by a time characteristic of each temperature

temperatures ranging from below the glass transition to above the rubber plateau. For each temperature, extension was recorded at times from a few seconds up to one hour. Experimenters have found that by choosing some reference temperature, conventionally the glass transition temperature T_g, and shifting these curves along the time axis, they can be made to overlap and form a *master curve*–whose time axis might extend from fractions of a second to centuries! Moreover, the time shift, $\log(a_t)$ required to compensate for a given temperature shift from T_g to T tends to be similar for all amorphous polymers, and can be expressed by the Williams–Landel–Ferry equation:

$$\log a_t = \frac{- C_1(T - T_g)}{C_2 + T - T_g} \tag{12·8}$$

where $C_1 = 17 \cdot 4$ K and $C_2 = 51 \cdot 6$ K. Thus we find that polymers at temperatures near T_g at ordinary testing rates can behave like glassy polymers under rapidly applied stresses (such as impact loads or high-frequency oscillations), and as rubbers under essentially static loads after very long times.

12·8·4 Viscoelastic models

For a component such as a tension rod in which the stress is constant, uniform and uniaxial, it would be a straightforward matter for a designer to apply creep test data directly. The analysis of more complicated structures requires some *model* of the material through which its behaviour can be expressed in simple *constitutive* equations. A good model can also throw light on the underlying structural processes, and linear viscoelastic models illustrate this point well.

For small strains polymers exhibit *linear viscoelasticity*, strains and strain rates varying linearly with stress. Models for this behaviour can be constructed from the linear spring, which exerts a load proportional to its extension, and the linear dashpot, which exerts a load proportional to its rate of extension (a dashpot, as used in automotive shock absorbers, is a cylinder within which a leaking piston moves through a viscous fluid). All of the work done on a spring is stored and can be recovered, while all of that done on a dashpot is lost. Connecting each of these elements in turn (Fig. 12·17) to support the stress σ on a unit cube models, respectively, a linearly elastic solid with constitutive equation

$$\sigma = E\varepsilon$$

and a linearly viscous (Troutonian) fluid with constitutive equation

$$\sigma = \eta \frac{d\varepsilon}{dt}$$

In the Maxwell model [Fig. 12·18(a)] a spring and a dashpot are connected in series. Each element carries the same load, and their extensions add. Thus

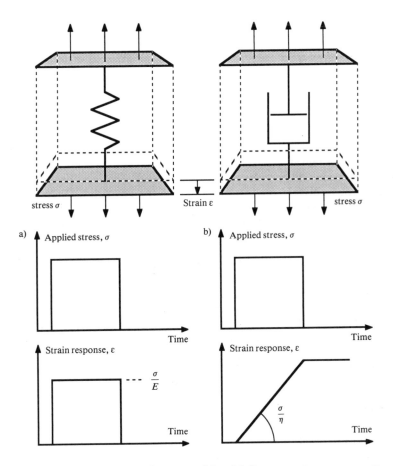

Fig. 12·17 Elements of viscoelastic models: (a) linear spring representing a linearly elastic solid, and (b) linear dashpot representing a linearly viscous fluid

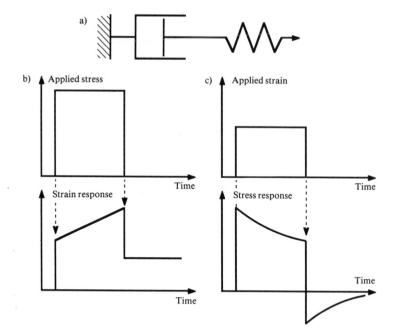

Fig. 12·18 Response of the Maxwell model (a), during creep (b), and relaxation (c)

under constant stress the spring extension will remain fixed, the dashpot will extend indefinitely at a constant rate, and when the stress is removed only the spring extension will be recovered [Fig. 12·18(b)]. This is a fair model for an elastoviscous polymer melt but a poor one for any solid. Under constant strain, the dashpot will extend at a rate which falls off exponentially with time as the initially stretched spring relaxes [Fig. 12·18(c)]. This is quite a good model for relaxation of a viscoelastic solid, although the modulus towards which it relaxes is zero. The ratio η/E, which is the time after which the stress has relaxed to $1/e$ of its initial value, is referred to as the *relaxation time*, t_r. Under cyclically applied loads or extensions the Maxwell model shows a 'transition' of sorts. At low frequencies the dashpot exerts negligible load and the material can support no stress: its modulus is zero, and remains so up to a frequency of about $1/t_r$ at which the dashpot starts to exert a significant resistance. At high frequencies the dashpot is effectively a rigid strut and the modulus is E. We can think of the Maxwell model as a rubbery thermoplastic (the spring representing a rubbery chain and the dashpot a temporary entanglement) undergoing its transition to a melt.

In the Voigt model [Fig. 12·19(a)] a spring and a dashpot are connected in parallel; each element undergoes the same extension but they share the load. Under constant stress the solid will start to flow viscously at an almost constant rate, the spring being unextended, but this rate will fall as the spring bears more and more of the load [Fig. 12·19(b)]: this models creep quite accurately. The Voigt model cannot be displaced instantaneously (the

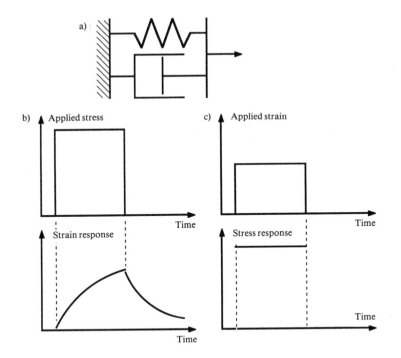

Fig. 12·19 Response of the Voigt model (a), during creep (b), and relaxation (c)

dashpot would resist with an infinite force), but even if the zero-strain rate condition for a relaxation test could be applied, the dashpot would then bear no load, the spring would bear a constant load, and there would be no relaxation [Fig. 12·19(c)]. The Voigt model represent a rigid glass (the spring again representing a rubbery chain, and the dashpot a mechanism for segmental diffusion) near the glass transition.

Since the strengths and weaknesses of the Maxwell and Voigt models complement each other, attempts have been made to combine them in three-element models: for example, a spring connected in parallel with a Maxwell model will prevent it collapsing into a liquid at low strain rates, providing a much better model for a transition between glassy and rubbery moduli. This *standard linear solid* remains relatively easy to analyse but quite restricted in its ability to simulate real behaviour. Models can be constructed which represent creep or relaxation response to any desired degree of accuracy, by connecting large numbers of Maxwell elements in parallel (Fig. 12·20), so that as each reaches its relaxation time others remain partially elastic. It is possible to identify elements (or groups of elements) in such models with specific *relaxation processes* in the polymer, such as coordinated crankshaft rotations of chain segments to allow rubbery extension or reptation (slip at entanglements) to allow flow; each has its own relaxation time and modulus. However, such realism must be paid for in mathematical complexity.

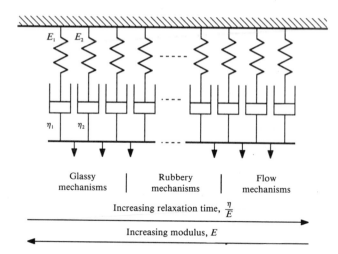

Fig. 12·20 Models for creep behaviour can be built from large numbers of Maxwell elements, each representing a relaxation process and the time which it needs to operate

12·8·5   Crack and craze growth

Plastics are relatively new materials, many of which were pressed into service during a period of rapid expansion in the availability of low-cost consumer goods. Their reputation suffered badly in the 1950s and 1960s due to poor design and inappropriate materials selection, compounded by hurried processing and a lack of performance data. Although caution generally prevailed against their use in highly stressed structures, the application of plastics in small, low-cost load-bearing components became associated with the development of cracks, particularly over long periods or under impact.

Pre-existing cracks or flaws are always present in engineering materials, if only at sub-microscopic sizes, and always have a powerful stress-concentrating effect at their tips. Under light, constant stresses a material at a crack tip may respond in one of two basic ways. It may flow to blunt and harden the tip, taming its stress-concentrating effect and preventing further damage; most semicrystalline polymers in benign environments behave like this, since they tend to draw and strengthen with orientation. On the other hand, glassy and amorphous polymers with little capacity to self-strengthen under stress tend to succumb to constant, steady *slow crack growth*. This tendency is amplified if some environment, even an apparently innocuous one, can reach the crack tip and plasticize vulnerable material there.

Amorphous polymers exhibit *crazing*. An isolated craze is a crack-like defect whose faces are tied together by a spongy mass of threads. Crazes form both at the tips of long cracks, reducing the stress-concentrating effect of the tip, and of short cracks, which they may prevent from extending further.

12·8·6 Impact and fatigue

As usual, time dependence complicates the behaviour of polymers as compared to metals. Amorphous polymers such as PMMA will present a greater resistance to a faster-running crack until some critical crack speed, perhaps a few metres per second, is attained at which resistance suddenly falls again-perhaps because heat generated by flow near the crack tip cannot be conducted away fast enough and causes local melting. The resulting weakness to impact is even seen in linear, highly crystalline polyethylenes: although it is almost impossible to drive cracks through this material at room temperature using steady loads, catastrophic fracture may ensue once a sharp, fast crack has been initiated by impact. The older themosets have notoriously little resistance to cracks under any conditions, but epoxies (and to a lesser extent polyesters), which can yield and flow very locally, may successfully blunt and arrest a running crack-if the crack speed drops to a low enough speed to allow the required flow process to occur.

12·9 Physical and chemical properties

12·9·1 Optical properties

For a material to be optically transparent, it must be uniform on a scale greater than that of a wavelength of light (0·1 μm). This is about a thousand times larger than a single mer, but smaller than a typical crystallite. Because crystallites have a different refractive index to that of the amorphous phase, refraction at interfaces diffuses light and semicrystalline polymers are generally white or translucent in their pure state. Amorphous polymers, on the other hand, are highly transparent to light. Melting of the crystalline phase in polymers such as LDPE is clearly visible as a sudden increase in transparency.

The transparency of amorphous thermoplastics has found them many applications. Polymethyl methacrylate (PMMA), with high transparency and moderate strength, has long been exploited for its optical properties, in both cast sheet form and in mouldings. Although tougher than glass, PMMA is not as hard, and is vulnerable to surface scratching and crazing. Impact-resistant grades have been developed by incorporating a compatible rubber with the same refractive index, but this further compromises hardness. The same is true of polystyrene, which is generally cheaper and easier to process but has a lower toughness. Polycarbonate has displaced PMMA in many applications for which toughness and impact-resistance are important (e.g., for aircraft windows and canopies), but it is slightly less transparent and tends to yellow and embrittle in ultraviolet (UV) light unless a *UV block* is incorporated in the plastic. PVC, with a significant degree of crystallinity, is much less transparent, but is widely used for applications such as packaging and bottle blowing on the basis of its lower cost.

A few thermosetting polyesters and epoxies are significantly transparent. A thermoset with an outstanding combination of optical and mechanical properties is polydiallyldiglycol carbonate (CR39), used for moulded lenses.

12·9·2   Electrical conductivity

Because the electrons which are shared covalently within chains are highly localized, most polymers are very good insulators; indeed, many early applications of both thermoplastics and thermosets exploited this property. Offset against this is their generally low resistance to diffusion, which can allow the absorption of water and other polar fluids and a corresponding increase in conductivity. Plasticized polyvinyl chloride, polyethylene and polytetrafluoroethylene are all notable examples of flexible insulators with low enough permeability to win them high volume applications in shielding wire and cables.

On the other hand, useful levels of conductivity can be produced in polymers by loading with a conducting filler such as graphite particles; this technique is used to produce anti-static foams and packaging for electronic components.

12·9·3   Density

Polymers are characterized, as mainly hydrocarbon compounds, by their low density. The lightest of the common thermoplastics are *polyolefins* consisting of carbon and hydrogen only, specifically polypropylene ($0·85$–$0·92$ Mg/m^3), while the densest are those containing heavier halide atoms (e.g., polytetrafluoroethylene, $2·1$–$2·3$ Mg/m^3). The degree of crystallinity has a strong secondary influence; because the chains pack more closely in crystallites, a higher degree of crystallinity is associated with higher density. For the polyethylenes there is a simple linear law-of-mixtures relationship between density and degree of crystallinity. The amorphous phase has a density of about $0·85$ Mg/m^3 and the crystalline phase $1·00$ Mg/m^3, so that branched LDPE of 65% crystallinity has a density of $0·93$ Mg/m^3 and HDPE of 90% crystallinity a density of $0·97$ Mg/m^3. Melting of PE is therefore associated with a large volume increase: after thermal expansion has taken place as well, the density of LDPE melt has fallen to just $0·75$ Mg/m^3. This can cause problems in moulding, since unless high pressure is maintained during cooling, contraction on recrystallization causes surface sinking and internal voids.

12·9·4   Diffusion, absorption and corrosion

Very broadly speaking, polymers resist agents which attack metals (aqueous acids and bases) but may succumb to organic solvents which metals resist. These may dissolve the polymer completely, or may cause *swelling* or plasticization. For molecules of a solvent to be able to penetrate and open up the structure of a thermoplastic, they must be attracted to sites on the chain at least as strongly as they are attracted to each other: thus 'like dissolves like'. Diffusion can itself be a problem: for example, obnoxious-smelling additives have been known to diffuse from gas pressure pipes into adjacent water pipes. Just as metals are most vulnerable in disordered regions at grain boundaries, it is the generally amorphous and open structure of polymers that allows smaller molecules to diffuse in relatively easily. Thus high crystallinity again confers benefits.

Since under the extremely high stresses near a crack tip the structure of a polymer opens up much further, a particularly insidious problem in some polymers is *environmental stress cracking* or *stress corrosion cracking*. Stress cracking agents (usually organic liquids) do not inflict much damage on the bulk material but may, even in extremely small quantities, cause a dramatic reduction in ductility and crack resistance under stress. This is probably due to plasticization of vulnerable material at the tip of pre-existing microcracks, allowing them to extend at stress levels which they would otherwise resist. A well-known example is the effect of aqueous detergent solution on LDPE, whose early use for moulded washing-up bowls did little to enhance the reputation of plastics in general.

12·9·5 Degradation, oxidation and ageing

Degradation is the deterioration of properties of a polymer. Degradation may occur during processing, or afterwards, under the onslaught of the service environment.

The seconds or minutes during which a polymer is processed into its final form are the most destructive it will ever experience. Temperatures high enough to melt a thermoplastic, and the high shear stresses inflicted by extrusion and injection moulding, will always break some covalent bonds. Chemical decomposition at processing temperatures is a particularly severe problem for unplasticized PVC; this can evolve large amounts of corrosive HCl and must be blocked by *stabilizers* in the formulation. Simple *chain scission*, breakage of backbone bonds, maintains the chemical structure but reduces the average molecular weight. The first sign of this, as Eqn (12·5) suggests, may be a reduction in melt viscosity, but mechanical properties, particularly toughness and impact strength, are also affected. This is a matter of concern to processors, who naturally want to exploit the re-formability of thermoplastics by reprocessing scrap. In practice the proportion of scrap must be kept low to avoid repeated degradation, and this can be monitored by measuring the melt viscosity.

Some polymers, for example, polyacetal, tend to depolymerize at high temperature by breakage of monomer groups from chain ends. A much more stable material is provided by copolymerizing with a little ethylene oxide, which forms much more stable chain ends; even if a chain end is lost, decomposition only continues until the next ethylene oxide group is reached. This *acetal copolymer* is the basis for an important engineering plastic.

Oxidation reactions, some extremely complicated, can occur in polymers either during processing or in service. Many of these have the effect of forming crosslinks which, as we have seen, tend to have an embrittling effect. This is a particular problem in rubbers, which rely on their tendency to form crosslinks for their mechanical properties. At normal service temperatures oxidation can be catalysed by light–particularly by sunlight–and can be reduced by the two-pronged strategy of incorporating both anti-oxidants and light-blocking pigments such as carbon black. Processes such as these in which degradation progresses over long periods of time in normal environments are known as *ageing*.

12·9·6   Thermal properties

The physics underlying thermal properties of solids has been introduced in Chapter 7. We are now familiar with the fact that the overall structure of polymers, even semicrystalline ones, is quite loose compared to that of a truly crystalline solid such as diamond. Each segment along a chain is strongly bonded only to its two immediate neighbours. We can therefore visualize both that a segment will only efficiently transmit vibration to these neighbours (i.e., there is poor 'thermal contact' between non-crosslinked chains), so that overall thermal conductivities k in polymers are low; and that its freedom for independent vibration is high, so that the specific heats C_p are high.

This freedom increases, of course, as each of the successive transitions takes place, unlocking another mode of vibration. Thus a convenient and accurate way of observing a transition such as the glass transition is to plot the heat which has flowed into a sample of the polymer (the total heat, or *enthalpy*) as a function of temperature. The glass transition is marked by a quite clearly defined increase in slope [Fig. 12·21(a)]. Also observable on such a plot is the melting point of any crystalline phase, whose disruption absorbs latent heat at constant temperature [Fig. 12·21(b)], although this is usually obscured by the small crystalline content and by its complicated and imperfect structure. A very useful investigative tool in polymer science, *differential scanning calorimetry* (DSC), is based on these changes in slope of the enthalpy–temperature curve.

The ratio $k/\rho C_p$ is known as the *thermal diffusivity*, characterizing the rate at which a body of the material will cool to the temperature of its environment. For all polymers, despite their low density ρ, this ratio is very low. This has very important economic consequences for the processing of thermoplastics, since long times are needed for mouldings and extrusions to cool to a temperature at which they will support their own weight.

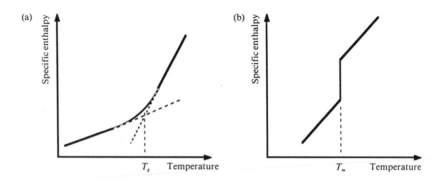

Fig. 12·21 (a) The glass transition and (b) the melting point of crystalline phase, as seen on enthalpy versus temperature curves

12·10 Multi-phase and cellular plastics

The mixing of two metals to form an alloy, which may combine the properties of its constituents or allow new, enhanced properties to emerge, has been discussed in detail in Chapter 10. True alloys, however, are rare between amorphous polymers and virtually unknown between crystalline ones. Genuinely homogeneous blends, like random copolymers, have a single glass transition temperature T_g; because every chain segment finds itself in a well-mixed environment, this lies between the T_g values of the two consitutents. Mutual solubility between different chain structures, however, is usually low, and even where simple blending is effective the mixture needs little encouragement to separate into distinct phases again. Multi-phase blends have the physical properties of a simple mixture: two transition temperatures, for example. But we have also seen in precipitation-hardening alloys and steels the way in which the mechanical properties of metals can be *enhanced* by introducing a secondary phase, provided that this is properly bonded to the matrix or primary phase. In a similar way, the immiscibility of most polymer blends can, with care, be turned to advantage, and this has become a widely exploited route for developing new polymeric materials without the need to develop new base polymers.

The original and best-known examples of such blends are high-impact polystyrene (HIPS) and acrylonitrile–butadiene–styrene (ABS). In HIPS an uncrosslinked diene rubber, usually butadiene, is dissolved in styrene monomer and the blend is polymerized. Since the phases begin to separate during polymerization, vigorous stirring is used to isolate crosslinked rubber particles within a thermoplastic polystyrene matrix (otherwise the rubber surrounds isolated polystyrene enclaves and crosslinks permanently). Some grafting (Section 12·2·3) takes place and improves bonding at the interface.

In ABS the two phases are a single-phase copolymer, styrene acrylonitrile (SAN), and a lightly crosslinked butadiene elastomer onto which SAN has been grafted. The copolymer phase differs from the thermosetting elastomers which we considered earlier, having much smaller elastomer blocks. The elastomer phase again forms a fine dispersion of particles which are securely bonded into the copolymer matrix by their SAN interface. Among other polymers which benefit from rubber toughening is unplasticized PVC, which illustrates a higher level of complexity in that the toughening phase may itself be ABS or another rubber-toughened thermoplastic compatible with the PVC matrix. In each case the role of the rubber particles, which are usually at least $0·1 \mu m$ in size, is to blunt cracks by sustaining large extensions, and to distribute damage over a larger volume of the matrix. The stress–strain curves of PS and HIPS are dramatically different, the rubber-toughened plastic yielding at about one-third of the fracture stress for PS, but showing a ductility of perhaps 40% instead of fracture at a strain of less than 1%.

A special case of a multi-phase polymer is that in which the second component is a gas–usually air. These are the foamed or *cellular* polymers, widely used for packaging, padding, cushioning and sound and thermal insulation. The properties of a polymer foam depend not only on those of the solid from which it is made, but also on the size, proportion and shape and openness of the voids.

The earliest foamed material was natural rubber, in which air was enfolded by a mechanical whisking process. Current technology relies on the incorporation of a *blowing agent* which generates gas (usually on heating) within the polymer. In an *open-cell* structure this gas usually diffuses out quite rapidly and is replaced by air. The polymers now most widely used in cellular form are the polyurethanes. These yield a vast range of properties varying from rigid *closed cell* blocks for thermal insulation (e.g., in refrigerators and freezers) to the flexible *open-cell* types used in furnishing. Closed-cell foams acquire their good thermal insulating properties from the enclosed gas, the cell walls serving only to prevent convective heat transfer.

The high stiffness-to-weight ratio of some rigid structural polymer foams is increasingly being exploited in *sandwich* structures, in which they form a lightweight core. The outer surfaces may be of a distinct material, for example a fibre-reinforced polymer, but a commonly used process is the injection moulding of a plastic containing a blowing agent. The melt solidifies rapidly on the mould surface to give a solid skin, but the interior of the moulding remains hot for long enough for the blowing agent to expand it into a foam. The result is a light and strong sandwich structure produced in a single operation.

12·11   Adhesives

Adhesives are always applied in liquid form, since part of their function is to fill the surfaces to be adhered (the *substrates*). This demands good wetting qualities and low viscosity. For some applications, such as pressure-sensitive adhesive tapes and films (e.g., 'cling-film'), highly plasticized amorphous rubbery polymers are used which remain permanently in a highly cohesive liquid state. Alternatively, *hot melt* adhesives are essentially thermoplastics which are applied at high temperatures and pressures in a sort of moulding operation and resolidify on cooling; the EVA (ethylene–vinyl acetate copolymer) adhesives used for bookbinding and packaging are an example. Other adhesives are solutions of polymers which set by evaporation or diffusion of a volatile solvent, and then perhaps crosslink.

Most uncrosslinked systems tend, as we have seen, to be vulnerable to temperature and solvent action, and the development of strong and stable adhesives for use at high temperatures and in adverse environments has focused on thermosets, in particular the epoxies. With good surface preparation these materials can attain remarkable bond strengths, and have won use in demanding structural applications in the automotive and aerospace industries. For high-strength structural joints, the toughness of the cured adhesive layer is as important as the strength of the adhesive-to-substrate interface, and for the strong epoxy adhesives use has been made of rubber-toughening.

Another well-publicized class of polymer adhesives is the cyanoacrylates, which are based on a liquid monomer stabilized by small amounts of a weak acid. On encountering a slightly basic surface the inhibitor is neutralized, and polymerization and cure occur with prodigious speed to give a strong bond. However, there is little crosslinking and the temperature and solvent resistance of the joint are not exceptional.

12·12 Speciality polymers

We have seen some of the ways in which the chemistry and architecture of polymer molecules is reflected in their processing and mechanical properties. Since the mid-1970s a major interest of the polymer producers has been the exploitation of this knowledge to *design* polymers with highly specific outstanding properties. This movement has resulted in the development of a group of polymers which, although so far negligible in terms of tonnage and extremely expensive in comparison with commodity thermoplastics, offer immense promise to the future of engineering.

12·12·1 High-performance polymers

The suitability of a plastic for engineering use normally depends on a good balance of several properties, rather than on excellence in one. Particularly important in many applications are a combination of high modulus, good temperature resistance (which, according to the concept of time-temperature correspondence, will imply good long-term stability) and good toughness. Notable new contenders in this field derive outstanding properties from strong bonding and high rigidity in their chain backbones; these include the polyimides (PI) and polyether ether ketone (PEEK). PEEK is a semicrystalline polymer with a melting point of 330°C. Both PI and PEEK are of great interest as matrix materials for carbon fibre composites; PEEK has also been used as a hot-melt structural adhesive in the aircraft industry.

12·12·2 Liquid-crystal polymers

We have seen that it is the tendency of polymer chains with flexible backbones to wriggle themselves into an entangled amorphous mass in the melt state which underlies many of their solid properties. Even 'crystalline' polymers contain large intercrystalline amorphous regions. There is currently great interest in a class of linear polymers whose backbones are so rigid that even in the melt state, in which thermal agitation has overcome inter-molecular forces enough for them to slip freely against each other, individual chains remain mutually aligned. Entanglements are rare if not completely absent. In the absence of shear forces which would impose orientation on a large scale, these polymers, in the liquid state, form small regions of almost uniform orientation separated by sharp boundaries; these give them their descriptive name: *liquid-crystal* (LC) *polymers*. In the solid form this structure is retained to produce an extremely anisotropic material in which the covalently bonded chain backbones are aligned in the same direction.

Lyotropic LC polymers can only be processed in solution, so that manufacturing methods are restricted to the production of sheets and films. This class contains the most industrially important LC polymer so far developed, the 'aramid' (aromatic polyamide) polyparaphenylene terephthalamide (PPTA) fibre 'Kevlar', which is spun from a solution of the polymer in concentrated sulphuric acid. The remarkable properties of this polymer arise from the aromatic components which confer rigidity to the backbone, and the strong hydrogen bonding between side-groups on adjacent chains, which raises the crystalline melting temperature to above the decomposition temperature. Axial moduli of approaching 200 GN/m^2 have been achieved, with strengths of about 3·5 GN/m^2. A similar lyotropic LC polymer, poly-paraphenylene benzobisoxazole (PBO), can better these values by 50% or so and is currently entering commercial production.

Thermotropic LC polymers can be processed in melt form. Although their properties do not generally match those of the lyotropics, this is largely offset by the greater ease of processing. Because entanglements are virtually absent from the melt, they are much less viscous than similar amorphous polymer melts. Moreover, because the crystalline structure persists in the melt state, there is little contraction on or after solidification, adding dimensional stability to the portfolio.

12·12·3   Functional polymers

Finally and briefly let us consider a group of materials which are at the forefront of research in polymer physics and chemistry. These are *functional polymers* which have been designed to fulfil specialized physical or chemical functions: they 'do something', beyond serving passively as the material of a structure. Although many functional polymers are still at the research stage, a few are already well-established commercially.

Electroactive polymers can be macroscopically aligned, to give a polarizing effect or a colour change, by an external electrical or magnetic field. Unlike the liquid-crystal compounds familiar from display devices, the rigid LC segments are connected to a polymer backbone through flexible spacers so that they are only free to align when these spacers are in the melt state. They can therefore be written on by a finely focused laser beam and used as optical information storage media. *Photoactive* polymers can be constructed in a similar way by fixing light-activated charge-transfer groups to the main chain; these materials are already in use for xerography, in copiers and in laser printers.

In Section 12·9·2 we noted that polymers could be given a useful level of electrical conductivity by loading with a conductive filler. Recent developments have developed *intrinsically* conducting or semiconducting polymers: notably polyacetylene, which can be modified to acquire a conductivity approaching that of copper at room temperature. As well as offering some potential for lightweight accumulators for vehicle and aerospace applications, these materials have been cited as potential elements in 'molecular computers' with electronic components a few nanometres in size. Also useful in the electronics field as well as for transducers are *piezoelectric* polymers, which develop an electric field when stressed, and *pyroelectric* polymers, which do so when heated. Polymer *electrolytes* have also been developed.

Problems

12·1 The C–C bond length is 0·154 nm. What is the fully extended length of a molecule of polyethylene whose molecular weight is 10^5? Compare this length to the typical dimensions of lamellae and spherulites.

12·2 What is the molecular weight of PTFE with degree of polymerization (DP) 10^5? How does this compare with the molecular weight of polyethylene with the same DP?

12·3 Explain why a glassy thermoplastic is not as hard as an inorganic glass.

12·4 Suggest why rubber tends to crystallize when it is stretched. (Think of what happens to the shape of the molecules.)

12·5 An experimental grade 'C' of linear polyethylene is produced by mixing a monodisperse polymer A, with molecular weight 10^5, with an equal quantity by weight of monodisperse polymer B having molecular weight 4×10^5. What are the number-average and weight-average molecular weights of grade C?

12·6 What would be the ratio of the melt viscosities (at low shear strain rates) of polymer grades C and A in the previous question? Which grade would be more suitable for processing by extrusion, and why?

12·7 The glass transition temperature of polyethylene is well below room temperature, but the elastic modulus of high-density polyethylene is nearly as high as that of a glassy polymer. Explain why this is so.

Self-Assessment questions

1 Adjacent molecules in a thermoplastic polymer are held together by
  A) primary bonds   B) secondary bonds   C) strong bonds
  D) weak bonds   E) covalent bonds   F) van der Waals or hydrogen bonds.

2 Side-groups are attached to the main-chain atoms in a thermoplastic polymer by
  A) primary bonds   B) secondary bonds   C) strong bonds
  D) weak bonds   E) covalent bonds   F) van der Waals or hydrogen bonds.

3 The molecular weight of vinyl chloride is 62·5. Thus the molecular weight of a polyvinyl chloride with degree of polymerization 20,000 is
  A) 320   B) $3 \cdot 1 \times 10^{-3}$ C) $1 \cdot 25 \times 10^6$
  D) 20,000   E) 62·5

4 Which of the following polymers is most similar in its overall properties to high-density polyethylene?
  A) PVC   B) high-impact polystyrene   C) PTFE
  D) isotactic polypropylene   E) atactic polypropylene.

5 The glass–rubber transition is caused by the onset of
  A) scission of backbone bonds   B) free rotation at backbone bonds

C) melting of crystallites   D) breakage of crosslinks
E) slippage between
   adjacent chains.

6 The glass transition involves significant changes in

A) degree of entanglement   B) specific heat   C) elastic modulus
D) density of crosslinking   E) tacticity   F) latent heat.

7 The large extensions possible in rubbers and elastomers are associated with

A) stretching of primary bonds
B) stretching of secondary bonds
C) bending of primary bonds
D) spring-like extension of crumpled polymer chains.

8 Melt-spun polymer fibres have higher tensile moduli than their parent materials because

A) they can contain only very small flaws
B) polymer chains within them are highly oriented
C) they contain more amorphous material
D) they contain inorganic fillers.

9 Stress relaxation in polymers is

A) a steady increase in strain under constant applied stress
B) a steady reduction in stress under a constant applied strain
C) a steady reduction to the original length after applied stress has been removed
D) largely reversible
E) largely irreversible.

10 Thermosets are generally more brittle than thermoplastics because

A) they are susceptible to degradation
B) they are always used below their glass transition temperature
C) crosslinking inhibits plastic flow to blunt a crack tip
D) they contain brittle fillers.

11 Thermosetting polymers

A) have network structures
B) have very few crosslinks
C) cannot be remoulded
D) have very temperature-sensitive mechanical properties
E) are produced by the process of step-growth polymerization.

12 The presence of bulky side-groups in a long-chain polymer increases the rigidity of the material.

A) true   B) false.

13 Only polymers which are highly crystalline exhibit a glass transition temperature.

A) true   B) false.

14 The combination of a spring and a dashpot in series is known as the Maxwell model.

A) true   B) false.

15 The Voigt model accurately represents the behaviour of polymer melts.

A) true   B) false.

16 Plasticizers in polymers

A) increase the melting point   B) depress the glass transition temperature

C) enhance crosslinking   D) are generally organic liquids

E) are generally monatomic gases.

17 Polymers with a glassy structure are generally hard and brittle.

A) true   B) false.

18 The degree of crystallinity of a polymer is decreased by

A) the presence of random side-branches   B) absorption of water

C) fast cooling from the melt   D) slow cooling from the melt.

19 A molecular structure consisting of two repeat units, A and B, part of which is schematically illustrated below, is

A) a branched polymer   B) a block copolymer   C) a graft copolymer

D) alternating copolymer   E) a random copolymer.

```
        B
        B
        B
        B
  - AAAAAAAAAAAAA -
        B
        B
        B
        B
```

20 Isotactic polypropylene is

A) a polyolefin   B) a semicrystalline thermoplastic

C) a glassy polymer   D) an amorphous thermoplastic

E) a thermoplastic elastomer.

21 The density of a polyethylene correlates well with

A) the degree of polymerization   B) the molecular weight   C) the melt viscosity

D) the degree of crystallinity   E) the absolute temperature.

22 The effect which increasing the temperature of a thermoplastic polymer has on its elastic modulus is similar to that of

A) increasing crystallinity   B) decreasing the rate of loading

C) increasing the rate of loading   D) decreasing the density of crosslinks.

23 Crosslinks between chains in a rubber have the effect of

A) resisting permanent creep under load

B) reducing the elastic modulus   C) preventing softening on heating

D) increasing crystallinity   E) preventing oxidation.

24 The viscosity of polymer melts falls as

A) molecular weight   B) flexibility of the chain
increases   backbone increases
C) pressure increases   D) shear strain rate increases
E) temperature increases.

25 Polystyrene is modified to give high-impact polystyrene (HIPS) by incorporating

A) a plasticizer   B) an anti-oxidant
C) a blowing agent   D) ethylene oxide co-monomer
E) a diene rubber.

26 Liquid-crystal polymers have the property that

A) their chain backbones are extremely rigid
B) they are electrically conductive
C) they have elastic moduli typical of metals
D) they are photoactive.

Answers

1 B, D, F	**2** A, C, E	**3** C	**4** D
5 B	**6** B, C	**7** D	**8** B
9 B, D	**10** C	**11** A, C, E	**12** A
13 B	**14** A	**15** B	**16** B, D
17 A	**18** A, C	**19** C	**20** A, B
21 D	**22** B	**23** A, C	**24** D, E
25 E	**26** A, C		

ELECTRICAL CONDUCTION IN METALS

13·1 Role of the valence electrons

Many readers will have learned quite a lot about electricity and electric currents without knowing just why it is that some materials will conduct readily while others are insulators which can become statically charged. In the foregoing chapters we have seen how all matter is built of charged constituents–positive and negative–and obviously conduction of electricity must be associated with motion of those charges. Since the protons in the nucleus of an atom are firmly fixed they can only move when the whole atom moves. Now in the case of electrical conduction in metals, we know that no matter is transported when conduction occurs so that the motion of protons cannot be involved and the loosely bound electrons must be responsible for the passage of current. On the other hand, we know in the case of electrolytic conduction that atoms from the electrolyte are released at cathode and anode so that atoms (actually *ions*) are moving during conduction and both positive and negative charges are involved in the process.

For the moment we confine ourselves to metals and we note first that when discussing metallic bonding we pointed out that only the *valence* electrons could be readily removed to take part in bonding. Similarly, we would expect only the valence electrons to be able to take part in conduction. Thus the number of conducting electrons per atom is determined by the atomic structure. Looking at Table 13·1 we can see that copper, silver, and gold have only one such electron per atom, zinc and cadmium have two, while aluminium has three. The univalent metals sodium and potassium are, of course, too reactive to be of engineering importance while if there are more than three valence electrons the character of the bonding changes and with it the electrical properties.

Table 13·1

Chemical symbol	Resistivity (Ω m)	Number of valence electrons per atom
Cu	$1 \cdot 8 \times 10^{-8}$	1
Ag	$1 \cdot 6 \times 10^{-8}$	1
Au	$2 \cdot 4 \times 10^{-8}$	1
Cd	$7 \cdot 5 \times 10^{-8}$	2
Zn	$6 \cdot 0 \times 10^{-8}$	2
Al	$2 \cdot 7 \times 10^{-8}$	3

Table 13·1 also gives the resistivities of these elements measured at 20°C and we see that the number of valence electrons alone does not determine its value; it does not follow that more electrons lead to higher conductivity. We must therefore look in more detail at the mechanism of conduction.

13·2   Electrons in a field-free crystal

We begin by considering the behaviour of the valence electrons when there is no potential gradient in the metal and no current is flowing. Having understood this, we then consider how the behaviour is modified when a potential drop is applied across the solid.

The valence electrons in a metal are not only able to leave their parent atoms but they can also wander freely through the lattice of ions. One might expect them to collide with each immobile ionic core, and it was once thought that this would cause a large resistance to their flow. However, this conclusion was drawn without taking account of the wave nature of electrons as we shall now show.

An electron moving in a given direction at a constant velocity behaves as a plane wave. The wave interacts with each ionic core and is thereby 'scattered'. Each ion then becomes the 'source' of spherical secondary wavelets in phase with the incident wave (Fig. 13·1), exactly as was assumed in the discussion of Bragg diffraction at the end of Chapter 6. It was concluded there that the secondary wavelets reinforced one another when the condition

$$n\lambda = 2d \sin \theta$$

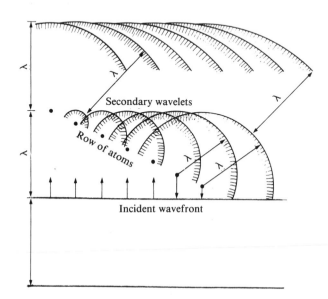

Fig. 13·1 A plane wave cannot be diffracted by a row of atoms when $\lambda > 2d$

was satisfied. In this equation, the expression $2d \sin \theta$ represents the difference in total path length for two adjacent wavelets. But if $\lambda > 2d$, as in the case for electrons of sufficiently low speed, there is no value of θ for which the path difference can be as much as λ, so that reinforcement can only occur where the path difference is zero. This can only occur in the direction of travel of the original wave which therefore proceeds, undiminished in intensity by its encounter with the ions.

A low-energy electron, therefore, is able to move undeflected by a perfectly regular lattice, and behaves as though the ions were not there! We can thus calculate the wave functions for the conduction electrons and determine the expected energy levels by ignoring the presence of the lattice. (A more exact treatment shows that this is not strictly valid. However, the elementary approach outlined here gives the main features of metallic conduction, while the more advanced method merely enables the finer details to be understood.)

Naturally, in a real metal there are many electrons, all moving in different directions and having different wave functions, but we begin by asking what are the possible wave functions for a single electron in the metal. Then we can add the other electrons, taking care not to violate Pauli's principle.

If, as explained in earlier chapters, the electrons are spread uniformly throughout the lattice they cannot be standing waves since in standing waves there are local maxima and minima in the probability distribution. They must therefore be travelling waves since these have uniform intensity everywhere.

Now each electron must be confined within the solid but it is not localized near any atom or group of atoms. It can be regarded as 'orbiting' all the atoms in the solid, moving in a curved path of very large radius so that it is (almost) a plane wave. The path will link every atom in the solid so that the overall path length is very great.

An illustration of this in two dimensions is shown in Fig. 13·2. The path must eventually close upon itself for the same reason as in the hydrogen atom (Chapter 3), namely, that if it were not closed it would eventually lead off to infinity. A mathematical description of this picture, in one dimension for simplicity, is to introduce boundary conditions (see Section 2·6), which describe the conceptual model. The main point is that the long, meandering path of the electron must eventually close up on itself, so at the junction the wave functions must be equal and we can write

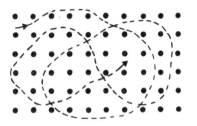

Fig. 13·2 The valence electrons in a metal move along a path which encompasses the whole crystal

$$\psi(x) = \psi(x + L) \tag{13.1}$$

where L is the length of the meandering path and x is measured along the path. This is called a 'periodic' boundary condition.

Since the electron is free to wander anywhere in the solid unaffected by the ions, its potential energy with respect to the lattice must be zero, i.e., $V = 0$ in Schrödinger's equation. Thus, in one dimension, we want to solve

$$\frac{d^2\psi}{dx^2} + \frac{8\pi^2 m}{h^2} E\psi = 0 \tag{13.2}$$

and the solution has to obey the boundary condition of Eqn (13.1). We assume a solution [see Eqn (2.9)]

$$\psi = A \exp(jkx) \tag{13.3}$$

in which A is a constant. Differentiating twice and substituting in Eqn (13.2) we obtain

$$E = h^2 k^2 / 8\pi^2 m \tag{13.4}$$

which is the expression for the kinetic energy of the electron [see Eqn (2.5)]. But we require, for the boundary condition,

$$A \exp(jkx) = A \exp(jk(x+L))$$
$$= A \exp(jkx) \exp(jkL)$$

so that

$$\exp(jkL) = 1 \tag{13.5}$$

or

$$\cos(kL) + j \sin(kL) = 1$$

using de Moivre's theorem. This is fulfilled if $kL = 2n\pi$, i.e.

$$k = 2n\pi/L \tag{13.6}$$

where n is an integer.

Substituting this in Eqn (12.4)

$$E = n^2 h^2 / 2mL^2 \tag{13.7}$$

and we see that the energy is quantized, n being the quantum number, and that if the path length, L, is large the quantum steps in energy will be small. Normalizing the solution [see Eqn (2.21)] we have

$$\int_0^L A \exp(jkx) \cdot A \exp(-jkx) = 1$$

i.e.

$$\int_0^L A^2 \, dx = 1$$

and

$$A = 1/\sqrt{L} \tag{13·8}$$

Thus, the full solution is

$$\psi = \frac{1}{\sqrt{L}} \exp(jkx) \tag{13·9}$$

The probability distribution is given by $|\psi|^2 = A^2 = 1/L$, which is independent of position x, so there is an equal probability of finding an electron anywhere, which fits our picture of 'free' electrons in a metal.

A full treatment would show that, just as in the hydrogen atom, there will be one quantum number for each dimension, since the solid is three-dimensional, so we require three quantum numbers to describe the energy levels. These are, in fact, so closely spaced that the electron can have almost any energy and the range of permitted energy forms an essentially continuous *band*. The energy level diagram is shown in Fig. 13·3, where the spacing between the levels is greatly exaggerated for clarity.

Now, if we add all the other valence electrons one by one to the solid we shall gradually fill these energy levels from the lowest up. Only two electrons, with opposite spins, may have the same set of quantum numbers so that even where several sets of quantum numbers give the same energy there is only a finite number of electrons in each energy level.

Eventually each of the available electrons will have been allocated to an energy level, and all those levels up to a given value, known as the *Fermi level*, will be filled (Fig. 13·3).

This is the Fermi energy which appears in the Fermi–Dirac probability distribution given in Eqn (7·11b). At absolute zero, $kT = 0$ and so any value of energy above the Fermi level, which makes $(E - E_F)$ finite and positive will give $\exp(\infty)$ in the denominator of Eqn (7·11b) and the probability of occupancy of levels above E_F is, at absolute zero, itself equal to zero.

We now summarize the important results obtained so far and which will be used in subsequent sections:

(a) The valence electrons in a metal behave as if the ion cores were absent and the solid were composed of free space.

(b) The spacing between the energy levels available to the valence electrons

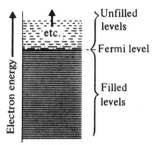

Fig. 13·3 The energy levels available to the valence electrons

is so close that their energy may be regarded for practical purposes as continuously variable.

(c)   In spite of this there is a finite number of available levels, and these are filled by the valence electrons, up to a highest level called the Fermi level. The energy of this level is typically about 4 eV above the lowest level in the band.

13·3   'Electron gas' approximation

Since the kinetic energy of the valence electrons in a metal is almost continuously variable the electrons may be treated as if they were particles in empty space. This important deduction enables us to simplify greatly the calculation of electrical resistivity. Before proceeding with that calculation it is useful to think a little more about the electron in this light. The electrons have a wide range of kinetic energies, as we have seen, owing to Pauli's principle. Thinking of them as particles, we see that this implies a correspondingly wide range of velocities. However, if there is no overall motion of electrons in any direction (that is, no net current) there must be as many electrons moving in one direction as in the opposite direction. The velocity may thus have both positive and negative values but the *average* velocity is zero.

If we plot the fraction, f, of the total number of electrons with a velocity, v, against the value of v, we obtain a curve like that in Fig. 13·4. All velocities between $-v_{max}$ and $+v_{max}$ are represented. The value v_{max} is that corresponding to the energy, E_F, of the Fermi level, mentioned in the previous section. By putting $\frac{1}{2}mv^2_{max} = E_F$ it is found that, when E_F is 4 eV, v_{max} is about $1\cdot2 \times 10^6$ m/s–a very high velocity.

We conclude that the valence electrons in a metal may be pictured as particles moving randomly around at a very high average speed, although also with a wide range of speeds. This suggests that the electrons are very much like a gas of atoms, in which the atoms move about randomly with high thermal energy. The analogy must not be allowed to cloud the fact that the reason why the electrons have high energy (velocity) is quite different from that for the gas atoms. The latter are in rapid motion because of thermal agitation. In the case of the electrons, the effects of thermal agitation are secondary: the velocity of an electron with kinetic energy $\frac{3}{2}kT$ is very tiny indeed compared to the average velocity in Fig. 13·4.

There is one further important effect that must be considered and this is that the electrons fail to 'collide' with the lattice atoms only when the lattice is perfectly regular. If any irregularity is present the moving electron may be scattered and its direction altered. In a real metal there are many such irregularities (see Chapter 8). The most important, however, is due to the vibration of atoms about their mean positions as a result of thermal agitation. Since these vibrations are random they upset the regularity of the lattice structure and the electrons can thus 'collide' with them. We shall see later how it is we know that this is normally the most commonly occurring kind of collision and that the presence of impurity atoms, grain boundaries, and so on, is often of secondary importance.

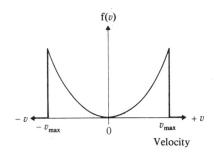

Fig. 13·4 The idealized distribution of velocities among the valence electrons

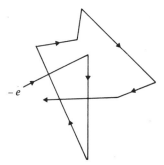

Fig. 13·5 Each valence electron takes a zig-zag path, due to frequent collision with vibrating atoms

In between these collisions the electrons move undisturbed in a straight line (Fig. 13·5). Their paths are thus random zig-zag patterns threading the lattice.

It should be noted that the speed of the electron may also change when it makes a collision–the loss or gain in its kinetic energy is transferred to or from the vibrating atom with which it collides. However, the overall distribution of electron velocities is still as shown in Fig. 13·4 since for every electron which gains in velocity by collision there is another which loses an equal velocity.

We see from the above that the analogy between the electron and a gas is quite close. Just as in a gas, it is possible to define an average path length between collisions, which is usually called the *mean free path*, and a corresponding *mean free time*, namely, the average time spent in free flight between collisions. It is found that the mean free path is about 400 Å in copper at room temperature so that, on average, the electron passes about 150 atoms before colliding with one.

13·4 Electron motion in applied electric fields

Having established the behaviour of electrons in the absence of an electric field, we now turn to the effects produced by applying a field. Since the

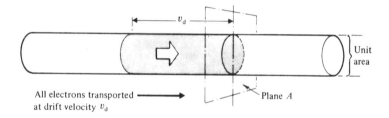

All electrons transported ⟶
at drift velocity v_d

Plane A

Fig. 13·6 Calculation of the current carried by electrons moving with a drift velocity v_d

electrons act between collisions as if they were in free space they are readily accelerated by a potential difference across the crystal, and a large current flows.

The field creates a drifting motion of the whole cloud of electrons in the direction opposite to the field–we often say that the electrons acquire a drift velocity, v_d. This gives rise to an electric current whose magnitude we now calculate in terms of v_d.

Let the current be flowing in the x direction and imagine a cylinder of unit cross section whose axis is also parallel to the x direction (Fig. 13·6). The current flowing down it is equal to the amount of charge crossing any plane, say, plane A, in unit time. This must be just equal to the amount of charge contained in the cylinder in a distance v_d upstream of the plane A. This volume is shown shaded in the figure.

If there are n electrons per unit volume, each of charge, $-e$, the total charge in the shaded volume is just $-nev_d$. They are moving in the $+x$ direction but by convention the current is in the $-x$ direction, and therefore has a negative sign. Thus we obtain

$$-J = -nev_d \quad \text{or} \quad J = nev_d \tag{13·10}$$

Since J is the current per unit cross-sectional area it is called the current density. When the current density in a piece of copper is say 1 A/cm² the drift velocity, v_d is about $7·4 \times 10^{-7}$ m/s (see Problem 13·2) so that v_d is only a very small fraction of the average random velocity of the electron gas. It is therefore permissible to assume that the distribution of velocities among the electrons is scarcely affected by the flow of current. All that happens is that each electron acquires an additional velocity equal to v_d; if the original velocity was v the new value is $(v + v_d)$. Note that if v is negative there is a reduction of the magnitude of the velocity. We show in Fig. 13·7 the new graph of $f(v)$ which is the same shape as before but shifted horizontally by an amount v_d in the direction *opposite* to the field ξ. The size of v_d has been exaggerated to clarify the illustration.

13·5   Calculation of v_d

So far we have assumed that each electron acquires a drift velocity, v_d, without enquiring why it is that the field does not accelerate the electron without

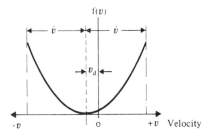

Fig. 13·7 Distribution of velocities among the valence electrons in the presence of an electric field

limit. To understand this point we must look again at the details of the electron motion. In Fig. 13·5 the path of the electron was depicted as a zig-zag course due to collisions with vibrating atoms. In the interval between collisions the electron experiences a force due to the electric field. Figure 13·7 shows how an electron already travelling in the $-x$ direction is thereby accelerated, while an electron travelling in the $+x$ direction is decelerated. We show in Fig. 13·8 how the path becomes curved when the electron is travelling in some other direction–the curvature is greatly exaggerated for clarity. In each case, however, the force, whose magnitude is $-e\xi$, is in the same direction. During free flight the electron is accelerating; the rate of change of v_d is then equal to the force divided by the electron mass so that in free flight

$$\left(\frac{dv_d}{dt}\right) = -\frac{e\xi}{m} \qquad (13\cdot11a)$$

On the other hand, since the system is in equilibrium we know that, averaged over a significant period of time, $dv_d/dt = 0$. This means that the rate of change of v_d given by Eqn (13·3a) is counterbalanced by the rate at which the drift velocity is lost by collision with the lattice vibrations. The latter can be calculated as follows.

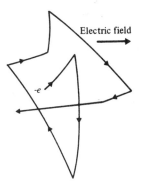

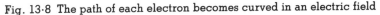

Fig. 13·8 The path of each electron becomes curved in an electric field

The essentially random nature of a collision is not affected by the fact that the electron has acquired a little extra velocity while in free flight. The direction of the electron velocity immediately after a collision is thus entirely random, as it would have been if the electric field were not present. Immediately after collision, then, the electron's velocity is completely random, and all memory of the drift velocity appears to have been lost. (Note that the *energy* associated with the drift velocity is not lost. It is in part transferred to the lattice.) At the beginning of the next free flight we may assume that $v_d = 0$.

At each collision the change in v is thus equal to v_d. If we let τ be the average time spent in free flight between collisions, then each electron collides $1/\tau$ times per second, each time losing a velocity v_d.

The total amount of velocity lost in one second is therefore $v_d \times 1/\tau$, and this is the rate of change of v_d.

Thus the average collision loss is

$$\left(\frac{dv_d}{dt} \right)_{loss} = \frac{v_d}{\tau} \tag{13·11b}$$

Note that, unlike the gain in velocity, the loss occurs only at the time of the collision. The rate of loss is therefore zero for the time, τ, between collisions and very large for the negligible time it takes to complete the collision. The average rate is given by Eqn (13·11b).

We have already noted that the total rate change of v_d is zero so we must add Eqns (13·11a) and (13·11b) and equate to zero, giving

$$\frac{dv_d}{dt} = \left(\frac{dv_d}{dt} \right)_{gain} + \left(\frac{dv_d}{dt} \right)_{loss} = 0$$

Thus

$$\frac{v_d}{\tau} = \frac{\xi e}{m} \quad \text{and hence} \quad v_d = \frac{e\xi\tau}{m} \tag{13·12}$$

Thus v_d is proportional to ξ, and it is customary to call the drift velocity in unit electric field the *mobility* of the electron, μ.

Using Eqns (13·10) and (13·12) we obtain the equation for current density

$$J = \frac{ne^2\xi\tau}{m} \tag{13·13}$$

Since the resistivity, ρ, of the metal is just ξ/J (see Problem 13·3) we have

$$1/\rho = \frac{ne^2\tau}{m} = ne\mu \tag{13·14a}$$

where

$$\mu = \frac{e\tau}{m} \tag{13·14b}$$

The two quantities which determine the resistivity are therefore the electron density, n, and the mean free time, τ. Since the electron density cannot vary with temperature the whole of the temperature dependence of ρ must be due to changes in τ. In the next section we shall consider how τ depends on temperature and also how it can be influenced by the defect structure of a metal.

For the moment let us return to consider the implications of the increase in the random velocity which occurs between each collision. Naturally this cannot be a cumulative process and the electrons actually lose some of this extra energy during the collision transferring it to the vibrating atoms whose vibration therefore increases. This corresponds to an increase in temperature of the solid. Indeed this is just what is observed; the passage of current through a metal causes heating. Since the amount of energy transferred to the lattice increases as v_d^2, the heating effect is proportional to the square of the current.

13·6 Phonon scattering

So far, in discussing collisions between electrons and atoms we have treated each atom as vibrating independently. This was the basis of the theory of resistivity developed by Drude in 1900 and subsequently refined by Lorentz. However, as was pointed out in Chapter 7, a single atom in a solid, when it vibrates, exerts forces on its neighbours and the atomic vibrations must be treated in terms of lattice waves (see Section 7·10).

As with electron waves, we can allocate a wave vector, $q = 2\pi/\lambda$, to the lattice wave. Using the wave-particle duality concepts which we applied to electromagnetic and to electron waves, we represent the lattice waves as a stream of 'particles' each carrying a momentum $hq/2\pi$ and with an energy quantized in units of $h\nu$ where ν is the frequency of the lattice wave. These 'particles' are the *phonons* described in Chapter 7 and the interaction of an electron wave with a lattice wave can be regarded as a 'collision' between an electron and a phonon. In the collision process either the phonon energy will be transferred to the electron and the phonon disappears, or else the electron will lose energy to the lattice thereby creating a phonon. In thermal equilibrium at a fixed temperature the rate of creation equals the rate of destruction and the phonon population is constant.

In such an interaction, since energy must be conserved, we may write

$$E(k) - E(k') = \pm h\nu/2\pi \qquad (13\cdot15)$$

where $E(k)$ is the electron energy of wave vector k before the interaction and $E(k')$ afterwards. Also, by conservation of momentum

$$hk'/2\pi - hk/2\pi = \pm hq/2\pi \qquad (13\cdot16)$$

where the positive sign will apply if the electron has taken up a momentum $hq/2\pi$ from the lattice and the minus sign if momentum is transferred to the lattice.

We saw in Section 13·5 that the effect of an electric field in (say) the $-x$

direction is to increase the velocity of the electron in the $+x$ direction and to decrease it in the $-x$ direction. Through electron–phonon 'collisions' the $+x$ electron interacting with a lattice wave can lose its extra velocity by creating one or more phonons, whilst the $-x$ electron can regain the velocity it lost by destroying one or more phonons. These processes result in a change in velocity and direction of travel for an electron at each 'collision' and are referred to as phonon scattering of the electrons, each 'collision' being a scattering event.

13·7   Dependence of resistivity on temperature

The reader is probably familiar with the fact that the resistivity of a metal increases linearly with temperature. It is possible to demonstrate this using the theory outlined in this chapter and we now do so. To simplify the derivation we merely deduce the form of the relationship and the constant of proportionality (that is, the temperature coefficient of resistance) will not be calculated.

In Eqn (13·14) the only temperature-dependent quantity is the mean free time. This is connected with the probability of a phonon–electron interaction (i.e., a 'collision' or 'scattering event'). If the probability is small a long time will elapse before a collision occurs and the mean free time will be long. It is, in fact, a standard result of kinetic theory (see the end of this chapter) that if the mean free time is τ, then the probability of a collision occurring in time δt is $\delta t / \tau$ or, putting it another way, $1/\tau$ is the probability per unit time of a collision occurring.

The probability of a phonon–electron interaction occurring will be proportional to the product of the numbers of electrons and of phonons present. Now the number of electrons is constant and independent of temperature but the number of phonons is given by Eqn (7·18). At ordinary temperatures, $h\nu$ is very small ($< 4 \times 10^{-4}$ eV) whilst the value of $k_B T$ at room temperature is 0.025 eV (k_B is Boltzmann's constant). Thus the fraction inside the exponential bracket in Eqn (7·18) is very small and we can replace the exponential by the first two terms in the exponential series, i.e., by $(1 + h\nu/kT)$. Thus Eqn (7·18) can be approximated to

$$n_q \simeq k_B T / h\nu$$

We see, therefore, that the probability, p of an interaction will be given by $p \propto n_e n_q \propto n_e k_B T / h\nu$, i.e.

$$p = 1/\tau = CT$$

where C is a constant. But from Eqn (13·6) resistivity ρ is proportional to $1/\tau$ and so we deduce that the resistivity of a metal will be proportional to temperature over the range of temperature for which $kT \gg h\nu$. This precludes low temperatures (below, say, 20K) where, experimentally, ρ is no longer proportional to T. At extremely low temperatures, a metal may exhibit superconductivity (see Section 13·9).

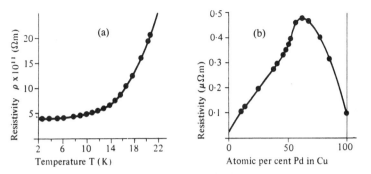

Fig. 13·9 Resistivity of (a) sodium at low temperature and (b) palladium–copper alloy at room temperature

13·8 Structural dependence of resistivity

When measurements are made at low temperatures the resistivity tends to a finite value [Fig. 13·9(a)] as the temperature is lowered. This is because the number of phonons becomes so small at these temperatures that electron and phonon collisions become less important than those with crystal defects such as impurities, grain boundaries, and so on. Since these are fixed in number the mean free time no longer depends so strongly on temperature.

This result suggests that it might be possible to increase the resistance of a metal at ordinary temperatures and at the same time reduce its temperature coefficient of resistivity, by introducing many defects. This is achieved in some disordered solid solutions whose resistance may be much higher than that of either constituent. An example of this is shown in Fig. 13·9(b) for the copper–palladium alloys. The copper–nickel alloys (for example, 'Constantan') which are used for making electrical resistances show similar behaviour, though not entirely for the same reason.

On the other hand, the behaviour is quite different in a two-phase alloy in which the structure at intermediate compositions is eutectic. In this case the material is essentially a mixture of two phases with different resistivities, and from an electrical viewpoint it may be regarded as a complicated network of resistances in series–parallel with one another. It can be shown that the resistance of such a structure varies linearly with the volume fraction of either constituent. Thus, if the resistivities of the two phases, α and β, are ρ_α and ρ_β then the resistivity of the two-phase material is

$$\rho = \rho_\alpha V_\alpha + \rho_\beta V_\beta$$

where V_α and V_β are the volume fractions of the α and β phases.

13·9 Superconductivity

The electrical resistance of many metals drops suddenly to zero at some, very low, temperature called the *critical temperature*, T_c. This was first observed in mercury at 4·2K by Kammerlingh Onnes in 1911 and since that time it has been found that, if magnetic elements are excluded, about half of all elemental metals and thousands of alloys and intermetallic compounds, including metallic oxides, borides and nitrides as well as some semiconductors, exhibit the effect at low enough temperatures. In the zero-resistance state they are called *superconductors*. Niobium is the pure metal with the highest T_c at 9·2K whilst others have values all the way down to 0·001K.

Until 1986 the highest known critical temperature was 23K for a complex alloy of niobium. Then Mueller and Bednorz published their results for a new class of tertiary metal oxides which showed much higher critical temperatures, for which they were awarded the 1987 Nobel Prize for Physics. An example is $YBa_2Cu_3O_7$ with a critical temperature of 92K. The great significance of this development is that it is no longer necessary to use expensive liquid helium to cool below T_c since liquid nitrogen is at 77K and costs about the same as a cheap beer! At the time of writing the highest reported critical temperature is about 120K. The way is now open for many practical applications of superconductivity at economic prices.

In the superconducting state the electrical resistivity is zero so that, if a current is induced in the superconductor, it will continue to flow, without diminution, so long as the material remains at a temperature below T_c. As the value of the induced current is increased, a *critical current density*, J_o, is reached at which the material loses its superconductivity and reverts to normal behaviour.

Associated with superconductivity there is an equally dramatic magnetic effect, discovered in 1933, called the *Meissner effect*. If the superconductor material, in the normal state, is placed in a magnetic field and then cooled through the critical temperature it is found that, at T_c, the whole of the

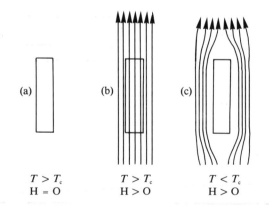

$$
\begin{array}{ccc}
\text{(a)} & \text{(b)} & \text{(c)} \\[2mm]
T > T_c & T > T_c & T < T_c \\
H = O & H > O & H > O
\end{array}
$$

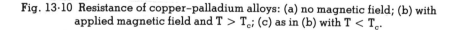

Fig. 13·10 Resistance of copper–palladium alloys: (a) no magnetic field; (b) with applied magnetic field and $T > T_c$; (c) as in (b) with $T < T_c$.

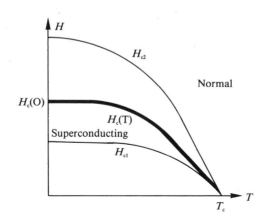

Fig. 13·11 Critical fields of a superconductor

magnetic flux is suddenly ejected from the material and it behaves as a perfect diamagnet (see Chapter 16) as illustrated in Fig. 13·10. However, as the strength of the applied magnetic field is increased, a *critical field*, H_c, is reached at which the superconductivity is destroyed and the material reverts to its normal state. This effect was an important clue in the development of the theory of superconductivity given later. It is related to the critical current density mentioned above since an electric current sets up a magnetic field. The magnitude of the critical field is temperature dependent, being lower the nearer the temperature approaches T_c from below, as illustrated in Fig. 13·11. The value of H_c quoted for a superconducting element is usually $H_c(0)$, the value of field at which the curve cuts the H_c axis at $T = 0$.

As either the temperature approaches T_c or the applied magnetic field approaches H_c, the magnitude of the critical current density approaches zero. J_c is not strictly an intrinsic property of the superconductor and is found to depend on the metallurgical state of the material, i.e., crystallite size, defect density and so on.

13·9·1 Type I and Type II superconductors

The flux density, B, in a material in a magnetic field, H, is given by Eqn (16·1)

$$B = \mu_0 (H + M)$$

where μ_0 is the permeability of free space and M the magnetization of the material. Due to the Meissner effect, $B = 0$ in a superconductor and hence $H = -M$; in other words the superconductor can be thought of as having a negative magnetization equal to the applied field. In Fig. 13·12 negative magnetization is plotted against field for typical elemental and typical alloy superconductors. For the former, the magnetization rises in magnitude up to the critical field and then abruptly drops to zero; this is called *Type I*

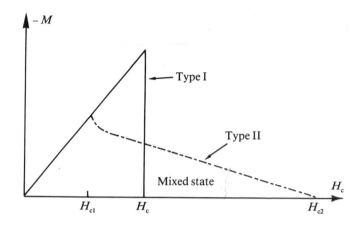

Fig. 13·12 Magnetization of Types I and II superconductors

behaviour. In a few elemental and most alloy superconductors it is found that, above a certain value of field, H_{c1}, magnetic flux can penetrate the material without destroying the superconductivity. As the applied field is increased, the magnitude of the negative magnetization decreases until, at a higher field, H_{c2}, the material reverts completely to the normal state. The material is called a *Type II* superconductor and the value of H_{c2} may be up to 2,000 times the value of H_{c1}. Between these values of field the material is in a mixed state in which the specimen is penetrated by quantized filaments of magnetic field each of which has a strength given by $hc/2e$. The filament contains a cylindrical core of 'normal' material surrounded by a cylinder around which a vortex of supercurrent flows. Each cylinder is referred to as a vortex and adjacent vortices repel each other. As the magnetic field increases, the number of filaments per unit area increases and the filaments pack together in a two-dimensional lattice. The interstitial material remains super-conducting until the field H_{c2} is reached, when the core regions are forced into contact and the whole specimen reverts to the normal state. The new high-temperature superconductors are Type II with very high values of H_{c2} and $H_{c2} \gg H_{c1}$.

A list of representative superconducting materials is given in Table 13·2.

13·9·2   *Theoretical position*

For a long time after its discovery, superconductivity was considered the most mysterious and difficult phenomenon to account for theoretically and it was not until the work of Bardeen, Cooper and Schrieffer in 1957 that a satis-factory theory (the BCS theory) was produced. The essential basis of this is that there is an attractive force between electrons that have about the same energy which, in the right circumstances, causes them to move in pairs. The criterion of superconductivity is that this attraction should exceed the repulsion of like charges which always occurs between electrons. It should be noted that the repulsive force between electrons in the solid is not as great as

Table 13·2

Element	$T_c(K)$	$H_c(0)$ (Tesla)	Alloy	$T_c(K)$	H_{c2} (4·2K) (Tesla)
Aluminium	1·19	0·01	$Pb_{0.9}Mo_{5.1}S_6$	14·4	60·3
Indium	3·41	0·029	V_3Ga	14·8	23·9
Mercury	4·15	0·041	NbN	15·7	10·0
Tantalum	4·48	0·084	Nb_3Sn	18·0	26·4
Vanadium	5·3	0·132	Nb_3Ga	20·2	32·7
Lead	7·18	0·090	$Nb_3(Al_{0.7}Ge_{0.3})$	20·7	41·5
Niobium	9·46	0·196	Nb_3Ge	22·5	36·4

High-temperature alloy†	$T_c(K)$	H_{c1} (Tesla)		H_{c2} (Tesla)	
		$H//c$	$H//a,b$	$H//c$	$H//a,b$
$YBa_2Cu_3O_7$	92	~0·1	~0·01	~50	~200

†Note: $H//a,b$ is a field in the ab basal plane of the crystal and $H//c$ is a field along the c-axis, where the crystal lattice constants are a, b and c. The ab plane contains the Cu and O atoms.

the force between two isolated electrons because the lattice of positive ions acts to 'screen' the electrons from each other to some extent, reducing the repulsion.

The attractive forces arise because, in a lattice of positive ions, an electron will produce a small distortion of the lattice by attracting the positive ions towards itself, making the lattice slightly more dense in its vicinity. To a passing electron this will look like a local increase in positive charge density and it will be attracted towards it. Thus the electrons tend to form pairs through interaction with the lattice. Clearly if the lattice is vibrating through the usual thermal effects the pairwise interaction will be obscured, but at very low temperatures where thermal agitation is small, the attractive force can assert itself. It can be shown that the attractive force will be a maximum for electrons having opposite spins and moving with equal velocities in opposite directions.

The effect of this force is to produce an ordering of the electron gas; just as atoms of a gas are ordered on condensing into a crystalline solid, so the electrons are ordered in pairs, whose movements are correlated with each other over a finite distance called the *coherence length*, on passing into the superconductive state. This ordering is associated with a lowering of the total energy of the electron gas by an amount ΔE, and in order to return to the 'normal' state an electron must be supplied with this amount of energy. It can come from an increase in temperature above T_c, so that the thermal energy kT is greater than ΔE, or from magnetic energy due to a field greater than H_c.

In terms of the energy level concept, introduced in Section 13·2, ΔE represents an 'energy gap' at the Fermi level, in which there are no allowed energy states for electrons. There is no possibility of scattering of electrons since there are no states to be scattered into, unless the electrons gain sufficient energy to cross the gap, and this is not possible at temperatures below T_c. Thus the normal mechanism of resistance to current cannot operate and the resistance of the material will be zero. This will remain true until the drift velocity of the electrons which carry the current exceeds a critical value corresponding to their having gained extra energy greater than ΔE from the applied field. Hence there will be a limit to the supercurrent density that can be carried. The critical velocity corresponding to this critical current density is given approximately by

$$v_c = \frac{\lambda_F \Delta E}{2h}$$

where λ_F is the electron wavelength for an electron at the Fermi level.

The width of the energy gap is a function of temperature, being a maximum at $T = 0$ and falling to zero at $T = T_c$ in the manner shown in Fig. 13·13. Just below T_c it rises steeply from zero, varying as

$$\Delta E(T) = 3·2kT_c(1 - T/T_c)^{\frac{1}{2}}$$

and at lower temperatures varies more slowly, flattening off to a constant value $\Delta E(0) = 1·76kT_c$ at about $T = T_c/2$.

The coherence length, c_0, can be related to the energy gap through the approximate relationship

$$c_0 \simeq \frac{h^2}{m\pi\Delta E\lambda_F}$$

and is of the order of 1 μm in most elemental superconductors. This means that the electron pairing is extended over thousands of atomic spacings so that atomic scale defects and impurities are too small to have much effect on superconducting behaviour. However, in the 123 high-temperature super-conductors the coherence length is of the order of a few nanometres, which means that the pairing behaviour is almost on an atomic scale and the super-conducting behaviour will be much more dependent on atomic scale defects.

At the time of writing there is considerable doubt about the mechanism by which pairing of electrons comes about in the 123 superconductors. There is evidence that the mechanism may not involve phonons but is, rather, a combination of charge and magnetic interactions. It is generally agreed that

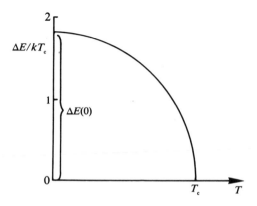

Fig. 13·13 Variation of energy gap with temperature

the crucial role is played by the interaction between copper and oxygen ions, but the exact form of the interaction is, as yet, unknown. The only certain thing is that the next decade will see a tremendous level of activity in super-conductor research and an upsurge of interest in its applications.

*13·10 Collision probability and mean free time of electrons

The probability of an electron undergoing collision within time δt is propor-tional to dt, so we may let it equal $p\delta t$. Take a group of N_0 atoms at time $t = 0$. Some of these will collide only a short time later, others will be in free flight for longer. To calculate the *mean* free time we must know how many are left (i.e., have not collided) after an arbitrary time t; for the moment we assume that there are N of these.

Between times t and $(t + \delta t)$ a further small number, δN, will have collided. This number δN will be equal to the number N left in the group, multiplied by the probability of collision, $p\delta t$, for each electron, that is

$$\delta N = -Np\delta t \tag{13·17}$$

The minus sign enters because δN represents a decrease in N.

We now find the relationship between N and t by integrating Eqn (13·17) between times $t = 0$ (when $N = N_0$) and t (when $N = N$), thus,

$$\int_{N_0}^{N} \frac{dN}{N} = -\int_{0}^{t} p \, dt$$

Performing the integration we find that

$$N = N_0 \exp(-pt) \tag{13·18}$$

Now Eqn (13·17) tells us the number of electrons, δN, which collide between t and $(t + \delta t)$, and which therefore have a time of free flight equal to t. The average flight time, τ, is given by

$$\tau = \frac{1}{N_0} \int_{0}^{t=\infty} t \, dN$$

with the help of Eqn (13·17) this becomes

$$\tau = -\int_{0}^{\infty} \frac{N}{N_0} pt \, dt$$

Using Eqn (13·18) this may be integrated by parts as follows

$$\tau = -\int_{0}^{\infty} pt \exp(-pt) \, dt$$

$$= \left[t \exp(-pt) \right]_{0}^{\infty} - \frac{1}{p} \int_{0}^{\infty} p \exp(-pt) \, dt$$

$$= \left[t \exp\left(-pt\right) \right]_0^\infty + \left[\frac{1}{p} \exp\left(-pt\right) \right]_0^\infty$$

Since the first term is zero at both $t = 0$ and $t = \infty$ we finally discover that

$$\tau = \frac{1}{p}$$

so that the collision probability is just $1/\tau$ per unit time.

Problems

13·1 Show that the following two definitions of the resistivity ρ are equivalent

(a) Resistance of a bar $= \rho \times \dfrac{\text{length of bar}}{\text{cross-sectional area of bar}}$

(b) $\rho = \dfrac{\text{Electric field}}{\text{Current density}}$

13·2 The density of copper is $8{\cdot}93 \times 10^3$ kg/m³. Calculate the number of free electrons per cubic metre and hence deduce their drift velocity when a current is flowing whose density is 1 A/cm².

13·3 The Fermi energies and interatomic distances in several metals are given below. Check in each case that the electrons of highest energy have a wavelength greater than twice the distance between atomic planes, and hence are not diffracted by the lattice.

Metal	Na	K	Cu	Ag
Fermi energy (eV)	3·12	2·14	7·04	5·51
Interplanar spacing (Å)	3·02	3·88	2·09	2·35

13·4 What is the kinetic energy of an electron which has a wavelength just short enough to be diffracted in each of the metals listed in Problem 13·3?

13·5 From the data given in Table 13·1 and Appendix 3, calculate the mobility of an electron in each metal at room temperature.

13·6 What is the mean free time between collisions in the metals Cu, Ag, and Au? Their resistivities are given in Table 13·1 and their densities are respectively 8·93 g/cm³, 10·5 g/cm³, and 19·35 g/cm³.

13·7 Deduce the maximum random velocity of electrons in Cu, Ag, and Au if their Fermi energies are respectively 7·04 eV, 5·51 eV, and 5·51 eV. Using these velocities and the mean free times obtained in Problem 12·6, find approximate values for the electron mean free path in the three metals.

13·8 When about 1 atomic percent of a monovalent impurity is added to copper, the mean free time between collisions with the impurities is about 5×10^{-14} s. Calculate the resistivity of the impure metal at room temperature.

Using the electron velocity deduced in Problem 13·7, find the mean distance between collisions with impurities. Compare this with the average distance between the impurity atoms and explain why the two distances are different.

13·9 If the electron mean free path were independent of temperature (as a result, say, of a large impurity content) and the electrons behaved as if they constituted a perfect classical gas, show that the resistivity would be proportional to $T^{1/2}$.

13·10 In Chapter 9 the changes in a metal which result from plastic deformation were described. Suggest what changes you would expect in the conductivity of copper as a result of cold working. How could you restore the resistivity to its original value?

Self-Assessment questions

1 The conductivity of a metal is determined only by the number of valence electrons per atom.

A) true   B) false.

2 In a perfectly regular metal lattice the valence electrons behave as though the ions were not there.

A) true   B) false.

3 The energies of the valence electrons in a metal are quantized because

A) the electrons are each localized near an atom
B) the electron wave is a plane wave
C) the electron's path must eventually close on itself.

4 The spacing between the energy levels of the valence electrons is small because

A) there is a large number of valence electrons
B) the average length of closed path traversed by the electron wave is very large
C) Pauli's principle applies to the electrons.

5 The Fermi level is

A) the highest occupied energy level at absolute zero of temperature
B) the average of the available energy levels
C) the highest available energy level.

6 The valence electrons in a metal may be treated as an 'electron gas' because

A) they exert a pressure at the boundaries of the solid
B) there is a very large number of them
C) the kinetic energy of each electron is almost continuously variable.

7 The velocity of an electron at the Fermi level, if the Fermi energy is 5·0 eV, is

A) 3×10^{10} m/s   B) $1·33 \times 10^6$ m/s   C) $1·33 \times 10^3$ m/s.

8 Electrons will 'collide' with the lattice atoms if the atoms are

A) very large
B) close together
C) displaced from their regular positions.

9 An applied electric field causes an electron gas cloud to

A) drift in the direction of the field
B) drift in the direction opposite to the field
C) become smaller.

10 If a metal contains a density of 10^{28} valence electrons per metre^3 and carries a current of 10^6 A/m, the drift velocity is

A) $6·25 \times 10^{-4}$ m/s   B) $6·25 \times 10^4$ m/s   C) $6·25$ m/s.

11 The distribution of velocities amongst the electrons is hardly affected by the flow of current.

A) true   B) false.

12 The current density in a metal having 10^{28} valence electrons/m^3 with a mean free time of 10^{-14} s, in a field of 10^4 V/m, is

A) $2 \cdot 8 \times 10^{10}$ A/m^2 B) $2 \cdot 8 \times 10^6$ A/m^2 C) $2 \cdot 8$ A/m^2.

13 The mobility of an electron in a metal is the drift velocity per unit applied field.

A) true   B) false.

14 The mobility of an electron in a solid in which it has a mean free time of 10^{-14} s, is

A) $17 \cdot 6$ m^2/V s   B) $5 \cdot 69 \times 10^2$ m^2/V s   C) $1 \cdot 76 \times 10^{-3}$ m^2/V s.

15 The mean free time of an electron in a metal is

A) proportional to   B) inversely proportional to
C) proportional to the square of

the temperature.

16 The resistivity of a metal is a function of temperature because

A) the electron velocity varies with temperature
B) the electron gas density varies with temperature
C) the number of phonons varies with temperature.

17 The resistivity of all normal metals as temperature is lowered

A) tends to zero   B) tends to a constant value
C) at first decreases and then increases.

18 A disordered solid solution between two metals has a resistivity which is

A) higher than that of either pure component
B) lower than that of either pure component
C) equal to that of the pure component which has the highest resistivity.

19 A two-phase polycrystalline alloy contains 10% of the α phase and 90% of the β phase by volume. If the resistivities are $1 \cdot 7 \times 10^{-8}$ Ωm and $7 \cdot 0 \times 10^{-8}$ Ωm respectively, the resistivity of the alloy is

A) $8 \cdot 7 \times 10^{-8}$ Ω m   B) $6 \cdot 47 \times 10^{-8}$ Ω m   C) $2 \cdot 23 \times 10^{-8}$ Ω m.

20 The collision probability per unit time for an electron in a metal is equal to the reciprocal of the mean free time.

A) true   B) false.

Answers

1 B	2 A	3 C	4 B
5 A	6 C	7 B	8 C
9 B	10 A	11 A	12 A
13 A	14 C	15 B	16 C
17 B	18 A	19 B	20 A

SEMICONDUCTORS

14·1 Introduction

If one examines the electrical resistivities of the elements, particularly of the elements in Groups I B, II B, III B, and IV B, it will be seen that they tend to increase with increasing valency. At first sight, this is surprising since increasing valency means an increasing number of loosely bound electrons in the outer shells of the atoms, and an increase in the number of electrons available might be expected to lead to an increase in electrical conductivity, especially in view of Eqn (13·14). The resistivity of copper (Group I B) is about $1\cdot7 \times 10^{-8}$ ohm-metre at room temperature while that of germanium (Group IV B in the same period of the Table) is in the range 10^{-3} to 10^4 ohm-metre, depending on its purity. This phenomenon is directly related to the interatomic bonding in these materials.

14·2 Bonding and conductivity

It will be recalled that the bonding in a metal is visualized in terms of a fixed array of positive ions held in position by a 'sea' of negative charge. The sea consists of the valence electrons which are distributed throughout the solid, i.e., they are shared by all the atoms. In the Group IV materials covalent bonds are formed whose distinguishing feature is that the electrons are *localized* in the bonds. This means that the wave function of a bonding electron is confined to a region of space between adjacent atoms, as depicted in Figs. 5·9 and 5·12. In fact, the maximum probability density for the bonding electrons is just midway between the atoms. We show this in the diamond lattice in Fig. 14·1, where for clarity the three-dimensional structure has been 'unfolded' and drawn in two dimensions. Each bond contains two electrons with opposite spin directions–more than this is not possible owing to Pauli's exclusion principle.

One must not think, however, that these electrons are immobile, condemned forever to stay in the same bond. The probability distributions are smeared out so that those of neighbouring bonds overlap slightly, and this makes it possible for a pair of electrons in adjacent bonds to change places with one another. Indeed, all the electrons are continually doing so, moving through the network of bonds at great velocity. In spite of this rapid motion, Pauli's principle still ensures that at any one time there are no more than two electrons in any particular bond.

Now although the electrons are moving through the network of bonds,

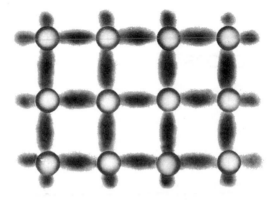

Fig. 14·1 A two-dimensional representation of the covalent bonds in a crystal of C, Si, or Ge; the electron clouds are concentrated between the atoms

they are not strictly to be regarded as travelling waves. Travelling waves have a constant intensity at every point, while we know that in covalent bonds the electron clouds are concentrated midway between the atoms. The electron density (Fig. 14·1) is more like the distribution in a set of standing waves, with nodes at the ion centres and antinodes between them as suggested in Fig. 14·2.

The electron waves are therefore standing waves, not running waves. However, this does not mean that the electrons cannot change places with one another, since each standing wave is composed of two running waves travelling in opposite directions. Because the amplitudes of the two oppositely travelling waves are equal, there is exactly as much charge being transported in one direction as the opposite one. In this way, the charge density at each point remains unchanged, and Pauli's principle is satisfied. From this we deduce that no net flow of charge in any direction is possible, i.e., a covalent crystal cannot carry an electric current.

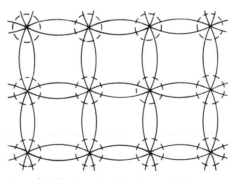

Fig. 14·2 The electron distribution in the diamond structure is like a system of standing waves

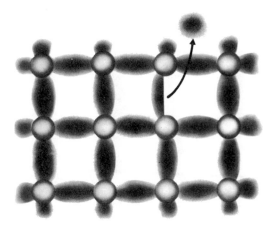

Fig. 14·3 Thermal vibration causes an electron to be freed from one of the bonds

14·3 Semiconductors

The above argument suggests that a crystal with covalent bonding should be a good insulator, and in some cases this is substantially correct. Our prototype covalent solid, diamond, has a resistivity of 10^{12} Ω m at room temperature, and polymers such as polyethylene and polyvinyl chloride have resistivities of about 10^{15} Ω m in their pure state. However, two elements, silicon and germanium, both of which are of great technological importance, have much lower resistivities. Why is this? If we compare the bond strengths of diamond, silicon, and germanium, we find that they decrease as the atomic weight increases. (This has already been discussed in Section 5·11.) It is thus easier to break the bonds in germanium than in silicon, while it is hardest of all to break them in diamond. If the bond is sufficiently weak, the thermal vibration of the atoms may occasionally be sufficient to sever a bond, even when the temperature is not very high. Naturally, the higher the temperature, the more likely it is that any one bond will be broken, and hence the more bonds will be broken at a given time.

Now breaking a bond does not affect the atoms which it joins together–each still has three bonds to keep it in place. But breaking a bond does entail releasing one of the electrons which form it, as shown in Fig. 14·3. Thus at any temperature a small proportion of the large number of bonding electrons will be set free. Being no longer confined in the bonds, these electrons may wander freely through the lattice and conduct an electric current. At room temperature the freed electrons in a piece of silicon constitute only about one in 10^{13} of the total number of valence electrons. This is in complete contrast to the situation in a metal, in which all the valence electrons are free to conduct current.

The number of free electrons is essentially fixed at a given temperature, although the process of bond rupture by thermal agitation is continuous. The rate at which electrons are freed is exactly balanced by the rate at which those

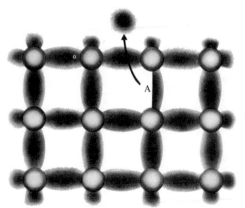

Hole is created at A

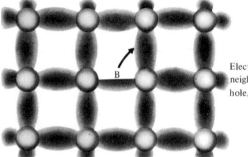

Electron moves from neighbouring bond to fill hole, which now appears at B

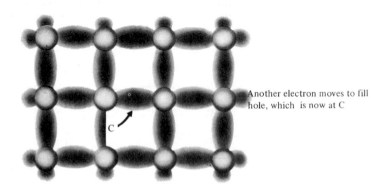

Another electron moves to fill hole, which is now at C

Fig. 14·4 As the bonding electrons move about, the hole is filled and appears to move

already free are trapped by an empty bond and lose their freedom.

We shall postpone the calculation of the conductivity generated by the presence of free electrons to the next section, for we must now take note of a curious phenomenon which occurs in the bonds.

The freed electrons leave behind them gaps in the bonds, called holes. Since the bonded electrons are in constant motion, these holes are soon filled by nearby electrons. This necessarily causes the holes to appear in the bonds vacated by the itinerant electrons, and the holes appear to move (Fig. 14·4). In fact the holes move about in much the same random way and at much the same velocity as the electrons in the filled bonds.

We shall now show that the presence of these holes enables the bonded electrons to carry a current in such a way that it appears as if it is conducted by positive charges carried by the holes!

14·4 Conduction by holes

We have seen that the hole moves by virtue of an electron jumping into it from an adjacent bond. We have also stressed that, when all the bonds are full, the electron distribution is not affected by an electric field. However, when a field is applied and a hole is present, the probability of an electron jumping into the hole from an adjacent bond will be increased if its jump is assisted by the field; conversely, the probability of a jump against the field is lowered. Thus, if the field favours the motion of electrons from left to right (say) in Fig. 14·4, the hole will tend to drift from right to left and thus its motion is influenced by the field. Furthermore, its drift direction in the field is opposite to that of electrons and so it behaves as if it carried a *positive* charge. Its speed of movement between bonds may be divided into two components: a random velocity arising from the random electron motion; and a drift velocity which comes from the fact that the field makes the electron velocity slightly greater in one direction than in the other. Thus, a hole experiences an acceleration in an electric field equal but opposite in direction to that of a free electron, that is, it behaves like a free electron with *positive* charge. Its drift velocity is limited by collisions in just the same way as is the drift velocity of an electron, because its movement is determined by the drift of real electrons in the opposite direction.

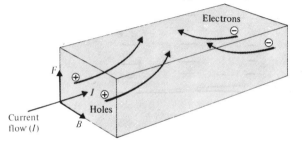

Fig. 14·5 The forces on electrons and holes moving through a magnetic field

Since the drift velocity is a direct result of the application of an electric field, the hole makes a direct contribution to the conductivity of the crystal. If there are p holes present the current they carry is p times the current carried by one hole: it is as if there were p positively charged 'electrons'.

The validity of this point of view has been demonstrated in experiments in which both magnetic and electric fields are applied. Consider a charged particle which is accelerated to a drift velocity v_d in an electric field $\mathscr{E}$. If a magnetic flux B is also present, lying perpendicular to the direction of $\mathscr{E}$, then the magnetic force F on the particle, as given in Equation 1.13, is $F = Bev$ in a direction perpendicular to both v and B (Fig. 14·5). Note that if the sign of e is changed, the direction of motion due to the electric field is reversed, i.e., v becomes negative. The force F, however, remains in the same direction because *both* e and v have changed signs. The particle, whether hole or electron, tends to move in the direction of F and by detecting this transverse motion electrically it is possible to determine the sign of its charge. Such experiments when carried out on semiconductors confirm the existence of positive holes and the phenomenon illustrated in Fig. 14.5 is known as the *Hall effect*.

14·5   Energy bands in a semiconductor

We have seen that there are two kinds of electrons in a semiconductor: those in the bonding system and those that are 'free'. The free electrons move about in the crystal and contribute to electrical conductivity in just the same way as the free electrons in a metal. They can be treated as plane waves and their energies must be quantized in a large number of closely spaced energy levels as were the electrons in a metal. The band of energies formed by these levels is called the *conduction band*. Since the electrons in the covalent bonds also move about at varying speeds, they also have a range of permitted energies. Detailed consideration of the quantization conditions shows that these levels are also closely spaced and form an energy band called the *valence band*.

The energies of the levels in the valence band must all be below those of the conduction band, for we have seen that an electron has to acquire energy from the thermal vibrations in the lattice in order to leave a bond and become free. This situation is shown on the energy band diagram in Fig. 14·6. There is, in fact, a minimum value for the amount of additional energy which a

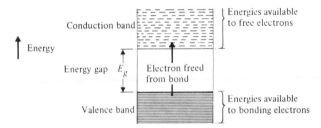

Fig. 14·6 The energy diagram for a semiconductor, showing the energy gap

bonding electron must acquire in order to leave the bond. The absolute minimum for this energy is that which the most energetic bonding electron needs in order just to become free, with no kinetic energy left over to carry it away. This will be represented in the energy level diagram by a gap between the highest level in the band of energies occupied by the bonding electrons, that is, the valence band, and the lowest level in the conduction band, as shown in Fig. 14·6. This range of energies which is prohibited to the electrons is called the forbidden energy gap, or often just the energy gap, and is a direct measure of the amount of energy an electron needs to leave the bonds. It is different for different materials. In germanium it is 0·65 eV, in silicon 1·15 eV, while in diamond it is 5·3 eV.

The arrow in Fig. 14·6 represents the transition of an electron from the valence band to a state in the conduction band, i.e., from a bond to freedom. This leaves vacant a state in the valence band, i.e., a 'hole'. A downward arrow would represent an electron falling back into a lower level, and hence recombining with a hole.

This diagram gives us another way of looking at conduction by electrons. Clearly, electrons in the conduction band can accelerate readily because empty, higher energy levels are available to accommodate them. If a hole (i.e., a vacant state) exists at the top of the valence band, an electron in a lower level can be given energy to fill it and the accelerating electron carries a current. On the other hand, if there were no vacant levels in the valence band no electrons could be accelerated.

14·6 Excitation of electrons

When an electron near the top of the valence band (i.e., in a bond) collides with a vibrating atom, it can absorb some of the atom's energy and jump into the conduction band. This is just another (more accurate) way of describing how thermal vibrations can break a bond and create an electron–hole pair. The probability of this occurring is just proportional to the probability of an atom having thermal energy greater than E_g, the amount of energy needed to excite the electron. This probability was stated in Chapter 7 to be very nearly proportional to $\exp(-E_g/kT)$. The rate of generation of electron–hole pairs is therefore proportional to the same factor.

In equilibrium, the rate of generation of electron–hole pairs must be exactly equal to the rate at which holes and electrons recombine to form fully occupied bonds again. This latter rate must be proportional both to the number of electrons per unit volume, n, and to the number of holes per unit volume, p. It is therefore proportional to the product np. By equating the rates of generation and recombination, we therefore find that

$$np = N_c^2 \exp(-E_g/kT) \tag{14·1}$$

where N_c^2 is a proportionality constant.

Since $n = p$, we then have

$$n = p = N_c \exp(-E_g/2kT) \tag{14·2}$$

This equation can be interpreted as follows. In the discussion of thermal energy distributions in Chapter 7 it was remarked that electrons, because they obey Pauli's principle, cannot be distributed exponentially over the different energy states. A thorough analysis of the problem shows that in an intrinsic semiconductor the probability of an electron being in a quantum state at the bottom of the conduction band is just $\exp(-E_g/2kT)$. This factor also appears in Eqn (14·2). Of course, not all the electrons are in the lowest quantum state in the conduction band, but the vast majority have only a little higher energy. Hence, if we regard the quantity N_c as an *effective number of states* at the bottom of the band, then the total number of electrons in the band is the number of states N_c multiplied by the probability $\exp(-E_g/2kT)$ that a state is occupied. This is

$$n = N_c \exp(-E_g/2kT)$$

as before.

Because N_c is not a number of actual states, it is not a constant, independent of temperature. However, its temperature dependence is small compared to that of the exponential factor in Eqn (14·2) and may safely be ignored for our purposes. It also varies but little among the common semiconductors from the typical value of 2.5×10^{25} m^{-3} at 300K.

The electrons in the conduction band and the holes in the valence band may carry current in a similar way to that described for metals in Chapter 13. Thus we can use Eqn (13·10) to write down the current density J_n carried by n free electrons per cubic metre whose drift velocity is v_n:

$$J_n = nev_n$$

Note that J_n has a positive sign, since both v_n and e are negative. Similarly the current density J_p due to the motion of holes in the valence band may be written in terms of their drift velocity $+v_p$

$$J_p = pev_p$$

where p is the number density of positive holes. The total current density J is then the sum of these

$$J = nev_n + pev_p \tag{14·3}$$

The number of free electrons n and the number of holes p per unit volume are usually equal (they are not always so; see Section 14·7) and given by Eqn (14·2), so that we may write the current density

$$J = N_c(ev_n + ev_p) \exp\left(-\frac{E_g}{2kT}\right) \tag{14·4}$$

Assuming that both the free electrons and the holes behave like the free electrons in a metal, we may use the results of Chapter 13. Thus v_n and v_p will both be proportional to the electric field $\mathscr{E}$ (Eqn 13·12) and we may divide Eqn (14·3) by $\mathscr{E}$ to obtain the conductivity:

$$\frac{J}{\mathscr{E}} = \sigma = N_c(e\mu_n + e\mu_p) \exp\left(-\frac{E_g}{2kT}\right) \tag{14·5}$$

where μ_n and μ_p are the drift velocities of electrons and of holes respectively in a unit electric field. As before we call μ_n and μ_p the *mobilities* of electrons and of holes respectively. In general they will not be equal to one another, but we may expect them both to vary inversely with temperature as we showed for electrons in Chapter 13. However, the exponential in Eqn (14·5) varies more rapidly with temperature and completely dominates the temperature dependence of the whole expression. Neglecting the temperature dependence of μ_p and μ_n, we see that σ rises approximately exponentially with temperature, in marked contrast to metals, for which it falls with rising temperature. The reason is that for semiconductors the *number* of available charge carriers grows exponentially as the temperature rises, while for metals it is roughly constant. For semiconductors, the rising temperature gives more and more valence electrons enough energy to jump into the conduction band, freeing them and the corresponding holes to conduct an electric current.

The experimental values of the resistivity of silicon measured at different temperatures T are plotted on a log scale against $1/T$ in Fig. 14·7. Exactly as predicted by Eqn (14·5) for conductivity (which is the reciprocal of resistivity), the points lie very closely on a straight line, the slope of which can be seen by differentiation to be equal to $E_g/2k$. Using the value for Boltzmann's constant k we can find from the graph that for silicon $E_g = 1·15$ eV, as stated earlier.

There are more accurate ways than this of determining energy gaps, and the values for several materials are given in Table 14·1. Among these, silicon and germanium have been the materials most used for transistors, diodes, etc., though nowadays some compound semiconductors are being used for special purposes. Compound semiconductors are discussed in Section 14·9.

Among these materials a wide range of properties is available–different energy gaps, different electron and hole mobilities–which makes each suitable for a different application. In this chapter we shall just consider diodes and transistors, whose operation depends upon the controlled introduction of very small amounts of impurities into an initially very pure crystal.

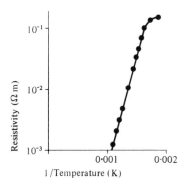

Fig. 14·7 The dependence of the resistivity of silicon on temperature

Table 14·1
Energy gaps, electron densities, and conductivities

Material	E_g (eV)	Calculated value of n at 300K (m^{-3})	Measured value of σ at 300K (Ω^{-1} m^{-1})
Copper	None	$8\cdot5 \times 10^{28}$	5×10^7
Aluminium	None	$1\cdot8 \times 10^{29}$	$3\cdot3 \times 10^7$
Grey tin	$0\cdot08$	$5\cdot4 \times 10^{24}$	10^6
Germanium	$0\cdot65$	$9\cdot3 \times 10^{19}$	$2\cdot2$
Silicon	$1\cdot15$	6×10^{15}	5×10^{-4}
Diamond	$5\cdot3$	$1\cdot1 \times 10^{-20}$	10^{-12} †
LiF	11	2×10^{-70}	10^{-11} †

†Conductivity in these materials does not occur by the motion of electrons since other processes occur more readily.

14·7   Impurity semiconductors

Each germanium or silicon atom in the semiconducting crystal has four valence electrons which it shares with its neighbours to form four covalent bonds. From another viewpoint, these electrons fill energy levels in an energy band (the valence band) so that the band is just filled at the zero of absolute temperature.

Suppose that a pentavalent atom (Group V) were to be substituted for one of the silicon atoms. An arsenic or antimony atom would be suitable, but a phosphorus atom would fit better than either of these into the silicon lattice, since its atomic radius is closely similar to that of silicon. Four of the five valence electrons belonging to this atom are required to form the four covalent bonds with the neighbouring silicon atoms. The fifth electron is bound to the atom by the extra positive charge on the nucleus, but, as we shall shortly show, it requires very little energy to break this bond, and at room temperature the thermal energy in the lattice is more than sufficient to do so. The extra electron is then free to move through the silicon lattice.

Let us now build up the same picture using the energy band model. The freed electron has been excited to an empty state in the conduction band.

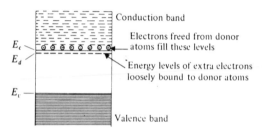
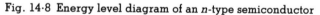

Fig. 14·8 Energy level diagram of an n-type semiconductor

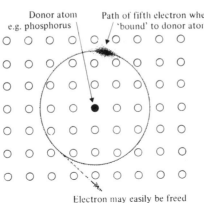

Fig. 14·9 Motion of the fifth electron around a donor atom in the lattice of a semiconductor

Since the energy required to remove it from its parent nucleus was very small, it must have come from an energy level *just below* the bottom of the conduction band–in the energy gap! (Fig. 14·8). Thus the introduction of the Group V atom has created a new energy level, near to the conduction band, which is normally vacant at room temperature because the electron is donated to the conduction band–the impurity is called a *donor* for this reason.

In order to indicate that the donors are localized, we make the horizontal axis of the band diagrams represent distance through the crystal: the donor levels are then shown as short lines.

We may show that the binding energy of an electron in this level is small by regarding the extra electron and the excess positive nuclear charge as if they were like a hydrogen atom. In both situations a single, negatively charged electron moves in the electric field of a single, fixed positive charge (Fig. 14·9). But in the present case the electron moves through a silicon lattice which we may regard approximately as a medium with a large relative permittivity (dielectric constant) ϵ_r. The expression derived in Chapter 3 for the hydrogen atom can be used now to derive the binding energy, but we must replace ϵ_0 by $\epsilon_r \epsilon_0$ giving the result

$$\text{Binding energy} = \frac{me^4}{8\epsilon_0^2 \epsilon_r^2 h^2} \tag{14·6}$$

This is the energy difference between the donor level and the bottom of the conduction band, i.e., it is equal to $(E_c - E_d)$ (Fig. 14·8).

Since the relative permittivity of silicon is 11·7, the binding energy is smaller than the ionization energy of hydrogen by a factor $11·7^2$, so that

$$(E_c - E_d) = \frac{13·6}{11·7^2} = 0·1 \text{ eV}$$

Naturally this result is only approximate [$(E_c - E_d)$ is actually nearer 0·05 eV] but it serves to show that the binding energy is indeed small compared with the width of the energy gap, which in silicon is $1 \cdot 15$ eV. The correspond-ing values for germanium are $(E_c - E_d) = 0 \cdot 01$ eV and $E_g = 0 \cdot 65$ eV.

The density of free electrons in *pure* silicon, n_i, (often called the *intrinsic* value) can be calculated from Eqn (14·2). Using the value for N_c given earlier, the density of electron–hole pairs at $T = 300$K is found to be

$$n_i = N_c \exp(-E_g/2kT)$$
$$= 2 \cdot 5 \times 10^{25} \times \exp(-1 \cdot 15/0 \cdot 026 \times 2)$$

Thus

$$n_i = 6 \cdot 3 \times 10^{15} \, \text{m}^{-3} \tag{14·7}$$

On the other hand there are 5×10^{28} atoms per cubic metre and about 2×10^{29} valence electrons per cubic metre in solid silicon. This confirms our earlier statement that only a few electrons are thermally excited across the band gap.

If now only one silicon atom in 10^{10} were replaced by a donor impurity, the density of donors would be about $5 \times 10^{18}/\text{m}^3$, and the same number of extra electrons would be available to conduct an electric current. This is far in excess of the number of carriers in the intrinsic material, so the conduction process is dominated by the presence of the electrons, which are called the majority carriers. Since the current carriers are almost exclusively negative, the material is termed an *n*-type impurity semiconductor.

We have glossed over one point in the argument. Not all the donor elec-trons appear in the conduction band. Just as in the case of intrinsic electrons, the number of electrons in the conduction band must be calculated from the probability of a donor electron being excited across the gap between the donor level E_d and the conduction band edge E_c. This probability is cal-culated in more advanced textbooks (see Further reading) with the result that, at room temperature, nearly all the donor atoms are found to be ionized, unless the density of donors becomes too large (greater than about $10^{25}/\text{m}^3$ in germanium).

For present purposes we may therefore treat a piece of *n*-type semiconduc-tor containing N_d donors per unit volume as a conductor in which the free electron density is approximately equal to N_d. As the density of electrons is increased by the presence of donors, so the density of holes is decreased. This is because a hole has a greater probability of meeting an electron with which it can recombine. The number of holes therefore reduces so that the net recom-bination rate, which is proportional to the product np, is decreased also. The generation rate, on the other hand, is not affected by the presence of donors, since it depends solely on the transfer of thermal energy from atoms to bound electrons. The number of donor atoms is so small that the effect they have is negligible. Since the generation rate must equal the recombination rate, we must again have the relation:

$$np = N_c^2 \exp(-E_g/kT)$$

exactly as in Eqn (14·1). But in the present case n and p are not equal, for n is approximately equal to N_d, so that we can now find an approximate value for p:

$$p = \frac{N_c^2}{N_d} \exp\left(-E_g/kT\right)$$

Note that this result is quite different from that for an intrinsic semiconductor.

14·8 Impurity semiconductors containing Group III elements

By contrast with the action of Group V elements, small traces of a Group III element such as boron or aluminium create a deficiency of electrons. Each impurity atom has only three valence electrons, while four are required for bonding. The incomplete covalent bond contains a hole which may, if it escapes from the impurity atom, take part in conduction.

The escape of the hole occurs by the capture of a valence electron from a neighbouring silicon atom, i.e., from the valence band on the energy diagram. For this reason the Group III impurity is called an *acceptor* atom. The captured electron now has a slightly higher energy than the other valence electrons, since it has upset the neutrality of the impurity atom, which has now become a negative ion. We therefore allocate to the captured electron an energy level called an acceptor level just above the top of the valence band (Fig. 14·10). Since the energy gap between the top of the valence band and the acceptor level is very small, electrons from the valence band are readily excited across it and most of the acceptor levels will be filled at normal temperatures. A corresponding number of holes is created in the valence band and these may conduct an electric current.

As with electrons in *n*-type silicon, the number of holes created by only a small concentration of acceptors may greatly exceed the number of holes in intrinsic silicon [given in Eqn (14·7)] and conduction thus occurs primarily by the motion of these positive holes. The material is called a *p*-type impurity semiconductor, and in contrast to *n*-type material the majority of carriers are now holes.

The energy gap between the valence band and the acceptor levels may be calculated in a similar way to that for donor levels. Since the captured electron gives an acceptor atom a charge $-e$, it has an 'attraction' for holes. So a hole can become bound to an acceptor as if it were an 'inverted'

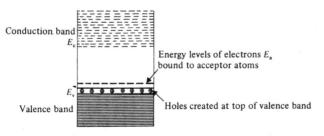

Fig. 14·10 Energy diagram of a *p*-type semiconductor

hydrogen atom–a positive charge orbiting a fixed negative charge. Equation (14·6) can again be used to obtain an approximate value for the binding energy, which is thus about 0·05 eV in silicon.

Many other impurities may be put into silicon or germanium to create new energy levels at almost any point in the energy gap, but the Group III and Group V elements are by far the most important for practical purposes. Many kinds of lattice defects can also act as donor or acceptor levels, or as traps for current carriers. It is therefore important to prepare the materials for diodes and transistors with the greatest care, to reduce both unwanted impurities and lattice defects below very low levels. As a result of the development of new technologies (Chapter 15) for doing this, silicon and germanium have become two of the purest obtainable elements.

14·9   Compound semiconductors

It was noted in Section 6·6 that covalent bonding, with the four-fold coordination exhibited in the diamond structure, is found in compounds of Group III and Group V atoms like GaAs and InP, and also in some II–VI compounds like ZnS. As in silicon, there are just enough valence electrons to fill the covalent bonds, leaving none (in the perfect crystal) in the conduction band to carry a current. They are therefore semiconductors, with many properties like silicon. Additionally, the absence of a Group V (or VI) atom from the lattice has the effect of removing a valence electron, creating a hole, while the absence of a Group III (or II) atom creates a free electron. The imperfect crystal behaves as if doped with donor or acceptor impurities, making it somewhat more difficult to control their electrical properties.

Further differences show up in the optical properties. As explained in Section 18·6, absorption of light only occurs in a semiconductor if the photon energy $h\nu$ of the light is equal to or greater than the energy gap E_g. This is because the absorption of a photon creates an electron–hole pair. Conversely, the recombination of an electron with a hole can, in the right circumstances, result in the emission of a photon of light with energy E_g. This is called direct recombination. However, in silicon or germanium, recombination almost invariably occurs indirectly, by first trapping the hole or electron at an impurity. This stems from a peculiar property possessed by the energy bands in these materials, which causes free electrons with energy near E_c to have either much greater momentum (in Si) or much less momentum (in Ge) than holes with energy near E_v. Thus so-called radiative recombination, involving the emission of a photon, almost never occurs, since momentum cannot be conserved during the process. For, if a free electron is to 'fill' a hole in silicon, it must first lose its momentum, so that recombination occurs in two stages. First, the electron becomes trapped at an impurity atom, losing its momentum to the atom as it does so. Subsequently, a hole happens to pass by, whereupon recombination occurs, the energy being transferred to the lattice, in the form of phonons.

Indirect recombination dominates in materials with energy bands similar to those in silicon or germanium, e.g., AlAs, GaP. They are said to have an indirect band gap. Materials such as GaAs or InP, in which radiative recombination occurs easily, are said to have a direct band gap. Both of these can be used as the basis for sources of infrared light. For fibre optic communications, the four-component semiconductor GaInAsP is frequently used. Indium antimonide ($E_g = 0·18$ eV) is used for the detection of far-infrared radiation, while cadmium sulphide is a good detector for visible light and is used in photographic light meters.

14·10 The *p-n* junction: a rectifier

A piece of semiconductor in which a section of *p*-type material is joined to one of *n*-type material is called a *p-n* junction. It is on the properties of this junction that the semiconductor diode and the transistor depend. Such a junction is not usually made by joining two separate pieces of material, but by taking a single crystal of silicon and introducing donor and acceptor impurities (see Chapter 15) into the appropriate parts of the solid. We shall now consider the electrical properties of such a junction.

Let us imagine what happens to the electrons and holes in two pieces of semiconductor, one *p*-type and one *n*-type, as they are placed together. The initial distribution of holes and electrons is shown in Fig. 14·11(a).

On one side of the junction is a material containing electrons moving freely and randomly at high velocities through the lattice, while on the other side is a material with free holes moving at similar velocities. It is natural that the excess carriers on either side will be able to cross the junction into the 'foreign' material, just as two gases will diffuse into one another. But, unlike gases, the free electrons and holes can combine self-destructively with one another; far from the junction on either side the normal, undisturbed carrier densities must persist. This means that the concentration gradients of free electrons and holes across the junction persist in spite of the diffusion, as shown in Fig. 14·11(c). Moreover, the diffusion rate, which is proportional to the concentration gradient (cf. Chapter 7) is high at this point, and the carrier density close to the junction is thereby depleted. This depletion is so marked that there is a thin layer on either side of the junction which is virtually empty of free carriers–the densities are several orders of magnitude lower than in the rest of the crystal. This region is illustrated in Fig. 14·11(b), and is termed the *depletion layer*. In practice it is about 10^{-4} cm thick.

Although there are very few charge carriers in the depletion layer, the region is actually highly charged [Fig. 14·11(d)] owing to the presence of the ionized impurity atoms. Thus the donor atoms in the *n*-type material are positively charged, while the acceptors on the other side have negative charge. These charges are normally neutralized by the presence of the majority carriers. Since the whole crystal has to be neutral, the numbers of positive and negative charges must be equal. The charge distribution on these impurities is roughly as shown in Fig. 14·11(e).

This charge distribution is rather like that on a parallel-plate capacitor each of whose plates is situated roughly mid-way between the edge of the depletion layer and the junction. The charge per unit area on each 'plate' is equal to the

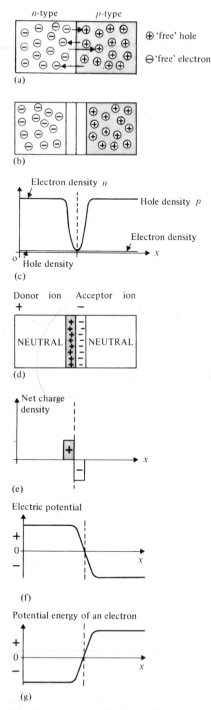

Fig. 14·11 The function of a depletion layer in a p–n junction

density of donors (or acceptors) multiplied by the width W of the depletion layer. The voltage V_0 across this 'capacitor' is a function of the carrier densities on either side of the junction, and an expression for V_0 will be calculated shortly.

This voltage is equal to the work required to be done in moving unit positive charge across the junction, and the voltage distribution across the junction, shown in Fig. 14·11(f), therefore gives the potential energy distribution for a unit positive charge. Since electrons are negatively charged, their potential energy distribution is the opposite, as shown in Fig. 14·11(g). In the light of this diagram, the energy band diagram for the p–n junction can be redrawn as in Fig. 14·12(b), where the energy bands are shown bent to take account of the potential variation across the junction.

It is clear that this potential distribution prevents the free passage of electrons and holes across the junction without work being done on them. Only a few, those with exceptional amounts of thermal energy, can cross this barrier in the 'uphill' direction. The holes in the p region are thereby hindered in moving across to the n region, and similarly, motion of electrons towards the p region is impeded. Equilibrium is achieved when the flows of both

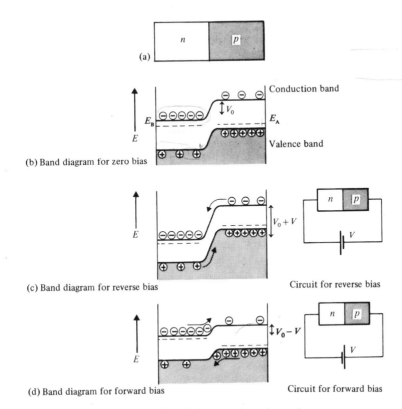

Fig. 14·12 The energy band diagram for a biased p–n junction

electrons and holes in the two opposite directions just balance one another.

To calculate the voltage difference across the junction we shall make use of the fact that the two halves of the crystal are in thermal equilibrium. The condition for thermal equilibrium between two bodies, which was derived in Section 7·7, is that the probability that quantum states of a given energy are occupied is the same in both bodies. This was originally stated for the quantum states of atoms, but it must also be true for electrons (and holes) if they are in thermal equilibrium with the atomic lattice. Since the free electrons are present in the conduction band because they have acquired thermal energy from the lattice, they must be in thermal equilibrium, and therefore we may assume that the probability of occupation of electron states in the conduction band in Fig. 14·12(b) is independent of position.

The probability distribution for electrons is the Fermi–Dirac expression of Eqn (7·11b), but it was pointed out in Section 7·6 that, provided $(E - E_F) \geq 2kT$, the relative probabilities for different energies are given almost exactly by the Boltzmann factor. We have not discussed the Fermi level for a semiconductor but we have a clue available from impurity semiconductors. We know that the impurity energy level, E_D, is just below the conduction band edge for electrons in an n-type semiconductor and that release of electrons from E_D to the conduction band is due to thermal agitation. Thus near absolute zero the occupation probability of the donor level E_D should be approaching 1, i.e., we find $E \leq E_F$ in Eqn (7·11b). On the other hand since the conduction band is practically empty, E_F must be less than E_C. Thus we can deduce that E_F must be somewhere in the vicinity of E_D for an n-type semiconductor and for normal doping levels this is so. Similarly it is in the vicinity of the acceptor level, E_A, for the p-type material.

In order for the occupation probability of any quantum level to be independent of position for thermal equilibrium, the Fermi energy, E_F, must be the same everywhere. Thus, if it coincides with E_A in p-type and with E_D in n-type materials, we have the basis for Fig. 14·12(b), which has been drawn with E_A and E_D at the same level. We can now see that, the probability of occupation for an energy level, E_C, at the conduction band edge in the n-type material will be proportional to $\exp\left[-(E_C - E_D)/kT\right]$ and in the p-type material will be proportional to $\exp\left[-(E_C - E_A)/kT\right]$. But, from Fig. 14·12(b), $E_C - E_A = [V_0 + (E_C - E_D)]$ so, in the p-type material, the probability is proportional to $\exp\left[-(E_C - E_D)/kT\right]\exp\left(-V_0/kT\right)$. The actual numbers of electrons close to E_C in each material will therefore be in the ratio

$$n_n/n_p = \exp\left(-V_0/kT\right) \tag{14·8a}$$

where n_n is the number of electrons on the n side and n_p the number on the p side. Thus

$$V_0 = kT\ln\left(n_p/n_n\right) \tag{14·8b}$$

This is the equation that determines the height of the potential barrier which is set up in the p–n junction.

We have seen that the product np is always a constant at a given temperature and is given by n_i^2 [Eqn (14·1)]. On the p side, therefore, $n_p p_p = n_i^2$ and p_p will be the hole density (which equals the acceptor impurity density N_A).

Similarly, on the n side, $n_n p_n = n_i^2$ and n_n will be given by the donor density, N_D. Thus, with a little manipulation, Eqn (13·8b) can be rewritten as

$$V_0 = kT \ln (N_A N_D / n_i^2)$$

(14·8c)

In Eqn (14·7) the value of n_i^2 is given as $6·3 \times 10^{15}$ m^{-3} at $T = 300$K. If N_A and N_D each have a typical value of 10^{22} m^{-3} we find $V_0 = 0·7$ V.

14·11   Current flow through a biased p-n junction

The dynamic equilibrium set up by the interdiffusion of holes and electrons is upset when a voltage is applied across the ends of the crystal. To understand the new situation we must first decide what form the energy band diagram takes when the voltage is applied. Let us take the case where the n-type region is made electrically positive with respect to the p-type region. The first thing to note is that, compared to the bulk of the crystal, the depletion layer has a high resistance because of the absence of free carriers. Thus, remembering that the depletion layer is effectively in series with the rest of the crystal, we see that the applied voltage drop appears almost entirely across the depletion layer. This gives a new energy band diagram as shown in Fig. 14·12(c). The effect of the applied voltage is to increase the potential difference across the junction. This *reduces* the flows of holes into the n region and of electrons into the p region, while making little difference to the reverse flows since the latter are limited mainly by the scarcity of holes in the n region and electrons in the p region. The result is a small net current flow from p to n, nearly independent of applied voltage.

On the other hand, reversing the sign of the bias voltage V reduces the height of the potential drop across the junction [Fig. 14·12(d)]. Once again the flow of holes and electrons from n and p regions is scarcely affected–remember that these flows are limited by the total numbers of minority carriers in the crystal and not by the rate at which they cross the depletion layers. But the flow in the other direction–up the potential hill–is increased, because the hill now causes less hindrance to the motion of the large number of majority carriers. This current can increase continuously as the applied voltage increases, so that the junction responds differently to voltages applied in opposite senses.

Just how the current depends upon voltage can be determined by reconsidering Eqn (14·8a). That equation expresses the fact that the two halves of the crystal are in thermal equilibrium and the passage of a current through it does not upset this equilibrium. However, the potential energy difference between electrons on either side of the junction is no longer eV_0, but $e(V_0 \pm V)$. Reference to Fig. 14·12 shows that this can be written just as $e(V_0 - V)$, if we give the voltage V a positive sign when the n region is made electrically negative and a negative sign when it is positive with respect to the p region.

This change in the potential energy difference means that Eqn (14·8a) must be modified, giving

$$n_p' = n_n' \exp\left[-e(V_0 - V)/kT\right] \tag{14.9}$$

where the concentrations n_p' and n_n' have been primed to indicate that they are different from the normal values given by Eqn (14.8a). A similar equation holds for holes:

$$p_n' = p_p' \exp\left[-e(V_0 - V)/kT\right] \tag{14.10}$$

Since the crystal is electrically neutral outside the depletion layers any change in n_p must be associated with an equal change in p_p. But since p_p is very much greater than n_p, the *percentage* change in p_p is negligible compared to the *percentage* change in n_p (the same can be said of n_n and p_n). Hence the interpretation of Eqn (14.9) is that the dominant change occurring is that n_p, the concentration of electrons in the p region, is dramatically altered. Equation (14.10) similarly tells us that p_n is also changed. In forward bias (V positive) the value of n_p is increased. The extra electrons have come from the n region, because the potential barrier has been lowered by the applied voltage. But these extra electrons can travel only a short distance (about 0.1 mm on average) before recombining with holes so that, far from the junction, the electron concentration is back to its normal value. The concentration gradient is illustrated by the circled charges in Fig. 14.12(d), and it is this gradient which gives rise to the flow of current, by diffusion, at the edge of the depletion layer where the voltage gradient is very small. According to Fick's Law (Chapter 7) the magnitude of the current is proportional to the concentration gradient, which in turn is proportional to the *excess* concentration of carriers, $(n_p' - n_p)$.

Using Eqns (14.8a) and (14.9) and remembering $n_n' \approx n_n$ we find that

$$n_p' - n_p = n_p[\exp(eV/kT) - 1]$$

So that the current can be written

$$I = I_s[\exp(eV/kT) - 1] \tag{14.11}$$

where I_s is a constant, independent of the applied voltage.

Figure 14.13 shows that the I–V characteristic for a real junction does indeed have this form–at least for low values of applied voltage. The reverse current is so small that for practical purposes the junction behaves very nearly as a one-way conductor, that is, as a rectifier.

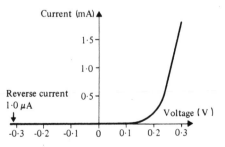

Fig. 14.13 The voltage–current relationship for a p–n junction

The advantage of semiconductor rectifiers and transistors over their vacuum valve equivalents lies largely in their reliability. There is little that can go wrong with the simple *p-n* junction and it deteriorates only if so much current is passed that the crystal becomes too hot. When this happens, kT increases and with it the current, causing further heating and eventually destruction of the device.

14·12 Junction transistors

The transistor is a device which can be used to amplify small alternating voltages or currents. It amplifies by converting power injected from a d.c. source (e.g., a battery) into a.c. power at the frequency of the input signal. In the following simplified description we shall show how power amplification is achieved.

The transistor can be regarded as two *p-n* junctions connected back-to-back as shown in Fig. 14·14 where the current carriers, holes, and electrons are shown. We now have two potential barriers, one between the (left-hand) *emitter* and (centre) *base* regions and one between the base and the (right-hand) *collector* region. If we apply forward bias, reducing the barrier height between emitter and base regions, holes pass into the base region from the emitter. (We say they are *injected* into the base.)

If the base regions were thick, they would soon recombine with the electrons there. But the base is made especially thin so that, before this can happen, they find themselves swept into the collector, since the potential drop across the base–collector junction encourages this.

Now quite a small forward bias across the emitter–base junction injects a large number of carriers into the base region. At the collector junction a fairly large reverse-bias voltage is applied which assists the flow of electrons into the collector from the base while opposing the flow of current in the opposite direction. Thus the current injected from the emitter enters at low voltage, and the same current (if no electrons are lost by recombination) is collected at a higher voltage.

If the emitter current is varied in accordance with the magnitude of an a.c. electrical signal, the collector current also varies. The input power is equal to the a.c. emitter current multiplied by the emitter–base voltage, which is

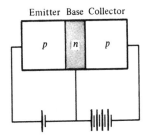

Fig. 14·14 The junction transistor, with its biasing voltages

small; the output power is greater, being equal to the a.c. emitter current multiplied by the (larger) collector–base voltage. The transistor is thus a power amplifier.

14·13   Field effect transistors (FETs)

The *p–n* junction and junction transistor described above depend for their operation on the presence of both holes and electrons and are referred to as *bipolar* devices. There is also a type of transistor which depends on only one type of carrier, either holes or electrons, for its operation. It could be described as a *monopolar* transistor, although it is generally called an FET. Its principle of operation is variation of the number of electrons (or holes) in a semiconductor by application of an external field–this is the 'field effect'.

To understand how it works let us consider a parallel-plate capacitor made up of two rectangular metal electrodes placed a short distance apart as in Fig. 14·15(a). When we apply a voltage, V, between the plates a charge, Q, is stored in the capacitor, of value $Q = CV$, where C is its capacitance (see Section 17·3). The charge is in the form of a net positive charge, $+Q$, on one plate and a net negative charge, $-Q$, on the other, but what are these actual charges? They must, in fact, be electrons and ions in the metal itself. Recalling our picture of bonding in a metal (Chapter 5) it is made up of an array of positive ions held in fixed positions with a mobile 'sea' of electrons acting as a glue. The electrons will move in an applied electric field, but the ions will not. Now the applied voltage produces an electric field between the capacitor plates. Referring to Fig. 14·15(b) this field will try to move the electrons away from the surface (deeper into the plate) at the top and to attract extra electrons to the surface at the bottom (Remember electrons move in the *opposite* direction to the electric field.) Thus the positive charge on the top plate is due to pushing back the electrons to reveal the fixed positive ions. The negative charge on the lower plate is due to pulling extra electrons to the surface. As this happens, immediately after closing the switch [Fig. 14·15(a)], electrons are displaced round the circuit, from the top plate to the bottom plate and a transient, charging current flows.

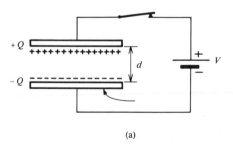
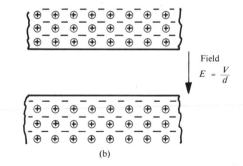

(a)   (b)

Fig. 14·15

We now determine how many electrons are likely to be moved. Let our capacitor have plates of area 1 cm² and spacing between them of 1 μm (10^{-4} cm). The capacitance will be given by Eqn (17·3) and has the value $8·84 \times 10^{-10}$ F or 884 pF. If the battery has an e.m.f. of 10 V we find $Q = 8·84 \times 10^{-9}$ C which is equivalent to $5·5 \times 10^{10}$ electrons and this is the number of ions 'revealed' per cm² on the positive plate and the number of extra electrons per unit area on the negative plate. Now a typical metal has $\sim 5 \times 10^{15}$ atoms/cm² in its surface and so that, on average, an electron is displaced to the layer of atoms below the surface–a distance of about 3 Å–for one in every 5,000 atoms. We see, therefore, that the electric field does not penetrate much below the surface of the metal.

Now suppose the plates of the capacitor are made of n-type semiconductor containing 10^{15} donor impurity atoms/cm³. The ionized donors provide the fixed positive charge and there will be, approximately, the same density of free electrons as there are ionized donor atoms. Each plane of the semi-conductor will, on average, contain $(10^{15})^{2/3} = 10^{10}$ impurity atoms/cm². Using the same figures as before, to obtain the necessary $5·5 \times 10^{10}$ electrons for the negative charges and $5·5 \times 10^{10}$ positive charges, and assuming half the donor atoms to be ionized, we would have to penetrate about 10 atomic layers below the surface. We conclude that the electric field must penetrate the semiconductor to a depth determined by the impurity density, for a given applied voltage and capacitance. This is the basis of the field effect.

An actual FET is illustrated in schematic form in Fig. 14·16(a) whilst Fig. 14·16(b) shows its actual structure on a silicon slice. The gate electrode is the top plate of the capacitor and the semiconductor slab–shown as p-type in Fig. 14·16(a)–is the bottom plate, the insulator between them being silicon dioxide (SiO_2). The gate electrode may be metal and, for this reason, the device is called a MOSFET, meaning metal-oxide-semiconductor FET. Nowadays the gate electrode is usually made of polycrystalline silicon, but the terminology MOS is still used. The transistor operates by passing a current through a shallow channel in the semiconductor substrate, just below the oxide-semiconductor interface, between two contacts called the *source* and *drain* electrodes. These are, in fact, heavily doped n-type regions. When a voltage is applied to the gate it sets up a field which penetrates the semiconductor and causes the number of electrons carrying the current in this channel to be increased or decreased, according to the polarity of the voltage. Thus the source-drain current is modulated by the gate voltage. A positive gate

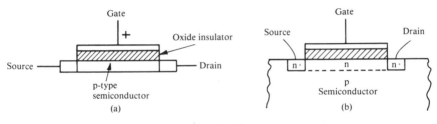

Fig. 14·16

potential will induce negative charge in the semiconductor and will repel holes, which are majority carriers in the p-type substrate. Thus a depletion layer, in which the free charge density is reduced below that of the bulk semiconductor, is produced in the surface channel. If an increasingly large, positive potential is applied to the gate then, at the surface, the concentration of minority carrier electrons will eventually become comparable to the majority carrier hole density. Electrons from the n^+ source and drain contacts will move into the channel under the attraction of the positive gate potential. Eventually, for a sufficiently positive gate potential, the electron density locally at the surface will exceed the hole density and the surface region is said to be inverted. This inversion region, referred to as an n-channel, bridges the n-type source and drain contacts and current flows between them. A transistor operating in this way is called an n-MOS transistor; with an n-type substrate and a negative gate voltage a p-channel will be formed in just the same way and the device is a p-MOS transistor. In each case the source and drain contacts are of the opposite type to the substrate, as in the example quoted with n^+ contacts in a p-type semiconductor. Thus the source/substrate and drain/substrate areas are, in fact, n–p junctions and, unless forward biased, will prevent current flowing from source to substrate or drain to substrate. We choose the source of drain voltage so that these junctions are always reverse-biased and virtually no current flows between source and drain until the gate 'turns on' the inverted channel. In this way the current is confined to the channel region.

The gain of the transistor is given in terms of source–drain current change for a given gate voltage change, in units of A/V. Typically, MOS transistors have forward gains of about 0·25 mA/V and a gate-to-semiconductor (input) capacitance of about 0·02 pF. The gain is about an order of magnitude lower and the capacitance about an order of magnitude higher than in a corresponding bipolar transistor. This combination of gain and input capacitance limits the frequency response of MOS transistor circuits to 10–20 MHz except in special devices. However, these disadvantages are offset by the fact that MOS transistors are physically very small and that their input impedance is purely capacitive. This latter point means that no current is consumed in applying bias to them and the resulting circuits have much smaller power demands than do bipolar circuits. It is partly this feature which has made possible the development of digital watches and small pocket calculators. A second advantage of the capacitive input is that it is possible to store signals in the form of charge on the gate capacitance, providing a built-in memory. The bipolar transistor has no built-in memory. That the MOS transistor occupies such a small area has been a key factor in the production of large-scale integrated (LSI) and very large-scale-integrated (VSLI) circuits, in particular the microprocessor chip.

Problems

14·1 Using the data in Table 14·1, calculate the mobilities of electrons and holes (assuming them to be equal) in grey tin, silicon, and germanium. Calculate also the mobility which would be necessary for diamond to have an *intrinsic* conductivity of $10^{-12}/\Omega$ m if the hole and electron mobilities were equal.

14·2 How much *n*-type impurity is necessary to explain the conductivity of $10^{-12}/\Omega$ m in diamond if the mobility is $0·18$ m^2/Vs for electrons and intrinsic conductivity can be neglected?

14·3 Calculate the carrier densities in intrinsic silicon and germanium at 100°C using the values $N_c = 2·5 \times 10^{25}/$m^3 and $E_g = 0·75$ eV in germanium and $1·1$ eV in silicon. Assuming the hole and electron mobilities to be inversely proportional to temperature, estimate their values at 100°C (see Problem 14·1 for the values at 300K) and hence calculate the resistivities of silicon and germanium at this temperature.

14·4 Calculate the total number of valence electrons per cubic metre in germanium (density $5·32$ g/cm^3) and find what fraction of them is available for conduction at 300K.

14·5 Calculate the intrinsic conductivity at 300K of the compound semiconductor gallium antimonide (GaSb), for which the energy gap is $0·70$ eV. Take $N_c = 2·5 \times 10^{25}$ m^{-3} and the electron and hole mobilities to be $2·3$ m^2/Vs and $0·010$ m^2/Vs respectively.

How much *n*-type impurity is required to give GaSb a conductivity of 100 Ω^{-1} m^{-1}? Would the same quantity of *p*-type impurity suffice?

14·6 Since Sb is a Group V element, an excess of Sb in GaSb would make the compound *n*-type. Using the data given in Problem 14·5, calculate the excess atomic percentage of Sb required to give a conductivity of 100 Ω^{-1} m^{-1}. The density of GaSb is $5·4 \times 10^3$ kg/m^3.

14·7 A piece of *p*-type silicon contains 10^{24} acceptors per cubic metre. Calculate the temperature at which intrinsic silicon has the same carrier density. Hence deduce qualitatively the manner in which the conductivity of the *p*-type material depends upon temperature.

14·8 What effect has an increase in temperature on the *V–I* characteristic of the diode shown in Fig. 14·13? Using Eqn (14·9), calculate the current through the diode under $0·3$ V reverse bias at 60°C, given that Fig. 14·13 is correct at 20°C.

14·9 How many holes and electrons would you expect to find in a piece of germanium containing an equal number N of donor and acceptor atoms? Calculate the conductivity of a piece of germanium containing 3×10^{22} donors and 8×10^{21} acceptors per cubic metre. The electron mobility in Ge is $0·39$ m^2/Vs. (Assume all impurity atoms to be ionized and neglect intrinsic carriers.)

14·10 Assuming that in intrinsic silicon at room temperature 10% of the free electrons recombine with holes each second, calculate the approximate rate of release of energy which results. Does this liberation of energy produce a rise in the temperature of silicon?

14·11 In order for an electron to cross from the top of the valence band to the bottom of the conduction band energy must be supplied to it. If the source of energy is a photon calculate the minimum frequency which the photon must have. What bearing does your result have on the optical properties of germanium?

14·12 The resistivity of many polymers (e.g., polyethylene, PVC) is very high. Draw the energy level diagram for electrons in polymers.

14·13 As described in the text, the depletion layers at a p–n junction form a kind of capacitor. The capacity may be calculated approximately by assuming all the charge in each layer to be concentrated at the mid-plane of the layer. By writing expressions for the capacity and the charge per unit area of the junction, deduce the expression

$$V_0 = eN_d \, \omega^2 / \epsilon_r \, \epsilon_0$$

for the voltage V_0 across the junction. Hence deduce the width W of the depletion layer in a silicon p–n junction, given $N_d = 10^{24}\,\mathrm{m}^{-3}$, $V_0 = 1\cdot0\,\mathrm{V}$, and $\epsilon_r = 11\cdot7$.

Self-Assessment questions

1 A perfect covalently bonded crystal at the absolute zero of temperature will be a perfect insulator because

A) the valence electrons are tied to the parent atoms
B) electrons cannot move from bond to bond
C) the electrons in the bonds form standing waves.

2 If n electrons are liberated from bonds the electrical conductivity is proportional to $2n$ because

A) there are two electrons in each bond
B) for each electron liberated a hole is created
C) liberated electrons move twice as fast as those in bonds.

3 The positive hole acquires a drift velocity in an electric field

A) in the same direction as the field
B) in the opposite direction to the field
C) perpendicular to the field.

4 The positive hole moves through the crystal by

A) escaping from the bonds
B) attaching itself to an electron in a bond
C) reciprocal motion of electrons in the bonds.

5 The total current through a semiconductor is given by

A) the sum   B) the difference   C) the product

of the currents due to the holes and the electrons.

6 Free electrons and bonding electrons in a semiconductor have bands of permitted energy levels with an energy gap between them. The free electrons occupy the conduction band and the bonding electrons the valence band. Which of the following statements are correct?

A) the conduction band is nearly full
B) the valence band is nearly full
C) holes are found mainly in the conduction band
D) holes are found only in the valence band

E) a hole reaches the conduction band by combining with an electron
F) an electron reaching the conduction band leaves a hole in the valence band
G) the bonding energy of an electron is equal to the energy gap
H) an electron recombines with a hole by falling from the conduction to the valence band
I) an electron recombines with a hole by the hole rising to the conduction band
J) when an electron and a hole recombine energy must be absorbed
K) when an electron and a hole recombine energy must be liberated
L) electrons in the conduction band can acquire a net acceleration from a field because there are empty energy levels available
M) an electron in the valence band cannot be accelerated by the field unless there are empty energy levels available
N) holes cannot be accelerated by the field unless there are empty energy levels available.

7 In a pure semiconductor having an energy gap of 0.65 eV and $N_c = 2.5 \times 10^{25}$ m^{-3} the density of holes at a temperature of 300K is

A) 1.3×10^{14} m^{-3}　　　B) 8.6×10^{19} m^{-3}　　　C) 9.8×10^{24} m^{-3}.

8 A pure semiconductor has $E_g = 0.7$ eV, $N_c = 2.5 \times 10^{25}$ m^{-3}, $\mu_n = 2.3$ m^2/Vs, $\mu_p = 0.01$ m^2/Vs. Its conductivity at 300K is

A) $12\ (\Omega\ \mathrm{m})^{-1}$　　　B) $0.13\ (\Omega\ \mathrm{m})^{-1}$　　　C) $6.4 \times 10^{-6}\ (\Omega\ \mathrm{m})^{-1}$.

9 The conductivity of a pure semiconductor is

A) proportional to temperature
B) rises exponentially with temperature
C) decreases exponentially with increasing temperature.

10 The slope of a graph of ln(conductivity) against reciprocal temperature is

A)　$-\dfrac{E_g}{2k}$　　　B)　$+\dfrac{E_g}{2k}$　　　C)　$\dfrac{E_g}{k}$

11 The following questions concern impurity semiconductors and consist of an assertion and a reason. Give the answer A, B, or C, as follows

A) assertion correct, reason correct
B) assertion correct, reason incorrect
C) assertion wrong and therefore reason irrelevant.

i) A group V impurity gives p-type conductivity *because* it has five valence electrons.
ii) A group III impurity gives p-type conductivity *because* it cannot take part in the covalent bonding system.
iii) A group V impurity is called a donor *because* it donates an extra electron to the covalent bonding system.
iv) A group III imparity is called an acceptor *because* it accepts an extra electron from the bonding system.
v) Boron is a donor *because* it donates a hole to the valence band.
vi) The energy difference between a donor level and the edge of the valence band is small *because* it easily accepts an electron.
vii) In an n-type semiconductor the current carriers are almost exclusively electrons *because* the product of electron and hole densities is a constant for constant temperature.
viii) In an n-type semiconductor the impurity atom density approximately equals the density of electrons in the conduction band *because* all the donor atoms are ionized.

12 Silicon contains 5×10^{28} atoms/m³. If it is doped with 2 parts per million of arsenic the electron density is approximately

A) $4 \times 10^{23}\,\text{m}^{-3}$ B) $10^{23}\,\text{m}^{-3}$ C) $2 \times 10^6\,\text{m}^{-3}$.

13 For the material of question 12, using $N_c = 2 \cdot 5 \times 10^{25}\,\text{m}^{-3}$ and $E_g = 1 \cdot 15\,\text{eV}$, the density of holes at 300K will be

A) $5 \times 10^{28}\,\text{m}^{-3}$ B) $10^{23}\,\text{m}^{-3}$ C) $3 \cdot 0 \times 10^8\,\text{m}^{-3}$.

14 The donor atoms in an n-type semiconductor at normal temperatures

A) carry a positive charge   B) carry a negative charge
C) are neutral.

15 The acceptor atoms in a p-type semiconductor at normal temperatures

A) carry a positive charge   B) carry a negative charge
C) are neutral.

16 In a p–n junction there is a charge difference across the depletion layer, due to the fixed impurity ions, such that

A) the p side is positive and the n side negative
B) the n side is positive and the p side negative
C) both sides are positive but the n side has the bigger positive charge.

17 The potential energy of an electron in a p–n junction is

A) low on the n side and high on the p side
B) high on the n side and low on the p side
C) the same on each side but with a minimum at the junction.

18 In equilibrium with zero bias

A) no holes or electrons cross the junction
B) only electrons cross the junction
C) equal numbers of holes and electrons cross the junction in opposite directions.

19 The height of the potential barrier at a p–n junction is determined solely by the densities of the impurity atoms on each side of the junction.

A) true   B) false.

20 Forward bias across a p–n junction is when an external battery is connected with

A) the positive terminal to p, negative to n
B) the positive terminal to n, negative to p.

21 In a silicon p–n junction the current at 300K for a reverse bias voltage V such that $eV \gg kT$ is 10 µA, the current with 0·2 volt forward bias will be

A) 23 mA   B) $3 \cdot 35 \times 10^{-9}\,\text{A}$ C) $-10^{-5}\,\text{A}$.

22 The transistor is a power amplifier.

A) true   B) false.

23 The field effect transistor is basically a capacitor.

A) true   B) false.

24 The law of mass action $np = n_i^2$ cannot apply in the source–drain channel of a MOSFET.

A) true   B) false.

25 When a negative gate voltage is applied to the n-channel MOSFET described in the text, the source–drain current will

A) increase   B) decrease   C) reverse

D) remain negligibly small.

Answers

1 C	2 B	3 A	4 C
5 A	6 B, D, F, G, H, K, L, N		

[*Note*: (M) is untrue because an electron in a full band can accelerate by changing places with an electron at higher energy, the latter therefore being accelerated, (N) is true because holes actually *are* empty levels.]

7 B	8 A	9 B	10 A
11 i) C; ii) B; iii) B; iv) A; v) C; vi) C; vii) A; viii) A			12 B
13 C	14 A	15 B	16 B
17 A	18 C	19 B	20 A
21 A	22 A	23 A	24 B
25 D*			

(**Note*: this channel would become p-type, but the n^+ source–drain contacts act as reverse-biased diodes.)

SEMICONDUCTOR PROCESSING

15·1 Introduction

The present age of microelectronics based on the silicon 'chip'–more properly integrated circuit–would not have been possible without the contributions of materials science. These have been important at every stage in the development of integrated circuit technology and in this chapter we discuss the principles and methods involved. Because the present technology is based dominantly on silicon we will confine our discussion mainly to this material.

15·2 Purification

Silicon is the most common element on earth, mainly found in the form of its oxides, e.g., sand and quartz. These can be reduced in a furnace with carbon, at about 1,450°C, to produce 98% pure metallurgical-grade silicon. We saw, in Chapter 14, that as little as one part per million of a dopant impurity in silicon has a profound effect on its electrical properties; it follows that a typical requirement for the starting material might be less than one impurity atom per 10^9 silicon atoms. Thus the commercially produced silicon requires considerable purification before use.

The metallurgical-grade silicon is purified to form electronic-grade polycrystalline silicon in which all impurities, with the exception of carbon, are at concentrations below one part per billion (1 ppb). The accurate measurement of trace quantities of carbon is difficult, the lower detection limit being 300 ppb; the carbon content is below this figure.

The standard purification process, used by industry over the past two decades, is hydrogen reduction of trichlorosilane on a heated silicon substrate. The first step is to react pulverized metallurgical-grade silicon with anhydrous hydrogen chloride in a fluidized bed at 300°C using a catalyst. The reactions are

$$\text{Si (solid)} + 3\text{HCl (gas)} \rightarrow \text{SiHCl}_3 \text{ (gas)} + \text{H}_2 \text{ (gas)} + \text{heat}$$

and

$$\text{Si (solid)} + 4\text{HCL (gas)} \rightarrow \text{SiCl}_4 \text{ (gas)} + 2\text{H}_2 \text{ (gas)} + \text{heat}$$

Since the reactions are exothermic, heat removal is essential to ensure maximum yield of trichlorosilane. Under correct operating conditions approximately 90% of the product is trichlorosilane with the remainder being mainly silicon tetrachloride together with chlorides of the various impurities.

At room temperature trichlorosilane is a liquid and has a boiling point of 31·8°C. It can be separated from the unwanted chlorides by fractional distillation. This process reduces the electrically active impurity concentrations to less than 1 ppb.

Electronic-grade silicon is produced from the highly pure trichlorosilane by a chemical vapour deposition process similar to that used for epitaxial growth (see Section 15·6). The trichlorosilane is vaporized and diluted with high-purity hydrogen. This is passed into a deposition reactor in which a long, thin rod of silicon is heated electrically to over 1,000°C. The deposition reaction is

$$H_2 + HSiCl_3 \rightarrow Si + 3HCl$$

with the silicon being deposited on the heated rod. The reaction byproducts exit the reactor as a high-temperature gas stream containing H_2, HCl, $HSiCl_3$, $SiCl_4$ and a small amount of H_2SiCl_2. The chlorosilanes are condensed and the trichlorosilane is recovered for recycling to the deposition process. Silicon tetrachloride is taken off as a useful byproduct and the HCl can be removed in anhydrous form for recycling to the trichlorosilane production process. The hydrogen is separated and recycled to the deposition reactor. A schematic diagram of the reactor is given in Fig. 15·1 and a diagram of the process is shown in Fig. 15·2. Present-day commercial reactors are capable of producing rods 2 m long and 20 cm in diameter, which takes about 200 hours of deposition.

The major shortcomings of the trichlorosilane technology are its relatively poor silicon and chlorine efficiency and the high power consumption of the deposition reactors. As much as 70% of the silicon and 90% of the chlorine

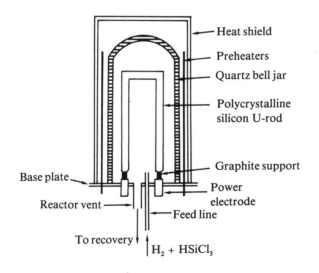

Fig. 15·1 Schematic diagram of quartz deposition reactor

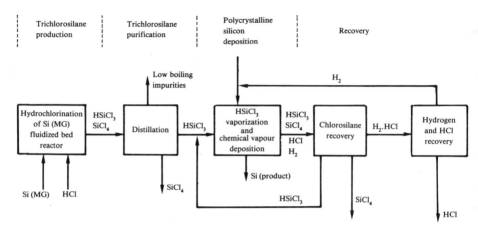

Fig. 15·2 Process flow block diagram for the production of polycrystalline silicon
via the hydrogen reduction of trichlorosilane

can exit the system in the form of silicon tetrachloride. An alternative
technology, which can achieve a virtually closed loop with only minor
byproduct streams, is based on the pyrolysis of silane (SiH_4). The process
steps are trichlorosilane production, silane production via multiple
redistributions of chlorosilane and silane pyrolysis to produce electronic-
grade silicon. The principal reactions are:

Hydrogenation of $SiCl_4$
$$3SiCl_4 + Si + 2H_2 \rightarrow 4HSiCl_3$$

Catalysed redistribution of $HSiCl_3$
$$2HSiCl_3 \rightarrow H_2SiCl_2 + SiCl_4$$

Catalysed redistribution of H_2SiCl_2
$$3H_2SiCl_2 \rightarrow SiH_4 + 2HSiCl_3$$

Silane pyrolysis

$$SiH_4 \xrightarrow{\text{heat}} Si + 2H_2$$

All chlorosilanes, silicon tetrachloride and hydrogen are recycled back
through the preceding processes, resulting in a closed-loop process with a
conversion efficiency approaching 100%. The pyrolysis can be carried out in
the same type of reactor as for the trichlorosilane process, depositing the
silicon on a thin, heated rod. A block diagram of the process is given in Fig.
15·3.

The world-wide production capacity of electronic-grade silicon in 1987 was
in excess of 11×10^6 kg/year and is projected to reach 20×10^6 kg/year in
1990.

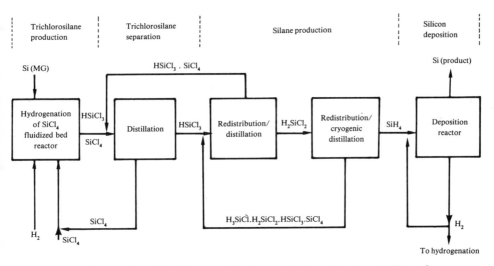

Fig. 15·3 Process flow block diagram for the production of polycrystalline silicon via the pyrolysis of silane

15·3   Crystal growth

The principle of crystal growth from the melt is that of controlled freezing under as near to thermal equilibrium conditions as can be achieved. In many techniques the freezing is started at a point and the solid–liquid interface is moved very slowly through the molten mass. In the case of silicon there is a considerable increase ($\sim 25\%$) in volume on crystallization and the method chosen must allow for this expansion. Thus containing the crystal in a sealed ampoule or crucible is not practical and the technique of pulling from the melt, named after Czochralski who developed it in 1917, is almost universally used.

A schematic diagram of a crystal puller is given in Fig. 15·4. An atmosphere of pure argon is used. To obtain a crystal with a known orientation, a small seed crystal is mounted with the desired growth plane accurately parallel to the melt surface. It is then dipped into the molten surface and the melt temperature is reduced until the molten silicon begins to freeze on to the seed. Pulling is started and more material freezes on to the crystal as it is withdrawn, the area of the solid–liquid interface increasing whilst the rod grows in length. This forms a neck whose diameter increases up to a constant value which is determined by the temperature gradients and heat losses, and the rate of pulling.

In the simplest view of the process the balance between heat input, H_i Joules to the system and heat lost, H_1 Joules from the system must equal the latent heat, L, of crystallization. If a length dx crystallizes in a time dt, then

$$(H_i - H_1)\,dt = L\rho_{si}A\,dx \tag{15·1}$$

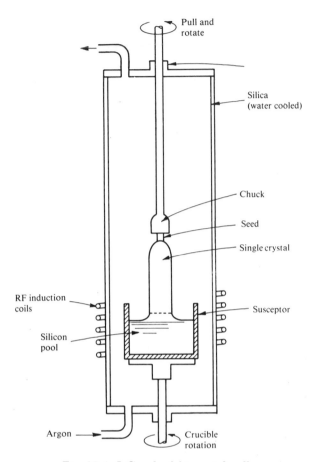

Fig. 15·4 A Czochralski crystal puller

where ρ_{Si} is the density of solid silicon and A is the area of the solid–liquid interface. Thus the rate of pulling will be given by

$$\frac{dx}{dt} = \frac{H_i - H_1}{L\rho_{Si}A} \qquad (15\cdot2)$$

and we see that the area of crystal produced, i.e., the diameter of the crystal rod, can be controlled by the rate of pulling. Typically growth rates vary between 10^{-4} and 10^{-2} mm/s, being finally determined on an empirical basis for any particular crystal puller.

The integrated circuit industry began with 2-inch diameter crystal rods, because that was the maximum diameter over which an acceptably low dislocation density could be obtained. It should be realized that dislocations have electrical effects in the lattice which seriously degrade the performance of a small area p–n junction in which a dislocation occurs. Thus for integrated circuit technology, in which thousands of transistors may be fabricated over a surface area of a few square millimetres, it is of extreme importance to produce crystals with as few dislocations as possible.

Dislocation-free crystals of silicon were first grown in 1960. They were not free from point defects in the form of vacancies or occasional impurity atoms but nevertheless were probably the most perfect crystals ever grown. Briefly the requisites for pulling a dislocation-free crystal are:

(a) initial seed diameter sufficiently small (1–2 mm) not to contain any dislocations itself;

(b) growth such that {111} glide planes make a large angle with the growth direction; this is because dislocations generated by plastic deformation concentrate in the {111} glide planes and if the growth axis makes a large angle to the planes, the dislocations tend to grow out to the surface. It is easy to see that the equivalent of plastic deformation must occur at the growth interface, as the solid is being pulled out of the liquid pool; there must be a point where the solidifying liquid behaves like a stressed solid;

(c) to favour climb of dislocations to the surface an adequate supply of vacancies should be available for their propagation. This is favoured by rapid growth in a large temperature gradient.

As a result of improvements in crystal growth technology, the semiconductor industry moved on to the use of 3-inch diameter crystal slices as standard substrates during the 1970s and 4-inch slices in 1980. At the time of writing, 6-inch diameter slices are becoming available.

The choice of crucible to contain the melt has an important bearing on the purity of the crystal. Silicon reacts with graphite to form silicon carbide and it alloys with refractory metals, so these materials cannot be used. The best crucible is vitreous silica which is used as a liner to a graphite crucible (a 'susceptor') heated by radio-frequency induction heating. Silicon slowly reacts with SiO_2 to form SiO, which is gaseous at the melting temperature. Thus the crucible is slowly attacked. The rate of attack is reduced by rotating the crucible and the crystal at the same rate and in the same direction during pulling. This has the effect of averaging out any thermal asymmetry in the radial directions (the reason for rotating the crystal in the first place) without introducing stirring of the melt, which would accelerate crucible erosion. A common contaminant of vitreous silica is boron, which dissolves in the melt as the crucible is eroded. This, of course, produces undesirable p-type doping. Oxygen also has an appreciable solubility in molten silicon and can enter the crystal lattice with an adverse effect on the electrical properties. However, these effects are small, and where a doped crystal is being grown they are unimportant.

15·4 Crystal doping

In IC technology the slice of crystal, into the surface of which the circuits are to be fabricated, is referred to as the substrate. The most popular substrate is p-type with a resistivity in the range 0·01–0·5 Ω-m, depending on the application. This corresponds to acceptor densities in the vicinity of 10^{21}–10^{22} m^{-3}, or between one impurity atom in 10^7 and 1 in 10^6 of silicon atoms. The introduction of this level of impurity requires successive stages in which a high-purity

Table 15·1

Impurity	Distribution coefficient	Dopant type
B	0·8	p
Al	0·002	p
Ga	0·008	p
In	0·0004	p
P	0·35	n
As	0·30	n
Sb	0·023	n

polycrystalline material is heavily doped to a resistivity of $\sim 10^{-4}$ Ω-m, by the addition of a sufficient quantity of boron to be accurately weighed. This is then powdered and a small quantity added to the melt from which the crystal is pulled. A similar procedure is required to produce n-type substrate material, the dopant usually being phosphorus, arsenic, or antimony.

Because the density (number of atoms per unit volume) of solid silicon is much less than that of liquid silicon a given number of silicon atoms per impurity atom will produce a greater *concentration*, in atoms *per unit volume*, of impurity in the liquid than in the solid. Thus there is a gradual enrichment of the dopant concentration in the melt pool as growth of the crystal proceeds. Thus, as the crystal pulling continues, there is produced a gradient of impurity density and hence resistivity, along the length of the crystal. The extent of this effect depends on the solubilities of the dopant impurity in the liquid and solid respectively and also upon its diffusion coefficient in the solid. These are summed up in a parameter called the *distribution coefficient*, the values of which for typical dopants in silicon are given in Table 15·1.

The nearer to unity the value of the distribution coefficient the less the enrichment effect. From this point of view, boron is the best dopant. Where the distribution coefficient is small, the melt volume should greatly exceed the volume of crystal to be grown. It is also possible to feed fresh material of appropriate composition continuously (e.g., pure silicon) into the melt to maintain it at the correct composition during growth.

15·5   Slice preparation

The single-crystal ingot has to be cut up into wafers for processing and these must have a perfectly flat, defect-free surface of an accurately oriented crystal plane.

After cutting off the seed and top ends of the ingot, its surface is ground by means of a rotating diamond wheel to reduce the diameter to the required value. Following this, a flat is ground along the length of the ingot. This is referred to as the primary flat and it is located, using an X-ray technique, relative to a specific crystal direction. It is used for mechanical location of a wafer in automatic processing plant and allows for orientation of integrated circuit devices with respect to crystallographic directions. A smaller, secondary flat is also cut, whose position with respect to the primary flat

Table 15·2

Type	Angle
{111} n-type	45°
{111} p-type	0° (i.e., no secondary flat)
{100} n-type	180°
{100} p-type	90°

identifies the type of wafer. The industry standards are given in terms of the angle between a radius to the centre of each flat as in Table 15·2.

After grinding, the ingot is ready to be sliced into wafers. The surface orientation is fixed by the angle of the diamond saw used for slicing. Usually trial wafers are cut, oriented by X-rays, and the angle of the saw adjusted until the correct orientation is obtained. For ⟨100⟩ orientation an accuracy of 1° is accepted; for ⟨111⟩ it is usually deliberately cut about 3° off orientation to aid epitaxial growth (see Section 15·6). The saw blade is annular with the cutting edge on the inside perimeter of the annulus. This allows the blade to be mounted in a rigid frame and ensures that it does not bend during the cutting. Saw blades up to 58 cm diameter with a 20 cm opening are available and blade thickness is 325 μm. The wafers are sliced with thicknesses of 0·5–0·7 mm, depending on their diameter, which means that up to one-third of the crystal is lost as sawdust.

The cutting process leaves the surface somewhat rough and it is followed by a mechanical, two-sided lapping operation performed under pressure using a mixture of alumina and glycerine as the polishing powder. Approximately 20 μm per side is taken off and produces a wafer with flatness uniform to within 2 μm. After lapping, the edges of the wafer are ground to a radius to round them off. This is necessary to avoid the edges chipping during handling and also serves to control the build-up of photo-resist at the edges during the photolithography process (see Section 15·8).

These processes leave a high density of defects due to mechanical damage extending to a depth of about 10 μm beneath the surface. To remove the damaged layer, the wafer is etched in a suitable solution.

After the first etching stage the surface is given a final polish to achieve a high degree of surface flatness, to an accuracy of between 5 and 10 μm across the whole face. A polishing pad of polyurethane felt is used, under high pressure, with a slurry of colloidal silica (particle size 100 Å) in an aqueous solution of sodium hydroxide. Typically about 25 μm of surface is removed. This is followed by a chemical clean with acid, base and/or solvent mixtures to remove the slurry residue.

Chemical etching of silicon has been extensively studied. Because it is a covalent material it is very resistant to chemical attack, but it will react with oxygen readily. However oxidation of the silicon surface results in the formation of a thin surface layer of silica, which is extremely resistant to the usual acids, but which can be dissolved in hydrofluoric acid. The simplest etch for silicon is a mixture of nitric and acetic acids, which oxidizes the silicon to SiO_2, and hydrofluoric acid which dissolves the oxide. Another factor to be considered is the different etch rate which applies to different crystal orientations. The (111) plane is close-packed in the diamond lattice and so contains most atoms. It will therefore be etched away more slowly than a (100) plane containing few atoms. If there is extensive mechanical damage of the surface, parts of it may no longer present (111) planes and the effect of the differential etching rates will be to bring out the defects in the form of bumps or holes in the surface. It is for these reasons that many different etches have been produced for silicon, and each processing company tends to develop its own formula.

The proper choice of etchant can either enhance or reduce the differential etching effect. In general an isotropic etchant will be used so that the silicon surface being etched is removed uniformly. This will be an etchant which reacts very quickly so that the material is dissolved away rapidly and differential effects are not able to develop. However a relatively recent technique has been developed which uses anistropic etching of (100) slices. Because the (111) silicon planes lie at an angle of 57·7°C to the surface and these planes are attacked at the slowest rate in an anisotropic etch (typically a potassium hydroxide–isopropanol mixture), a V-shaped groove is formed. Thus the available surface area in a given slice is increased enabling more, or larger-area, transistors, etc., to be packed into a given area of chip. This technique is usually referred to as 'VMOS' and it has been used mainly in the fabrication of high-power, field-effect transistors.

15·6 Epitaxial growth

Even after the treatment described above the crystal surface will still contain defects. It is possible to improve the crystal perfection by growing on to the surface a thin, single-crystal layer by the process known as epitaxy, from the Greek word meaning 'arranging upon'. In this, atoms are deposited on to the crystal surface under conditions such that, under the influence of thermal agitation, each atom can 'run around' until it finds its correct crystal lattice position where it forms bonds to the atoms in the solid surface. Thus, a new crystal is formed layer by layer so long as the supply of atoms continues. The source of silicon atoms will usually be a vapour, which may be produced by evaporating silicon from a molten pool in vacuum but, more usually, comes from a chemical reaction, as described below.

The use of an epitaxial layer has significant advantages in the fabrication of integrated circuits. The epilayer can readily be uniformly doped by the inclusion of dopant atoms in the silicon vapour. Typically a $10\,\mu$m thick n-type layer with a resistivity of 0·005 Ω-cm would be deposited on a 0·1 Ω-cm p-type substrate crystal slice. The resulting p–n junction over the whole surface area can act, if reverse biased, as an effective insulating layer between the epilayer and the substrate. This prevents currents of flowing via the substrate between adjacent components formed in the epilayer. The uniformly doped epilayer, isolated from the substrate, is an ideal starting point for the fabrication of diodes, transistors, etc., by diffusion, which is discussed in Section 15·7.

15·6·1 Chemical vapour deposition (CVD)

The universally used method for the preparation of epilayers is *chemical vapour deposition*. In this, the source of silicon atoms is silicon tetrachloride, $SiCl_4$, which is reduced by hydrogen on a hot silicon surface to deposit silicon atoms and gaseous hydrochloric acid. The reaction is

$$SiCl_4 + 2H_2 \rightleftarrows Si\ (solid) + 4HCl\ (gas)$$

A schematic diagram of a vapour-phase reactor is given in Fig. 15·5. The cleaned and etched crystal slices are placed on a graphite slab, which acts as a susceptor for radio-frequency induction heating, and heated to about 1,200°C. Initially, hydrochloric acid vapour carried in a hydrogen stream is passed over the slices, which etches away about 1 μm of their surface, cleaning them of all traces of contaminants. Growth is then started by switching off the acid gas and bubbling the hydrogen through a solution of silicon tetrachloride, which carries the silicon tetrachloride vapour to the hot substrates. The rate of epitaxial growth is about 1 μm/min. This rate depends upon the

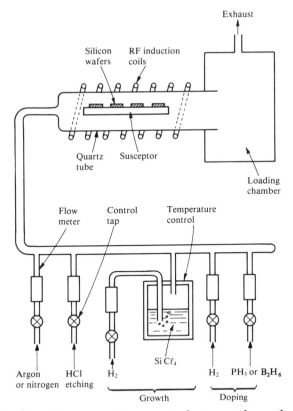

Fig. 15·5 A vapour phase reactor for epitaxial growth

vapour pressure of $SiCl_4$ in the reactor, determined by the temperature of the solution, which therefore has to be accurately controlled, usually in the range 0–30°C. Doping is achieved by mixing in with the $SiCl_4$ an appropriate gaseous source of dopant atoms diluted in a stream of hydrogen. For *n*-type doping the most usual dopant gas is phosphine, PH_3, which decomposes to phosphorus and hydrogen on the hot substrate surface. If a *p*-type epilayer is required, diborane, B_2H_6, is used. It may be noted that these dopant gases are highly toxic and stringent safety precautions are necessary when they are used.

The conditions for epitaxial growth of good, single-crystal material are quite critical. The substrate temperature must be sufficiently high that the arriving atoms can diffuse rapidly enough across the crystal surface to find their required crystal lattice positions. Their rate of arrival must be slow enough to allow this process to occur for each atom in turn, and this limits the permissible growth rate to a maximum of about 2 μm/min. Above this, a polycrystalline layer will be produced. For 1 μm/min deposition, the molecular fraction of $SiCl_4$ required in H_2 is about 0·01. If this fraction rises above about 0·25 a competing reaction sets in with the $SiCl_4$ reacting with solid silicon according to

$$SiCl_4 \text{ (gas)} + Si \text{ (solid)} \rightleftarrows 2 SiCl_2 \text{ (gas)}$$

Under these conditions the crystal surface is actually etched away and the 'growth' rate becomes negative.

Like the seed crystal used in the Czochralski growth technique, the substrate orientation determines the orientation of the epitaxial layer. Thus a $\langle 100 \rangle$ face will produce a $\langle 100 \rangle$ epitaxial layer, and so on. One of the advantages of epitaxial technology is that it allows the use of a buried layer. For example, to make an *n*–*p*–*n* transistor we require a heavily doped n^+ layer to make a low-resistance collector contact. This can be achieved by first diffusing an n^+ layer (see next section) into the *p*-type substrate and then growing an epitaxial *n*-type layer over it. The transistor is fabricated in the epitaxial layer, but clearly it has to be possible to locate the position of the n^+ region through the epitaxial layer. Fortunately the process of forming the n^+ layer causes a shallow pit, about 50–100 nm deep, in the wafer surface and this is duplicated at the surface of the epitaxial layer, providing it is not too thick.

Unfortunately, especially with CVD at higher rates, there is a shift in the epitaxial surface pit with respect to the buried layer. The extent of this depends on the orientation of the wafer, being greatest for $\langle 111 \rangle$, and on the deposition conditions. Current practice is to misorient $\langle 111 \rangle$ wafers about 3° towards the nearest $\langle 110 \rangle$ direction, which reduces the effect and gives about the same shift as occurs on a $\langle 100 \rangle$ wafer.

15·6·2   *Molecular beam epitaxy (MBE)*

It was mentioned above that the vapour of atoms, from which the epitaxial layer is grown, could be produced by evaporation in vacuum, a process now known as *molecular beam epitaxy*. Although the technique is very old-established, it was not adopted by the silicon semiconductor industry because of its intrinsically low deposition rate (0·01–0·03 μm/min) and the need for very high-grade vacuum equipment. However, it offers a number of advantages over the CVD process, the most important of which is that substrate temperatures can be as low as 400°C as opposed to up to 1,250°C for the CVD process. The significance of this in silicon technology is that, at the CVD temperatures, there is out-diffusion from localized, doped areas such as buried layers. As a consequence, devices cannot be packed as close together on the substrate as is required in very large-scale integration (VLSI). The use of MBE overcomes this problem as well as conferring greater flexibility in doping of the deposited layers.

Unlike the CVD process, MBE is not complicated by problems of gaseous transport, nor are there any chemical reactions. Furthermore, there are no foreign atoms, such as chlorine and hydrogen, to become incorporated with the epitaxial layer. The basic process consists of evaporation of silicon and one or more dopant species in an ultra-high-vacuum (UHV) chamber as illustrated in Fig. 15·6. To achieve UHV of 10^{-8}–10^{-10} Torr, the chamber must be made of stainless steel and be baked for several hours at 300°C under vacuum before use, to remove adsorbed gases from the walls of the chamber. Thereafter it is undesirable to let the chamber up to atmospheric pressure and loading of samples is done through a vacuum load lock. To avoid possible contamination by oil vapour, the vacuum pumps are usually ion or cryogenic pumps.

Because of its high melting point, silicon is evaporated by electron beam heating. In this, a high-energy electron beam is directed at the surface of a silicon ingot where it forms a molten pool from which the silicon evaporates. The ingot is held in a water-cooled copper hearth so that the body of the ingot

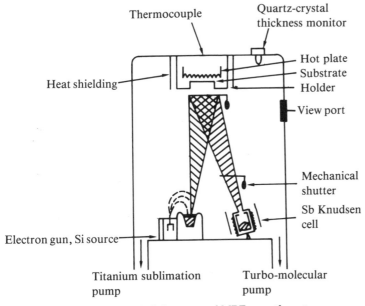

Fig. 15·6 Schematic of MBE growth system

remains solid and acts as a crucible for the molten silicon. Dopants are evaporated from a small furnace, known as a Knudsen effusion cell. This consists of a bottle-shaped silica crucible with a narrow neck. A resistance heater wound around it provides the heat necessary to melt the solid dopant material. The rate of deposition of each atom species on the substrate is determined by the vapour pressure above the liquid, which is a strong function of temperature (see Section 15·14). Thus the relative rates of evaporation of silicon and dopant can be regulated by control of the source temperature, which controls the atom flux reaching the substrate. Unfortunately, desirable dopants such as As and P evaporate too rapidly and B too slowly for controlled use and so Sb is used for n-type and Ga or Al for p-type doping. An advantage of this method is that the dopant concentration can be varied during deposition by varying the source temperature to produce graded doping profiles, which are desirable in some types of device.

The method is, of course, not confined to the deposition of silicon. With the increasing need for very fast digital electronic devices, ultra-high-frequency amplifiers and semiconductor lasers, there is a growing demand for devices and circuits based on compound semiconductors such as GaAs, GaAlAs, InP and InSb, all of which can readily be fabricated in an MBE system.

Because deposition of the epitaxial layer takes place in a vacuum, it provides the opportunity of *in situ* cleaning of the substrate surface before deposition. This is done by sputter cleaning (see Section 15·16) in which a beam of argon ions is directed at the substrate surface. This removes the native oxide and other adsorbed species, leaving an atomically clean surface on which to form the epitaxial layer. About 1 μm depth of material is removed and a short anneal at 800–900°C is sufficient to reorder the surface.

The high-vacuum feature also allows doping by ion implantation, which is discussed in Section 15·12 below.

15·7 Integrated circuit processing

We have now considered the production of a single crystal silicon slice with an epitaxially deposited layer on one surface. Let us say we wish to fabricate a transistor in the n-type epilayer which itself is used as the collector region for the transistor. It will be necessary to produce a p-type region for the base between two n-type regions which are the emitter and collector of the n–p–n transistor. The scientific principle involved is the law of mass action, as applied to semiconductors, quoted in Section 14·10, i.e., $np = n_i^2$. Since n_i, the 'intrinsic' carrier density is a constant for a given temperature, the introduction of more acceptor impurity atoms to increase p must automatically reduce n, and so if we can diffuse sufficient boron into the n-layer, it will be converted to p-type for the base region. If this is followed by diffusion of, say, phosphorus, part of the p-layer can be reconverted to n to complete the transistor. A schematic diagram of the sequence is shown in Fig. 15·7 in which (a) and (b) are the diffusion stages and (c) is the finished transistor.

The transistor must also be isolated electrically from adjacent com-

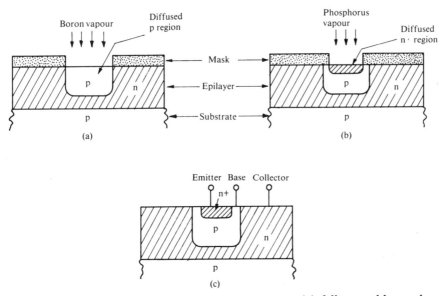

Fig. 15·7 The steps in integrated circuit processing: (a) diffusion of boron for p-type conductivity; (b) diffusion of phosporus for n-type conductivity; (c) the finished transistor

ponents, otherwise it would not operate correctly in a circuit. This can be achieved by an 'isolation diffusion' in which a border of p-type impurity is diffused in all around the transistor. This provides a p–n junction barrier around it which operates just like the barrier between the epitaxial layer and the substrate.

It is clear from Fig. 15·7 that a *masking process* is required to define the areas into which the impurities are to be diffused. A convenient barrier to the impurity vapours is silicon dioxide (silica), through which the rate of diffusion of the impurity atoms is negligibly small at the diffusion temperatures. The minimum oxide thickness for this purpose depends on the impurity atom but is generally in the region of 1 μm. The silica (SiO_2) layer can be produced by flowing a stream of oxygen over the hot silicon surface, producing the reaction

$$\text{Si (solid)} + O_2 \text{ (gas)} \xrightarrow[1100°C]{} SiO_2 \text{ (solid)}$$

The oxidation is more rapid in the presence of water vapour and, for masking oxide, the oxygen may be bubbled through water on its passage into the furnace. The oxide growth rate is around 0·5 μm/h. The oxidation process is discussed in more detail in Section 15·13.

15·8 Photolithography

After growing an oxide layer over the whole of the silicon surface, we require

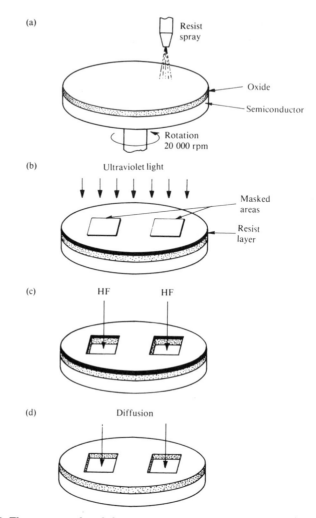

Fig. 15·8 The steps in photolithography: (a) 'spinning-on' photo-resist; (b) photo-resist exposure with masking; (c) etching 'windows'; (d) diffusing in dopant

to open 'windows' in the oxide to allow for impurity diffusion into the required area. This is relatively easily done by dissolving the silica in hydrofluoric acid (HF), which does not attack the silicon itself. However, it is necessary to protect the oxide from the acid in non-window areas (see Fig. 15·8) and it is also important that the positions and sizes of the windows be accurately defined. For this purpose organic *photoresists* have been developed, which are the emulsions of an organic monomer in a suitably volatile carrier liquid. Working in a yellow light this material is sprayed on the oxide surface which is set spinning at about 20,000 rpm to produce a uniform thickness of liquid on the surface by centrifugal force. The carrier liquid evaporates and the surface is gently baked to 150°C to solidify the layer of resist. The resist material is then in a condition such that it will polymerize if exposed to ultraviolet light. The polymerized material is not attacked by the hydrofluoric acid. The processing thus consists of preparing a mask, usually a film of chromium on glass, which masks the window areas so that they remain unexposed when the whole slice is irradiated with ultraviolet light. After irradiation the unexposed resist is washed off in a suitable solvent, referred to as a developer, which will not remove the polymerized material. Thus the windows to the oxide are opened for etching away in HF. After being etched, the polymerized resist is removed, using a mixture of sulphuric acid and hydrogen peroxide or–in present-day technology–by exposing it to an oxygen-rich ionized gas plasma, which 'burns' it off; this is known as plasma-etching. The whole sequence is illustrated in diagrammatic form in Fig. 15·8, and since the exposed areas are not washed off the resist is called a negative emulsion by analogy with photographic development. Because of the similarity it bears to processes used in producing etched plates for litho-graphic printing, the whole process is referred to as *photolithography*. There are many types of resist in use, including positive resists in which exposure makes them soluble in a solvent which does not attack the unexposed material.

The development of the photolithographic process has been crucial to the realization of integrated circuits. By using high precision mask-locating machines, together with high quality optics it has been possible to define and locate windows of a few square micrometres in area and line-widths down to 1–2 μm. Achieving this level of accuracy is expensive and difficult, and all operations have to be carried out in a totally dust-free environment. Present-day technology is moving away from wet processing towards electron beam lithography in which the mask-and-expose stage is replaced by 'writing' the pattern into a suitable resist material, such as polymethyl methacrylate, by means of an electron beam. This method is capable of defining line-widths of as little as 0·1 μm.

15·9 Impurity diffusion

The driving force for a diffusion process, as discussed in Section 7·21 is a con-centration gradient, the diffusion flux, F, being given by Fick's Law [Eqn (7·24)]

$$F = -D \frac{dn}{dx} \tag{15·3}$$

where dn/dx is the concentration gradient and D is the diffusion coefficient, measured in m^2/s. This coefficient has a value which depends on the mechanism by which the diffusion occurs. If we were to imagine an ideal, covalent lattice in which there were no vacancies or defects, it would obviously be very difficult for the foreign atoms to penetrate the lattice, and they could only do so into interstitial positions (see Chapter 8). However, due to thermal agitation there will always be several vacancies in the silicon

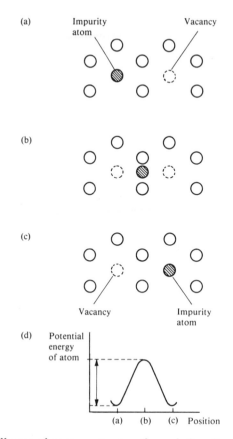

Fig. 15·9 The diffusion of an impurity atom through the silicon crystal: (a) initial position with adjacent vacancy; (b) the activated step; (c) new position exchanged with vacancy; (d) potential energy variation of impurity atom during the diffusion process

lattice, in which an atom has moved from its regular lattice site into an interstitial position (a Frenkel defect). If the interstitial diffuses away from the vacancy, the latter becomes a Schottky defect. Now, for an impurity atom to be electrically active, in the sense of providing a hole or an electron, it must substitute for a silicon atom on a regular lattice site. Thus we require that the impurity diffusing into the crystal lattice must occupy a vacancy. The process of diffusion, therefore, will be that of the impurity atom jumping from vacancy to vacancy, as illustrated in Fig. 15·9. Whether it is able to move from one vacancy to the next will depend on the availability of a vacant neighbouring site, and the probability of this being available will be the probability of an adjacent silicon atom jumping out of its lattice site to form a Frenkel defect. We therefore have two activated processes involved in the diffusion. We require to activate a silicon atom out of its normal position, requiring an energy E_i, and to activate the impurity atom from one site to the next, requiring an energy E_j, as illustrated in Fig. 15·9(d). The diffusion constant, D, is therefore given by

$$D = D_0 \exp\left[-(E_i + E_j)/kT\right] \tag{15·4}$$

where D_0 is a constant.

Whilst this describes the basic mechanism, the effects of electrostatic interaction between the impurity atom and charged defects also have to be taken into account. These lead to a lowering of the energy necessary to promote the diffusion.

Group III and V impurities diffuse with approximately the same average activation energy $E_D \simeq 3\cdot5$ eV with the prefactor $D_0 \simeq 10^{-4}$ m²/s.

15·10   Diffusion profiles

It is clearly important to be able to predetermine the depth of penetration and the concentration profile of impurity introduced into the semiconductor slice, when fabricating integrated circuits. To do this we require to make use of the *transport equation* for diffusion of atoms, which is derived from Fick's law (see Section 7·21).

Referring to Fig. 15·10, consider a rectangular elementary volume of unit area and thickness, dx. Let the flux entering it from the left-hand side be $F(x)$ and that leaving from the right-hand side be $F(x + dx)$. The rate of change in the number of atoms contained in the elementary volume will be the difference between the flux in and the flux out. Now the number of atoms in the elementary volume will be $n\,dx$, where n is the average concentration, so we may write

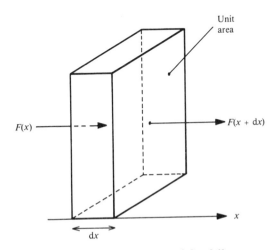

Unit area

$F(x)$

$F(x + dx)$

x

dx

Fig. 15·10 An elementary slice for derivation of the diffusion transport equation

$$\frac{\partial n}{\partial t} \cdot dx = F(x) - F(x + dx)$$

or

$$\frac{\partial n}{\partial t} = -\frac{\partial F}{\partial x} \tag{15·5}$$

where the negative sign arises from the fact that if flux in is greater than flux out, the gradient of flux in the positive x-direction must be negative.

Substituting for F from Eqn (15·3) we have

$$\frac{\partial n}{\partial t} = D \frac{\partial^2 n}{\partial x^2} \tag{15·6a}$$

which is the diffusion transport equation.

The appropriate solution of this equation will depend upon the boundary conditions, i.e., the values of $n(x, t)$ at particular positions x and times t.

One possible solution of this equation is

$$n(x, t) = t^{-1/2} \exp(-x^2/4Dt) \tag{15·6b}$$

as may be confirmed by differentiating and substituting in Eqn (15·6). This, however, represents only one particular solution of the diffusion equation and a general solution will be the sum of all such integral solutions. This may be written as

$$n(x, t) = \frac{1}{2\sqrt{\pi Dt}} \int_{-\infty}^{+\infty} f(x') \exp[-(x - x')^2/4Dt]\, dx' \tag{15·6c}$$

where x' is a dummy variable and $f(x')$ is the initial concentration distribution of the diffusion at time $t = 0$. Clearly the precise form of $f(x')$ must be known before the integral in Eqn (15·6c) can be evaluated. The simplest case is to assume a semi-infinite solid containing no impurities at time $t = 0$ into which an impurity is to be diffused from the surface. The concentration at the surface (at $x = 0$) is maintained constant by (say) a vapour of impurity atoms. If the surface concentration $n(0, t) = N_s$ then that at a depth x, after a time t, is given by

$$n(x, t) = N_s \operatorname{erfc}(x/2\sqrt{Dt}) \tag{15·7}$$

where erfc is the *complementary error function* whose value is tabulated in mathematical tables. This is the solution for an 'infinite source'.

If a finite source comprising a thin, uniformly doped region of thickness δ at the silicon surface is used, the initial impurity distribution can be approximated by a constant concentration N_0 over a distance δ. There is thus a finite source of $Q = N_0\delta$ atoms per unit area available for diffusion and as diffusion proceeds this source will be depleted, contrary to the previous case where the source was constant. The solution of Eqn (15·6) is then

$$n(x, t) = (Q/\sqrt{\pi Dt}) \exp(-x^2/4Dt) \tag{15·8}$$

which is a Gaussian distribution. As the diffusion proceeds, the surface source depletes with time as

$$n(0, t) = Q/\sqrt{\pi Dt} \tag{15·9a}$$

for $t > t_0$ where $t_0 = \delta^2/\pi D$.

If the background doping of the silicon is, say, P_b, then the depth at which $n(x, t) = P_b$, which will be the junction depth, is given by

$$x_j = (4Dt \ln [Q/P_b\sqrt{\pi Dt}])^{1/2} \tag{15·9b}$$

Writing $u = x/2\sqrt{Dt}$, Eqns (15·7) and (15·8) become, respectively, $N_s \text{erfc}(u)$ and $N_s \exp(-u^2)$. These two functions are plotted in Fig. 15·11 as n/N_s against u for a given time, t, when the surface concentrations are the same for each case. As time proceeds, the Gaussian distribution changes in accordance with Eqn (15·8). This is illustrated for three times $t_1 < t_2 < t_3$ in Fig. 15·12.

To obtain a nearly uniform distribution of impurity through the thickness of a slice a two-stage process is used. Initially an erfc profile is produced over a thin layer near the surface, from a vapour source. The vapour source is then removed and subsequent diffusion follows the Gaussian profile until, at a sufficiently long time, the concentration is fairly uniform through the slice.

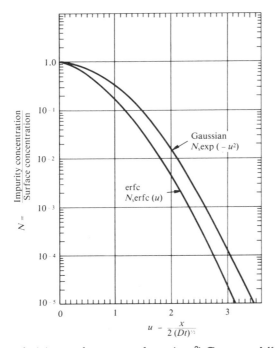

Fig. 15·11 The erfu (u) error function and $\exp(-u^2)$ Gaussian diffusion profiles

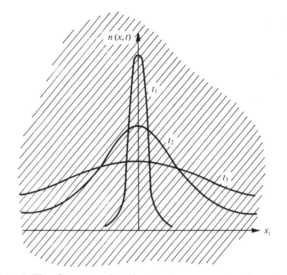

Fig. 15·12 The Gaussian distribution for successive times $t_1 < t_2 < t_3$

The initial diffusion is referred to as the deposition stage and the second process is called 'drive-in'.

15·11 Junction formation

In the formation of p–n junctions or transistors in an integrated circuit chip the initial slice will already be doped, typically being an n-type epitaxial layer. To produce a junction at some depth, x_j, below the surface we require to diffuse in boron as a p-type impurity. The actual junction will be formed at the depth where the in-diffused boron concentration equals the original n-type impurity concentration. This gives *compensation* where the two dopants effectively cancel each other out. The carrier concentrations n and p, which are equal, are given by $np = n_i^2$ and the material at this point behaves as if it were intrinsic. Assuming the p-type diffusion is by drive-in from a finite source of boron, the diffusion profile will be Gaussian as given by Eqn (15·8) and, assuming we know the original n-type impurity concentration, N_D, we can calculate the depth x_j of the junction as the point at which

$$N_D = N(x_j, t) = \frac{Q}{(\pi Dt)^{1/2}} \exp(-x_j^2/4Dt) \tag{15·10}$$

If, now, we wish to continue processing to produce an n–p–n transistor we will need a second diffusion stage in which an n-type impurity atom such as phosphorus is diffused in to form the emitter of the transistor. Because of its sharper fall-off, the erfc profile would normally be used at this stage with the diffusion being from a constant vapour source. This sequence is shown in

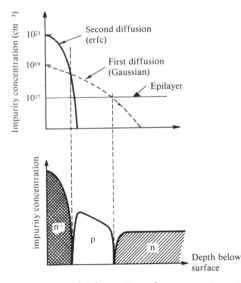

Fig. 15·13 Diffusion stages in the formation of an n–p–n transistor: (a) individual profiles for the successive diffusion; (b) resulting impurity concentration distributions

Fig. 15·13(a) with the resulting net impurity concentrations shown in Fig. 15·13(b).

It is clear that a given area of silicon slice may be subjected to several heating cycles corresponding to different stages in the oxidation and diffusion processes. For this reason it is better that the first in-diffused species should be the one with the lowest diffusion coefficient, requiring the highest temperature. The next stage should require a lower temperature and have a higher diffusion coefficient and the final stage the lowest temperature and highest coefficient. These requirements are met by the sequence arsenic (n), for which the diffusion coefficient is $1·15 \times 10^{-13}$ cm²/s at 1,200°C, boron (p) which has a similar diffusion coefficient, and therefore can be diffused slightly more slowly at (say) 1,150°C and phosphorus (n), which has the same diffusion coefficient of $1·15 \times 10^{-13}$ cm²/s at 1,090°C.

15·12 Ion implantation

In the discussion of diffusion of dopant atoms into silicon in the previous sections, a distinction was made between infinite and finite sources of dopant atoms. In the latter case, the deposition of a fixed number of dopant atoms from a chemical vapour is difficult to achieve and a much more controllable method is *ion implantation*. In this, dopant atoms are vaporized, accelerated and directed at the silicon surface in a vacuum chamber. They enter the crystal lattice and collide with silicon atoms, gradually losing energy and coming to rest at a depth below the surface determined by the acceleration energy. The actual number of dopant atoms delivered can be controlled quite closely by monitoring the ion current during implantation. The process causes disruption of the silicon lattice, due to ion collisions, which can be removed by subsequent heat treatment. Ion implantation thus satisfies the conditions necessary for an effective doping process. Implantation energies for doping of silicon range from 10 to 200 keV, producing penetration depths from 10 to 800 nm, depending on the type of atom implanted. Dose densities range from 10^{12} to 10^{18} ions/cm^2.

In the energy range 10–200 keV, used for implantation into silicon, the energy loss mechanism which dominates is that of ion collisions. The initial implanted ion energy of about 100 keV is much larger than the lattice binding energy of 10–20 eV so that we can infer that the implanted atom will come much closer to a silicon atom than the interatomic distance in the lattice. This allows the use of classical dynamics to describe the collision between pairs of nuclei, ignoring the relatively weak lattice forces. The collision will be elastic, with the energy lost by the incoming atom being transferred to the target atom, which recoils away from its lattice site. For a 'head-on' collision the energy loss, T_{max}, will be given by

$$T_{max} = \frac{4M_1 M_2}{(M_1 + M_2)^2} E$$

where M_1 and M_2 are the atomic mass numbers of the ion and target atoms respectively and E is the initial energy of the implanted ion.

For higher implantation energies, inelastic collisions between the incoming ion and electrons in the solid become increasingly important. In this process, the energy lost by the incident ions is dissipated through the electron cloud into thermal vibrations of the lattice. Detailed modelling of this process is very complex and corrections to energy loss arising from it are generally empirical.

Each implanted ion follows a random path as it moves through the target and its average total path length is called its range R, which is a mixture of vertical and lateral motion. The average depth reached by the implanted ions is called the projected range R_p and the distribution of ions about this depth is approximately Gaussian in both the vertical and lateral dimensions. Various theoretical models have been developed for the calculation of projected ranges and have resulted in reference tables from which the projected range, and spread in depth and width, can be found for a given atom type and implantation energy. At the time of writing, a computer program, TRIM, based on Monte Carlo calculations and developed in 1987 by J.F. Biersack and J.F. Ziegler, is widely used and examples of its output for boron, phosphorus and arsenic implantations in silicon are given in Table 15·3.

The ranges quoted in Table 15·3 are based on implantation into an amorphous material, but these figures may be modified in a single crystal by the phenomenon of *channelling*. For ions moving in certain crystallographic directions the atomic planes form a channel through which the ion may travel a relatively long distance, by glancing angle collisions with the atomic rows or planes. To avoid this, most implantatons into silicon are carried out with the

Table 15·3

Ion implantation ranges and spreads in silicon for different dopant atoms and energies

Ion	Energy (keV)	Projected range (Å)	Longitudinal spread (Å)	Lateral spread (Å)
B (mass 11)	10	428	258	271
	20	847	438	477
	50	2079	818	981
	100	3931	1195	1575
	200	6997	1588	2312
P (mass 31)	10	171	85	87
	20	305	140	143
	50	711	289	291
	100	1417	510	525
	200	2876	872	969
As (mass 75)	10	123	46	50
	20	194	68	74
	50	383	122	134
	100	681	201	219
	200	1279	345	376

wafer tilted so that the ion beam enters at an angle of 7° from the vertical.

As mentioned earlier, implantation damages the target and displaces many silicon atoms for each implanted ion. Indeed, at high dose rates the silicon may lose its crystalline character completely and become amorphous. However, the crystalline structure can be recovered and the dopant atoms placed on substitutional lattice sites, as required to make them electrically active, by a suitable annealing treatment. Raising the wafer temperature to 500°C during deposition induces self-annealing which maintains the crystal structure, but this may promote unwanted dopant diffusion elsewhere in the wafer.

Where the silicon has become completely amorphous, regrowth of the crystal occurs by solid-phase epitaxy, in which the amorphous/crystalline interface moves towards the surface at a velocity depending on the temperature, doping and crystal orientation. For example, in undoped silicon the interface velocity at 600°C is about 20 Å/min for a ⟨111⟩ direction and about 300 Å/min for a ⟨100⟩ direction.

If the implantation was not severe enough to create an amorphous layer, lattice repair occurs by the generation and diffusion of point defects. Paradoxically this process has a higher activation energy than solid-phase epitaxy and a temperature of about 900°C is required to remove all defects.

Whilst these annealing treatments can be carried out in a furnace, heating the whole slice for up to 30 minutes may cause unwanted diffusion of dopants. For this reason *rapid thermal annealing* (RTA) methods have been developed. These are generally based on heating by localized high-energy light irradiation, either from a laser, xenon tube, or tungsten–halogen lamps. RTA heating occurs because photons are absorbed by free carriers in the silicon, which then transfer their energy to the lattice. Thus the heating rate depends on the doping level and temperature, which determine the number of carriers, as well as the photon flux and the surface emissivity of the silicon.

There are three types of RTA, referred to as: adiabatic, with heating time $<0.1\ \mu s$; thermal flux, with heating from $0.1\ \mu s$ to 1 s; and isothermal, with heating times greater than 1 s. Adiabatic heating is done by means of a localized, high-energy, short-duration laser pulse which melts the surface to a depth of less than 1 μm. The surface recrystallizes by liquid-phase epitaxy, removing all traces of damage. Dopant diffusion in the liquid is very fast, so that uniform doping over the liquefied region is produced and any doping profiles or surface films are lost.

In thermal flux annealing, heating from one side of the wafer gives a temperature gradient across it. The surface is not melted and repair is generally by solid-phase epitaxy before any diffusion can occur. However, the rapid thermal quenching involved leaves many point defects which may condense to form dislocations, which have an adverse effect on the electrical properties of the material. Isothermal annealing uses tungsten–halogen lamps to raise the wafer surface temperature to 1,100°C for 30 s and produces much less diffusion than furnace heating.

Commercial equipments for ion implantation are available from a number of manufacturers. In Fig. 15·14 the layout of a typical implanter is shown, whilst in Fig. 15·15 a schematic diagram shows the features of the various parts. Beginning on the right-hand side of the diagram the Freeman ion source converts a gas containing the desired implantation species to an ionized plasma. For example, for boron the gas may be BF_3, which is ionized to B^+ and F^- ions. The ionization is effected by electrons emitted from a hot, tungsten filament, and a magnetic field causes them to spiral about the filament, increasing their path-length in the gas and hence their ionizing efficiency. The positive ions are extracted as a beam by means of suitable, negatively biased extraction and accelerating electrodes. The ion beam travels down a curved flight tube under the combined influence of an accelerating electric field and a variable magnetic field. The strengths of these two fields are adjusted so that only ions of the desired mass are able to follow the curve down the centre-line of the tube. An electrostatic lens system at the end of the tube directs the ion beam onto the target. The beam is scanned across the target by mechanical translation of the target holder in the x–y plane.

Implantation equipment is now a standard component of integrated circuit fabrication lines. Development of this technique has been so successful that it is used for almost all dopant stages in VLSI processing.

15·13   Thermal oxidation of silicon

As described earlier in the chapter, silicon is oxidized at various stages in the processing of an integrated circuit, to form diffusion barriers and electrically insulated areas. In the MOS (metal-oxide-semiconductor) technology, in which the transistors formed are field-effect devices, the oxide gate insulator is a crucial element in the operation of the transistor itself. As a consequence,

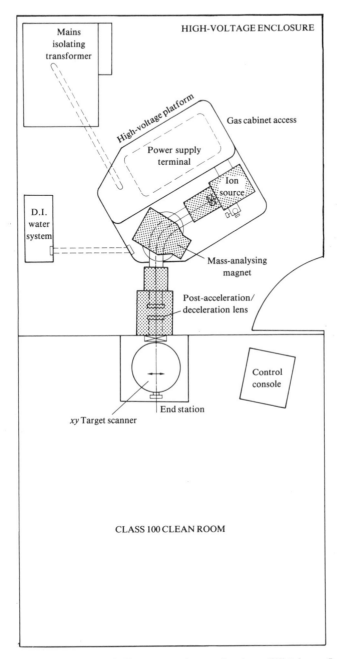

Fig. 15·14 The Imperial College ion beam facility (Whickam Ion Beam Systems Ltd)

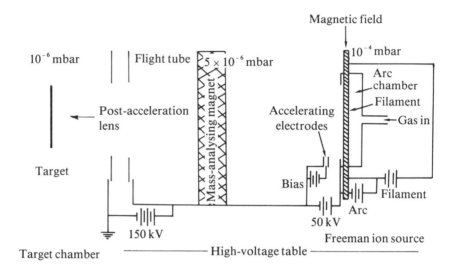

Fig. 15·15 Schematic diagram of the ion implantation equipment

the thermal oxidation of silicon has been closely and extensively studied.

Several experiments have shown that, in the course of thermal oxidation, oxygen diffuses through the oxide layer already grown to the Si/SiO$_2$ interface, where it reacts with the silicon surface to form more oxide. There are therefore three steps in the process, illustrated schematically in Fig. 15·16. The oxygen gas in the furnace will provide a flux F_1 mols/s to the surface, where F_1 will be proportional to the oxygen partial pressure. The oxygen will then diffuse through the oxide layer with a flux F_2, proportional to the oxygen concentration gradient across the oxide layer and to the diffusion coefficient, D_{eff}, for oxygen in SiO$_2$. Finally, there will be a reaction flux, F_3, at the silicon surface, determined by the chemical surface reaction rate constant, k_s. Under steady-state conditions, $F_1 = F_2 = F_3$ and the oxidation proceeds at a rate determined by the slowest of the three processes.

When the oxide layer is 'thick', meaning that the thickness x_0 is greater than D_{eff}/k_s (usually a few hundredths of a micrometre), solution of the full theoretical equations predicts that the thickness of the layer grows with time according to a square-law

$$x_0^2 = Bt \tag{15·11}$$

where B is called the parabolic rate constant and t is time. Experimentally this is found to be obeyed where the diffusion through the oxide layer is the rate-limiting factor. For short times and thin oxide layers the surface reaction is the rate-limiting factor; the theoretical solution predicts that the oxide growth is linear with time according to

$$x_0 = (B/A)(t + \tau) \tag{15·12}$$

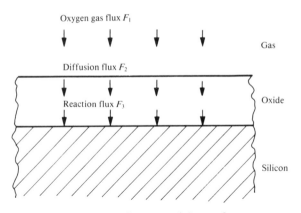

Fig. 15·16 Schematic diagram of the oxidation process

where B/A is called the linear rate constant and the parameter τ represents a shift in the time coordinate to correct for the presence of an initial layer of oxide on the silicon surface. If the thickness of the initial layer is x_i, then τ is given by

$$\tau = (x_i^2 + Ax_i)/B \qquad (15\cdot13)$$

Again the experimental results are found to follow the theoretical equation.

The linear and parabolic rate constants depend on temperature, following an activated law similar to that for the diffusion coefficient [Eqn (15·4)]. They are given by

$$B/A = C_1 \exp(-E_1/kT) \qquad (15\cdot14)$$

and

$$B = C_2 \exp(-E_2/kT) \qquad (15\cdot15)$$

where the values of the constants C_1 and C_2 and the activation energies E_1 and E_2 are given in Table 15·4 for dry and wet oxidation respectively.

The presence of water vapour considerably increases the rate of oxidation, by a factor of about ten times at 100°C. This is mainly due to an increase in the parabolic rate constant in the presence of water vapour, as indicated by the much lower activation energy, E_2, for wet oxidation.

The structure of the oxide formed on the silicon single-crystal surface is not itself crystalline, but rather is an amorphous, network structure. This is a

Table 15·4

Gas	C_1 (μm/h)	E_1 (eV)	C_2 (μm^2/h)	E_2 (eV)
Dry O$_2$	$6\cdot23 \times 10^6$	2·0	$7\cdot72 \times 10^2$	1·23
H$_2$O+O$_2$	$1\cdot63 \times 10^8$	2·05	$3\cdot86 \times 10^2$	0·78

more open structure than crystalline quartz, only 43% of the volume of oxide being occupied by atoms.

15·14 Metallization

Once the components have been formed on a chip, by the methods described in this chapter, they must be connected to each other to form a circuit and to the outside world. This is done by depositing thin films of metal to form the interconnection pattern on the chip and by bonding appropriate points in the pattern to very thin wires, to form the external connections.

15·14·1 Vacuum evaporation

The standard method for depositing metal on a chip is vacuum evaporation. In this technique the metal to be deposited is melted, usually by putting small lumps of it in an electrically heated 'boat' made of tungsten or molybdenum, in a good vacuum of 10^{-6} torr (1 torr is a pressure of 1 mm of mercury). Referring to Fig. 15·17, once the metal in the boat has formed a liquid pool, the shutter is opened and deposition starts. The rate at which atoms of metal

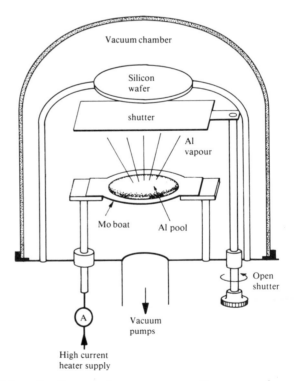

Fig. 15·17 Schematic diagram of vacuum evaporation process for metallization

leave a liquid surface in vacuum was first investigated, using mercury, by Hertz in 1882 whose work was later extended by Knudsen. They found that the evaporation rate was proportional to the difference between the equilibrium vapour pressure of the liquid material and the vacuum pressure above it. This theory finally leads to an expression for the mass evaporation rate per unit area, Γ, which is

$$\Gamma = 5 \cdot 834 \times 10^{-2} (M/T)^{1/2} P \qquad \text{(g/cm}^2\text{s)} \qquad (15 \cdot 16)$$

where M is the molecular weight of the evaporating species (in g), T is the temperature of the liquid (in K), P is the equilibrium vapour pressure (in torr) of the liquid metal at the temperature T. It is assumed that P is so much greater than the vacuum pressure that the latter can be neglected. Vapour pressures for most elements are published for practical ranges of temperature (see, for instance, *Handbook of Chemistry and Physics*, published annually by the Chemical Rubber Publishing Company in the United States). Taking aluminium as an example, its vapour pressure is 10^{-4} torr at 972°C (1,245K) and its atomic weight is 26·98, giving an evaporation rate of $8 \cdot 6 \times 10^{-7}$ g/cm² s. If we wished to deposit a square centimetre of film with a thickness of 0·1 μm, since the density of aluminium is 2·7 g/cm³, it would take 31 s.

15·14·2 Sputter deposition

It is sometimes required to deposit a thin film of metal, such as platinum, that has such a high melting point that it cannot easily be melted in the vacuum chamber. This can be done by the process known as *sputtering*. If a gas, such as argon at reduced pressure, is subjected to a strong electric field it will ionize to form a *plasma*. The process by which the plasma is formed is that a relatively small number of free electrons, which are bound to exist in any gas due to the ionizing effect of stray cosmic ray particles and photons of high energy, are accelerated by the electric field. They acquire kinetic energy such that, when they collide with a gas atom they cause it to ionize, producing an electron and a positive ion. The new electron is, in turn, accelerated and causes ionization of another gas atom by colliding with it. The process continues to build up until the whole body of gas is ionized. The positive ions in the plasma will be attracted to the negative electrode (cathode) and the electrons to the positive electrode (anode) and give rise to a current in the external circuit. When an ion reaches the cathode it will collect an electron, supplied round the external circuit from the anode, and again become a neutral atom which returns to the gas and is then re-ionized. Thus the cathode surface is bombarded continuously by positive ions which, if the applied electric field is sufficiently high, arrive at the surface with considerable momentum. This is transferred to the atoms of the electrode itself by collision of the ion with the surface and, as a result, the atoms may be kicked out of the metal surface. This release of atoms from the metal, which is akin to splashing water out of a pool by throwing stones into it, is called sputtering. A substrate placed facing the cathode, called the 'target' electrode, will be coated with a film of the metal sputtered off the target surface. Since this process does not involve melting, thin films of refractory metals like platinum or tungsten can be produced by making the target of the

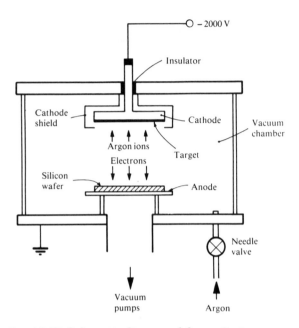

Fig. 15·18 Schematic diagram of the sputtering process

appropriate metal. A schematic diagram of a sputtering system is shown in Fig. 15·18. The anode is the bottom electrode which carries the silicon wafer and is electrically earthed. The top electrode carries the target material, which is to be sputtered, and is held at a high negative potential. A cathode shield, which is held at earth potential, surrounds the target and confines the ions to the surface to be sputtered. Compared with evaporation, sputtering is a relatively low-rate deposition system, typically requiring 10 minutes to deposit $0·1$ μm of platinum.

15·15 Metal-to-semiconductor contacts

It is necessary, in semiconductor devices, that the metal contact to the silicon exhibits low resistance, linear (non-rectifying) current–voltage characteristics over a wide range of current. Such contacts are referred to as 'ohmic'.

The basic characteristics of metal–semiconductor contacts can best be understood from a simple band diagram. It was mentioned in Chapter 14 that in a p–n junction, the Fermi energy must be constant throughout the material; the same will apply to a metal–semiconductor junction. In order to apply this principle, however, we must be able to 'define' the Fermi energy level with respect to some fixed energy reference for both the metal and the semiconductor. This can be done by defining the *work function* for each, which is the energy which must be supplied to an electron at the Fermi level in order for it to escape from the solid altogether, i.e., for it to be emitted from the solid surface. In the definition it is assumed that the electron appears just outside the solid surface with zero kinetic energy; of course to detect the presence of the electron we would have to have the solid in vacuum. Now an electron emitted from the semiconductor would be indistinguishable from one emitted from the metal, so we take their energy just outside their respective surfaces to be the same and use this as our energy reference level. We can now interpret the steps shown in Fig. 15·19. In Fig. 15·19(a) the metal and semiconductor are not in contact, the reference energy (called the vacuum energy level) is shown, the work function energies are ϕ_m for the metal and ϕ_s for the semiconductor and the diagram is drawn for ϕ_m less than ϕ_s. When the metal and semiconductor are brought into contact, since there are empty electron allowed energy levels in the conduction band of the semiconductor in contact with filled levels, containing electrons, below the Fermi level E_{FM} in the metal, electrons will flow from the metal into the semiconductor. The effect of extra electrons in the semiconductor conduction band will be to move the semiconductor Fermi level, E_{FS}, nearer to the edge of the band and so the Fermi levels of metal and semiconductor are moved towards each other. Figure 15·19(b) shows the equilibrium state in which there is an accumulation of electrons at the semiconductor surface and the Fermi level is the same throughout the two materials. There is no barrier to communication between the allowed electron levels in the metal and the electrons in the conduction band of the semiconductor and the contact is ohmic.

A similar sequence of argument shows that ohmic contact to a p-type semiconductor would require the metal work function to be greater than that of the semiconductor.

Whilst this theory predicts the general behaviour of metal–semiconductor contacts, the actual contact characteristics of a particular metal are difficult

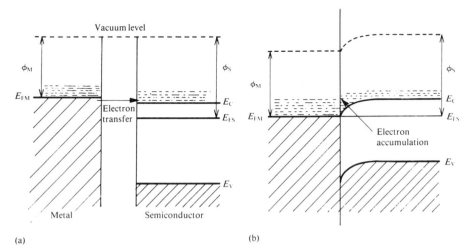

Fig. 15·19 Energy level diagrams for a metal-semiconductor contact: (a) metal and semiconductor before contact; (b) equilibrium state after contact

to predict. The reasons for this are that the work function of a material depends on the condition of its surface and is very sensitive to contamination by other materials. Furthermore, if any inter-diffusion between the metal and semiconductor occurs, their effective work functions may change drastically. Finally, the work function of a metal in thin-film form is not generally the same as its bulk value. Taking all these points together, it is impossible to predict just how good a contact a particular metal will make; generally the only approach is to try it experimentally.

15·15·1   Contact materials

Aluminium has been the most widely used contact material in integrated circuits. It is deposited by vacuum evaporation and the desired inter-connection pattern is defined by photolithographic techniques. To ensure low resistance, ohmic contact, the circuits are heat treated, typically at 550°C for 10 min. During heat-treatment the aluminium reduces some of the native SiO_2, which occurs naturally on the silicon surface, and Si–Al contact is made. At the heat-treatment temperature, dissolution of the silicon in the aluminium occurs, up to about 1·5 atomic percent at 550°C. Unfortunately the dissolution is anisotropic, the (111) face of Si dissolving more slowly than the (110) and (100) faces. The result of this is that shallow, faceted pits tend to appear under the aluminium contact during annealing and, also, the dissolution may spread under the oxide adjacent to the contact area. When the heat treatment is over, and the structure returns to room temperature, some silicon precipitates out of the aluminium epitaxially on to the silicon surface and the precipitate is heavily doped with aluminium. Since aluminium is a Group III material, the precipitates are very p-type. The consequences of these processes are undesirable: a dissolution pit can short out a shallow junction and precipitation may produce spots of high electrical field between the p-type precipitate and a differently doped silicon surface. Nevertheless, for many purposes, aluminium contacts can be satisfactory. However, the present generation of integrated circuits uses a more complex metallization technology, in which successive layers of different metals are used.

Two examples are given in Fig. 15·20, both of which are based on the formation of metal silicides, which form good ohmic contact to silicon but do not form pits or precipitates. A thin layer of platinum or palladium (0·05 μm thick) is sputtered on to the silicon and sintered at 600–700°C. This forms the silicides PtSi or Pd_2Si and the surplus metal is etched away. Unfortunately, aluminium reacts with the silicides and, at 400°C, may penetrate to the silicon. To prevent this a 'diffusion barrier' of chromium, which is partially oxidized, is next deposited and finally the aluminium 3% Cu deposited on top. In Fig. 15·20(a) is shown a contact to the n epilayer in a bipolar system whilst in Fig. 15·20(b) is shown the structure of a p-channel MOSFET. The inclusion of 3% Cu in the aluminium is to prevent a phenomenon known as *electromigration*. Since the aluminium inter-connections are both thin and narrow, quite small currents through them give rise to an enormous current density (amps per unit area). This means that a very high flux of electrons

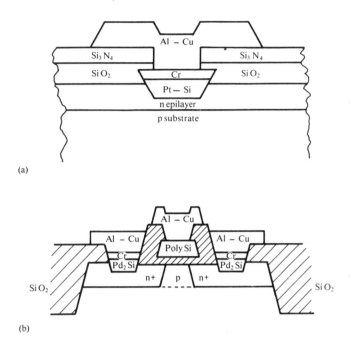

Fig. 15·20 Cross-sections of integrated circuits: (a) ohmic contact to an epilayer in a bipolar system; (b) p-channel MOSFET

flows through the metal, giving rise to an electron 'wind' which can literally 'blow' the metal atoms along in the direction of electron flow. As a result the conductor gradually thickens at the positive end and thins at the negative end until it eventually breaks. The copper atoms act as an obstacle that reduces the electromigration effect.

15·15·2   External connections

The layout of an integrated circuit invariably provides 'pads' of metal round the edge of the chip, to which external connections are made. In the 'classical' technology, these pads are squares of metal to which thin wires are bonded by 'ultrasonic welding'. In this technique a wedge-shaped tool presses the wire on to the pad and is vibrated up and down at an ultrasonic frequency, literally welding the metals together. Aluminium wire is used to connect to aluminium pads but aluminium wire cannot be soldered, and so a preferred route is to coat the pad with gold and use gold wire. Unfortunately gold does not adhere well to aluminium and so an intermediate 'glue' layer of chromium must be deposited first on the aluminium. To achieve thick pads, the gold is generally deposited by electroplating. When gold is used, thermocompression bonding becomes possible, in which the tool which presses the wire on to the pad is

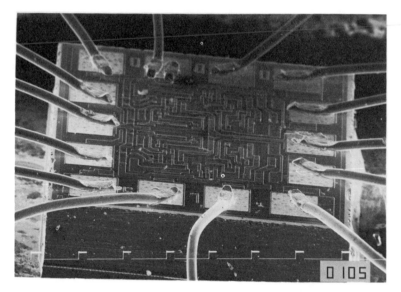

Fig. 15·21

heated. The combination of heat and pressure produces a thermocompression bond which is equivalent to a weld. The process is less violent than ultrasonic bonding and is less liable to failure by breakage of the wire.

In Fig. 15·21 is shown a photograph of a chip wire-bonded to a T05 'header'.

Problems

15·1 In a crystal puller with a molten silicon charge 99·9% of the heat input is lost by conduction and radiation. Calculate the heat input required to draw a 3-inch diameter crystal at a rate of 10^{-3} mm/s. The latent heat of crystallization of silicon is 49 kJ/mole, its density is $2·33 \times 10^3$ kg/m^3 and its atomic weight is 28.

15·2 An impurity diffusing into silicon has diffusion coefficient values of $9·46 \times 10^{-16}$ cm^2/s at 900°C and $1·07 \times 10^{-12}$/s at 1,200°C. Using Eqn (15·4), determine the activation energy E_D and the prefactor D_0.

15·3 A doped layer 0·01 μm thick on the surface of a p-type epitaxial layer is used as a source for a drive-in diffusion. If the p-type doping is 10^{23} boron atoms/m^3 and the drive-in source contains 10^{25} phosphorus atoms/m^3, determine the time taken to produce a p–n junction at 0·3 μm below the surface, given that the diffusion coefficient is $1·15 \times 10^{-13}$ cm^2/s. What will be the surface concentration of phosphorus atoms after this time? [Note: Eqn 15·19(b) will yield the answer, but an iterative calculation using a series of trial values of t is required.]

15·4 Using the values given in Table 15·4 determine the linear and parabolic rate constants for silicon at 1,100°C for both dry and wet oxidation.

15·5 Using the results of Problem 15·4 plot graphs of the growth of an oxide layer at 1,100°C for dry and wet oxidation. Assume an initial layer of oxide of 50 Å thickness and plot thickness against time up to a maximum thickness of 1 μm. The change over from linear to parabolic growth occurs at a thickness of 0·05 μm.

Self-Assessment questions

1 Electronic-grade silicon requires impurity levels below one part per million.

A) true B) false.

2 On freezing, the volume of silicon

A) increases B) decreases.

3 In crystal pulling the diameter of the crystal depends upon

A) temperature of melt B) temperature of solid
C) rate of pulling D) volume change of silicon upon freezing.

4 The dislocation content of a crystal may be reduced by increasing the rate of pulling

A) true B) false.

5 The concentration of dopant atoms during pulling of a doped crystal is greater in

A) the solid B) the liquid.

6 A single crystal slice is etched after polishing primarily to

A) remove surface oxide
B) remove mechanically damaged material
C) make the slice thinner.

7 Epitaxial growth is used to

A) improve crystal perfection
B) increase the doping level
C) change the type of dopant.

8 Producing a transistor by integrated circuit processing depends on the fact that, at a constant temperature

A) $n + p$ = constant B) np = constant C) $n = p$ = constant.

9 In integrated circuit processing the silicon slice is oxidized to

A) prevent the silicon evaporating
B) act as a barrier to impurity diffusion
C) prevent loss of dopant from the crystal.

10 The driving force for impurity diffusion is

A) concentration gradient B) temperature gradient
C) chemical potential.

11 The diffusion profile for a dopant in silicon is directly proportional to

   A) the surface concentration of impurity atoms
   B) the square root of diffusion coefficient
   C) the reciprocal of time.

12 For a finite source of impurity atoms the diffusion profile is

   A) an error function (erfc)   B) Gaussian.

13 In forming an *n–p–n* transistor it is best to use, as the first *n*-type diffusion,

   A) arsenic   B) phosphoros.

14 The rate at which oxide is formed on a silicon surface depends on the thickness of oxide already there

   A) true   B) false.

15 For a thin oxide layer the growth rate is

   A) constant   B) proportional to $(1/t)^{1/2}$.

16 The rate of evaporation of a molten metal in vacuum is proportional to its molecular weight.

   A) true   B) false.

17 Sputter deposition is a physical rather than a chemical process.

   A) true   B) false.

18 A single metal layer on a semiconductor will always make a satisfactory ohmic contact.

   A) true   B) false.

19 Electromigration is the movement of metal ions due to

   A) high temperature   B) high voltage   C) high current density.

Answers

1 B	2 A	3 A, B, C	4 A
5 B	6 B	7 A	8 B
9 B	10 A	11 A	12 B
13 A	14 A	15 A	16 B
17 A	18 B	19 C	

MAGNETIC MATERIALS

16·1 Introduction

The ability of certain materials–notably iron, nickel, cobalt, and some of their alloys and compounds–to acquire a large, permanent magnetic moment is of central importance in electrical engineering. The many applications of magnetic properties range from permanent magnets for magnetron oscillators or loudspeakers to soft iron and related materials for transformers and cover the use of almost every aspect of magnetic behaviour.

There is an extremely wide range of differing types of magnetic materials and it is important to know firstly, why these, and only these, materials should display strong magnetic properties and, secondly, to know what governs the differences between their behaviour. For example, why can one material carry a permanent magnetic moment while another has high permeability?

These two topics will be treated separately in this chapter, beginning with the origin of the magnetic moment.

16·2 Magnetic moment of a body

We know from experiments that if a magnetized solid is subdivided into smaller and smaller parts it is impossible to find a piece which carries a single magnetic charge, or monopole. Magnetic materials act always as dipoles, and for these materials it is possible to define a *magnetic dipole moment*.

In this respect the material behaves like an assembly of blocks (which we shall later identify as atoms) each carrying a circulating current, i (Fig. 16·1) and having a magnetic dipole moment, ia, where a is the cross-sectional area of the block, that is, the area enclosed by the current loop (see Problem 16·1). Thus if the solid is broken down into separate blocks, each block has a dipole moment.

Using this model we can calculate the magnetic moment of the whole body in terms of the dipole moment of each block. Now the net current on the surface of each interior block is zero since neighbouring currents are equal and cancel one another. The only place where no cancellation occurs is at the surface of the body. There is a surface current equal to the current i circulating round each layer of blocks, giving a total current in if there are n blocks along the length of the solid. This current encloses the total cross-sectional area, A, of the body so that the total magnetic moment is $inA = \Sigma ia$, the sum of the moments of all the blocks.

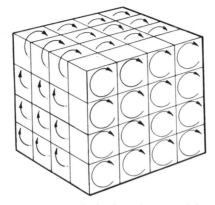

Fig. 16·1 The magnetic moment of a body is the sum of the magnetic moments of
the elementary units of which the body is composed

Clearly the magnetic moment is proportional to volume, so that the
magnetic moment per unit volume is a characteristic of the material and not
of the shape or size of the solid. The moment per unit volume is called the
intensity of magnetization (*M*).

Because we can write magnetic moment in terms of current × area its units
are ampere-metres^2, whence we obtain the units of the intensity of magnetization as amperes per metre.

Now, in the SI system of units the magnetic flux density, *B*, through a
magnetized body is equal to the intensity of magnetization multiplied by the
permeability of free space, μ_0; that is, $B = \mu_0 M$ (no applied field). If an
externally applied field, *H*, is present, we obtain

$$B = \mu_0(H + M) \tag{16·1}$$

So *H* and *M* share the same units while *B* is measured in webers per square
metre or teslas. However, μ_0 is usually quoted in henrys per metre so that we
have the dimensional relationship:

webers/m^2 = (henrys/m) × (amperes/m)

Hence,

webers = amperes × henrys

We shall meet the relationship between *B*, *H*, and *M* once again, later in
this chapter, but in the meantime we return to the problem of the origin of the
magnetic moment.

16·3   Atomic magnetic moment

The model of Fig. 16·1 suggests that a solid be divided into the smallest blocks
possible (atoms) so that our first task is to consider the magnetic moment of

single atoms and later to investigate the magnetic properties of a lattice of atoms.

We begin by considering an isolated atom. As we saw in Chapter 3, the magnetic quantum numbers, m_l and m_s, of an electron are both related to the magnetic moment of that electron. To obtain the total magnetic moment of the atom we have to add up the magnetic moments of the electrons in it, taking into account the fact that the direction of each moment is given by the sign of m_l and m_s. Thus, closed shells of electrons having equal numbers of electrons with positive and negative values of m_l and m_s have no net magnetic moment since each electron cancels another. This is shown in Table 16·1 where the dipole moment of each electron is represented by an arrow.

This then is the first important rule to note: the atomic magnetic moment comes only from incomplete electron shells.

Now the valence electrons together form an incomplete shell, so that we would expect all but the rare gases to exhibit magnetic properties. This is true for atoms in isolation, i.e., in the gaseous state, but when we come to consider solids we find only a few elements which are capable of permanently retaining a magnetic moment at room temperature.

If all the elements are studied at very low temperatures a lot more are found to be magnetic, while if we include compounds in our investigations yet more materials are found to be strongly magnetic.

However, not just any compound is found to be magnetic, as we show in Table 16·2. In this table the columns and rows represent the groups of the Periodic Table. At each intersection of a row and a column an example is given, where one exists, of a magnetic alloy or compound which is composed of an element from each of the two groups.

The table shows that, with one or two exceptions (which can be readily explained) all the magnetic materials contain *at least one transition element*.

The distinguishing feature of a transition element is that one of the inner electron shells is incomplete. This suggests immediately that the inner incomplete shell is the one which gives rise to the atomic magnetic moment, while the valence electrons are of no importance. This characteristic of the valence electrons is just a reflection of the rule which we learnt in Chapter 5 about bonding: atoms bond together in such a way as to form closed shells of electrons. As we saw above, a closed shell carries no magnetic moment.

The exceptions in Table 16·2 may now be explained if we note that the bonding is unusual in these cases: it results in an unfilled inner shell. [For example in copper (II) oxide (CuO), the copper atom must lose two electrons

Table 16·1

Directions of electron moments in a filled 2p shell

Electron	1	2	3	4	5	6
m_l	-1	-1	0	0	$+1$	$+1$
Moment due to m_l	↓	↓	—	—	↑	↑
m_s	$+\frac{1}{2}$	$-\frac{1}{2}$	$+\frac{1}{2}$	$-\frac{1}{2}$	$+\frac{1}{2}$	$-\frac{1}{2}$
Moment due to m_s	↑	↓	↑	↓	↑	↓

Table 16·2
Magnetic materials

Footnotes:

- *Antiferromagnetic
- †Compound of two Group IIIB elements
- ?Magnetic properties uncertain
- Dash denotes a disordered alloy

The table is arranged so that each entry is a representative magnetic compound; the column gives the group of the primary element and the row gives the group of the partner element (the diagonal, bottom row, giving the elements themselves). Reconstructed matrix:

Partner \ Primary group	IA	IIA	IIIA	IVA	VA	VIA	VIIA	VIII	VIII	VIII	IB	IIB	IIIB	IVB	VB	VIB	VIIB
—	UH_3																
IIIA			Gd–Sc														
IVA	TiH_2*																
VIA					CrV*												
VIIA	Mn–H		YMn_5			$MnCr$*											
VIII (Fe/Ru)	Fe–H	$FeBe_2$	PrRu	Fe_2Zr	VFe	Fe_3Cr	FeMn										
VIII (Co/Rh/Ir)	Co–H	Co–Be	$HoIr_2$	Co_4Zr	Nb–Co	$CrIr_3$	$MnRh$*	FeRh									
VIII (Ni/Pd/Pt)	Ni–H	Ni_3Mg	$GdPd_2$	Ni–Zr	Ni–V	$CrPt_3$	$MnPt_3$	$FePd_3$	CoPt								
IB (Cu/Ag/Au)			$AgDy$*		Au_4V		$MnAu_4$	FeAu	Co–Au	NiCu							
IIB (Zn/Cd/Hg)			GdCd	$ZrZn_2$			$MnHg$*	Fe_2Cd	Co–Zn	$Ni–Hg_3$							
IIIB (B/Al)			$CeAl_2$			$CrAl_2$*	MnB	FeB	Co_2B	Ni_3Al							
IVB (C/Si/Ge)			$NdGe_2$			$CrGe_2$	Mn_5Ge_2	FeC	Co–Si	Ni–Si							
VB (N/P/As/Sb/Bi)			TbN			CrAs	MnBi	Fe_2P	$Co_5–As_4$	Ni–Bi	Cu_2Sb*						
VIB (O/S/Se/Te)	KO_2*		EuS	Ti_2O_3	VSe*	CrTe	Mn_3O_4	Fe_3O_4	CoS_2	$Ni_2–Te$	CuO*						
VIIB (F/Cl/Br/I)			$GdCl_3$	$TiCl_3$*	VCl_3*	CrI_3	$MnBr_2$	$FeCl_2$*	CoF_2	NiF_2	AgF_2?						
ELEMENTS	—	—	Gd	—	—	Cr*	Mn*	Fe	Co	Ni	—	—	—	—	$GaAl_2$†	—	—

Group headers as printed across the top (GROUP): IA, IIA, IIIA, IVA, VA, VIA, VIIA, (TRANSITION ELEMENTS) VIII, then IB, IIB, IIIB, IVB, VB, VIB, VIIB.

to form an ionic bond with divalent oxygen and this leaves it with an unfilled $3d$ shell.]

16·4 Size of the atomic magnetic moment

Having established that the magnetic moment of an atom in the solid state is due only to an incomplete inner shell we now calculate what its magnitude should be. In Chapter 3 we noticed that the magnetic moment was related to angular momentum and for each electron this takes two forms: the orbital and the spin angular momenta. For convenience we shall treat these separately. But first we have to discuss the values of m_l and m_s for the electrons in the unfilled shell.

When we considered the filling of energy levels in multi-electron atoms in Chapter 4 no indication was given of the sequence of filling the energy levels in a subshell. The rule governing this was first deduced from the emission spectra of these elements, and is known as Hund's rule after its discoverer. Hund deduced from his spectroscopic measurements that levels of different m_l filled first with electrons having the same m_s value, that is, $m_s = +\frac{1}{2}$. Thus, the first level to fill is that with $m_l = -l$, $m_s = +\frac{1}{2}$; the second level to fill is that with $m_l = -(l-1)$, $m_s = +\frac{1}{2}$, and so on. When, however, the shell is just half full then according to Pauli's principle no more electrons with $m_s = +\frac{1}{2}$ are allowed so that the second half-shell fills in the sequence $m_l = -l$, $m_s = -\frac{1}{2}$ then $m_l = -(l-1)$, $m_s = -\frac{1}{2}$, and so on.

We may summarize this by saying that the electrons arrange themselves among the levels to give the maximum possible total spin angular momentum consistent with Pauli's principle. This is the usual form of Hund's rule and we may deduce from it that there must be an interaction between electrons when they have different wave functions, which tends to make their spin axes parallel. Moreover it seems that two electrons with different wave functions and with spins parallel have lower energy than two of the same wave functions with antiparallel spins. This cannot be explained without going deeply into the subject of quantum mechanics but we may note that this spin–spin interaction is a direct consequence of Pauli's principle combined with the coulomb repulsion of two like charges.

We can now return to the problem of calculating the atomic magnetic moment. Let us begin with the case of a single orbiting electron. The element scandium is an example of this situation, for there is but one electron in the $3d$ shell. According to the rules given above, the magnetic quantum numbers of this solitary electron are $m_l = -2$, and $m_s = \frac{1}{2}$. In Chapter 3 it is stated that the magnetic moment is proportional to the angular momentum and we must now derive the exact relationship between them.

We have seen that the magnetic moment, μ, of a current, i, flowing in a loop of radius r is given by

$$\mu = \pi r^2 i \tag{16·2}$$

In place of i we must put the equivalent current carried by the electron in its orbit. In the present case we wish to calculate the current flow around the circumference of the circular path of radius, $\bar{r}$, the mean radius of the wave function. The current carried around by the electron can be derived in the manner used earlier to obtain Eqn (13·10). Let us assume that all the electron's charge, $-e$, is moving around at a radius, $\bar{r}$, so that the amount of charge per unit length of the circumference is just $-e/2\pi\bar{r}$. The amount of charge passing a point on the circumference in unit time is then $-ev/2\pi\bar{r}$, where v is the velocity at which the charge is being transported. This velocity is just p/m, that is, the momentum divided by the mass. Thus we can write down the equivalent current, i, as follows

$$i = \frac{-e}{2\pi\bar{r}} \cdot \frac{p}{m} \tag{16·3}$$

The magnetic moment is obtained by combining Eqn (16·2) and Eqn (16·3):

$$\mu = -\pi\bar{r}^2 \cdot \frac{ep}{2\pi\bar{r}m} = -\frac{e}{2m} \cdot p\bar{r} \tag{16·4}$$

Now the product $p\bar{r}$ is equal to the angular momentum of the electron. Since the total orbital angular momentum is precessing about a fixed axis as described in Chapter 3, only its component along that axis contributes to the atomic magnetic moment. The perpendicular component is continuously rotating (see Fig. 3·10) and averages to zero.

Thus the angular momentum to be inserted in Eqn (16·4) to obtain the orbital magnetic moment μ_{orb} is equal to $m_l h/2\pi$, and the resultant expression for the magnetic moment is

$$\mu_{orb} = -m_l \frac{eh}{4\pi m} \tag{16·5}$$

Since m_l is an integer, we can regard the quantity $eh/4\pi m$ as a natural unit of magnetic moment. It is called the Bohr magneton, symbol β, and the atom of scandium thus has an orbital magnetic moment of two Bohr magnetons, since its sole $3d$ electron has $m_l = -2$.

The spin magnetic moment is not obtained simply by replacing m_l by m_s in Eqn (16·5) but the relationship is similar

$$\mu_{spin} = -m_s \frac{eh}{2\pi m} \tag{16·6}$$

so that a single electron with $m_s = \frac{1}{2}$ has a spin magnetic moment equal to one Bohr magneton.

16·5   3d transition elements

The second element in the first transition series is titanium. It has two $3d$ electrons and according to Hund's rule their magnetic quantum numbers are respectively $m_l = -2$, $m_s = \frac{1}{2}$ and $m_l = -1$, $m_2 = \frac{1}{2}$.

The total orbital magnetic moment of titanium is the sum of those of the two electrons, that is,

$$\mu_{orb} = \beta + 2\beta = 3\beta \qquad \text{(titanium)}$$

The spin magnetic moment is obtained in the same way but in this case the contributions of the two electrons are both one Bohr magneton:

$$\mu_{spin} = \beta + \beta = 2\beta \qquad \text{(titanium)}$$

In the same way the orbital and spin magnetic moments of vanadium, which has three $3d$ electrons, are both found to be 3β.

So far nothing has been said about the way in which the orbital and spin magnetic moments combine to give the total moment for the atom. This combination is just a little too complicated for consideration here and for our present purpose is unnecessary. For we are confining our attention to the $3d$ series of transition elements, which are different from the $4f$ series because in the former the orbital magnetic moment disappears in the solid state. This is because each atom in the crystal experiences an electric field due to the charge on all the neighbouring ions. This electric field alters the shape of the wavefunctions, and makes the total orbital angular momentum $m_l h/2\pi$ precess in a much more complicated way, with the result that, on the average, there is no component of orbital magnetic moment along any direction. The spin magnetic moment, on the other hand, is not affected by this electric field.

This 'crystal field', as it is called, does not affect the orbital moments of the magnetic electrons in the $4f$ transition metals, starting with lanthanum and cerium, since those electrons are deeper inside the atom and are shielded by the $5d$ and other electrons. Since these shells are full each acts like a charged sphere in shielding anything inside from external electric fields. The conclusion we draw from the above discussion is that the atomic moments of the $3d$ transition elements in the solid state are obtained by adding the spin magnetic moments of the electrons in the unfilled shell.

We have only to note that, at manganese, the $3d$ shell becomes half full, five electrons being present. Because of Pauli's principle the next electron to be added must have $m_s = -\frac{1}{2}$, so that its magnetic moment is opposite to, and subtracts from, those of the first five electrons. Thus iron has a spin moment of 4β, cobalt 3β, and so on. This is shown in Table 16·3.

16·6 Alignment of atomic magnetic moments in a solid

So far we have considered two effects arising from the close proximity of

Table 16·3
Spin distributions and magnetic moments in the first long period

Element:	K	Ca	Sc	Ti	V	Cr	Mn	Fe	Co	Ni	Cu	Zn
Number of $3d$ electrons	0	0	1	2	3	5	5	6	7	8	10	10
Spin directions	—	—	↑	↑↑	↑↑↑	↑↑↑↑↑	↑↑↑↑↑	↑↑↑↑↑ ↓	↑↑↑↑↑ ↓↓	↑↑↑↑↑ ↓↓↓	↑↑↑↑↑ ↓↓↓↓↓	↑↑↑↑↑ ↓↓↓↓↓
Magnetic moment in Bohr magnetons	0	0	1	2	3	5	5	4	3	2	0	0

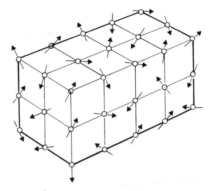

Fig. 16·2 In a paramagnet the atomic magnetic moments (represented by arrows) are all in different directions

atoms to one another in a solid. These were (1) the forming of closed shells by the valence electrons to give no resultant magnetic moment and (2) the disappearance of the magnetic moment connected with m_l in the case of the unfilled inner 3d shell.

We now come to a third effect which has to do with the directions adopted by the atomic 'magnets' relative to one another. This is governed by the interactions between the spins of electrons on neighbouring pairs of atoms which are quantum mechanical in nature. These interactions, when they are large, generally manifest themselves as a tendency for adjacent atomic magnets to remain either parallel or antiparallel to one another. This leads the classification of magnetic behaviour into the following groups:

(a) Materials in which the interactions are negligible compared to thermal energy which causes the directions of the atomic 'magnets' to be random, and fluctuating rapidly. Figure 16·2 shows a 'still' picture of this arrangement. An applied field can produce a small degree of alignment, but this disappears on removal of the field. This behaviour is termed *paramagnetic*.

(b) Materials in which the magnetic moments on alternate atoms point in opposite directions. As an example Fig. 16·3 shows the arrangement in metallic chromium.

(c) Other solids in which the atomic moments are all parallel (see Fig. 16·4 for iron). This is *ferromagnetism*.

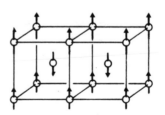

Fig. 16·3 Antiferromagnetic chromium which has the b.c.c. structure

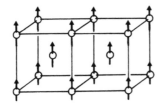

Fig. 16·4 Ferromagnetic iron, which has the b.c.c. structure

(d) In dilute solid solutions of magnetic atoms in a host material having closed shells, a randomly oriented arrangement rather like Fig. 16·2 can become 'frozen in' at very low temperatures. This is a result of competing interactions between atomic magnets whose spacing is not regular, and the material is termed a *spin glass*.

Because of their practical importance, we shall concentrate on the more highly ordered classes, (b) and (c).

16·7 Parallel atomic moments (ferromagnetism)

The most important of the ferromagnetic materials are all metallic, and, iron, cobalt, nickel, and gadolinium are the only elements which are ferromagnetic at room temperature. The origin of the force which aligns the moments in metals is beyond our scope, but it is known to arise from the wave mechanical nature of electrons and it is related to the force which aligns the spins of electrons in a partly filled atomic shell.

There is a further peculiarity about metallic bonding. It is found that in the case of iron, cobalt, and nickel the $3d$ electrons are able to wander through the lattice just like the $4s$ electrons. One effect of this is to reduce the effective magnetic moment of each electron so that instead of the expected values of the atomic magnetic moment calculated above, the measured values are as follows:

Metal	Number of Bohr magnetons per atom (calculated)	Number of Bohr magnetons per atom (measured)
Fe	4	2·22
Co	3	1·72
Ni	2	0·61

If one of these elements is bonded ionically or covalently in a compound, however, the moment is not reduced because none of the $3d$ electrons can move through the lattice.

16·8 Temperature dependence of the magnetization

The magnetic moment of a solid body is the sum of all the individual atomic

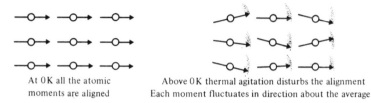

At 0 K all the atomic
moments are aligned

Above 0 K thermal agitation disturbs the alignment
Each moment fluctuates in direction about the average

Fig. 16·5 The effect of thermal agitation on the alignment of atomic moments in a ferromagnet

moments. Thus in metallic nickel, with an atomic moment of 0.61β and an atomic weight of 58.7, the magnetic moment per kilogramme is

$$\frac{6.02 \times 10^{26} \times 0.61\beta}{58.7} \text{ Am}^2/\text{kg}$$

since by Avogadro's law there are 6.02×10^{23} atoms in a mole.

The magnetic moment per unit volume (called the *intensity of magnetization, M*) is the above value multiplied by the density, which for nickel is $8.8 \times 10^3 \text{ kg/m}^3$. Thus

$$M = \frac{6.02 \times 10^{26} \times 0.61 \times 9.27 \times 10^{-24} \times 8.8 \times 10^3}{58.7}$$

$$= 5.1 \times 10^5 \text{ A/m}$$

Now the measured value of the intensity of magnetization is exactly equal to this at a temperature of 0K. But at room temperature the measurement yields a value of 3.1 A/m, while at 631K and above nickel is not ferromagnetic! This is explained by the thermal agitation in the lattice which disturbs the alignment of the atomic moments (Fig. 16·5) so reducing the component of each resolved along the direction of the net magnetization. At the *Curie temperature* (631K for nickel) and above the forces trying to align the moment are not strong enough even partially to overcome the thermal agitation and the atomic moments become randomly oriented–in fact the material behaves like a paramagnet. A graph relating M and T for nickel is shown in Fig. 16·6.

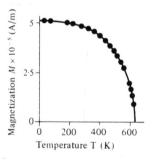

Fig. 16·6 The intensity of magnetization of nickel plotted against the absolute temperature

16·9 Antiparallel atomic moments

In chromium metal there are as many moments pointing in one direction as in the opposite direction, resulting in exact cancellation, so that the material has no magnetic moment. This behaviour is called *antiferromagnetic* and it occurs in many compounds as well as some metals. Table 16·2 lists some examples.

There is yet another class of magnetic materials which are like the anti-ferromagnets in that alternate atoms have magnetic moments pointing in opposite directions. But in this case alternate atoms carry different magnetic moments, as indicated in Fig. 16·7 by the sizes of the arrows. The moments, therefore, do not cancel one another and there is a residual intensity of magnetization. Such materials are termed *ferrimagnetic* and are of great engineering importance since, unlike all the usable ferromagnets they are insulators, being ionic-covalent compounds frequently incorporating oxygen. An example is the oxide of iron, Fe_3O_4, known to the ancients as Lodestone. Here some of the iron atoms are divalent, while the rest are trivalent and hence have a different magnetic moment since one more electron is missing from the $3d$ shell. The net magnetization in this case is due solely to the Fe^{2+} ions since the moments of the Fe^{3+} ions cancel one another.

Further ferrimagnets can be made by substituting other divalent $3d$ transition elements (e.g., Mn, Cr) for the Fe^{2+} ions. These particular magnetic oxides all have the same crystal structure and are called *ferrites*.

16·10 Magnetization and magnetic domains

In Section 16·5 it was explained that in the ferromagnetic elements the magnetic moments of all the atoms are parallel to one another at 0K. This would mean that any piece of iron, for example, would be spontaneously magnetized to saturation, that is, it would be a permanent magnet.

On the contrary, in practice a piece of iron is generally unmagnetized in the natural state. On application of a magnetic field, H, it becomes magnetized to an extent determined by its susceptibility to the magnetic field.

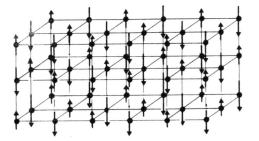

Fig. 16·7 An imaginary ferrimagnet in which alternate magnetic atoms have anti-parallel moments of different magnitudes. In real ferrimagnets the crystal structure is usually more complex

The intensity of magnetization, M, is related to H by the equation $M = \chi H$, where χ is the magnetic susceptibility. Since M and H share the same units, χ has no dimensions–it is just a number.

The magnetic flux density, B, in a magnetized solid has, as was explained earlier, two components and this is expressed in the relationship

$$B = \mu_0(H + M).$$

Using the relation $M = \chi H$ this can be transformed to read

$$B = \mu_0(H + \chi H) = \mu_0 H(1 + \chi)$$

The proportionality constant $(1 + \chi)$ is often lumped into one symbol: μ_r, the relative permeability; like χ it has no units.

Before going on to consider why M is not just a constant, independent of H, we shall study in more detail the observed behaviour of a piece of iron for example, when it is placed in a variable magnetic field.

Starting with the iron unmagnetized, we apply an increasing field and measure the flux density produced in the material. Referring to Fig. 16·8, the flux B rises along the line O–P–Q–R, reaching a saturation value, B_{sat}, beyond which any increase in the field has a negligible effect on B.

The intensity of magnetization at this point is equal to the value obtained from a graph like that in Fig. 16·6. If the experiment were performed at 0K it would also be equal to the value given in Section 16·6 corresponding in iron to $2·22\beta$ per atom.

If now the field is reduced again, B does not follow the same curve but decreases more slowly along the line R–B_r; reversing the field and increasing it in the opposite direction causes the flux to follow the line B_r–$(-H_c)$–S at which point the iron is again saturated but with the magnetization in the opposite direction. Reduction and re-reversal of the field carries the flux along the path S–$(-B_r)$–H_c–R and finally closes the loop. The closed path so described is called a *hysteresis loop* and it shows that when the field is removed after reaching the point R, the material retains a remanent flux density, B_r, and behaves as a 'permanent' magnet. In order to destroy the magnetization, it is then necessary to apply a reversed field equal to the coercive force H_c. The value of B_r and H_c may vary widely from material to material and some typical values are shown in Table 16·4.

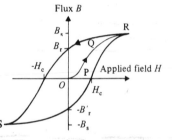

Fig. 16·8 The *B–H* curve of a ferromagnet or a ferrimagnet

Table 16·4

Properties of some typical magnetic materials at room temperature

Material	Use	B_r (T)	B_{sat} (T)	H_c (A/m)	μ_r	$(BH)_{max}$ (kJ/m³)
Fe (pure)		1·3	2·16	0·8	10^4	–
Fe (commercial)		1·3	2·16	7	150	–
Carbon steel (98Fe, 1Mn, 0·9C)	Permanent magnets	1·0	1·98	4×10^3	14	1·6
Silicon iron (Fe–3% Si)	Transformers	0·8	1·95	24	500 to 1,500	–
Mu-metal (5Cu, 2Cr, 77Ni, 16Fe)	Magnetic shielding	0·6	0·65	4	2×10^4	–
MnZn ferrite ($Mn_{0·48}Zn_{0·52}Fe_2O_4$)	High-frequency transformers	0·14	0·36	50	1,400	–
Alcomax (50·2Fe, 24Co, 13·5Ni, 8Al, 3Cu, 0·8Nb)	Permanent magnets	1·3	1·4	5×10^5	3	45
Barium ferrite ($BaFe_{12}O_{19}$)	Permanent magnets	0·40	0·41	$1·5 \times 10^5$	1	30

All the above features can be explained by the following theory.

It was suggested more than fifty years ago that the solid is indeed magnetized to saturation at all times but that it comprises a collection of small volumes called 'domains' each magnetized to saturation but each with its magnetization in a direction which is random with respect to all the others. The collection of randomly oriented domains gives the solid a zero overall magnetic moment. The boundaries between domains are called domain walls and subsequent research has led to ways of making these walls visible, confirming the theory. It has also been shown that in many cases the domains are not random but quite regular. As might be expected the domains in a single crystal are more regular than those in a polycrystalline solid. We show examples in Figs. 16·9 and 16·10.

The magnetization curve and the hysteresis loop can be interpreted in terms of domains. As an example, we take in Fig. 16·10 a single crystal of nickel, in which the process is clear. On first application of the field, in Fig. 16·10(b), the domain walls move so as to increase the volume of those domains whose direction of magnetization is nearest to the field direction. This is because, as in the examples sketched in Fig. 16·11, the lower domain has *lower magnetic energy*, and will therefore try to grow at the expense of regions with higher magnetic energy.

Referring back to Figs. 16·10(c) and (d), it can be seen that, as the unfavourably orientated domains shrink, they become more resistant to removal, and rotation of the magnetization direction within whole domains begins to play a larger part in the process. In some soft magnetic materials (i.e., those with high permeability) rotation of the magnetization is very easy, and occurs at much lower applied fields, simultaneously with domain wall

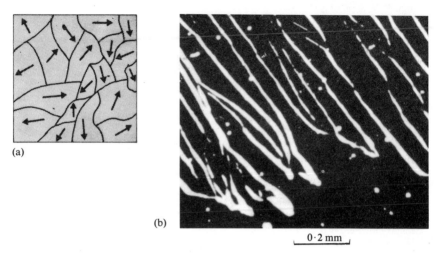

(a)

(b)

0·2 mm

Fig. 16·9 (a) Hypothetical domain arrangement in a demagnetized ferromagnet.
(b) Domains in a polycrystalline sample of nickel

motion–for example, in the amorphous Fe–Si–B alloys described in Section 16·11.

Saturation of the magnetization is only complete when all the unfavourable domains are removed–this occurs only at very high fields in the nickel sample of Fig. 16·10, fields several times the nominal saturation field as determined from the B–H loop.

This, then, explains qualitatively the shape of the initial magnetization curve (O–P–Q–R in Fig. 16·8) but does not explain why in most solids the same curve is not followed on reduction of the field. In the case of the single crystal in Fig. 16·10 the initial magnetization curve is followed quite closely in the reverse direction but most practical materials contain large numbers of imperfections which act as obstacles to rotation of the magnetization and domain wall motion. When the field is increased these obstacles are overcome by the energy supplied by the field but when the field is removed the defects prevent the domain walls from returning to their previous position. The situation is rather like that of a horse pulling a cart along a hilly road: when the horse is unhitched the cart does not roll back right to its starting point but merely to the bottom of the last hill.

Thus, on removal of the field the flux does not return to zero but to a remanent magnetization which can be quite a large fraction of the saturation value. To return the domain structure to a random distribution with zero net moment it is necessary to supply more energy, either from a reversed applied field or in the form of heat.

This model enables us to understand the difference between hard and soft magnetic materials in a qualitative way. In a soft material there are few obstacles to wall motion or rotation; the coercive force is low and the magnetization changes follow the applied field, as it increases or decreases, with little hysteresis. For example, in Table 16·4 it can be seen that careful

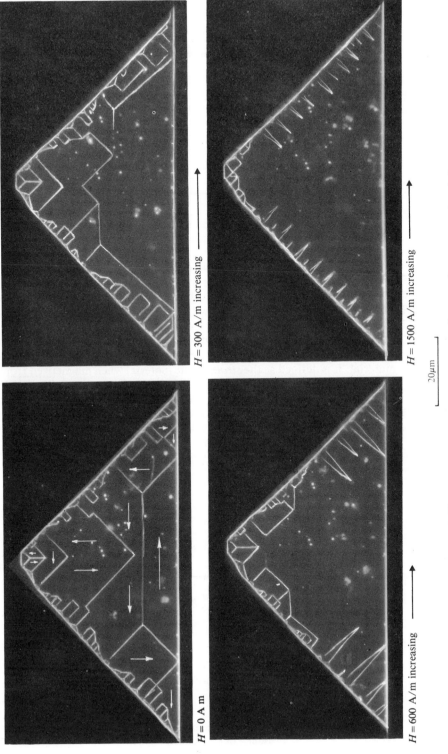

$H = 0 \text{ A m}$

$H = 300 \text{ A/m increasing}$

$H = 600 \text{ A/m increasing}$

$H = 1500 \text{ A/m increasing}$

20μm

Fig. 16·10 The magnetization process in a single crystal specimen of nickel

purification of iron both increases the relative permeability and decreases the coercive force compared to iron of a commercial grade.

Conversely, if many obstacles to domain wall motion exist the domains will lock in the fully magnetized position after magnetization has taken place, and both the coercive force and the remanent flux density will be large. Such a material would make a good permanent magnet and some permanent magnets are made from alloys in which a precipitate forms, the precipitated particles acting as obstacles.

The effectiveness of these obstacles to domain wall motion is dependent on: (i) two basic material parameters, the anisotropy constant and the saturation magnetostriction constant (both defined below); and (ii) the changes in these parameters with position, which is determined by the type and density of lattice defects.

The *anisotropy constant* is a measure of the energy required to rotate the magnetization direction uniformly away from a structurally preferred direction, such as a crystal axis, along which it normally lies in the absence of any external influence—for example, the {111} direction in the nickel sample of Fig. 16·10. An increase in the anisotropy constant generally makes the motion of domain walls harder, as the effectiveness of an obstacle to domain wall motion depends upon the ease with which the magnetization rotates when a domain wall passes through the obstacle.

The second parameter, the *saturation magnetostriction constant*, is a measure of the sensitivity of the state of magnetization to stresses internal to the material. It is defined as the *strain* of the material when magnetized from zero flux density to saturation, and it varies with the crystallographic direction of the magnetization. As the magnetization direction changes in a region through which a domain wall moves, there is an associated change in the lattice shape, i.e., there is a strain, which is proportional to the size of the magnetostriction constant. Domain wall motion thus involves less mechanical work, and hence is easier, when the magnetostriction constant is smaller.

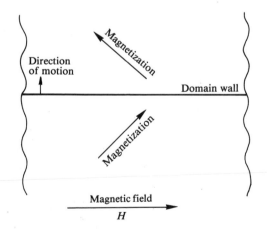

Fig. 16·11   Schematic diagram of a domain wall separating two domains, in the presence of a magnetic field H

These considerations lead to rather straightforward requirements for producing either a soft or a hard magnetic material. Hard, or permanent, magnetic properties require first and foremost a large anisotropy constant: the effects of magnetostriction are small enough to be neglected in this case. Additionally, defects must usually be specifically introduced, in order to create variations of anisotropy with position. A soft material, with high permeability, requires above all a small anisotropy constant, and secondly, a small magnetostriction constant. If both can be simultaneously reduced to near zero magnitude, the permeability obtainable ought to be highest. This is exemplified in a variety of ternary alloys of iron, one of which (Fe–Si–Al) is mentioned later.

It should be noted that these general rules cannot be allowed to overrule the requirement that a useful material must be able to sustain a high flux density, so that a high intensity of magnetization at saturation, M_s, is a prime necessity. Moreover, the intensity of magnetization in the absence of external influences should lie naturally along the direction required for the flux when the material is used. Thus a strip of material to be wound into a toroid for a transformer core would probably have its magnetization direction aligned with the axis of the strip, i.e., the circumferential direction for the toroid. In this way, the domain walls in a soft material are most readily moved by an external field lying parallel to the circumferential direction. In a similar way, a permanent magnet will have maximum flux density B_r at remanence, if the natural axis of the material is aligned with the flux direction required in use.

We shall now discuss a selection of practical materials in the light of these criteria.

16·11 Some soft magnetic materials

16·11·1 Silicon–iron

Iron containing a few percent of silicon has been the dominant soft magnetic material in volume production for over eighty years. Silicon, like carbon, is very soluble in b.c.c. Fe–up to about 15% at temperatures below 800°C. As the percentage of Si in Fe rises, the properties of the alloy change as follows.

(a) The Si assists in the removal of interstitial oxygen by forming SiO_2 during melting. This is beneficial, since interstitial oxygen increases both the coercive force H_c and the hysteresis loss in α-iron.

(b) The maximum permeability increases as the silicon content rises–only a little at first, then by a factor of about 2 between 3% and 6% Si. Simultaneously the coercive force H_c and the hysteresis loss decrease. The reasons underlying these improvements are that both the anisotropy constant and the appropriate saturation magnetostriction constant decrease, as indicated in Fig. 16·12. Whilst the lowest values are found above about 6% Si, it is not possible to reduce both simultaneously to negligible values in a binary alloy†.

†The addition of aluminium to FeSi provides the extra degree of freedom needed to select a composition with both constants zero. The resulting alloy, tradename Sendust, has a maximum relative permeability between 80,000 and 110,000, but is very brittle, and hence is used only in small quantities for specialist purposes.

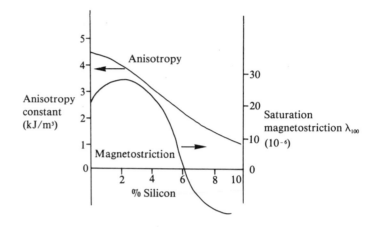

Fig. 16·12 Anisotropy constant K_1 and magnetostriction constant λ_{100}, as a function of the percentage of Si in Fe

(c)   The Curie temperature decreases, and with it the intensity of magnetization at saturation. This is disadvantageous, since it reduces the
maximum flux density at which the material may be used. Thus a
transformer or motor, for example, would need to be larger, in
proportion to the reduced flux density.

(d)   The electrical resistivity rises linearly from $1 \times 10^{-7} \Omega$ m to about $9 \times 10^{-7} \Omega$ m at 6% Si. This reduces the energy loss arising from the circula-
tion of eddy currents when the magnetic flux changes with time. It is
particularly important in machines and transformers operating with
alternating currents.

(e)   The material becomes more brittle and less easily worked, especially
above about 3% Si content. As a result, there are two distinct classes of
SiFe transformer steels, produced using different techniques, one with
silicon content up to 3% the other nearer 6%.

At Si concentrations up to 3% the alloy can be cold-rolled into thin sheets. As
ductility is greatest at low Si concentrations, the cheaper steels contain 1–2%
Si. Cold-rolling 3% Si steel induces a marked texture, or preferred orientation, of the crystallites. Two different textures can be produced, illustrated in
Figs. 16·13(a) and (b). Since the magnetization lies naturally along the ⟨100⟩
directions, the roof-top or Goss texture [Fig. 16·13(a)] has high permeability
along the rolling direction, while the four-square texture of Fig. 16·13(b)
gives similarly high μ_r both along and at right-angles to it. Unfortunately the
latter orientation is harder to produce, so that, for the bulk of applications
which require lower loss and higher permeability than is given by random
orientation, the roof-top texture is used.

   Recently, the techniques of hot- and cold-rolling which produce the four-
square orientation have been perfected to the point where the sheet is very
nearly monocrystalline. The saturation induction B_s is also raised by this

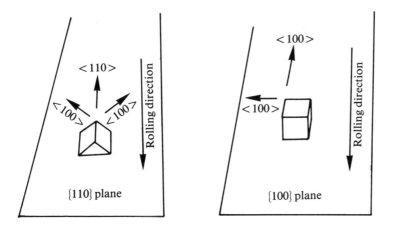

Fig. 16·13 Illustrating (a) roof-top or Goss texture, and (b) four-square texture, in a rolled SiFe sheet

more elaborate process, by about 5%, enabling a significant reduction in the amount of steel needed in a transformer.

Silicon steel with 6% Si content can be made economically by a technique of rapid cooling. A flowing stream of liquid is directed at the surface of a cooled rotating roller (Fig. 16·14) which rapidly removes the heat from the

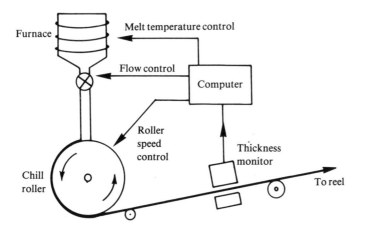

Fig. 16·14 The melt quenching process for producing fine-grained FeSi and amorphous metals

thin layer of metal in contact with it. The thin foil has a very small grain size, but is magnetically soft, because of the low anisotropy and magnetostriction. Total losses are higher than when the grain size is larger, but the cost of production is very much lower, at least in narrow widths.

16·11·2   Amorphous magnetic metals

Since anisotropy is related to the presence of crystalline order, the removal of that order should help to reduce the obstacles to domain wall motion, until only the effects of magnetostriction remain. This is indeed so; the anisotropy constant can be reduced to zero in iron which is made amorphous by rapid cooling, or quenching, at a rate of about 10^6 K/s. This inhibits the crystallization process, but since the glass transition temperature of most such metals is below room temperature, they are not useful materials. However, the introduction of sufficient boron, silicon, carbon or phosphorus into the alloy stabilizes the amorphous state. These impurities, known as metalloids, readily form metallic alloys with the host metal without segregating, and, when present in sufficient quantity, they get in the way of the normal diffusion of host atoms which allows recrystallization to occur.

Amorphous metallic alloys are readily made by the same melt-casting technique as was described above for 6% SiFe. The ribbon so produced cannot be made more than about 50 μm thick, otherwise its outer surface cools slowly enough to crystallize. Since the rate of production is fairly high (the foil speed is up to 50 m/s) the process is economic, though it is difficult to scale up for the production of wide sheet. This is because the nozzle's dimensions, and its distance from the roller, are critical in controlling the thickness. An alternative method is to quench the metallic vapour onto a cooled surface in a vacuum.

Amorphous metals of suitable composition prepared by these methods can exhibit any of the types of magnetism observed in crystalline materials: dia-, para-, ferromagnetism, etc. For example, amorphous $Pd_{80}Si_{20}$ (where the subscripts are percentages) is paramagnetic, while $Mn_{75}P_{15}C_{10}$ is anti-ferromagnetic, and Tb–Fe is a ferrimagnet which is used in the newly developed erasable optical disc computer memory.

Of most interest here, however, are the soft ferromagnetic glasses such as $Fe_{80}B_{20}$ or $Fe_{40}Ni_{40}P_{14}B_6$. Unfortunately, many of these show a slow variation in magnetic properties with time due to a stress relaxation process. This can sometimes be overcome by annealing at a temperature 50–100°C below the crystallization temperature. The latter lies at about 400 ± 50°C in most cases. After much experimentation, the stable alloy $Fe_{78}B_{13}Si_9$ has become the basis of commercial materials, and is available as ribbon in widths up to a few centimetres.

In these amorphous metals, the addition of the non-magnetic constituents fills the $3d$ electron energy band, so that the magnetic moment per atom at the temperature of 0K is reduced below the values given in Section 16·7 for the pure metals. The average moment per atom falls approximately linearly with the addition of the metalloid at a rate which depends on the electrons available in the metalloid for filling the $3d$ band. The rate of fall is about 3β for each P atom, 2β for each Si or C atom, and 1β for each B atom, when these are added to Co. In other magnetic hosts, notably Fe, the rate of fall is more variable from alloy to alloy: thus the B atom gives a fall of between $0·3\beta$ and $1·6\beta$, but most commonly 1β. Thus the moment of $Fe_{80}B_{20}$ at 0K can be calculated assuming about $1·1\beta$ per Fe atom alone: the B atoms effectively contribute nothing. At room temperature, the value is lower by an amount which depends upon the Curie temperature.

Because the nearest magnetic neighbours are on average further apart than in a crystal, the strength of the interaction which holds their magnetic moments parallel is somewhat weakened. This is reflected in a slightly lower Curie temperature than in the crystalline counterpart

Nevertheless, the ease with which the domain walls can be moved results in a much lower coercive force and consequently lower energy losses in an alternating magnetic field. Higher resistivity, in the range $0·2–5$ $\mu\Omega m$ also results from the random atomic arrangement, since electrons then collide more frequently with the lattice. This reduces eddy-current losses compared with the crystalline equivalent.

The attributes of low anisotropy and high electrical resistance give the commercial 'Metglas' $Fe_{78}B_{13}Si_9$ a four-fold advantage over crystalline FeSi in respect of total losses at a frequency of 50 Hz. However, the brittle nature of the material and the problems of handling large quantities of such thin foil restrict its use to fairly small transformers and motors. Nevertheless, experimental transformers capable of handling 100 kW of power and showing improved losses have been made, so these materials have a promising future.

16·12 Hard or 'permanent' magnetic materials

The requirement for a 'permanent' magnet is two-fold: it should support a high flux density, and it should not easily be demagnetized. These requirements can be combined in a single figure, the so-called energy product $(BH)_{max}$, which is the maximum value reached by the product of B and H at a point in the second quadrant of the $B–H$ loop, as shown in Fig. 16·15. It is a measure of the magnetic energy stored per unit volume of the material at that flux density and is inversely proportional to the volume needed for a given task.

It is possible to show theoretically that this product cannot exceed $\mu_0 M_s^2/2$, so that a high saturation intensity of magnetization is the first requisite. In this respect the rare-earth transition metals, with their large atomic moments, are very useful constituents. Since, in some applications, very large fields can be applied in the reverse direction to the flux, the coercive force H_c is another useful figure of merit.

High coercive force goes along with a high anisotropy constant, which in turn derives from a very anisotropic lattice, i.e., a large difference in the dimensions of the unit cell.

Cobalt, with its hexagonal lattice, is the one ferromagnetic element with this type of structure, while among the simple ferrites, hexagonal barium ferrite, $BaFe_{12}O_{19}$, is the most notable example. Of these, barium ferrite is currently the most used, in part owing to the very short supply, and hence high price, of cobalt (in spite of its relative abundance in the earth's crust).

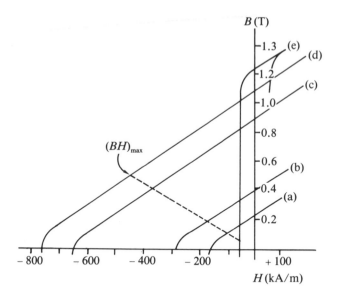

Fig. 16·15 *B–H* loops of some permanent magnet materials: (a) isotropic barium ferrite; (b) anisotropic Ba–Sr ferrite; (c) SmSo₅; (d) Nd–Fe–B; (e) Alcomax III (a cobalt-based alloy)

Barium ferrite, and its close relation strontium ferrite, have an unusually high coercive force among the cheaper magnet materials: typical partial *B–H* loops in the second quadrant of some commercial magnets are compared in Fig. 16·15. Note that all these loops have a slope of μ_0 at fields above $-H_c$.

The very elongated unit cell of barium ferrite gives it a large anisotropy constant, which holds the magnetization direction fixed until very large reverse fields are reached. On the other hand, being a ferrite, it has a lower intensity of magnetization compared to ferromagnetic alloys (see Table 16·4). This is why barium ferrite magnets are commonly shorter and fatter than their metallic counterparts. They are prepared by sintering a compressed powder of the mixed oxides, and are not monocrystalline. A high density of grain boundaries provides the obstacles needed to hinder the motion of domain walls. The coercive force H_c is found to rise as the grain size is reduced, in an almost inverse relationship. The theoretically predicted limit is not reached in practice, since it is difficult to ensure uniformity in the height of the barriers to wall motion using powder processing techniques. The magnet is normally moulded in its required final shape, since subsequent cutting is slow, owing to the mechanical hardness.

If the powder grains are randomly oriented, their local hexagonal axes and hence their preferred magnetization directions are all randomly distributed within a solid angle of 2π, centred on the average flux direction as in Fig. 16·16(a). In this case, the net total magnetic flux along the average direction must be lower than is achieved if the grains are all aligned as in Fig. 16·16(b).

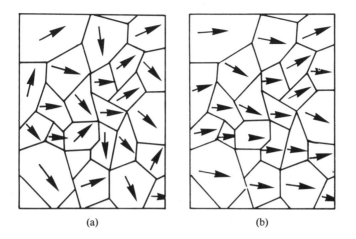

Fig. 16·16 Illustrating the magnetization directions in (a) randomly oriented grains and (b) aligned grains, of a barium ferrite magnet

This is apparent in a comparison of the B–H loops (a) and (b) in Fig. 16·15 for unaligned and aligned materials respectively. To align the grains, a large magnetic field is applied while the milled grains are compacted, in a slurry, before sintering.

In the search for materials with higher energy products, several intermetallic compounds of cobalt with rare-earth metals were discovered in the 1960s. $SmCo_5$ and Sm_2Co_{17} are the best of these, with $(BH)_{max}$ values close to 0·16 MJ/m^3 ($=$ MT A/m), near the theoretical limit of 0·19 MJ/m^3 for these materials. They, like cobalt, have hexagonal lattice symmetry, and must be carefully fabricated to form a microstructure which prevents the movement of domain walls. They are also made by sintering granules or powder in an inert atmosphere, and can only be cut subsequently with a silicon carbide or diamond saw.

The expense of separating samarium from other, more abundant, rare-earth metals of almost identical chemistry, and the scarcity of cobalt, limits the application of these materials. This has driven a search for alternatives, which has succeeded recently with the discovery of the ternary compound $Nd_2Fe_{14}B$, with a theoretical maximum energy product of 0·5 MJ/m^3. The lattice has tetragonal symmetry, though its structure has hexagonal features reminiscent of $SmCo_5$. Practical Nd–Fe–B magnets having energy products of more than 0·3 MJ/m^3 have been made, both by rapid solidification and by traditional powder metallurgy techniques, and are becoming commercially available. With their lower raw materials costs (Nd is more abundant than Sm), they will undoubtedly replace samarium–cobalt.

Problems

16·1 Calculate the couple on a circular current loop of area a carrying a current i when placed in a magnetic flux B which makes an angle θ with the axis of the loop. What is the corresponding couple on a short bar magnet whose magnetic moment (i.e., pole strength × distance between poles) is m? Hence deduce the magnetic moment of the current loop.

16·2 What are the magnetic moments, in Bohr magnetons, of the following ions? Ru^+, Nb^+, La^{2+}, Gd^{3+}, Pt^{2+}, Fe^{2+}, Fe^{3+}?

16·3 Calculate the intensity of magnetization at 0K of iron, in which the atomic magnetic moment is 2·22 Bohr magnetons and the density is $7·87 \times 10^3$ kg/m^3.

16·4 The chemical formula for Fe_3O_4 can be expressed in the form $(Fe^{2+}O)$ $(Fe_2^{3+}O_3)$, so that there are twice as many Fe^{3+} ions as Fe^{2+} ions. Calculate the intensity of magnetization of Fe_3O_4 at 0K, given that the density is $5·18 \times 10^3$ kg/m^3.

16·5 In manganese ferrite, $MnFe_2O_4$, the Mn^{2+} ions take the place of the Fe^{2+} ions in Fe_3O_4, while the dimensions of the unit cell are almost identical. Calculate the intensity of magnetization at 0K.

16·6 The magnetic moment of a paramagnetic material is directly proportional to the applied magnetic field. Aluminium is paramagnetic and has a susceptibility of 1·9 per cubic metre. Calculate the field required to give aluminium the same intensity of magnetization as nickel at room temperature (see Fig. 16·6). Is it possible to attain such a field in the laboratory?

16·7 A toroid of ferrite has the following dimensions: internal diameter 1·0 cm, outside diameter 2·0 cm, thickness 0·5 cm. Calculate the total flux through the specimen when it is magnetized circumferentially by a field of 100 A/m, if the intensity of magnetization is 12,000 A/m.

16·8 Why does Mu-metal (see Table 16·4) make a good magnetic shield? What other material in the table would also serve? Why is it not used in practice?

16·9 What effect should cold working have on the properties of Mu-metal? How would you restore its original properties after fabrication of a shield?

16·10 The coercivity of a material is raised by incorporating microscopic voids which impede the motion of domain walls. What other magnetic property is influenced by the voids?

16·11 According to the discussion in Chapter 3 of the gyroscopic nature of magnetic moments, one would expect that, when a magnetic field is applied to a magnetized body, its magnetic moment should precess around the field. Why does this not occur?

Self-Assessment questions

1 The magnetic dipole moment of a body is measured in units of

    A) A m^{-1} B) A m^2 C) Wb m^{-2} D) Wb m^2.

2 The intensity of magnetization of a body is defined as

    A) magnetic moment/volume   B) magnetic moment × area
    C) magnetic moment × volume.

3 The quantities B (magnetic flux density), H (magnetic field intensity) and M (intensity of magnetization) are related by the equation

A) $B = \mu_0(H + M)$ B) $M = \mu_0(H + B)$ C) $H = \mu_0(M + B)$.

4 The magnetic moment of a solid is the vector sum of the magnetic moments of its constituent atoms.

A) true   B) false.

5 The permanent magnetic moment of an atom is dependent on the quantum numbers

A) n and l B) m_l and m_s C) l and m_s.

6 The resultant permanent magnetic moment of two electrons each with $m_l = 0$ and with opposed spins is

A) twice the moment of one electron   B) zero.

7 A complete shell of electrons has

A) the maximum possible permanent magnetic moment
B) zero permanent magnetic moment
C) a moment dependent on the particular values of m_l and m_s.

8 The resultant orbital magnetic moment of two electrons, one with $m_l = +2$ and one with $m_l = -2$ is

A) zero   B) twice that of each electron
C) dependent on the values of m_s for the two electrons.

9 An element can form a strongly magnetic solid only if its atoms have

A) an incomplete valence shell   B) an incomplete inner shell
C) a vacant inner shell.

10 To work out the size of the magnetic moment due to the electrons in a subshell, we use Hund's rule which states that the electrons fill the states

A) in the order of decreasing m_l
B) so that the maximum possible value of spin angular momentum is achieved consistent with Pauli's principle
C) so that there is maximum cancellation of magnetic moments.

11 The magnetic moment of a single electron due to its orbital motion around the nucleus is

A) $-2m_l\beta$ B) $-\dfrac{m_l}{\beta}$ C) $-\dfrac{\beta}{2m_l}$ D) $-m_l\beta$.

12 The magnetic moment of a single electron due to its spin alone is

A) $-2m_s\beta$ B) $-\dfrac{m_s}{\beta}$ C) $-\dfrac{\beta}{2m_s}$ D) $-m_s\beta$.

13 In the $3d$ transition elements the 'crystal field' due to the charges on neighbouring ions in the solid causes

A) the spin magnetic moment to become negligible
B) the spin magnetic moment to be a maximum
C) the orbital magnetic moment to become negligible
D) the orbital magnetic moment to be a maximum.

14 In the table below, which of the elements are given the wrong value of magnetic moment in the solid state?

Element	A) Ca	B) V	C) Cr	D) Mn	E) Fe	F) Ni	G) Cu
No. of 3d electrons	1	3	5	5	6	8	10
Atomic magnetic moment in Bohr magnetons	1	3	4	6	4	2	1

15 If the atomic magnetic moments are randomly oriented in a solid its magnetic behaviour is termed

A) polycrystalline   B) paramagnetic
C) antiferromagnetic   D) polymagnetic.

16 A ferromagnetic material is one in which

A) one constituent is iron
B) the constituents are transition metal oxides
C) the atomic magnetic moments are parallel
D) the atomic magnetic moments are antiparallel and unequal.

17 The intensity of magnetization M of a ferromagnetic solid

A) increases with increasing temperature
B) decreases with increasing temperature
C) is independent of temperature.

18 The dependence of M on temperature is caused by

A) the presence of magnetic domains magnetized in varying directions
B) a permanent misalignment of the atomic magnetic moments
C) fluctuations in the directions of the atomic magnetic moments.

19 The Curie temperature is the temperature at which

A) the saturation intensity of magnetization becomes zero
B) the domains become entirely randomly magnetized
C) the atomic magnetic moment disappears.

20 A material with unequal, antiparallel atomic magnetic moments is termed

A) a ferrite   B) a ferrimagnet   C) an antiferromagnet.

21 Within each magnetic domain in a ferromagnet all the atomic magnetic moments are

A) antiparallel   B) demagnetized   C) parallel   D) random.

22 A piece of material has no net magnetic moment. It can only therefore be composed of domains magnetized in different directions.

A) true   B) false.

23 A piece of magnetic material has a net magnetic moment when no field is applied. It must therefore be ferromagnetic.

A) true   B) false.

24 If the domain walls in a magnetic material can be easily moved the material displays

A) high permeability   B) high flux density
C) permanent magnetic behaviour.

25 To increase the permeability of iron it is necessary to

A) introduce carbon   B) purify it   C) alloy it with cobalt.

26 Magnetic recording tape is most commonly made from
   A) small particles of iron   B) silicon–iron   C) ferric oxide.

27 Permanent magnets are sometimes made by the aggregation of particles which are
   A) smaller than a magnetic domain width
   B) non-magnetic particles in a magnetic bonding medium
   C) smaller than a domain wall thickness.

Each of the sentences in questions 28–35 consists of an assertion followed by a reason. Answer:

   A) If both assertion and reason are true statements and the reason is a correct explanation of the assertion
   B) If both assertion and reason are true statements but the reason is *not* a correct explanation of the assertion
   C) If the assertion is true but the reason contains a false statement
   D) If the assertion is false but the reason contains a true statement
   E) If both the assertion and reason are false statements.

28 The magnetic moment of an iron atom in Fe_3O_4 is less than that in metallic iron *because* the orbital moment is reduced by the crystal field.

29 A piece of iron may have no magnetic moment *because* it is antiferromagnetic.

30 Copper cannot be antiferromagnetic *because* it possesses a full $3d$ shell of electrons.

31 Valence electrons cannot contribute to the magnetic moment of an isolated atom *because* their spins are always opposed in pairs.

32 In the $4f$ transition series of metals, the crystal field does not cause the orbital magnetic moment to disappear in the solid *because* the $4f$ electrons are not deeply buried inside the atom.

33 The compound MnBi cannot be ferromagnetic *because* it contains no transition elements.

34 Ferrites are useful in high frequency transformers *because* they have a lower saturation flux density than does silicon–iron.

35 Silicon–iron has a high remanent flux density *because* defects prevent the domain walls from returning to their original positions when the magnetizing field is removed.

Answers

1 B	2 A	3 A	4 A
5 B	6 B	7 B	8 A
9 B	10 B	11 D	12 A
13 C	14 A, C, D, G	15 B	16 C
17 B	18 C	19 A	20 B
21 C	22 B	23 B	24 A
25 B	26 C	27 C	28 D
29 C	30 A	31 E	32 C
33 E	34 B	35 C	

DIELECTRICS

17·1 Introduction

Dielectric materials or insulators have the unique property of being able to store electrostatic charge. Very often the material is charged up by friction as in the classic school experiment of rubbing a glass rod with dry silk. Virtually all the modern plastic materials are good dielectrics: the charging up of nylon fabric through friction with other clothing gives rise to crackling sparks and sometimes to tangible arcs and shocks when the garment is removed in a dry atmosphere. One of the authors has seen a three-inch spark from a nylon garment in the dry atmosphere of the South African high veldt. Bearing in mind that an arc struck in dry air at 5,000 feet altitude needs a field of approximately 10^7 volts per metre, such a spark corresponds to a potential of 400,000 volts! Before attempting to account for this phenomenon we must first consider the dielectric in terms of the ideas of energy bands discussed in connection with semiconductors in Chapter 15.

17·2 Energy bands in dielectrics

Dielectric materials are invariably substances in which the electrons are localized in the process of bonding the atoms together. Thus covalent or ionic bonds, or a mixture of both, or van der Waals bonding between closed-shell atoms all give rise to solids (or gases) exhibiting dielectric (insulating) properties. The energy band diagram for a crystal will be just like that of a semiconductor with a valence band and a conduction band separated by an energy gap as was mentioned in Chapter 14. The gap is so large that, at ordinary temperatures, thermal energy is insufficient to raise electrons from the valence to the conduction band, which is, therefore, empty of electrons. Consequently, there are no free charge carriers and the application of an electric field will produce no current through the material.

This description applies to a perfect dielectric: in practice there will always be a few free electrons in the conduction band. These will be knocked there by stray high-energy radiation (such as cosmic rays) and irradiation by visible or ultra-violet light and will have a relatively long 'lifetime' before returning to the valence band since, once the electrons are free, the probability of their being recaptured by an empty bond is low because so few bonds are empty. However, when the energy gap exceeds about three electron volts the number of such electrons is so small that they are unable to give a significant current. In a good insulator the current, when a field of several hundred volts per cm is applied, will be of the order of 10^{-9} A or less.

It should be remembered that the energy band concept is strictly relevant only to a single crystal of material. This is because it assumes that every atom and its bonding system is the same as every other so that an electron requires precisely the same energy to be liberated from a bond anywhere in the solid. Many dielectrics used as insulators are highly disordered so that the environment of each atom tends to be a little different from that of its neighbours, as in the glassy network structure described in Chapter 11. However, it is still possible to consider the energy band picture to apply but with the band edges smeared out somewhat. This allows for the fact that slightly different energies may be needed to energize an electron from a bond at different places in the material.

In the electrostatic charging phenomenon mentioned in the introduction charge is stored on the surface of the material where it persists because there are no free carriers to neutralize it, and the charge itself becomes bound in the surface. Various mechanisms can be postulated whereby charge may be bound in the surface of the material but none is easy to demonstrate experimentally. In a material with a repetitive although not necessarily regular structure like glass, the surface must represent an interruption of the bonding system. There will be atoms which have been unable to complete a covalent or ionic bond with a neighbour because there is no neighbour with which to do so; there will be one or two valence electrons which are relatively loosely bound to their parent atoms and these may be transferred into the material used for the rubbing by mechanical work due to friction. The surface will then be left with a positive charge due to the loss of the negatively charged electrons. It is a demonstrable fact that a glass rod acquires positive charge on being rubbed. The positive charge will remain so long as it is unable to acquire replacement negative charge to compensate it. When there is water vapour in the air the H_2O molecules will readily ionize to OH^- and H^+ and the hydroxyl (OH^-) groups become attached to the surface to replace the lost negative charge. Thus the charge only persists in a dry atmosphere. In practice, in the process of fabrication of glass the surface usually becomes covered with hydroxyl ions and the frictional work tends to remove these rather than to remove electrons, as described above. This results in a positively charged surface just the same.

The case of polymer-based materials, such as resin, Bakelite, silk, nylon, and so on, must be somewhat different. This is evidenced by the fact that, on rubbing they acquire a negative charge. The structure of such materials comprises covalently bonded, long-chain molecules held together by van der Waals forces and there are no loosely-bound electrons in the surface. However, all these molecules have side-groups which are like electrostatic dipoles. The simplest case, that of the paraffins, is illustrated in Fig. 17·1.

Since the bonding electrons are concentrated mainly midway between the carbon and hydrogen ions the positive hydrogen ion represents a positive charge spatially displaced from the negative charge in the bond, so forming a dipole. This occurs with other kinds of side-groups so that water molecules, which are dipolar, can attach themselves by forming hydrogen bonds with the side-groups. The action of the mechanical work due to friction may then be to

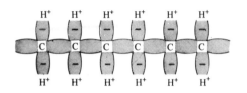

Fig. 17·1 Paraffin molecule showing polar side arms

ionize the water, that is, to 'wipe off' a hydrogen ion, leaving the negatively charged hydroxyl group attached to the end of the polymer side chain. This would then give a negative surface charge. Again, in a damp atmosphere the hydroxyl could readily capture a hydrogen ion and so neutralize the surface charge.

It is emphasized that the above mechanisms are only tentative explanations and that the actual processes involved in the frictional surface charging of dielectrics are not yet well understood. However, the general features of the explanations are well founded and lead to the conclusion that, since the sites for charge absorption are atoms or groups of atoms, it will be possible for a very large density of surface charge to be accumulated. For example, it can be shown that there are 10^{11} to 10^{12} 'dangling bonds' on one square centimetre of the surface of a covalently bonded insulator. Thus the maximum charge density could be in the region of 10^{12} electrons per square centimetre, that is, $1 \cdot 6 \times 10^{-19} \times 10^{12} = 1 \cdot 6 \times 10^{-7}$ coulombs per square centimetre. In order to estimate what this implies in terms of voltages we shall first recapitulate the basic laws of electrostatics.

17·3   Coulomb's law

Experiments on electrically charged bodies yield the following observations:

1. Like charges repel and opposite charges attract each other.
2. The force between charges is:
    (a) inversely proportional to the square of the distance between them;
    (b) dependent on the medium in which they are embedded;
    (c) acts along the line joining the charges;
    (d) is proportional to the product of the charge magnitudes.

These facts are summarized in Coulomb's law of force which gives the force, $\mathbf{F}$, on a charge q_2 due to a charge q_1 of like sign as

$$\mathbf{F} = \frac{q_1 q_2}{4 \pi \epsilon r^2} \mathbf{a} \text{ newtons} \tag{17·1}$$

where q_1 and q_2 are the charges in coulombs, r is the distance between them in metres, and ϵ is called the permittivity of the medium in which they are embedded. The vector $\mathbf{a}$ is a unit vector from q_2, pointing away from q_1 in the direction of the line joining the charges and reflects the fact that when the charges have the same sign the force is one of repulsion. The units of permittivity may be deduced from Eqn (17·1) thus:

$$\epsilon = \frac{(\text{coulombs})^2}{\text{newtons (metres)}^2}$$

The properties of a material as a dielectric enter into Coulomb's law only through this permittivity, which is also a measure of its ability to store charge. This follows from Coulomb's law of capacitance which states that the capacity of a body to store charge is defined by the equation

$$Q = CV \tag{17·2}$$

where $+Q$ is the charge on one surface of the body and $-Q$ is the charge on the opposite surface, V is the potential drop between the surfaces, and C is the capacitance of the body. The dimensions of C are given by Eqn (17·2) as

$$\text{Capacitance} = \frac{\text{coulombs}}{\text{volts}}$$

and the unit of capacitance is called the farad. The capacitance of a body is found experimentally to be an intrinsic property of the material which forms the body and of its geometry, that is, its physical shape. It is most conveniently defined in terms of a parallel plate capacitor. If we have two metal electrodes, each of area a square metres, separated by a distance, d metres and parallel to each other, filled with a material of permittivity, ϵ (Fig. 17·2), the capacitance of the system is given by

$$C = \frac{\epsilon a}{d} \tag{17·3}$$

It can be shown that the potential energy stored by a capacitor is given by

$$E = \frac{1}{2}\frac{Q^2}{C} = \tfrac{1}{2}CV^2 \quad \text{joules} \tag{17·4}$$

when it has Q coulombs of charge stored. From this we see that

$$\text{joules} = \text{newton metres} = \frac{(\text{coulombs})^2}{\text{farads}}$$

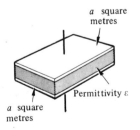

Fig. 17·2 Parallel-plate capacitor

Combining this with Eqn (17·1) shows that the dimensions of permittivity reduce to farads per metre. When the medium in the capacitor is empty space, the permittivity is written as ϵ_0 and has the value $\dfrac{1}{36\pi} \times 10^{-9}$ farads per metre.

The permittivity, ϵ, of a dielectric material may be related to the permittivity of vacuum by writing it is

$$\epsilon = \epsilon_r \epsilon_0 \qquad\qquad\qquad (17\cdot5)$$

where ϵ_r is the relative permittivity and is simply a number. It has exactly the same value as the dielectric constant, k, defined in the CGS system of units. Values of relative permittivity for a range of materials are given in Table 17·1. It should be noted that dry air has a value approximately equal to one, the relative permittivity of vacuum.

Table 17·1
Relative permittivity and loss factors of various materials

Material	Relative permittivity (ϵ) at 1 MHz	Loss factor (tan δ) at 1 MHz
Alumina	4·5–8·4	0·0002–0·01
Amber	2·65	0·015
Glass (Pyrex)	3·8–6·0	0·008–0·019
Mica	2·5–7·0	0·0001
Neoprene (rubber)	4·1	0·04
Nylon	3·4–3·5	0·03–0·04
Paraffin wax	2·1–2·5	0·003 ($\approx$ 900Hz)
Polyethylene	2·25–2·3	0·0002–0·0005
Polystyrene	2·4–2·75	0·0001–0·001
Porcelain	6·0–8·0	0·003–0·02
PTFE (Teflon)	2·0	0·0002
PVC (rigid)	3·0–3·1	0·018–0·019
PVC (flexible)	4·02	0·1
Rubber (natural)	2·0–3·5	0·003–0·008
Titanium dioxide	14–110	0·0002–0·005
Titanates (Ba, Sr, Ca, Mg, and Pb)	15–12,000	0·0001–0·02

We may now return to the question of the potential of a friction-charged surface. We may take the surface as one plate of a capacitor with the other being the nearest earthed surface, that is, the nearest surface at zero potential.

Let us take the example of the three-inch spark quoted in the introduction to the chapter. The cross-sectional area of the arc is small, say one square millimetre, and so only that tiny area of the nylon surface is discharged by the spark. We can calculate the charge stored on that area, knowing that the potential across the three-inch capacitor is 400,000 volts, and from the capacity, C, which is given by

$$C = \frac{\epsilon_0 a}{d}$$

where $a = 1$ mm^2 and $d = 0.075$ m. Thus

$$C = 10^{-9} \times 10^{-6}/36\pi \times 0.076 = 1.2 \times 10^{-16} \text{ farads}$$

The charge stored is then

$$Q = CV = 1.2 \times 10^{-16} \times 4 \times 10^5 = 4.6 \times 10^{-11} \text{ coulombs}$$

This charge would correspond to Q/e single electronic charges, that is to $(4.6 \times 10^{-11})/(1.6 \times 10^{-19}) = 2.9 \times 10^8$ electronic charges in an area of one square millimetre. This is an electron charge density of 2.9×10^{10} per square centimetre. Now, in a nylon fabric there are about 10^{15} dipolar side-chains per square centimetre of surface so that this rough calculation suggests that on average one side-chain in 10^5 acquires a single electronic charge as a result of friction.

17·4 A.C. permittivity

Permittivity, $\epsilon = \epsilon_r\epsilon_0$, has been defined as a property of the medium by means of Coulomb's law. If a parallel plate capacitor has a capacitance, C_0, in air, and the space between the plates is then filled by a medium of relative permittivity ϵ_r then, neglecting fringing effects, the capacitance becomes $C = \epsilon_r C_0$.

When an alternating electromotive force, V, is applied across an ideal capacitor an alternating current, i, will flow, its value being determined by the reactance of the capacitor, the value of which is given by $1/\omega C$, where $\omega = 2\pi f$, f being the frequency. Now a fundamental property of a capacitor is that the current that flows to and from it due to its charging and discharging successively is 90° out of phase with the alternating voltage applied across it. This means that the current and voltage are related as shown in Fig. 17·3. In real capacitors, containing a dielectric, the phase angle is found not to be exactly 90° but is less than 90° by some small angle, δ. This is due to the flow of a small component of current which is in phase with the applied voltage as the current would be in a resistor. This resistive current shown in Fig. 17·3, combines with the capacitive current to give a total current which is $(90 - \delta)°$ out of phase with the applied voltage. The resistive current is due to the dielectric actually acting as a very poor conductor and it is often referred to as leakage current. The magnitude of the leakage current is a property of the dielectric and can be represented in a circuit by showing the capacitor to have

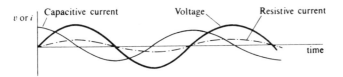

Fig. 17·3 Voltage–current phase relationships in a capacitor

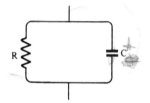

Fig. 17·4 Equivalent circuit of lossy dielectric

a resistance in parallel with it, as in Fig. 17·4. The value of the resistance is determined by the size of the leakage current.

To allow for this phenomenon mathematically we represent the relative permittivity by a complex number so that

$$\epsilon_r = \epsilon' - j\epsilon'' \qquad (17\cdot6)$$

Here ϵ' represents the part of relative permittivity that increases capacitance and ϵ'' represents the 'leakage' or 'loss'. The derivation of this expression is given, for those familiar with j-notation in a.c. theory, in Section 17·16 at the end of the chapter. It is shown there that the 'loss angle', δ, is given by

$$\tan \delta = \frac{\epsilon''}{\epsilon'} \qquad (17\cdot7)$$

and $\tan \delta$ is called the loss factor of the dielectric. This can range in value from 10^{-5} for the very best insulators to 0·1 for rather poor dielectrics. Some typical loss factors are included in Table 17·1.

17·5   Electric flux density

When a dielectric material is placed in an electric field already existing in a homogeneous medium such as air, it has the effect of changing the distribution of the field to a degree depending upon its relative permittivity, that is, the electric field intensity is a function of the medium in which it exists. We represent this situation mathematically by defining an electric flux density **D**, in the dielectric by the equation

$$\mathbf{D} = \epsilon \mathbf{E} \qquad (17\cdot8)$$

where electric field **E** is a vector, since it has magnitude and direction, and so therefore is the flux density. The units of flux density may be derived from Eqn (17·8) and are coulombs per square metre. Now, electric field at a point is defined as the force per unit charge on a positive test charge placed at that point. Thus, from Eqn (17·1),

$$\mathbf{E} = \frac{q}{4\pi\epsilon r^2} \, \mathbf{a} \qquad (17\cdot9)$$

will be the field at a point distant r from a charge q.

Combining Eqns (17·9) and (17·8) we have

$$\mathbf{D} = \frac{q}{4\pi r^2}\,\mathbf{a} \qquad\qquad (17\cdot10)$$

and we see that, since ϵ does not appear in the equation, the flux density aris-ing from a point charge, q, is independent of the medium and is a function of the charge and its position only. If we take q to be at the centre of a sphere of radius, r, the flux density will be the same at all points on the surface. The total flux, Ψ, crossing the surface will be the flux density multiplied by the area of the sphere, hence, using Eqn (17·10),

$$\Psi = 4\pi r^2\,\frac{q}{4\pi r^2} = q \text{ coulombs}$$

Thus we see that the total flux crossing the surface of a sphere with a charge at its centre is equal to the value of the charge.

This statement is generalized in Gauss' law which states that, for any closed surface containing a system of charges, the flux out of the surface is equal to the charge enclosed.

17·5·1 Electrostatic potential

In Section 17·3 we introduced potential drop V in connection with Coulomb's law of capacitance. A potential drop between two points A and B is the difference in the electrostatic potentials at the points A and B respec-tively. Formally this is defined as the work done on a unit charge in moving it from A to B when an electric field exists between the two points (Fig. 17·5). Now the force on a unit positive charge in a field E is, by definition, equal to the field, i.e., $F = E$. Work done is force times distance so that, if the charge moves a distance dl, the work done is E dl. However, the path between points A and B may not necessarily be parallel to the field E, as in the figure. If there

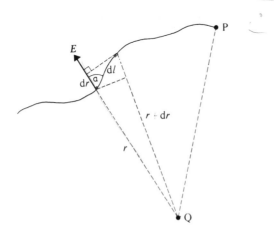

Fig. 17·5 Illustration of calculation of potential

is an angle, say α, between the direction of E and dl, then the work done is $E \cos \alpha \, \mathrm{d}l$.

For a unit positive charge, in a field, E, we define potential difference mathematically as

$$V_A - V_B = -\int_B^A E \cos \alpha \, \mathrm{d}l \tag{17·11}$$

where V_A is the potential at A, V_B is the potential at B, and dl is an element of the path of length l between A and B.

Since only difference in potential has been defined, the potential at a given point can only be stated with respect to some other point of known, or arbitrarily fixed, potential. We usually meet this difficulty by defining as zero the potential of a point an infinite distance from the region in which we are interested. Thus the potential at some point P is given by

$$V_P = -\int_\infty^P E \cos \alpha \, \mathrm{d}l \tag{17·12}$$

Suppose that the field at P is due to a charge Q and we bring a unit charge up from infinity to the point P by any path. From the diagram, it is clear that, treating dl as a straight line, d$l \cos \alpha = \mathrm{d}r$, and the field due to a point charge Q is given by Eqn (17·9). Since it is the magnitude only of the field that we require, the unit vector in Eqn (17·9) is dropped. Thus

$$V_P = -\int_\infty^{r_P} \frac{Q}{4\pi\epsilon r^2} \, \mathrm{d}r$$

i.e., $V_P = \dfrac{Q}{4\pi\epsilon r_P}$

This will be recognized as the expression used for the potential energy of an electron in Eqn (2·6), in which, instead of unit charge, we are considering an electron of charge $-e$ in the field due to a charge Q equal to $+e$. Thus the potential energy of the electron is

$$V_P = \frac{-e^2}{4\pi\epsilon r_P}$$

Since potential energy is equal to work, which is given by force times distance, the units are newton-metres, i.e., joules. Thus potential energy, as defined above, has the dimensions of energy.

The total energy of a unit positive charge at a point P in the field due to a system of charges at various distances from it will be given by

$$V_P = \sum_{i=1}^n \frac{Q_i}{4\pi\epsilon r_i}$$

where we have charges $Q_1, Q_2, Q_3 \ldots Q_n$ at distances $r_1, r_2, r_3 \ldots r_n$ from the point P.

In general, differentiating Eqn (17·12), we may write the relation between field and potential. However, because of the vector nature of field we can only write the relation for a specific direction. Thus for, say, a field E_x, in the x-direction

$$E_x = -\frac{\partial V}{\partial x}$$

Putting this equation in words: the field in a given direction is equal to the negative value of the gradient of the potential in that direction.

17·6 Polarization

When connected to a battery an air capacitor will charge until the free charges on each plate produce a potential difference equal and opposite to the battery voltage as illustrated in Fig. 17·6(a). A dielectric increases the charge storage capacity of a capacitor by neutralizing some of the free charges which would otherwise contribute to the potential difference opposing the battery voltage. More charge can, as a result, flow into the capacitor which then has an increased storage capacity given by $\epsilon_r C_0$, where C_0 is the original capacity in air. We visualize this effect as arising from alignment of electrostatic dipoles in the dielectric under the influence of the field between the capacitor plates

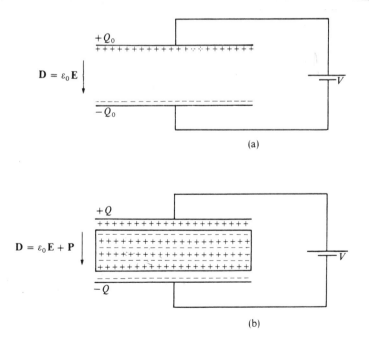

(a)

(b)

Fig. 17·6 (a) Charges on the plates of an air capacitor, (b) modification of charges on capacitor plates due to insertion of solid dielectric

as shown in Fig. 17·6(b). The dipoles form long chains with a positive charge at one end and a negative charge at the other. The positive charge will be adjacent to the negative capacitor plate and will neutralize some of the charge on it. Similarly the negative end of the dipole chain will neutralize some of the charge on the positive capacitor plate.

For an applied battery voltage V, the charge carried by the air capacitor will be $Q_0 = C_0 V$ and, on insertion of the dielectric, $Q = \epsilon_r C_0 V$ where ϵ_r is defined by Eqn (17·6).

Since we are not, in this argument, considering loss mechanisms, ϵ_r may be replaced by ϵ' (ϵ'' assumed zero) and we may write

$$\frac{Q}{\epsilon'} = C_0 V \qquad \text{or} \qquad V \propto \frac{Q}{\epsilon'}$$

The implication is that, of the total charge Q, only a fraction, Q/ϵ', contributes to neutralization of the applied voltage, the remainder $Q\,[1 - (1/\epsilon')]$ being bound charge neutralized by the polarization of the dielectric. We define the polarization of the dielectric in terms of this bound charge. The polarization, $\mathbf{P}$, is equal to the bound charge per unit area of the dielectric surface and is measured in coulombs per square metre–the same units as the flux density, $\mathbf{D}$; like $\mathbf{D}$, the polarization is a vector quantity.

Thus we may imagine the electric flux density in a dielectric to be due to two causes: the flux density, which would be set up in the space occupied by the dielectric by an applied electric field, and the polarization of the dielectric which results from the electric field. Thus we may write,

$$\mathbf{D} = \epsilon_0 \mathbf{E} + \mathbf{P} \tag{17·13}$$

In the example of the parallel plate capacitor discussed above, $\mathbf{D}$, $\mathbf{E}$, and $\mathbf{P}$ are all parallel to each other so that the magnitudes D, E, and P may be used in Eqn (17·13), that is, $D = \epsilon_0 E + P$. In Eqn (17·8) we defined D such that

$$\mathbf{D} = \epsilon' \epsilon_0 \mathbf{E} \tag{17·14}$$

and using this with Eqn (17·13) we have

$$\mathbf{P} = \mathbf{D} - \epsilon_0 \mathbf{E} = \epsilon_0 \mathbf{E} (\epsilon' - 1) \qquad \text{and} \qquad \mathbf{P} = \chi \mathbf{E} \tag{17·15}$$

where $\chi = \epsilon_0 (\epsilon' - 1)$ is called the dielectric susceptibility of the medium and is given by

$$\chi = \frac{\text{bound charge density}}{\text{free charge density}}$$

The measurement of polarization (which is the analogue of magnetization of a magnetic material) is based on Eqn (17·15) and consists in the measurement of χ or, in practice, of ϵ'. This is dealt with at the end of the chapter.

17·7 Mechanisms of polarization

Permittivity is essentially a macroscopic, or averaged out, description of the properties of a dielectric. To understand exactly what is happening in the material when an electric field is applied, we have to link the permittivity to atomic or molecular mechanisms which describe the processes of polarization of the material.

On the macroscopic scale we have defined the polarization, **P**, to represent the bound charges at the surface of the material.

Two point electric charges, of opposite polarity, $+Q$ and $-Q$, separated by a distance, d, represent a dipole of moment, μ, given by

$$\mu = Qd \qquad (17\cdot16)$$

The moment is a vector whose direction is taken to be from the negative to the positive charge.

We now have, in the polarized dielectric, P bound charges per unit area and if we take unit areas on opposite faces of a cube separated by a distance l, the moment due to unit area will be

$$\mu = Pl \qquad (17\cdot17)$$

For unit distance between the unit areas $l = 1$ and we have $\mu = P$ per unit volume. Thus the polarization, P, is identical with the electric moment per unit volume of the material. This moment may be thought of as resulting from the additive action of N elementary dipoles per unit volume, each of average moment $\bar{\mu}$, therefore

$$P = N\bar{\mu} \qquad (17\cdot18)$$

Furthermore, μ may be assumed to be proportional to a local electric field inside the dielectric which is not necessarily the same as the applied field E. If this is denoted by E_{int}, being the value of the field acting on the dipole, we define

$$\mu = \alpha E_{int} \qquad (17\cdot19)$$

where α is called the *polarizability* of the dipole, that is, the average dipole moment per unit field strength. The dimensions of α are

$$\frac{\text{coulombs} \times \text{metres}}{(\text{volts/metre})} = \epsilon \times (\text{métres})^3$$

Thus we have

$$\mathbf{P} = (\epsilon - 1)\epsilon_0 \mathbf{E} = N\alpha \mathbf{E}_{int} \qquad (17\cdot20)$$

This is referred to as the Clausius equation.

Since α is defined in terms of dipole moment, its magnitude will clearly be a measure of the extent to which electric dipoles are formed by the atoms and molecules. These may arise through a variety of mechanisms, any or all of which contribute to the value of α. Thus, for convenience, we regard the total polarizability to be the sum of individual polarizabilities each arising from one particular mechanism, i.e.,

$$\alpha = \alpha_e + \alpha_a + \alpha_d + \alpha_i$$

where the terms on the right-hand side are the individual polarizabilities which are illustrated in schematic form in Fig. 17·7. We will now discuss each of these in turn.

17·8 Optical polarizability (α_e)

An atom comprises a positively charged inner shell surrounded by electron clouds having symmetries determined by their quantum states. When a field is applied the electron clouds are displaced slightly with respect to the positive charges because the force, due to an electric field, on a negative electron is in the opposite direction to the force on a positive ion core. Thus there is, on average, a displacement between the ion core and the centre of gravity of the orbiting electrons as illustrated in Fig. 17·7(a). We now have positive and negative charges separated by a small distance which constitute an electric dipole having a dipole moment μ_e. The strength of this induced moment is proportional to the intensity of the local electric field in the region of the atom in accordance with Eqn (17·7), i.e., $\mu_e = \alpha_e \mathbf{E}_{int}$, where α_e is called the optical polarizability; it is sometimes also referred to as the electronic polarizability.

17·9 Molecular polarizability $(\alpha_a$ and $\alpha_d)$

Consider a diatomic molecule made up of atoms A and B as shown in Fig. 17·7(b). Because of the interaction between them there is a redistribution of

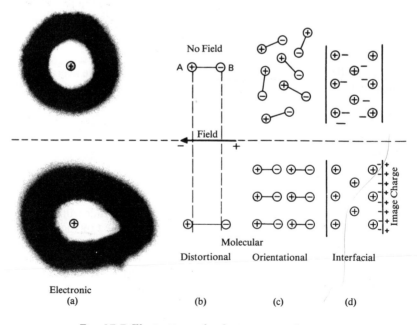

Fig. 17·7 Illustrations of polarization mechanisms

electrons between the constituent atoms which should, generally, be axially
symmetrical along AB. It may be expected that the diatomic molecule will
possess a dipole moment in the direction AB, except where the atoms A and B
are identical, when the dipole moment should vanish for reasons of
symmetry. Molecules having a large dipole moment are described as 'polar',
an example being hydrochloric acid in which there is a displacement of charge
in the bonding between H^+ and the Cl^- ions. This gives rise to a configuration in which a positive charge is separated from a negative charge by a small
distance, thus forming a true dipole.

Under the influence of an applied field the polarization of a polar substance will change by virtue of two possible mechanisms. Firstly, the field
may cause the atoms to be displaced, altering the distance between them and
hence changing the dipole moment of the molecule. This mechanism is called
atomic polarizability, α_a. Secondly, the molecule as a whole may rotate
about its axis of symmetry, so that the dipole aligns itself with the field. This
is *orientational polarizability* (α_d) [Fig. 17·7(c)].

HCl is, of course, a gas or liquid in which each molecule is complete in itself
and is bound to the others only by van der Waals forces. In an ionically
bonded solid, such as NaCl, the ions Na^+ and Cl^- are arranged in a regular
extended lattice. The unit cell of the lattice is the face-centred cubic structure
pictured in Fig. 6·9(b) and, referring to that diagram, we may take the dark-
coloured spheres as Na^+ and the light-coloured ones as Cl^-. Considering
each set separately, a moment's thought will show that the centre of gravity
of each set lies at the centre of the unit cell. The crystal is centro-symmetric
and thus the centres of gravity of the positive and of the negative sets of
charges coincide at the centre and the whole array will have no resultant
electric dipole moment. When an electric field is applied, all the positive
charges will tend to move in one direction and all the negative ones in the
opposite direction. Because of the powerful bonding forces these movements
will be small, but they result in the centres of gravity of the positive and negative sets of charges being slightly displaced from each other. As a result the
array acquires a small dipole moment and the material exhibits atomic
polarizability. Because of the fixed positions of the atoms with respect to
each other, however, the orientational mechanism described above will be
impossible.

17·10 Interfacial polarizability (α_i)

In a real insulating material there inevitably exists a number of defects such as
lattice vacancies, impurity ions, and so on, together with some free electrons.
Under the influence of an applied field some or all of these may migrate
through the material towards the electrode of appropriate polarity. If they
reach the electrode and are able to discharge there, the ions by acquiring electrons from the electrode and electrons by escaping into the electrode, the
result is a loss current through the dielectric. However, if not all of them can
discharge there results a pile-up of charge, of opposite types, in the vicinity of
each electrode. This gives the dielectric a dipole moment and constitutes a

separate mechanism of polarization, known as *interfacial polarizability*, as illustrated in Fig. 17.7(d).

Any or all of the above mechanisms may contribute to the behaviour in an applied field. As has been described earlier, they are lumped together in a phenomenological constant, α, the polarizability, defined by Eqn (17.19).

17.11 Classification of dielectrics

3 6

In general any or all of the above mechanisms of polarization may be operative in any material. The question is how can we tell which are the important ones in a given dielectric? We can do this by studying the frequency dependence of permittivity.

Imagine, first of all, a single electric dipole in an electric field. It will, given time, line itself up with the field so that its axis lies parallel with the field; if the field is reversed, the dipole will turn itself round through 180° so that it again lies parallel with the field. When the electric field is an alternating one the dipole will be continually switching its position in sympathy with the field. For an assembly of dipoles in a dielectric the same will apply, the polarization alternating in sympathy with the applied field. If the frequency of the field increases a point will be reached when, because of their inertia, the dipoles cannot keep up with the field and the alternation of the polarization will lag behind the field. This corresponds to a reduction in the apparent polarization produced by the field, which appears in measurements as an apparent reduction in the permittivity of the material. Ultimately, as the field frequency increases, the dipoles will barely have started to move before the field reverses, and they try to move the other way. At this stage the field is producing virtually no polarization of the dielectric. This process is generally called *relaxation* and the frequency beyond which the polarization no longer follows the field is called the relaxation frequency.

Considering now the various mechanisms of polarization we can predict, in a general way, what the relaxation frequency for each one might be.

Electronic polarizability relies on the position of electrons relative to the core of an atom. Since the electrons have extremely small mass they have little inertia and can follow alternations of the electric field up to very high frequencies. In fact, relaxation of electronic polarizability is not observed until the visible or ultraviolet light range of the frequency spectrum.

In atomic polarizability we require individual ions to change their relative positions. Now we know that these atoms vibrate with thermal energy and the frequencies of the vibrations correspond to those of the infrared wavelengths of light. Thus the relaxation frequencies are in the infrared range.

Orientational polarizability refers to actual reorientation of groups of ions forming dipoles. The inertia of these groups may be considerable so that relaxation frequencies may be expected to occur in the radio frequency spectrum.

In the case of interfacial polarizability a whole body of charge has to be moved through a resistive material and this can be a very slow process. The relaxation frequencies for this mechanism can be as low as fractions of a cycle per second.

In Fig. 17·8 we show a curve of the variation of relative permittivity, ϵ', with the logarithm of frequency over the entire spectrum, as it might be expected to occur for a hypothetical material showing all these effects. A slight extension of the above description is necessary in connection with this diagram.

Relaxation arises from the *inertia* of the system of charges. In many instances there will also be a *restoring force*, acting on the charges, which opposes the force due to the applied electric field. For example, in electronic polarization the attraction between the outer electrons and the inner ionic core of the atom will resist any attempt to pull the electron further away from the core. Similarly, in an ionic solid the positions of the ions are determined by a balance of attractive and repulsive forces and any attempt to displace the ion brings these forces into play to restore it to its original position. Thus these systems combine inertia and restoring force and in any mechanical system where this is true, resonance is possible. The classical example is a weight suspended on a coil spring.

In the presence of heavy damping, such as would arise if the weight were immersed in a pot of oil, the resonance effect is damped out and the small force is insufficient to produce significant movement of the weight. In this case the system exhibits relaxation since, if the frequency of alternation of the force is raised it will produce progressively less movement. The same considerations apply in the dielectrics; in the electronic and atomic mechanisms the damping is generally small and we observe resonance, rather than relaxation, at the characteristic frequencies. Resonance causes an increase in the displacements of the charges so that the alternating polarization, and therefore the permittivity, at first rises in value. As the applied field frequency rises above the resonant frequency it passes through an antiphase condition, where it opposes the vibrations of the charges, and the polarization falls to a very low value. Finally, as the frequency is raised still further, complete relaxation occurs and the polarization resumes a steady value in which the particular mechanism is no longer operative. This process gives the characteristic resonance shape to the permittivity curves shown for electronic and atomic polarization mechanisms. In most systems the damping is sufficiently large that orientational and interfacial polarization mechanisms do not show resonance behaviour.

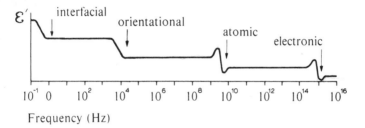

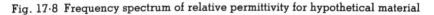

Fig. 17·8 Frequency spectrum of relative permittivity for hypothetical material

The basic electronic and atomic polarizabilities, α_e, α_a, and α_d, lead to a general classification of dielectrics. All dielectrics fall into one of three groups:

(a) non-polar materials which show variations of permittivity in the optical range of frequencies only; this includes all those dielectrics having a single type of atom, whether they be solids, liquids, or gases;

(b) polar materials having variation of permittivity in the infrared as well as the optical region; the most important members of this group are the ionic solids, such as rock-salt and alkali halide crystals in general;

(c) dipolar materials which, in addition, show orientational polarization; this embraces all materials having dipolar molecules of which one important common one is water. The chemical groups O–H and C = O are dipolar and may impart dipolar properties to any material in which they occur.

17·12 Piezoelectricity

'Piezo' is derived from the Greek word meaning 'to press', and the piezo-electric effect is the production of electricity by pressure. It occurs only in insulating materials and is manifested by the appearance of charges on the surfaces of a single crystal which is being mechanically deformed. It is easy to see the nature of the basic molecular mechanism involved. The application of stress has the effect of separating the centre of gravity of the positive charges from the centre of gravity of the negative charges, producing a dipole moment. Clearly, whether or not the effect occurs depends upon the symmetry of the distributions of the positive and negative ions. This restricts the effect so that it can occur only in those crystals not having a centre of symmetry since, for a centro-symmetric crystal, no combination of uniform stresses will produce the necessary separation of the centres of gravity of the charges. Crystals may be divided into 32 classes on the basis of their symmetry. Of these, 20 show the property of piezoelectricity because of their low symmetry. This description makes it clear that the converse piezoelectric effect must exist. When an electric field is applied to a piezoelectric crystal it will strain mechanically. There is a one-to-one correspondence between the piezoelectric effect and its converse, in that crystals for which strain produces an electric field, will strain when an electric field is applied.

In a piezoelectric crystal the polarization, **P**, is related to the mechanical stress, **T**, or, conversely, the electric stress, **E**, is related to the mechanical strain, **S**. We define a *piezoelectric coefficient d* relating polarization to stress and strain to field by

$$d = \left(\frac{\partial P}{\partial T}\right)_E = \left(\frac{\partial S}{\partial E}\right)_T \tag{17·21}$$

where the suffix E indicates that the field is held constant and the suffix T that the stress is held constant. In words, the piezoelectric coefficient is given by the rate of change of polarization with stress at constant field, or the rate of change of strain with field at constant stress. The units of d will be coulombs per newton or metres per volt.

Because the polarization, field, stress and strain are all vector quantities, the value of d will depend on the relative directions of the quantity involved as well as their magnitudes. There are two types of stress, linear and shear, along each of three axes, giving six possibilities. Thus, in general,

$$P_i = \sum_j d_{ij} T_j \; (i = 1,2,3; j = 1, \ldots, 6) \tag{17·22}$$

and

$$S_j = \sum_i d_{ij} E_i \; (i = 1,2,3; j = 1, \ldots, 6) \tag{17·23}$$

where d_{ij} are the piezoelectric coefficients. There will be 18 of them (three possible values of i times six possible values of j) but, with $d_{ij} = d_{ji}$ there are 15 independent ones. How many of these are non-zero depends on the symmetry of the crystal.

An alternative piezoelectric coefficient, g, may be defined as

$$g = \left(\frac{-\partial E}{\partial T} \right)_P = \left(\frac{\partial S}{\partial T} \right)_T \tag{17·24}$$

where the suffix P indicates constant polarization and T constant stress. The dimensions of g will be $m^2 \, C^{-1}$.

The relationship between g and d can be seen, by inspection of Eqns (17·21) and (17·24), to be

$$d = \varepsilon g \tag{17·25}$$

In practical applications the important property of a piezoelectric is its effectiveness in converting electrical to mechanical energy or vice versa. This is given by its *coupling coefficient* k which is defined by

$$k^2 = \frac{\text{Electrical energy converted to mechanical energy}}{\text{Input electrical energy}}$$

or   (17·26)

$$k^2 = \frac{\text{Mechanical energy converted to electrical energy}}{\text{Input mechanical energy}}$$

The magnitude of k is proportional to the geometric mean of the piezoelectric coefficients d and g and is a measure of the ability of the material both to detect and to generate mechanical vibrations.

Piezoelectric crystals are widely used to control the frequency of electronic oscillators. If a crystal is cut in the form of a thin plate it will have a sharp mechanical resonance frequency determined by the dimensions of the plate. In a suitable circuit, this resonance can be excited by an applied alternating voltage the frequency of which it then controls, giving a very stable electronic oscillator working a fixed frequency. Such circuits are universally used to provide the fixed frequency 'clock' pulses in computers and watches and to control the frequencies of radio transmitters.

17·13 Piezoelectric materials

Historically, quartz was the first piezoelectric material to be used in practical devices. This was because large single crystals occur in nature and they are relatively cheap. It is still universally used in quartz crystal electronic oscillators because its piezoelectric coefficients vary only slowly with temperature. Other materials have higher piezoelectric activities but are also pyroelectric (see next section) and are therefore very temperature-sensitive. Table 17·2 lists some practical piezoelectric materials together with their piezoelectric coefficients and permittivities. Two modern developments have been a key feature in the rapid growth of piezoelectric devices and applications, namely ceramic and plastic piezoelectrics.

Table 17·2

Material	Formula	Piezoelectric coefficients (C/N)	Relative permittivity
Quartz	SiO_2	$d_{11} = -2\cdot25 \times 10^{-12}$ $d_{14} = 0\cdot85 \times 10^{-12}$	$e_{11} = e_{12} = 4\cdot58$
Ammonium dihydrogen phosphate (ADP)	$NH_4H_2PO_4$	$d_{36} = 5 \times 10^{-11}(0°C)$	$e_{11} = 44\cdot3$ $e_{33} = 20\cdot7$
Potassium dihydrogen phosphate (KDP)	KH_2PO_4	Similar to ADP	
Lithium niobate	$LiNbO_3$	$d_{33} = 1\cdot6 \times 10^{-11}$ $d_{15} = 7\cdot4 \times 10^{-11}$	$e_{11} = 85\cdot2$ $e_{33} = 28\cdot7$
Lithium tantalate	$LiTaO_3$	$d_{33} = 8 \times 10^{-12}$ $d_{15} = 2\cdot6 \times 10^{-11}$	$e_{11} = 53\cdot5$ $e_{33} = 43\cdot4$
Rochelle salt	$NaKC_4H_4O_6 \cdot 4H_2O$	$d_{14} = 2\cdot33 \times 10^{-9}$ $d_{25} = -5\cdot6 \times 10^{-11}$ $d_{36} = 1\cdot16 \times 10^{-11}$	$e_{11} = 3,000$ $e_{33} = 11$
Lead zirconate titanate (PZT)	$PbTi_{0\cdot48}Zr_{0\cdot52}O_3$	$d_{31} = -9\cdot4 \times 10^{-11}$ $d_{33} = 2\cdot23 \times 10^{-10}$	$e = 730$
Polyvinylidene fluoride (PVDF)	$(CH_2-CF_2)_n$	$d_{31} = 1\cdot82 \times 10^{-11}$ $d_{32} = \sim 3 \times 10^{-12}$	$e = 160-200$

17·13·1   Ceramics

Ceramic oxide compositions, based on the ferroelectric oxides listed in Table 17·3 can generally be produced by thoroughly mixing the constituents as oxides and calcining the mixture at a temperature which gives substantial interdiffusion of cations. The calcine is then finely ground and hot-pressed into the wanted shape. After application of electrodes the material is poled (see later) to give it a uniaxial polarization. The most successful ceramic piezoelectric to date is lead zirconate titanate (PZT) having the general formula $PbTi_{1-z}Zr_zO_3$, z having a value around 0·52. This gives a saturation polarization of 47 microcoulombs per square metre and a d coefficient in the region of 10^{-10} C/N, i.e., 50 times that of quartz. To get an idea of what these figures mean, consider a rod of 1 mm^2 cross-sectional area and length 1 cm; its capacitance between electrodes on each end, using the value of relative permittivity in Table 17·2, will be $6·45 \times 10^{-13}$ F. Suppose we apply an impulse stress of 10 N (hit it with a hammer!), the charge on the electrodes would be 10^{-9} C and the voltage between its ends would be 1550 V. This sensitivity has been widely exploited in record player pick-ups, microphones, force transducers for measuring pressure and weights, and even so-called electronic cigarette or gas lighters, which use the voltage across the crystal to generate a spark. Unfortunately the piezoelectric activity has a rather large temperature dependence, which limits its use in some applications.

17·13·2 Plastics

In 1969 Kawai (*Jpn J. Appl. Phys.*, **8**, 975) reported a strong piezoelectric effect in polyvinylidene fluoride (PVDF). This is a polymer having the basic monomeric unit CH_2–CF_2 and is similar to PTFE in that it is chemically very inert. In 1972 Bell Laboratories discovered that it could also be pyroelectric (see Section 17·14).

Piezoelectricity in polymers arises because many of them have regions where the polymer chains are ordered and form localized crystalline phases surrounded by amorphous regions. At least four different crystalline phases have been identified in PVDF and in the untreated form 50–90% of the volume is crystalline, mainly in the α form. In this phase the polymer chains have a relatively high electric dipole moment but the crystal structure is such that adjacent polymer chains are aligned antiparallel and overall the material is non-polar. The β phase has all the fluorine atoms facing in one direction and is piezoelectric. This phase is typically produced by mechanically stretching a film of the material at temperatures between 50 and 100°C, producing extensions in length of 400–500%. The stretching can be performed biaxially or uniaxially and produces regions of the polar β phase which are randomly oriented throughout the material.

In order to exhibit piezoelectricity the material must now be poled. This can be done by depositing aluminium electrodes on the top and bottom of the sheet, heating to about 100°C and applying a large field of 8×10^7 V/cm. This has the effect of reorienting the dipoles of the β phase in the direction of the applied field. Finally the specimen is brought down to room temperature with the electric field still applied, locking in the dipole orientation.

The importance of PVDF in practical applications is that it can be produced in large areas, relatively cheaply. It has been used for large-area ultrasonic receivers, particularly in underwater sonar systems. One novel application has been to use it as the analogue of human skin since its response to pressure emulates the sense of touch.

17·14 Pyroelectricity and ferroelectricity

In accounting for the piezoelectric effect we found it necessary to consider the symmetry of the crystal. When under stress, the centres of gravity of the positive and negative charges separated, forming an electrostatic dipole and hence a polarization of the crystal. There are many materials in which the symmetry is such that the centres of gravity of the positive and negative charges are separated even without a stress being applied. These will exhibit *spontaneous polarization*, which means that there must be permanent electrostatic charge on the surfaces of the crystal, with one face positive and another negative, depending on the direction of the polarization vector. However, the observer would not, in general, be aware of these charges because the atmosphere normally contains sufficient free positive and negative ions to neutralize the free surface charge by being attracted to, and adsorbed on, the surface.

The spontaneous polarization will be a strong function of temperature, since the atomic dipole moments vary as the crystal expands or contracts. Heating the crystal will tend to desorb the surface neutralizing ions, as well as changing the polarization, so that a surface charge may then be detected. Thus the crystal appears to have been charged by heating. This is called the *pyroelectric effect*. It was observed in the natural crystal tourmaline as early as the 17th century. In the 18th and 19th centuries many experiments were made to characterize the phenomenon and it was these that led to the discovery of piezoelectricity by the Curies in 1880.

In 1920 it was discovered, by Valasek, that the polarization of Rochelle salt, which was known to be pyroelectric, could be reversed by the application of an electric field. By analogy with ferromagnetic materials, which have a spontaneous magnetic polarization that can be reversed by applying a magnetic field, materials of this type were described as *ferroelectric*.

Of the 20 piezoelectric crystal classes, 10 are characterized by the fact that they have a unique polar axis and possess spontaneous polarization; these are the pyroelectric crystals. A crystal is said to be ferroelectric when it has two or more orientational states of polarization, in the absence of an applied electric field, and can be switched from one to another of these states by the application of an electric field. Most pyroelectrics are also ferroelectric.

For the majority of ferroelectrics there exists a prototype crystal phase at high temperature which is sufficiently symmetrical not to exhibit spontaneous polarization in the absence of a field. As the crystal is cooled it undergoes a phase transition to a ferroelectric phase at a temperature, T_c, called the *Curie temperature*. Below T_c, in the absence of an applied field, there are at least two directions along which spontaneous polarization can develop. To minimize the depolarizing fields, different regions of the crystal polarize in each of these directions, each volume of uniform polarization being called a domain. The result is a domain structure which reduces the net macroscopic polarization nearly to zero and the crystals consequently exhibit very small, if any, pyroelectric effects until they are 'poled'. Poling of a crystal is done by applying a strong electric field across it as it cools through the Curie temperature. This has the effect of forcing the polarization into the

Table 17·3

Name	Formula	Curie temperature, T_c (°C)	Spontaneous polarization P_s (C cm^{-2}) (at T°C)	
Barium titanate	$BaTiO_3$	135	26·0	(23)
Lead titanate	$PbTiO_3$	490	>50	(23)
Potassium niobate	$KNbO_3$	435	30·0	(25)
Lithium niobate	$LiNbO_3$	1,210	71·0	(23)
Lithium tantalate	$LiTaO_3$	665	50	(25)
Potassium dihydrogen phosphate (KDP)	KH_2PO_4	−150	4·75	(−177)
Guanidinium aluminium sulphate hexahydrate (GASH)	$C(NH_2)_3Al(SO_4)_2 \cdot 6H_2O$	None	0·35	(23)
Triglycine sulphate (TGS)	$(NH_2CH_2COOH)_3H_2SO_4$	49	2·8	(20)
Rochelle salt	$NaKC_4H_4O_6 \cdot 4H_2O$ upper lower	24 −18	0·25	(5)
PLZT (ceramic)	$Pb_{0.88}La_{0.08}Zr_{0.35}Ti_{0.65}O_3$	~97	~47	(20)

direction of the field throughout the crystal, eliminating the domain structure. On removing the field at room temperature the single direction of polarization will remain in many crystals, because the energy contained in the depolarizing field is less than the energy required to switch the polarization direction to form domains. Crystals are usually poled before being used in practical applications.

There are hundreds of ferroelectric materials and a comprehensive list is given in the book by Lines and Glass†. Table 17·3 lists some of the more commonly used materials.

17·14·1 Molecular mechanisms

A large number of possible mechanisms of ferroelectricity have been discussed and, to date, no completely satisfactory quantitative theory has been forthcoming. It is clear from careful X-ray analysis, however, that the dipole moment is associated with distortion of molecular groups, while the cooperative alignment of these dipoles, to give spontaneous polarization, is a function of the interatomic bonding in the crystal.

In the case of the uniaxial crystals, and of sulphates in the multiaxial group, the dipole moment is due to the deformation of atomic groups such as SO_4, SeO_4, AsO_4, etc., in which the undeformed state is a symmetrical arrangement of the oxygen ions around the central sulphur (or other) ion. At the Curie temperature the crystal strains spontaneously, the central ion is slightly displaced, and the atomic groups acquire a dipole moment. The spontaneous strain is due to ordering of the hydrogen bonds (present in all the uniaxial ferroelectrics), corresponding to the order–disorder phase change in the crystal. The ordered bonds act to align the induced dipoles causing

†*Principles and Applications of Ferroelectrics and Related Materials*, M.E. Lines and A.E. Glass, Oxford University Press (1977).

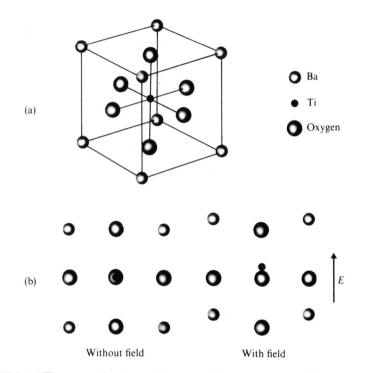

(a)

Ba

Ti

Oxygen

(b) E

Without field   With field

Fig. 17·9 (a) The perovskite crystal structure of barium titanate; (b) sectional view
of the structure through a (100) face without and with applied field

spontaneous polarization. The determining factor in the order–disorder
phase change is the (thermodynamic) free energy associated with the crystal
structure, and is outside the scope of the present text.

The case of the multiaxial ferroelectrics of perovskite and pyrochlore
structure differs somewhat. Here the basic molecular structure is an
octahedral arrangement of oxygen ions around a central ion such as Ti in
BaTiO$_3$ as shown in Fig. 17·9(a). In Fig. 17·9(b) is shown a sectional view of
the unit cell as seen through one face and it is easy to see that the centres of
gravity of the positive and negative charges coincide. Below the Curie temperature the crystal strains spontaneously with the Ti^{4+} and Ba^{2+} ions moving
upwards with respect to the O^{2-} ions, separating the centres of gravity of the
charges and producing a dipole moment. The alignment of the moments,
however, is not due to hydrogen bonds but to coupling between the oxygen
ions. This coupling is partially ionic and partially covalent, and the directional property of the covalent bond is responsible for the alignment.

17·14·2   Dielectric behaviour

The total permittivity of normal dielectrics decreases with decreasing temperature. In ferroelectric materials the permittivity and susceptibility
increase with decreasing temperature, going through a sharp maximum at the

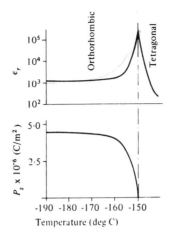

Fig. 17·10 Permittivity and polarization of KDP as a function of temperature

Curie temperature and thereafter falling further, as temperature decreases. Above the Curie temperature the dielectric susceptibility follows a Curie–Weiss law of a type which is also encountered in magnetism, that is,

$$\chi \simeq \frac{C}{T - T_c} \tag{17·27}$$

where C is called the Curie constant. For ferroelectrics containing hydrogen bonding, C is of the order of 100K. The temperature dependence of permittivity and of spontaneous polarization of potassium dihydrogen phosphate (KDP) are shown in Fig. 17·10(a) and (b) and for Rochelle salt in Fig. 17·11(a) and (b) over the region of their Curie temperatures.

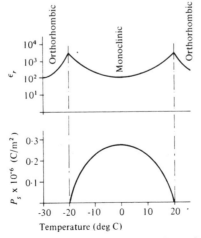

Fig. 17·11 Permittivity and polarization of Rochelle salt as a function of temperature

The high permittivity, typically of the order of 10,000 at the Curie temperature, is readily understandable. At this temperature the ions are on the point of moving into or out of the position corresponding to spontaneous polarization; consequently an applied field will be able to produce relatively large shifts, with big changes in dipole moment, corresponding to a high permittivity. The falling susceptibility below T_c corresponds to an increasing degree of spontaneous saturation of polarization in the material.

The high relative permittivities of ferroelectric materials are potentially of considerable practical importance, and much effort has been put into taking advantage of them in capacitors. Piezoelectric behaviour and chemical instability as well as the rapid variation of permittivity with temperature have, however, limited their application.

17·15 Pyroelectric devices

The electric field developed across a pyroelectric crystal can be remarkably large when it is subjected to a small change in temperature. We define a *pyroelectric coefficient p* as the change in flux density in the crystal due to a change in temperature, i.e.

$$p = \frac{\partial D}{\partial T}$$

the units of which are coulombs per square centimetre per degree. For example, a crystal with a typical pyroelectric coefficient of 10^{-8} C cm^{-2} K^{-1} and a relative permittivity of 50 develops a field of 2,000 V cm^{-1} for a 1K temperature change.

At equilibrium, the depolarization field due to the polarization discontinuity at the surfaces of the pyroelectric crystal is neutralized by free charge. This usually arises from free (mobile) electrons and holes in the crystal itself, where it can be regarded as a wide-band-gap semiconductor, or from an external circuit connected to the electrodes, or both. When the crystal temperature changes so that an excess of charge appears on one of the polar faces, a current will flow in the external circuit, the sense of the current flow depending on the direction of the polarization change. After the initial surge, the current dies away exponentially with time and eventually falls to zero until another temperature change comes along.

Pyroelectric devices can be used to detect any radiation that results in a change in temperature of the crystal, but are generally used for infrared detection. Because of its extreme sensitivity a temperature rise of less than one-thousandth of a degree can be detected. The detector must be designed so that the heat generated in the crystal by the radiation does not flow away too quickly and a typical design is shown in Fig. 17·12. Such detectors are widely used in burglar alarms, which detect the thermal radiation from a human body. By using a pyroelectric as the sensitive screen in a television camera tube, infrared images can be formed from the differing heat radiation from the scene being viewed, so that the operator can 'see' in the dark. These are used in a wide variety of satellite and military applications.

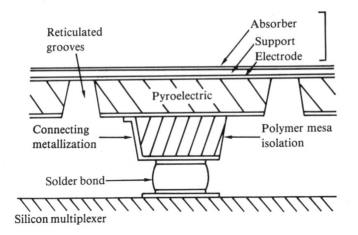

Fig. 17·12 Thermal isolation using a mesastructure

*17·16   Complex permittivity

We assume a capacitor of value C_0 in air to be filled with a medium of relative permittivity ϵ_r. When an alternating voltage V is applied across it the current, i, will be 90° out of phase with the voltage and will be given by

$$i = j\omega \epsilon_r C_0 V \qquad (17·28)$$

provided that the dielectric is a 'perfect' one. This current is just the normal capacitor charging and discharging current, which is 90° out of phase with the applied voltage and consumes no power. In general, however, an in-phase component of current will appear, corresponding to a resistive current between the capacitor plates. Such current is due entirely to the dielectric medium and is a property of it. We therefore characterize it as a component of permittivity by defining relative permittivity as

$$\epsilon_r = \epsilon' - j\epsilon'' \qquad (17·29)$$

Combining this with Eqn (17·28) we have

$$\begin{aligned} i &= j\omega(\epsilon' - j\epsilon'')\,C_0 V \qquad \text{or} \\ i &= \omega\epsilon''\,C_0 V + j\omega\epsilon' \end{aligned} \qquad (17·30)$$

Thus the current has an in-phase component $\omega\epsilon''\,C_0 V$ corresponding to the observed resistive current flow. The magnitude of ϵ'' will be defined by the magnitude of this current.

It is conventional to describe the performance of a dielectric in a capacitor in terms of its *loss angle*, δ, which is the phase angle between the total current, i, and the purely quadrature component i_c.

If the in-phase component is i_L, then

$$|i| = (|i_L|^2 + |i_c|^2)^{1/2} \qquad (17·31)$$

and

$$\tan \delta = \frac{|i_L|}{|i_c|} = \frac{\omega \epsilon'' C_0 V}{\omega \epsilon' C_0 V} = \frac{\epsilon''}{\epsilon'} \tag{17.32}$$

This is the result quoted as Eqn (17·7).

17·17 Measurement of permittivity

A wide variety of methods exists for the measurement of permittivity, the particular method adopted being determined by the nature of the specimen and the frequency range in which the measurement is to be made.

The measurement of the real part of the relative permittivity, ϵ', is generally done by measuring the change in capacitance of a capacitor, brought about by the introduction of the dielectric between its electrodes. The imaginary part, ϵ'', is found by measurement of tan δ, the loss factor resulting from the introduction of the dielectric.

The most usual type of measurement employs a bridge circuit working in the range 10^2 to 10^7 Hz. The commonest circuit is the Schering bridge illustrated in Fig. 17·13. In this, C represents the dielectric-filled capacitor and R the dielectric loss. It should be noted that, as shown in Fig. 17·4, the loss is represented by a resistance in *parallel* with the capacitor. However, this can always be simulated by a series resistor R_s which will produce the same relationship between current and voltage as a parallel resistor R_p in parallel with a capacitor C_p. The relationships between the components are then

$$R_p = \frac{1}{\omega^2 C_s^2 R_s} \quad \text{and} \quad R_s = \frac{1}{\omega^2 C_p^2 R_p}$$

In the bridge circuit, C_2 and C_3 are standard calibrated capacitors and the resistors R_1 and R_2 are usually made equal. In use, C_2 and C_3 are adjusted until there is zero current through the detector D.

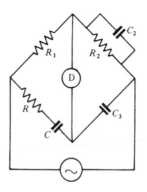

Fig. 17·13 The Schering bridge

Initially, without the dielectric present, $C = C_0$ and $R = 0$; if, in addition, $R_1 = R_2$, the balance equations are

$$C_0 = C_3 \quad \text{and} \quad C_2 = 0$$

With the dielectric inserted, the equations become

$$C = C_3 \quad \text{and} \quad R = \frac{R_1 C_2}{C_3} \tag{17.33}$$

Since $\tan \delta = \omega CR$ for a series R–C circuit, we have $\tan \delta = \omega C_2 R_1$ and, since R_1 is fixed, the dial of C_2 may be calibrated directly in values of $\tan \delta$. Using $\epsilon''/\epsilon' = \tan \delta$ and $C/C_0 = \epsilon'$, the value of the complex permittivity for the dielectric is found.

Problems

17·1 State Coulomb's law of force between electric charges. A positive sodium ion and a negative chlorine ion are a distance 5 Å apart in vacuum. Calculate the attractive force between them if each has a charge of magnitude equal to the charge on the electron. What will be the value of this force if the ions are embedded in water of relative permittivity 80? What is the relevance of your calculation to the solubility of rock salt in water?

17·2 A flat plate of ebonite of area 10 cm^2 is rubbed so that its surface is charged electrostatically. If the charge is equivalent to 10^{14} electrons per square centimetre, what will be the potential difference between the ebonite and a metal plate of the same area placed in air a distance 1 cm from it? At what distance would be expect a spark to jump from the ebonite to the plate if the breakdown strength of the air is 10^9 V/m?

17·3 Explain why a complex permittivity is attributed to a lossy dielectric material. Show that the flux density due to a point charge is independent of the medium in which it is embedded.

In a parallel plate capacitor the distance between the plates, each of area 1 cm^2 is 1 mm. The dielectric has a relative permittivity of 10 and it holds 10^{-9} coulombs of charge. What is the flux density in the dielectric? What is the potential between the plates?

17·4 Summarize the atomic and molecular mechanisms of polarization indicating, with reasons, in what range of frequency you would expect relaxation for each one.

17·5 Explain what is meant by the term ferroelectric when applied to a dielectric material. How would you determine whether or not a material was: (a) piezoelectric, (b) pyroelectric, (c) ferroelectric?

17·6 A coaxial transmission line comprises an inner conducting cylinder of radius a and an outer conductor of radius b. If the inner conductor carries a charge $+q$ per unit length, the outer will have a charge $-q$ per unit length induced on its inner surface. Use Gauss' law to determine the field existing in the electric medium, of permittivity ϵ, between the conductors.

17·7 Suppose the inner conductor in Problem 17·6 to be replaced by an electron beam having a radius a. The charge density in the beam is uniform throughout its volume and has a value ρ. Determine the field at a radius R, where $R < a$. If the beam current is 1 mA and the beam velocity arises from the electrons being accelerated through a potential of 1,000 V, what is the field at the surface of the electron beam if its diameter is 10^{-3} cm?

17·8 For the coaxial line of Problem 17·6 obtain an expression for the potential difference between the conductors. Hence find the capacitance per unit length of a line having an insulator of permittivity 2·56 and an outer conductor whose diameter is 10 times that of the inner.

17·9 An electrolyte containing equal concentrations of positive and negative ions in water (relative permittivity = 80) is placed in a cell between parallel electrodes. The cell is placed in an electric field perpendicular to the plane of the electrodes so that the charges separate to opposite sides. Assuming that the density of positive charge is uniform over half the width of the cell and that of the negative charge uniform over the other half, sketch a graph of the potential between the plates as a function of x, where x is a distance perpendicular to their plane and the distance between the plates is d.

Self-Assessment questions

1 In a dielectric material nearly all the electrons are localized on atoms or in bonds.
   A) true   B) false.

2 A dielectric can be regarded as a semiconductor with an energy gap which is
   A) small   B) large.

3 A dielectric surface positively charged by friction retains its charge in a dry atmosphere because there are no free electrons in the solid to neutralize it
   A) true   B) false.

4 A dielectric surface can only be charged negatively by friction if there are no free electrons in the solid.
   A) true   B) false.

5 The force between charges is independent of the medium in which they are embedded.
   A) true   B) false.

6 The permittivity of a medium has the units
   A) Farad metre   B) Farad $(\text{metre})^{-1}$ C) $(\text{Farad metre})^{-1}$
   D) is a number.

7 The capacitance of a parallel plate capacitor is inversely proportional to the voltage between the plates.
   A) true   B) false.

8 If the distance between the plates of a parallel plate capacitor is initially small and then is doubled the capacitance is
   A) doubled   B) halved   C) increased by a factor of 4.

9 The charge held by a parallel-plate capacitor is proportional to the area of its plates for a fixed applied voltage.

 A) true   B) false.

10 Permittivity is represented as a complex number to take account of losses in the dielectric.

 A) true   B) false.

11 The loss factor of a capacitor in which the dielectric has permittivity given by ϵ' − $j\epsilon''$ is equal to

 A) $\tan^{-1}(\epsilon''/\epsilon')$ B) ϵ'/ϵ'' C) ϵ''/ϵ'.

12 An equivalent circuit for a lossy capacitor comprises a resistor and capacitor in parallel. If the loss were large the resistor would be small.

 A) true   B) false.

13 List the true statements in the following question. The loss angle of a capacitor represents

 A) the phase angle, θ, between total current and total voltage
 B) $(90° - \theta)$
 C) the phase angle between the resistive component of current and the applied voltage
 D) the phase angle between the total current and its capacitive component.

14 List the true statements in the following question. The electric flux density at any point in a medium distant r from a point charge q is

 A) a vector
 B) proportional to the electric field at the same point
 C) dependent on the permittivity of the medium
 D) proportional to the distance from the charge
 E) has units of coulombs per square metre.

15 The electrostatic potential at any point P in an electric field E due to a point charge Q is the work done on a unit positive charge in bringing it to the point P from infinity. The potential is

 A) a vector quantity
 B) positive if the charge Q is negative
 C) positive if the charge Q is positive
 D) independent of the path followed
 E) given in joules per metre
 F) proportional to the permittivity of the medium
 G) equal to the gradient of the field.

16 The polarization in a dielectric is

 A) the free charge per unit volume of the dielectric
 B) the bound charge per unit volume of the dielectric
 C) the bound charge per unit area of the dielectric.

17 The dielectric susceptibility is the polarization per unit electric field.

 A) true   B) false.

18 The polarizability of a dipole is the average dipole moment per unit field strength.

 A) true   B) false.

Each of the sentences in questions 19–25 consists of an assertion followed by a reason. Answer:

A) If both assertion and reason are true statements and the reason is a correct explanation of the assertion

B) If both assertion and reason are true statements but the reason is *not* a correct explanation of the assertion

C) If the assertion is true but the reason contains a false statement

D) If the assertion is false but the reason contains a true statement

E) If both the assertion and reason contain false statements.

19 Displacement of the electron cloud round a nucleus by an electric field is called optical polarizability *because* its relaxation frequency is in the optical range.

20 Optical polarizability arises in all substances *because* the electron cloud round a nucleus experiences a force in an electric field in the same direction as does the nucleus.

21 Molecular polarizability occurs in all substances *because* diatomic molecules can possess a dipole moment.

22 HCl is a polar molecule *because* it comprises a positive charge separated from a negative charge by a small distance.

23 Rock salt cannot acquire a dipole moment in an electric field *because* the crystal is centro-symmetric.

24 Orientational polarization generally occurs in solids *because* the atoms can move freely.

25 Interfacial polarization in a solid has a very low relaxation frequency *because* it requires the movement of a whole body of charge through a resistive medium.

26 A piezoelectric material is one which generates bound electrostatic charge on its surface when it is mechanically deformed.

   A) true   B) false.

27 A pyroelectric material must also be piezoelectric.

   A) true   B) false.

28 Pyroelectricity is only found in centro-symmetric crystals.

   A) true   B) false.

29 All ferroelectric materials exhibit pyroelectricity.

   A) true   B) false.

30 Poling of a ferroelectric material creates a domain structure.

   A) true   B) false.

31 The Curie temperature for a ferroelectric is the temperature above which its spontaneous polarization disappears.

   A) true   B) false.

32 Barium titanate is a ferroelectric because its lattice strains spontaneously above the Curie temperature.

   A) true   B) false.

Answers

1 A	2 B	3 A	4 A
5 B	6 B	7 B	8 B
9 A	10 A	11 C	12 A
13 B, D	14 A, B, E	15 C, D	16 C
17 A	18 A	19 A	20 A
21 A	22 B	23 A	24 B
25 A	26 B	27 A	28 B
29 A	30 B	31 A	32 B

OPTICAL MATERIALS

18·1 Introduction

The evidence that materials selectively absorb or reflect different parts of the visible spectrum is all around us, in the natural and artificial colours of every day objects. This selectivity naturally extends beyond the range of wavelengths we can see, into both the ultraviolet (shorter wavelengths) and the infrared (longer wavelength) parts of the electromagnetic spectrum. We shall cover the whole of this broader spectral range, from about 200 nm to 16 μm in wavelength, in the course of this chapter.

The optical behaviour displayed by materials can be thought of merely as a special case of the response of a solid to the application of an electric field. This is because light is an electromagnetic wave which consists of electric and magnetic fields travelling together through space. The difference from behaviour discussed in the preceding chapter is that the electric field strength is alternating in sign and amplitude (usually sinusoidally) at a very high frequency indeed: between, say, 10^{14} and 10^{16} Hz, corresponding to wavelengths ranging from the infrared to the ultraviolet.

The electric field in an electromagnetic wave can exert forces on, and its energy can be absorbed by, either the electrons or the atomic nuclei in a material, since both carry electrostatic charges. At the far infrared end of the spectrum, where the frequency is lowest, the rate of change of the electric field is small enough that the nuclei in a crystal lattice can respond to the alternating force exerted on them.

As the frequency of the incoming wave is raised and approaches the natural frequency of vibration of the lattice, the amplitude of the induced nuclear motion increases and the energy absorbed increases correspondingly, reaching a maximum at the natural, or resonant, frequency. Beyond that frequency, the induced motion and the energy absorption decrease due to the inertia of the nuclei. An optical wave of frequency substantially above the resonant frequency is transmitted with very little absorption of the lattice if the quantum of energy hf is below the threshold for exciting transitions of the electrons between allowed energy levels.

Thus in insulators and semiconductors with a gap in the electronic energy bands which is greater than the photon energy, we can neglect the free electrons in discussing infrared spectra, and can obtain information on the crystal lattice absorption alone. This is the subject of Section 18·3. In metals (and very narrow band gap semiconductors), however, the electrons dominate even the infrared response (see Section 18·5).

The absorption and scattering of ultraviolet and visible radiation on the other hand is primarily due to the excitation of electrons into higher energy levels, although nuclear motion, as so often, can influence the effects we observe.

The study of absorbed or emitted radiation as a function of wavelength is called spectroscopy, and is a powerful tool for the materials scientist to use in characterizing solids, liquids and gases. For the electronic energy levels which can thus be studied are not only characteristic of the lattice and even the ionic species present, but also contain information about the immediate surroundings in which the atoms are situated. This last point can be understood if we recognize that (a) the number of occupied energy levels depends on the state of ionization of an atom and that (b) the solutions to Schrödinger's equation must be slightly altered if the potential distribution in and around the nucleus is modified by the presence of neighbouring ions.

18·2 Absorption spectroscopy

The degree of absorption of light by a medium can be determined from the ratio of the transmitted and incident intensities for a slab of material, if allowance is made for the energy reflected at the front and back surfaces. The intensity decreases with distance inside the absorbing medium according to an exponential law. This is simply due to the fact that in any small thickness δx, a fraction $\alpha \delta x$ of the intensity of the light is absorbed, where α is called the absorption coefficient, and has the dimensions (length)$^{-1}$.

Putting this mathematically, the fall δI in intensity I over the distance δx is given by

$$\delta I = -\alpha \delta x I$$

the minus sign being necessary because I decreases as x increases. In the limit $\delta x \rightarrow 0$, the intensity I will obey the differential equation

$$dI/dx = -\alpha I$$

The solution of this equation gives the intensity I at position x as

$$I = I_0 \exp(-\alpha x)$$

where I_0 is the intensity at the position where $x = 0$. By taking natural logarithms,

$$\alpha x = -\ln(I/I_0) \tag{18·1}$$

This result is known as Lambert's law, from which it can be seen that the absorption coefficient α is the slope of a graph of $-\ln I$ against x.

When the material being studied is in solution, or is a compound or mixture, the concentration of the absorbing species also influences the strength of absorption. In many cases, the absorption coefficient α is directly proportional to the concentration C, i.e., $\alpha = KC$, where K is a constant. Combining this with Eqn (18·1) leads to the Lambert–Beer law

$$\ln(I/I_0) = -KCx$$

It is usual practice to use base-10 logarithms in place of the natural logarithm, i.e.,

$$\log (I/I_0) = -aCx$$

where the coefficient a is called the absorptivity of the species whose concentration is C. When C is expressed in moles per litre, and x is in centimetres, this coefficient is known as the molar absorptivity, and is given the symbol ε, so that

$$\log (I/I_0) = -\varepsilon Cx$$

Thus ε is expressed in units of litre mole^{-1} cm^{-1}.

Beer's law strictly applies only at low concentrations, though deviations at high concentration are often small enough to be ignored.

If a light source with a broad spectral output (a 'white' light) is concentrated on a semi-transparent material, the absorptivity can be measured for each wavelength. The simplest method of measuring the wavelength dependence of absorptivity is to pass the transmitted light through a prism which separates the various wavelengths, and the resulting spectrum can be recorded on a photographic film, as in Fig. 3·15, or can be scanned by a suitable small detector which produces an output current or voltage proportional to intensity. Commercial spectrometers for these measurements are readily available.

The radiant energy absorbed by the material may be

(a)   re-radiated without change in wavelength, i.e., scattered into another direction of propagation;

(b)   re-radiated at a different wavelength, if the absorbing charge subsequently makes a transition to a different energy level to that from which it was first excited;

(c)   lost to kinetic energy of atomic motion (i.e., to thermal energy) if the atom(s) concerned 'collide' with their neighbours.

Which of these processes occurs depends on their relative probabilities: we often refer to *transition probabilities* between energy levels.

All the above processes, including (a), can give rise to a reduction in the intensity transmitted through a material, and hence show up in an absorption spectrum.

Before discussing some of the features to be found in absorption spectra, let us study the absorption due to induced nuclear motion alone. We shall find that it displays features which are universally found even in electronic spectra.

18·3   A model for crystal lattice absorption

Crystal lattice absorption dominates at infrared frequencies in ionic insulators and semiconductors, for example LiF, MgO, InSb, and ZnS.

The motion of ions in a crystal lattice was discussed in Chapter 7 and illustrated in Fig. 7·7. When alternate ions have equal and opposite charges, we have a simple two-dimensional model of a CsCl lattice. Forces on neighbour-ing ions due to the electric field are opposite in direction, and in the normal mode of vibration which is thus excited they oscillate in antiphase motion as illustrated in Fig. 7·7(b). The natural frequency of this mode lies in the infrared, and incident radiation at this same frequency will be most strongly absorbed because the amplitude of the motion it can create is a maximum at the natural frequency. The degree of absorption at this frequency is limited by the damping of the oscillatory motion.

To calculate how the absorption of energy is expected to vary with the fre-quency of the incoming radiation, we can use the above model in which we assume for simplicity that the negative ions (say) are fixed, while the positive ions each of mass M move under the constraints of the 'springs' attached to them. (This approximates to the actual motion if the negative ions are very much more massive than the positive ions.) These exert a restoring force proportional to the displacement x. Then each positive ion with charge q obeys the equation of damped simple harmonic motion, under the influence of an oscillatory driving force $F = qE_0 \cos \omega t$, where E_0 is the amplitude of the alternating electric field strength. The equation of motion, which simply states that the mass $\times$ acceleration equals the sum of the forces acting, is

$$M\ddot{x} = -kx - \beta\dot{x} + qE_0 \cos \omega t \tag{18·2}$$

where $-kx$ is the restoring force, and $-\beta\dot{x}$ is the damping force, which is assumed to be proportional to the velocity of the ions.

The resulting alternating displacement of the ions follows the driving force, but with a phase difference we shall call ϕ, i.e., the solution of Eqn (18·2) has the form

$$x = A \cos (\omega t + \phi) \tag{18·3}$$

By substitution of this equation into Eqn (18·2) and solving for A and ϕ we obtain the results

$$A = \frac{qE_0}{M} \left[\frac{1}{(\omega_0^2 - \omega^2)^2 + \gamma^2\omega^2} \right]^{1/2} \tag{18·4}$$

and

$$\tan \phi = -\gamma\omega/(\omega_0^2 - \omega^2) \tag{18·5}$$

where

$$\omega_0 = \sqrt{k/M} \quad \text{and} \quad \gamma = \beta/M$$

Here ω_0 is the natural frequency of oscillation, and γ depends on the strength of damping. The rate P (J/s) at which energy is absorbed from the driving force is the damping force multiplied by dx/dt, the velocity

$$P = (\beta\dot{x}) \, dx/dt = \beta \, (\dot{x})^2$$

The mean of P over many cycles thus depends on $\langle \dot{x}^2 \rangle$, the mean of the square of the velocity, thus using Eqn (18·3) the mean power $\langle P \rangle$ is given by

$$\langle P \rangle = \beta \langle \dot{x}^2 \rangle = \frac{\beta}{2} \, (\omega A)^2 \tag{18.6}$$

A plot of the mean absorbed power $\langle P \rangle$ against angular frequency ω is shown in Fig. 18·1 and it represents the shape of the absorption spectrum line of this model 'oscillator'. The similarity to the measured shape of the transmission spectrum of lithium fluoride shown in Fig. 18·2 is obvious.

However, as shown in Chapter 7, in a crystalline solid there is no single natural frequency ω_0, but a band of frequencies, the 'optical' phonon band. Each frequency in the band corresponds to a mode of oscillation of the lattice with a different periodicity, i.e., wavelength. Only one of the modes absorbs strongly–the one for which the mode wavelength coincides with the *optical* wavelength at the same frequency. In all other modes, the ionic motion cannot remain in phase with the force exerted by the optical wave as it travels

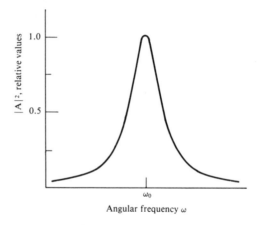

Fig. 18·1 Absorption versus frequency predicted from Eqn (18·4)

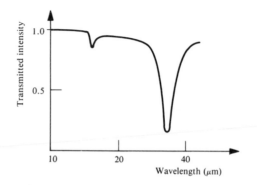

Fig. 18·2 Transmission spectrum of a thin slice of lithium fluoride. Relative transmitted intensity is plotted against wavelength

through the material. The phase angle ϕ between the two changes continuously from 0 to π and on to 2π as the waves proceed. The work done by the optical wave on the ions is the product of a force $F = F_0 \cos \omega t$ with displacement $x = x_0 \cos (\omega t + \phi)$. The work done, $W = F x_0 \cos \omega t \cos (\omega t + \phi)$ varies with ϕ and is even negative if $\frac{1}{2}\pi < \phi < \frac{3}{2}\pi$! In fact, when averaged over all phase angles 0 to 2π the net energy absorbed is just zero. For absorption to occur both the frequencies *and* the wavelengths of the optical and lattice waves must coincide. Since only one frequency f_0 is absorbed, the spectrum of light transmitted through the crystal and recorded on a photographic plate shows a narrow dark line at the wavelength $\lambda = c/f_0$ corresponding to that frequency, where c is the velocity of light. We call this a *line spectrum*, for obvious reasons. The width at half-intensity of the absorption line, shown in profile in Fig. 18·1, can be found from the condition that the amplitude A in Eqn (18·4) must be $1/\sqrt{2}$ of its peak value at half-intensity. This occurs when $(\omega_0^2 - \omega^2) = \gamma\omega$, i.e., $\Delta\omega \simeq \frac{1}{2}\gamma$ since $\gamma \ll \omega_0$, so that the *total* width of the line is $2\Delta\omega = \gamma = \beta/M$. Thus the damping coefficient β in the equation of motion alone determines the absorption line width. In practice this damping, or energy loss from the vibrational motion (a phonon) is governed by 'collisions' with other phonons and/or with defects.

In molecular solids (which includes the vast range of organic solids), vibrations of one molecule couple only weakly to the next, owing to the

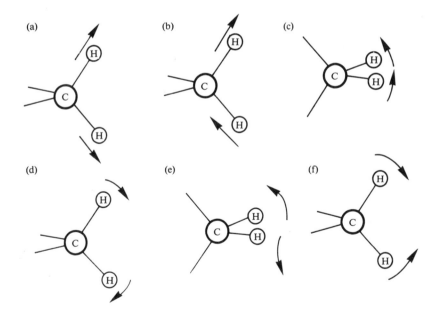

Fig. 18·3 Vibrational modes of a CH_2 group. (a) Symmetric stretching; (b) asymmetric stretching; (c) wagging; (d) rocking; (e) twisting; (f) scissor

relative weakness of intermolecular bonds. The concept of lattice vibrations or phonons must then be replaced by treating each molecule, or even each group within a molecule, as a separate vibrator.

For example, the methylene group as illustrated in Fig. 18·3 can vibrate in several characteristic ways, each with a characteristic natural frequency which causes an absorption at the corresponding optical wavelength. A list of wavelengths for some modes of this and other groups is given in Table 18·1. Note that bending vibrations of the group have lower characteristic frequencies than stretching modes of the same group, because a bond displays a lower stiffness when bent than when stretched or compressed. The natural frequency is proportional to the square root of the stiffness, i.e., the force constant, k, as seen in Eqn (18·5).

Table 18·1
Wavelengths of infrared absorption peaks of vibrational modes

Group and mode		Wavelength (μm)	Strength
—C—H	Stretch	3·03	Strong
$=C\begin{smallmatrix} /H \\ \backslash H \end{smallmatrix}$	Stretch	3·23–3·25	Medium
$\backslash\!\!\diagup\!\!CH_2$	Rocking	14	Weak
—OH	Stretch	2·7–3·1	Variable
—O—H	Bend	7·09–7·94	Strong
N—H	Stretch	2·8–3·03	Medium
N—H	Bend	6·3–6·7	Weak

The absorptivity due to a fundamental vibrational mode is proportional to the square of the rate of change of the dipole moment with displacements of the atoms. Thus symmetric motions, as in Fig. 18·3(a), which produce almost no change in the dipole moment, do not show up in infrared absorption spectra. But asymmetric motions such as in Fig. 18·3(b) are said to be infrared 'active', and produce significant absorption.

18·4 Electronic absorption in insulators

The absorption spectra of different materials, while at first sight being very dissimilar, are made up of features which broadly fall into two categories: absorption *lines* such as we have already studied, and absorption *edges*, which must wait until Section 18·6.

Frequent mention is also made in textbooks on spectroscopy of *absorption bands*, which are no more than a series of overlapping lines.

Absorption of visible and ultraviolet energy by electrons also frequently gives line spectra. That this must be so can be understood from what we have already learned about electronic energy levels in materials. If energy levels are discrete and narrow, like those of the hydrogen gas whose emission spectrum is shown in Fig. 3·15, absorption of energy occurs over a narrow band of frequencies centred on the one given by Planck's law

$$hf = E_1 - E_0 \tag{18·7}$$

where E_1 and E_0 are the energies of the final and initial levels respectively. Thus line spectra are observed in the visible and ultraviolet regions of the spectrum, as well as in the infrared. This can be seen in Fig. 18·4 which shows transmission spectra of some organic dye molecules. The wavelength of the principal absorption maximum determines the colour, and its width the 'purity' of the colour: narrow absorption lines give bright, 'pure' colours. It can be seen from Fig. 18·4 that the absorption peaks in these molecules, which have alternate single and double bonds ('conjugated' bonds), shifts towards longer wavelengths (lower frequencies) as the molecule lengthens. The positions of these peaks and those of many other conjugated molecules can be predicted remarkably well with the aid of a very simple model. The N electrons in the conjugated bonds are assumed to be free to move right along the molecule, whose length L along the zig-zag bonds is calculable from known bond lengths. The 'electron in a box' model discussed in Section 2·6 can then be employed to predict energy differences between the highest filled level E_0 and the lowest unfilled level E_1. The resulting equation for the absorption peak λ_{max} is

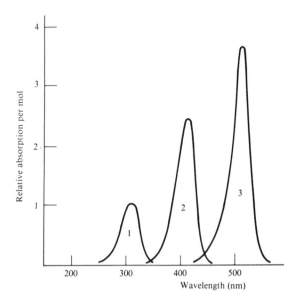

Fig. 18·4 Ultraviolet spectra of the dyes $Me_2N = CH - (CH = CH)_n - NMe_2$, where Me is a methyl group; and n is given in the figure

$$\lambda_{max} = \frac{8m\,cL^2}{h(N+1)} \qquad\qquad (18\cdot8)$$

This equation predicts the values of 170 nm, 270 nm and 350 nm for the three dyes in Fig. 18·4. While agreement with the observed values is only approximate, it is surprisingly good for such a simple model especially as λ_{max} is sensitive to the molecular surroundings, a fact which the equation above does not predict. However the model can predict the relative strengths of absorption fairly well. These and other data convince us that electrons are indeed rather free to move within organic molecules having conjugated bonds: the molecules behave rather like minute metallic conductors. This knowledge is useful for predicting other properties of organic materials. Naturally the λ_{max} of dye molecules can be predicted much more accurately by other, better, models, and nowadays the absorption peak of a dye can be routinely predicted to within 10 nm or less before it has been synthesized. It is notable that the shapes of the absorption lines in Fig. 18·4 are not dissimilar to the one predicted for lattice absorption in Fig. 18·1. A classical model for the absorption of electrons can indeed be quite useful.

In more complex organic molecules and in inorganic insulators there may be many more energy levels, producing many absorption lines which may overlap with one another. Furthermore, the width of an absorption line often increases substantially when atoms or molecules are condensed to form solids or liquids. The effect is analogous to that which broadens electron energy levels into bands in a solid, and the resulting absorption bands can be quite broad. Atomic vibrations in molecules and solids can also broaden electronic absorption lines.

The identification and separation of these lines and bands is a major task of spectroscopists, and the detailed shape of an absorption spectrum can often be used as a 'fingerprint' for the identification of unknown substances.

18·5 Electronic absorption in metals

The 'free electrons' in organic molecules display narrow absorption lines because their electronic energy levels are separated by energies comparable to those of photons. In contrast the electron energy bands in metals allow the conduction electrons freedom to respond to electromagnetic waves of almost any frequency. It is this strong response of the conduction electrons which prevents a metal from transmitting light except when thinned to less than 100 nm.

The seeming paradox, that metals both absorb strongly and reflect strongly, is no contradiction. The strong reflecting power results from the high conductivity–a perfect conductor would be a perfect reflector. This is because an accelerating charge radiates an electromagnetic energy, and the electrons are being rapidly accelerated backwards and forwards. Most of the incident energy is re-radiated backwards, while little penetrates the surface. Having penetrated, that little is absorbed, just because the conductivity is finite, so that kinetic energy acquired by electrons is readily transferred to the lattice by collision, as described in Chapter 13 in the discussion of conduction. The strength of the absorption at a frequency f is a consequence of the large concentration of electrons which lie within an energy hf (typically 1–2 eV) below the Fermi level. Strong absorption and reflection persist as the frequency is raised throughout the infrared, visible and near ultraviolet parts of the spectrum, until in the far ultraviolet, around a wavelength of about 100 nm the metal becomes significantly more transparent. It is only at these high frequencies, above 10^{16} Hz, that the inertia of the electrons makes itself felt and they fail to move sufficiently to either absorb or reflect the incoming wave.

It is obvious from the differences in the colours of metals, that the strength of absorption varies across the optical part of the spectrum. This is due to variations in the absorption by electrons lying well below the Fermi level, and is due to the fact that most common metals have partially empty d shells. Absorption is usually strongest at higher photon energies, i.e., at the blue end of the spectrum. The metal therefore reflects the complementary colour–yellow if the absorption is weak, but reddish, as in copper, when the absorption is rather stronger.

18·6 Electronic absorption in semiconductors

Because semiconductors have many fewer free carriers, free carrier absorption such as occurs in metals is very weak. Lattice absorption can occur in the infrared, as we have seen in Section 18·3. As long as the photon energy hf is less than the energy gap E_g between the valence band and the conduction

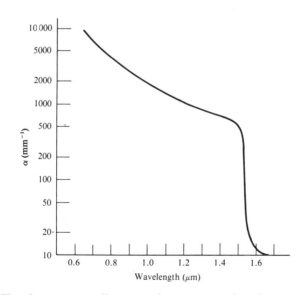

Fig. 18·5 The absorption coefficient α of germanium plotted against wavelength

band, little or no absorption by electrons can occur. The only vacant levels at an energy which the valence electrons can reach are the few holes at the top of the valence band.

Hence significant absorption only occurs at frequencies f such that $hf > E_g$, the energy gap. Figure 18·5 shows such a case: the spectrum of germanium. This shows features which can be related to the details of the energy bands in germanium, which are not nearly as simple as was described in Chapter 14. The rise in absorption at higher frequencies seen in Fig. 18·5 is called an absorption edge. This is because, on the low frequency (longer wavelength) side of the main 'edge' the absorption rises exponentially with wavelength over a long range. Thus the shape differs noticeably from the $1/(\omega_0^2 - \omega^2)^2$ dependence given by Eqn (18·4) when $\gamma\omega$ is small. From Fig. 18·5 it is clear that germanium absorbs strongly like a metal at visible wavelengths, and as might be expected it looks metallic to the eye.

From the energy gaps of other semiconductors given in Table 14·1 the wavelengths at which the absorption edges occur can be calculated using $hc/\lambda = hf = E_g$. Thus silicon, like germanium, appears metallic with an absorption edge at the infrared end of the spectrum, whilst cadmium sulphide is yellow (absorption edge at 540 nm) and diamond is of course visibly transparent because it only absorbs ultraviolet light at wavelengths shorter than 230 nm.

18·7   The refractive index

From the absorption of light, we now turn to another property of transparent solids which is of interest to us: the refractive index. The theory of electromagnetic waves tells us that the refractive index is equal to the square root of the relative dielectric permittivity of a material, determined at the frequency of oscillation of the electromagnetic wave. This, of course, is controlled by the motion of the charges in a material in response to the electric field of a light wave. It can therefore be predicted from the mathematical model developed in Section 18·3. The change in the refractive index with frequency of wavelength is usually termed *dispersion* of the index.

The equations of Section 18·3 enable us to calculate the polarizability (see Chapter 17) of the moving charges, which is just $qA \cos \phi/E_0$ for each charge. A and $\cos \phi$ are to be found from Eqns (18·4) and (18·5) respectively.

Equation (17·20) gives us a relation between permittivity ε and polarizability α which, however, needs modification to relate the internal electric field experienced by moving charge to the applied field $E_0 \cos \omega t$. When this is done, and n^2 is substituted for ε, Eqn (17·20) leads to the result

$$\frac{n^2 - 1}{n^2 + 2} = \frac{N\alpha}{\epsilon_0} = \frac{NqA \cos \phi}{3\epsilon_0 E_0} \tag{18·9}$$

Inserting A and $\cos \phi$ from Eqns (18·4) and (18·5) gives the result

$$\frac{n^2 - 1}{n^2 + 2} = \frac{Nq^2}{3M\epsilon_0} \left[\frac{(\omega_0^2 - \omega^2)}{(\omega_0^2 - \omega^2) + \gamma^2\omega^2} \right] \tag{18·10}$$

JUST
KNOW,
THIS!

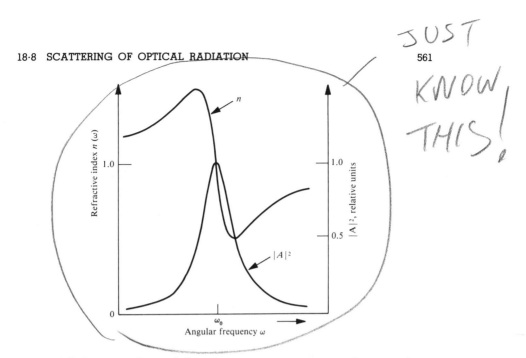

Fig. 18·6 Refractive index and absorption variations across an absorption line, predicted by Eqn (18·10)

The resulting shape for $n(\omega)$ is shown in Fig. 18·6 together with the associated power absorption, for an arbitrary density N of moving charges. This reproduces fairly closely the observed behaviour in at least some materials but, as with the absorption spectrum, broadening is common in solids and liquids. The refractive index changes with frequency in a characteristic way as the absorption line is crossed. Note that, except in a narrow range near the absorption peak, the index increases with the frequency. The exceptional behaviour close to the absorption peak is often called anomalous dispersion–a rather outdated description now that the theory is so well established!

18·8 Scattering of optical radiation

Besides direct absorption, there are various mechanisms by which a beam of radiation is scattered within a material, which contribute to the loss of energy in the beam.

The simplest of these, Rayleigh scattering, is an important contributor to the energy loss in highly pure, transparent materials such as are used for optical fibres. Rayleigh scattering is essentially due to small, random fluctuations in the refractive index from point to point which in an amorphous substance arise from the density and composition fluctuations consequent upon a random structure. Such variations also occur in crystalline materials, due to thermal motion of the lattice, and other lattice defects. Variations of index can be particularly large in liquids and gases, and Rayleigh scattering is responsible for the blue colour of the sky. The reason for this is that the scattered energy is inversely proportional to the fourth power of the wavelength and so is much stronger–by about 2^4–at the blue end of the visible spectrum.

We now explain how this comes about, by considering the scattered radiation to be due to induced dipoles located at the randomly sited points where the index differs from the average values.

The dipole moments of these induced dipoles are proportional to the incident light amplitude just as in the previous section. So the dipole moment p varies cosinusoidally in sympathy with the driving force, i.e.

$$p = p_0 \cos \omega t \qquad (18\cdot11)$$

where $\omega = 2\pi f = 2\pi c/\lambda$ is the angular frequency of the incident radiation of wavelength λ. The total scattered amplitude is the sum of many wavelets originating at random distances from a point of observation, so that their relative phases are random. But the total scattered intensity will be proportional to the intensity of each scattered wavelet.

Now according to electromagnetic theory, a moving charge radiates electromagnetic energy with an amplitude proportional to its acceleration (a steady current will not radiate) so that the wavelet has amplitude proportional to d^2p/dt^2. From Eqn (18·11)

$$\frac{d^2p}{dt^2} = p_0\omega^2 \cos \omega t = \frac{p_0 c^2}{\lambda^2} \cos \omega t$$

Hence the scattered *intensity*, being proportional to the square of this amplitude, is proportional to $1/\lambda^4$. To compare this result with experiment, we shall use the example of attenuation of light in optical fibres designed for long-distance communication. Because of their current importance, they deserve more than a brief mention.

18·9 Optical fibres

The scattering and absorption of light in glasses, and particularly in silica, has received much attention because of the cost reductions obtainable in telephone systems by the use of optical fibres. These fibres, which consist primarily of silica, have a cylindrical core of 5–50 μm diameter doped with GeO_2 to give a slightly higher refractive index. The light is guided in the core by total internal reflection at the interface with the pure silica cladding.

In 1970 the best fibres then made attenuated light of 850 nm wavelength by a factor of 100 over a 1 km length. Careful study and control of the sources of absorption had by 1978 reduced the attenuation by 1 km of fibre to a factor of less than 1·2 at a wavelength of 1·3 μm. The wavelength dependence of the measured attenuation typical in recent fibres is shown in Fig. 18·7.

Rayleigh scattering is clearly the main limiting factor for wavelengths up to about 1 μm. At longer wavelengths the curve is reminiscent of spectral absorption 'lines'. Indeed, it has been shown that the first few absorption lines are 'overtones', or harmonics, of an absorption line at about 2·8 μm

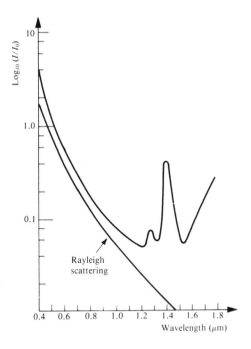

Fig. 18·7 Ratio of transmitted intensity I to incident intensity I_0 typical of 1 km of silica fibre, plotted against wavelength. The lower curve shows the estimated effect of Rayleigh scattering alone

wavelength which results from the vibration of O–H bonds inadvertently incorporated in the silica. The exclusion of water from the fibre is thus vital to achieve the lowest possible attenuation of the guided light. Exclusion of other absorbing impurities such as Na, Fe, Cu is even more important, and indeed the allowed impurity levels rival those achieved in semiconductors for transistor manufacture.

The magnitude of the Rayleigh scattering in any given material must depend upon the amplitude of both density and composition fluctuations. The variations in density of a glass can be attributed to thermally driven fluctuations arising from the Brownian motion in the liquid glass before it freezes. The amplitude of the fluctuations must therefore be dependent on the freezing temperature, being greater for a higher freezing point. A high melting point material such as silica might, therefore, be expected to show a higher Rayleigh scattering loss than a lead glass of much lower melting temperature. But the reverse is the case, since in multi-component lead glasses local composition fluctuations cause bigger refractive index changes than does the non-uniform density. Hence silica is the preferred material for long-distance optical fibre communication. At still longer wavelengths than those in Fig. 18·7, the attenuation is predominantly due to absorption by induced vibration of the constituent atoms in the glass. Each interatomic bond

Table 18·2
Characteristic stretching frequencies of bonds in glasses for optical fibres

Bond	Wavelength (μm)
Si–O	9·0
B–O	7·3
P–O	8·0
Ge–O	11·0

oscillates at a different characteristic frequency and hence absorbs most strongly at the corresponding wavelength, as indicated in Table 18·2.

18·10   Luminescence

Luminescence is the generic term used to describe the emission of electromagnetic radiation after the prior absorption of energy, often in the form of radiation. Thus photoluminescence, which includes fluorescence and phosphorescence, is the emission of light after photons have been absorbed. Fluorescence occurs via energy transitions of electrons which do not involve a change of electron spin, while phosphoresence involves a change of spin, and is in consequence much slower than fluorescence. Typically, fluorescence is over within about 10^{-5} to 10^{-6} seconds of the absorption process. Phosphorescence decays more slowly, in 10^{-4} to 10 seconds. In both cases the lifetime is defined as the time required for population of the excited states to decay to $1/e$ of its original value after the source of excitation is switched off. In this section we shall concentrate on fluorescence.

A typical energy level diagram showing fluorescence transitions is given in Fig. 18·8. The states S_0 and S_1 both have zero spin, and are split into several vibrational energy levels. Having zero spin, they are called singlet states, while the energy levels labelled T_1 are associated with an electronic state having a spin quantum number of one, called a triplet state.

In a fluorescence process, a molecule or solid in the lowest vibrational level of the singlet ground state S_0 is excited by an incoming photon into a higher vibrational level of the first excited singlet state S_1. The vibrational energy of the excited state disappears rapidly: it is communicated to neighbouring atoms or molecules in about 10^{-12} s, a time much shorter than the excited state lifetime of 10^{-8} to 10^{-6} s quoted above. The excited state may then decay with emission of a photon of fluorescent radiation, usually to an upper vibrational level of the ground state. Because of the similarity between the vibrational levels of the ground and excited states, the fluorescent spectrum is often somewhat like a mirror image of the absorption spectrum of the same material, as in the example in Fig. 18·9.

Phosphorescence occurs following an alternative decay process, via the triplet state T_1, which is also shown in Fig. 18·8. This process has lower probability as it requires a flip (a reversal of direction) of the spin of the

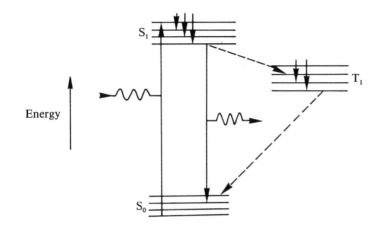

Fig. 18·8 Schematic energy level diagram showing fluorescence via singlet levels S_0 and S_1, and phosphorescence via a triplet state T_1

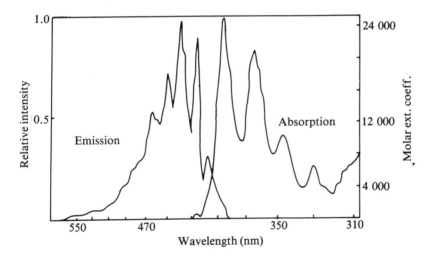

Fig. 18·9 Emission and absorption spectra for benzo(ghi)perylene in benzene

electron involved. Decay to the ground state must then occur with another spin flip, either by phosphorescent emission, or by emission of vibrational energy.

Note that the luminescent spectrum always has longer wavelength (lower energy) than the exciting radiation. If both fluorescence and phosphoresence occur in the same material, the latter has the longer wavelength for a similar reason.

18·10·1   Applications of luminescence

Fluorescence is important for two quite different reasons. Inorganic 'phosphors', as they are confusingly called, are used for producing light in fluorescent tubes and in lasers, and will be covered in Sections 18·10·2 and 18·11. Fluorescence in organic materials, too, is used in lasers, but in addition it is most useful as an analytical tool for the scientist and engineer. This is because it is more sensitive to structural changes than is absorption spectroscopy, and can often be used in the analysis of trace contaminants in both industrial and biological environments, down to levels of parts per trillion in some cases. It is relatively easy to filter the fluorescent light from the incident excitation of a shorter wavelength, so that the few photons produced by a trace impurity can be detected against a 'dark' background. Thus the drug LSD (lysergic acid diethylamide) can be detected at levels of 1 ng/ml in a 5 ml sample of blood plasma or urine.

Fluorescence occurs in aromatic molecules, and in those which contain stable arrangements of conjugated double bonds. Such molecules have delocalized electrons with excited singlet states at energies of 2–3 eV which fluoresce readily. Substituents for hydrogen which withdraw electrons from the conjugated system (e.g., $-Cl$, $-Br$, $-I$, $-NO_2$, $-COOH$) will tend to inhibit the fluorescence, while substituents which help to delocalize the electrons enhance fluorescence (e.g., $-NH_2$, $-OH$, $-F$, etc.). Thus aniline fluoresces, while nitrobenzene does not. The importance of such molecules in industrial chemistry and as drugs and carcinogens is what makes fluorescence spectroscopy of such value.

The reader is advised to consult a textbook of analytical chemistry for further details on this topic.

18·10·2   Phosphors

Phosphors are inorganic photoluminescent solids used for the generation of light. They consist of an inactive host (often an insulating oxide or wide-band-gap semiconductor, e.g., $CaWO_4$, ZnS), which is doped with a small percentage of luminescent atoms. The latter are commonly transition-metal ions, e.g., Ag^+, Cr^{2+}, Nd^{3+}, because these have suitable excited states between 2 and 3 eV above the ground state.

The fluorescent light tube is perhaps the best known application of photoluminescence. In the tube, a mercury vapour discharge produces

Table 18·3

Some important lamp phosphors

Host	Host formula	Activator	Emission colour
Zinc silicate	Zn_2SiO_4	Mn	Green
Calcium halophosphate	$Ca_5(PO_4)_3(F,Cl)$	Sb, Mn	Blue to pink
Calcium tungstate	$CaWO_4$	None	Deep blue
Yttrium oxide	Y_2O_3	Eu	Red

(Source: *Lamps and Lighting*, Thorn-EMI Lighting Ltd, 1983)

ultraviolet light efficiently, predominantly at the two wavelengths of 185 and 254 nm. This irradiates a phosphor coated on the tube walls, consisting of one or more of the combinations of host and active dopant listed in Table 18·3. These materials are chosen for their efficiency in converting ultraviolet light into visible light, and because their performance deteriorates little with rising temperature and age. Their lack of toxicity and ease of preparation in fine-particle form are also important to the manufacturer.

Other means of 'pumping' luminescent atoms into their excited state will also cause luminescence, including exposure to an energetic electron beam. This, called cathodoluminescence, is the mechanism at work in the cathode ray tube. For colour tubes the three phosphors normally used are Cu^+-doped ZnS (green), Ag^+-doped ZnS (blue) and Eu^{3+}-doped YVO_4 (red).

18·11   Principles of lasers: optical amplification

This chapter would hardly be complete without a discussion of **L**ight **A**mplification by **S**timulated **E**mission of **R**adiation–the fundamental mechanism underlying the LASER. Stimulated emission is, in fact, just another form of luminescent emission.

To understand stimulated emission, let us recall that spontaneous emission of a photon occurs when an electron falls from a higher energy level to a lower one, while absorption of a photon causes a quantum jump from a lower to a higher energy level. But what if an electron in a high energy level collides with a photon? Clearly the electron makes a quantum jump, perhaps to a higher level by absorbing the photon. But it is equally probable that the jump will be downward with the emission of a photon of energy equal to that of the photon which stimulated the jump. This is the process called stimulated emission. The three processes are compared diagrammatically in Fig. 18·10.

The proof that the stimulated emission and absorption processes are equally likely lies in the symmetry of the equations which describe the interaction between photon and electron.

One more feature of stimulated emission which is most important is this:

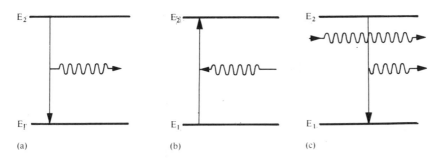

Fig. 18·10 A representation on an energy level diagram of (a) spontaneous emission, (b) absorption, (c) stimulated emission

the electromagnetic wave which is the emitted photon has the same phase everywhere as the stimulating photon wave. In fact the two photons can be represented as a single wave with twice the intensity of a single photon: the two photons are, of course, indistinguishable. We sometimes say that the two photons are *coherent*, meaning simply that they have the same phase.

Now, here is the possibility of an optical amplifier, for we have two photons where there was one. What is required to observe it is a material containing plenty of atoms in excited states, and rather few in a lower energy level (not necessarily the ground state) into which the excited electrons may go on receipt of a stimulus from a suitable photon.

Now a material which has more electrons in higher energy levels than low can hardly be in thermal equilibrium–the ratio of the population of the higher level to that in the lower in equilibrium is given by the exponential Boltzmann factor exp $-\Delta E/kT$. To make this ratio greater than unity is impossible unless energy is continuously put into the material to 'pump' electrons into the higher levels. It is not surprising that energy is needed, for we cannot expect to amplify the intensity of an optical beam by breaking the first law of thermodynamics. We can relate the populations of the upper and lower energy levels E_2 and E_1 to the amplification of the intensity, in the following way. Suppose a plane wave of a frequency f and intensity I is passing through the laser, in which there are N_2 atoms per unit volume in the upper level and N_1 in the lower energy level. If the probability of a stimulated transition of one atom from level 2 to level 1 is W_s, then the probability of absorption causing an atom to move from level 1 to level 2 is also equal to W_s. The net flow of atoms from 2 to 1 is thus

$$(N_2 - N_1)\, W_s$$

But W_s must also be proportional to the intensity I of the plane wave. The net power P generated per unit volume is then

$$P = (N_2 - N_1)\, W_s\, hf = (N_2 - N_1)\, w_s\, I$$

where w_s is just a proportionality constant. If there were no energy loss by other means (never the case) the intensity *increase* per unit length is equal to P so that we can write

$$\frac{\mathrm{d}I}{\mathrm{d}z} = (N_2 - N_1)\, w_s\, I$$

where z is the distance, and the solution of this equation is

$$I(z) = I(0)\, \exp \gamma z$$

where $I(0)$ is the intensity at $z = 0$ and $\gamma = (N_2 - N_1)\, w_s$ is called the *gain* of the laser medium. (Note that it has the same dimensions as the absorption coefficient of the medium.) We see that gain exists only as long as N_2 is greater than N_1. If the atoms are pumped directly from the lower to the higher level, the pump power needed would be very high, since the lower level contains few atoms available to absorb the pump energy. Less pump power is needed if level 1 is not the ground state (as in Fig. 18·11), for two reasons. One

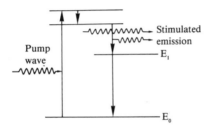

Fig. 18·11 Energy level diagram of a four-level laser

is that the density of atoms N_0 in the ground state can be much higher than either N_1 or N_2, so that the pump power is readily absorbed. But in addition the lower lasing level population N_1 is held low since atoms can decay rapidly to the ground state. The energy levels and transitions in such a laser are indicated in Fig. 18·11. This scheme is called a *four-level* laser, since the pump more often than not excites the electron into a fourth level, higher than E_2 (the upper lasing level) just as in fluorescence.

The conditions in which the four-level laser will produce gain can be determined from the equations describing the rate at which the populations N_2 and N_1 change with time.

If the atoms are pumped up to the level E_2, at rate R per unit volume per second, then the rate of increase of N_2 is less than R because

(a) atoms decay to level 1 by spontaneous emission at a rate N_2/τ_2 where τ_2 is the mean lifetime of an atom in level 2 before this decay process occurs, and because

(b) the net rate of stimulated emission given earlier as $(N_2 - N_1)\,W_s$ also reduces N_2.

The rate of increase of N_2 is thus

$$\frac{dN_2}{dt} = R - \frac{N_2}{\tau_2} - (N_2 - N_1)\,W_s$$

In similar fashion the rate of increase of N_1 is given by

$$\frac{dN_1}{dt} = \frac{N_2}{\tau_2} - \frac{N_1}{\tau_1} + W_s\,(N_2 - N_1)$$

where τ_1 is the natural lifetime of level 1, by decay to the ground state of the atom.

Since in the steady state both dN_2/dt and dN_1/dt are zero, the above two simultaneous equations give the result:

$$N_2 - N_1 = \frac{R(1 - \tau_1/\tau_2)}{W_s + 1/\tau_2}$$

This equation tells us the vital condition for achieving gain: $N_2 - N_1$ is only positive if $\tau_2 > \tau_1$, i.e., the natural lifetime of level 2 must be greater than that of level 1 in the four-level laser.

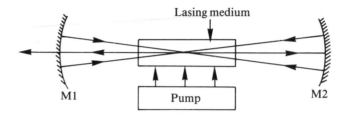

Fig. 18·12 Mirrors M1 and M2 reflect the amplified beam in a laser back upon itself to sustain the oscillation. The pump is a source either of light or electrical power

The pump energy for a laser may be provided in several ways, though the obvious way (employed in the first lasers) is to use an intense source of light, such as a flash lamp. Naturally this method only produces pulses of amplified light, but it is still used today in some types of pulsed lasers.

Alternatively, an electrical discharge in a gas will excite atoms to high energy levels–after all, such a discharge radiates spontaneously. However, a neon or a sodium lamp is not called a laser, although stimulated emission probably occurs to a small extent in the discharge.

Oddly, we reserve the acronym LASER for what in electronic parlance is actually an *oscillator*. The difference is that, in an oscillator, a portion of its output is fed back into the input to make it continue oscillating at a fixed frequency. In electronic terms, the neon or sodium lamp produces a very noisy output–that is, one which is continually changing phase, and hence cannot be at a single, fixed, frequency. The constancy of the output phase and frequency is, perhaps, the most important characteristic of the laser, and is what gives it its unique qualities as a light source.

The feedback of output to input is done by mirrors, as shown in Fig. 18·12. One of these is made partially transmitting to enable a fraction of the light to

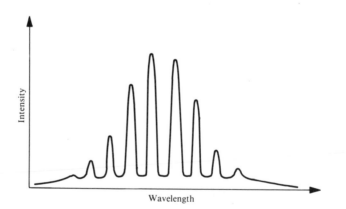

Fig. 18·13 Typical output intensity versus wavelength for a laser

be used externally. Since oscillation can only be sustained if the fed-back optical wave arrives back in phase with itself after a complete round-trip between the mirrors, the only wavelengths λ which will oscillate must satisfy the equation

$$N\lambda = 2L \qquad\qquad\qquad (18\cdot12)$$

where N is an integer. Thus the mirror separation L also plays a role in determining the spectral distribution of the output light.

A typical distribution is given in Fig. 18·13. Each narrow emission line is centred on a wavelength which satisfies Eqn (18·12), whilst the envelope of the peaks is determined by (though not equal to) the natural spectral width of the gain γ of the lasing material defined earlier.

18·12 Laser materials

It is clear that a material which fluoresces is likely to be a candidate as a lasing material. We shall select three examples to illustrate good laser materials: the solid-state neodymium laser, the liquid dye laser and the semiconductor laser.

18·12·1 The neodymium laser

The neodymium laser is a fairly straightforward example of a four-level laser. The low-lying energy levels of the Nd^{3+} ion are shown in Fig. 18·14. In the electronic ground state the Nd^{3+} ion has three electrons in its $4f$ shell having parallel spins (total spin $S = 3/2$) and with orbital angular momentum $L = 3 + 2 + 1 = 6$ (see Chapter 4). The letter I in the figure, the sixth in the series PDFGHIJ . . ., signifies that $L = 6$. There are four possible combinations of L and S, giving the following values of J: $6 - 3/2, 6 - 1/2, 6 + 1/2$ and $6 + 3/2$, corresponding to the values of the subscripts in the figure.

In the first excited electronic state an electron is promoted to the next $4f$ level where $l = 0$ giving $L = 2 + 1 = 3$, i.e., an F state, but its spin is unchanged, so that the total spin is still $S = 3/2$. Only the lowest two levels are illustrated.

Three possible lasing transitions are shown, of which one ($0\cdot914\ \mu$m) ends on the ground state, and therefore has a highly populated lower level. It consequently provides less gain than the other two. The laser may be pumped at $0\cdot914\ \mu$m, $0\cdot805\ \mu$m, or at one of several shorter wavelengths at which the ion absorbs strongly. The excited ion relaxes rapidly into the lowest level of the first excited state, just as any fluorescent species.

When the ion is in a solid, these energy levels are little affected by the surrounding anions, because the $4f$ electrons are shielded from their fields by the filled $5s$, $5p$, $5d$ shells, as described in Chapter 16. Hence by putting a small quantity (about 5% by weight) of Nd_2O_5 into a suitable rod of silicate glass, an excellent lasing rod can be made. Large rods, pumped by a powerful xenon flashlamp, can produce brief (less than 1 ns) but 'giant' pulses of light at $1\cdot314\ \mu$m or $1\cdot064\ \mu$m, with peak powers of a few gigawatts. Alternatively,

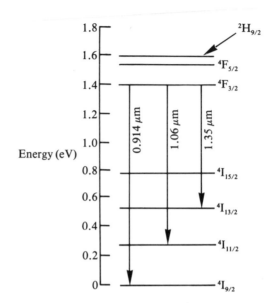

Fig. 18·14 Energy levels of the Nd^{3+} ion involved in lasing

the glass can be made in optical fibre form, pumped down the core by a few tens of milliwatts of infrared light of wavelength 0·914 μm from a cheap semiconductor laser to give an output at the more useful wavelength of 1·3 μm (see Section 18·12).

Nd^{3+} ions in crystals can, however, produce about ten times as much optical gain as the same ions in a glass. This is because the surroundings experienced by the Nd ions in a glass vary from site to site in the random structure, so detuning the energy levels just far enough from their average positions that many of them provide little or no gain at the frequency of an incoming photon. However, Nd^{3+} ions in a crystal of yttrium aluminium garnet (YAG), where each is surrounded uniformly by eight oxygen ions, have energy levels which all coincide, and all will contribute to the gain. Nd-YAG laser rods are readily obtainable, and are capable of producing higher power pulses than Nd-glass.

18·12·2   Dye lasers

Organic dyes are also four-level lasing materials, having a closely spaced set of vibrational levels associated with each electronic level, as shown in Fig. 18·15. Because there are so many lower levels, lasing is possible over an almost continuous set of frequencies, making the laser *tunable*. Tuning is usually achieved by putting a diffraction grating into the optical path, in such a way that only one diffracted wavelength reaches the laser mirror, and no spontaneously emitted photons at neighbouring frequencies can be amplified. By rotating the grating, the wavelength selected can be varied. The

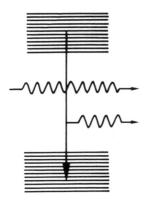

Fig. 18·15 Vibrational splitting of the electronic levels in an organic dye may allow stimulated emission over a range of wavelengths

dye is usually pumped by another laser operating at a shorter wavelength. Fairly high pump powers are needed.

18·12·3   Semiconductor lasers and LEDs

The semiconductor laser currently is of greater economic significance than other types, owing to its high efficiency, small size, ease of pumping, and range of applications in communications and data recording and retrieval.

It makes use of the radiation which, in some semiconductors, results from the recombination of an electron in an energy level close to the bottom of the conduction band with a hole (an empty level–see Chapter 14) close to the top of the valence band. The transition is illustrated in Fig. 18·16. The energy released in this process equals the energy gap E_g of the semiconductor, so that the wavelength is given approximately by $hc/\lambda = E_g$.

The commonest semiconductor usable in this way is gallium arsenide, with $E_g = 1·4\,\text{eV}$, for which $\lambda = 0·85\,\mu\text{m}$. Silicon and germanium cannot be used,

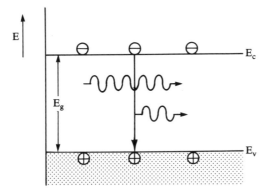

Fig. 18·16 Stimulated emission across the energy gap of a semiconductor

since their energy bands are such that electron–hole recombination can only produce radiation with the help of a phonon of suitable momentum. The lack of such phonons makes this a highly improbable process.

The energy level scheme is thus like that of the dye laser, but the pumping mechanism is quite different. The laser is made in the form of a p–n junction diode, through which a high current is passed. As a result, a big fraction of the high concentration of electrons present on the n-type side of the junction is injected across the depletion region into the p-type side. There they recombine with the large population of holes, which is created and maintained by the p-type doping.

Thus far, this description applies equally to a light-emitting diode (LED) and to a laser. The LED has no optical feedback from a pair of mirrors, as does a laser. To make the device into a laser, mirrors are made by cleaving the GaAs crystal along a [110] plane at either end. As the refractive index of GaAs is high (near 3·5), the natural reflectance of the surface is also relatively high, being close to 30%.

In order to raise the population of conduction electrons in the p-type region to the level required for producing gain (about 2×10^{24} m^{-3}), both sides of the junction must be very heavily doped, and the forward bias voltage applied to the diode must be equal to the energy gap voltage E_g/e, which is nearly 1·4 V in GaAs.

The efficiency of energy conversion is good. Each electron–hole pair produces a photon having energy E_g. The electrical energy used to produce the photon is also E_g, since it equals the electron's charge e, multiplied by the voltage E_g/e across the junction. Thus almost no energy is wasted in pumping electrons up into energy levels higher than the conduction band edge E_c. However, the overall efficiency is not 100%, as many electrons recombine outside the region between the mirrors, while others recombine at lattice defects without radiating photons.

Efficiencies of 20–40% are typical within the junction. Efficiency is further reduced because the high current density needed (typically 4000 A/cm^2) produces considerable heat. This must be removed by a heat-sink making direct contact with the junction. The heat-sink is usually much larger than the laser itself, which is only perhaps $0·1 \times 0·25 \times 0·05$ mm. Light output powers of up to 1 watt are available commercially at the time of writing (1989), and are rising annually.

Problems

18·1 Estimate the absorption coefficients of (a) a metal foil of thickness 50 nm which reflects 40% and transmits 20% of the light incident upon it, and (b) window glass, which absorbs 90% of the incident light in a thickness of 0·2 m.

18·2 By substitution of Eqn (18·3) into Eqn (18·2), obtain the results quoted in Eqns (18·4) and (18·5).

18·3 With the aid of Eqn (18·5) give a dimensioned sketch of the variation with frequency of the phase angle between atomic displacement and electric field of the incoming wave.

18·4 Remembering that an oscillating charge radiates energy in phase with its *acceleration*, show that the oscillator described by Eqns (18·3)–(18·5) re-radiates a wave lagging 90° behind an incident wave of the frequency ω_0.

18·5 Use the 'electron in a box' model of Section 2·6 to find an expression for the energy difference between two adjacent energy levels having quantum numbers N and $N+1$. Hence prove the result quoted in Eqn (18·8).

18·6 Describe some of the processes which give rise to attenuation of light in optical fibres.

18·7 Explain what is meant by the statement that the output of a laser is *coherent*. Why is the light from an incandescent source not coherent?

18·8 Find the separation between the emission lines in the spectrum of the output from

    (a) a helium–neon laser with mirror separation 0·5 m and wavelength 632·8 nm;
    (b) a gallium arsenide semiconductor laser operating at 850 nm wavelength and with mirror separation 0·3 mm.

Self-Assessment questions

1 The frequencies of infrared radiation are higher than those of visible light.
    A) true   B) false.

2 Crystal lattice absorption occurs predominantly in which part of the spectrum?
    A) visible   B) infrared   C) ultraviolet.

3 Electronic absorption occurs at infrared frequencies in
    A) insulators   B) metals   C) semiconductors
    D) organic molecules   E) glasses.

4 The following processes contribute to the intensity loss on transmission through a medium
    A) electronic absorption   B) electron–phonon collisions
    C) photon–phonon collisions   D) stimulated emission
    E) Rayleigh scattering.

5 Crystal lattice absorption in insulators gives rise to a spectrum of
    A) narrow lines   B) a broad band of wavelengths
    C) an absorption edge.

6 Metals become fairly transparent beyond wavelengths
    A) just longer than visible   B) just shorter than visible.

7 On the long wavelength side of an absorption edge the absorption varies with wavelength
    A) exponentially   B) logarithmically   C) parabolically.

8 The energy gap in a semiconductor determines the wavelength below which the absorption rises strongly.
    A) true   B) false.

9 The refractive index is which function of the permittivity ϵ measured at optical frequencies?

A) square   B) square root   C) cube.

10 Refractive index dispersion is the change in refractive index with

A) photon energy   B) wavelength
C) energy gap   D) absorption.

11 Anomalous dispersion occurs

A) far from an absorption line
B) at the edges of an absorption line
C) at the centre of an absorption line.

12 Rayleigh scattering is due to

A) fluctuations in absorption
B) stimulated emission
C) vibrations of O–H bonds
D) fluctuations in refractive index.

13 Rayleigh scattering varies with wavelength λ as

A) λ B) λ^{-1} C) λ^2 D) λ^{-2}
E) λ^3 F) λ^{-3} G) λ^4 H) λ^{-4}.

14 Density fluctuations in glasses cause stronger Rayleigh scattering than do composition fluctuations

A) true   B) false.

15 Stimulated emission occurs whenever

A) an electron makes a transition to a lower energy level
B) a photon is absorbed by an electron
C) both A) and B) occur simultaneously
D) a photon causes an electron to emit radiation.

16 Two photons are coherent when

A) they travel at the same speed
B) their wavelengths are the same
C) their phases are the same
D) they obey Planck's equation.

17 A laser requires mirrors because

A) they provide optical feedback
B) they invert the electron populations
C) they determine the wavelength at which lasing occurs.

18 The number of electrons in the upper level of a laser must be

A) higher than the equilibrium population
B) higher than the population in the lower level
C) lower than the population in the lower level.

Each of the sentences in questions 19–25 consists of an assertion followed by a reason. Answer:

A) If both assertion and reason are true statements and the reason is a correct explanation of the assertion.

B) If both assertion and reason are true statements but the reason is not a correct explanation of the assertion.
C) If the assertion is true but the reason contains a false statement.
D) If the assertion is false but the reason contains a true statement.
E) If both assertion and reason contain false statements.

19 Crystal lattice absorption cannot readily be observed in metals *because* it is masked by electronic absorption.

20 Pure silica shows a lower Rayleigh scattering loss than lead glass *because* it contains no elements other than Si and O.

21 A laser requires a pump energy source *because* stimulated emission is impossible without it.

22 Crystal lattice absorption occurs at a single frequency *because* only one phonon frequency travels at the same speed as the absorbed radiation.

23 Organic dye molecules are coloured *because* electrons in them are confined to the molecule.

24 Less pump power is needed in a two-level laser than in a three-level laser *because* it is easier to increase the population in the upper level.

25 Metals reflect strongly *because* they also absorb strongly.

Answers

1 B	2 B	3 B	4 A, C, E
5 A	6 B	7 A	8 A
9 B	10 B, A	11 C	12 D
13 H	14 B	15 D	16 C + B
17 A	18 B	19 A	20 A
21 C	22 A	23 B	24 E
25 B			

FURTHER READING

Chapter numbers are given in brackets.

General
Wyatt, O. H. and Dew-Hughes, D. (1974) *Metals, Ceramics and Polymers*, Cambridge.

Van Vlack, L. H. (1985) *Elements of Materials Science and Engineering* (5th edn), Addison-Wesley.

Van Vlack, L. H. (1982) *Materials for Engineering*, Addison-Wesley.

Chalmers, B. (1982) *The Structure and Properties of Solids*, Heyden.

Clauser, H. R. (1975) *Industrial and Engineering Materials*, McGraw-Hill.

Wert, C. A. and Thomson, R. M. (1970) *Physics of Solids* (2nd edn), McGraw-Hill, New York.

Easterling, K. (1988) *Tomorrow's Materials*, Institute of Metals, London.

Crane, F. A. A. and Charles, J. A. (1984) *Selection and Use of Engineering Materials*, Butterworths.

Cotterill, R. (1985) *The Cambridge Guide to the Material World*, Cambridge University Press.

Structure of matter
Porter, D. A. and Easterling, K. E. (1981) *Phase Transformations in Metals and Alloys*, Van Nostrand Reinhold, (9).

Verhoeven, J. D. (1975) *Fundamentals of Physical Metallurgy*, Wiley, (8, 9).

Budworth, D. W. (1970) *An Introduction to Ceramic Science*, Pergamon Press, (11).

Holloway, D. G. (1973) *The Physical Properties of Glass*, Wykeham, (11).

Henderson, B. (1972) *Defects in Crystalline Solids*, Arnold, (8, 9).

Vinson, J. R. and Chou, T. W. (1975) *Composite Materials and their Use in Structures*, Applied Science.

Ashby, M. F. and Jones, R. H. (1986) *Engineering materials 2 An introduction to microstructure, processing and design*, Pergamon Press (9, 10, 11).

Mechanical properties
Le May, I. (1981) *Principles of Mechanical Metallurgy*, Arnold, (9, 10).

Davidge, R. W. (1979) *Mechanical Behaviour of Ceramics*, Cambridge, (9, 11).

Martin J. W. and Hull, R. A. (1973) *Strong Materials*, Wykeham, (9, 10, 11).

Kelly, A. (1973) *Strong Solids* (2nd edn), Oxford University Press, (8, 10, 11).

Hull, D. (1981) *An Introduction to Composite Materials*, Cambridge University Press, (11).

Harris, B. (1986) *Engineering Composite Materials*, Institute of Metals, London, (11).

Thermodynamics
Tabor, D. (1969) *Gases, Liquids and Solids*, Penguin, London, (7).

Warn, J. W. R. (1969) *Concise Chemical Thermodynamics*, Van Nostrand, New York, Student edition, (7).

Swalin, R. A. (1972) *Thermodynamics of Solids* (2nd edn), John Wiley, New York (7).

Electrical and magnetic properties
Sze, S. M. (1985) *Semiconductor Devices–physics and technology*, Wiley, New York, (13–15).

Seymour, J. (1988) *Electronic Devices and Components* (2nd edn), Longman, London, (13, 14, 16).

Cullity, B. D. (1972) *An Introduction to Magnetic Materials*, Addison-Wesley, (16).

Crangle, J. (1977) *The Magnetic Properties of Solids*, Arnold, London, (16).

Lines, M. E. and Glass, A. E. (1977) *Principles and Applications of Ferroelectrics and Related Materials*, OUP (17).

Data books
Hodgman, C. D., (ed.), *Handbook of Chemistry and Physics*. Chemical Rubber Publishing Company, New York. (Published annually.)

Anderson, J. C., *et al.* (1982) *Data and Formulae for Engineering Students* (3rd edn), Pergamon, Oxford.

APPENDIX 1   UNITS AND CONVERSION FACTORS

SI units

(1) *Basic units*

quantity	*unit*	*unit symbol*
length	metre	m
mass	kilogramme	kg
time	second	s
electric current	ampere	A
thermodynamic temperature	Kelvin	K
luminous intensity	candela	cd
amount of substance	mole	mol

(2) *Derived units*

quantity	*unit*	*unit symbol*
force	newton	$N = kg\,m/s^2$
work, energy, heat	joule	$J = N\,m$
power	watt	$W = J/s$
pressure	pascal	$Pa = N/m^2$
	electrical units	
potential	volt	$V = W/A$
resistance	ohm	$\Omega = V/A$
charge	coulomb	$C = A\,s$
capacitance	farad	$F = A\,s/V$
electric field strength	—	V/m
electric flux density	—	C/m^2
	magnetic units	
magnetic flux	weber	$Wb = V\,s$
inductance	henry	$H = V\,s/A$
magnetic field strength	—	A/m
intensity of magnetization	—	A/m
magnetic flux density	tesla	$T = Wb/m^2$

(3) *Unit conversion factors*

Length, volume

$$1\ mil\quad = 2\cdot54 \times 10^{-5} m$$
$$1\ in\quad = 2\cdot54\ cm \qquad 1\ in^3 = 16\cdot39\ cm^3$$
$$1\ ft\quad = 0\cdot3048\ m \qquad 1\ ft^3 = 0\cdot02832\ m^3$$
$$1\ mile\quad = 5{,}280\ ft = 1\cdot609\ km$$
$$1\ \mu m\quad = 10^{-6}\ m = 39\cdot37\ \mu in$$
$$1\mathring{A}\quad = 10^{-10}\ m$$
$$1\ gal\quad = 0\cdot1605\ ft^3 = 4{,}546\ cm^3$$
$$1\ US\ gal = 0\cdot1337\ ft^3 = 3{,}785\ cm^3$$
$$1\ l\quad = 10^{-3}\ m^3 \quad = 1{,}000\ cm^3$$

Mass
   1 lb   = 0·4536 kg
   1 ton   = 2,240 lb = 1,016 kg
   1 tonne = 1,000 kg

Density
   1 lb/in^3 = 27·68 g/cm^3
   1 lb/ft^3 = 16·02 kg/m^3
   1 g/cm^3 = 10^3 kg/m^3

Force
   1 pdl = 0·1383 N
   1 lbf = 32·17 pdl = 4·448 N
   1 tonf = 9,964 N
   1 kgf = 2·205 lbf = 9·807 N
   1 dyn = 10^{-5} N

Power
   1 hp   = 550 ft lbf/s = 0·7457 kW
   1 ft lbf/s = 1·356 W

Torque
   1 lbf ft   = 1·356 N m
   1 tonf ft = 3,037 N m

Energy, work, heat
   1 ft lbf = 1·356 J
   1 kWh   = 3·6 MJ
   1 Btu   = 1,055 J = 252 cal = 778·2 ft lbf
   1 cal   = 4·187 J
   1 hp h   = 2·685 MJ
   1 erg   = 10^{-7} J

Pressure stress
   1 lbf/in^2 = 0.07031 kgf/cm^2 = 6,895 Pa
   1 tonf/in^2 = 157·5 kgf/cm^2 = 15·44 MN/m^2
   1 kgf/cm^2 = 0·09807 MN/m^2
   1 lbf/ft^2 = 47·88 N/m^2 = 47·88 Pa
   1 ft H$_2$O   = 62·43 lbf/ft^2 = 2,989 Pa
   1 in Hg   = 70·73 lbf/ft^2 = 3,3896 Pa
   1 mm Hg = 1 torr   = 133·3 Pa
   1 Int atm = 1·013 × 10^5 Pa   = 14·70 lbf/in^2
   1 bar   = 10^5 Pa   = 14·50 lbf/in^2

Temperature
   1°C = 1·8°F
   $TK = T°C + 273·15°C$

Dynamic viscosity
   1 poise (g cm s) = 0·1 kg/m s = 0·1 Ns/m^2
   1 kgf s/m^2 = 0·9807 kg/m s
   1 lbf s/in^2 = 6,895 kg/m s

Kinematic viscosity
   1 ft^2/s = 0·09290 m^2/s
   1 in^2/s = 6·452 cm^2/s

Thermal conductivity
   1 cal/cm s K $=$ 418·7 J/m s K

Electrical units
The conversion factors which follow are from the CGS system to the SI system. (Note: in the CGS system 1 e.m.u. $=10^{10}$ e.s.u. of charge.)

capacitance	1 e.s.u.	$= \frac{1}{9} \times 10^{-11}$ F
charge	1 e.m.u.	$= 10$ C
current	1 e.m.u.	$= 10$ A
electric field strength	1 e.s.u.	$= 3 \times 10^4$ V/m
electric flux density	1 e.s.u.	$= (1/12\pi) \times 10^{-5}$ C/m^2
electric polarization	1 e.s.u.	$= \frac{1}{3} \times 10^{-5}$ C/m^2
inductance	1 e.m.u.	$= 10^{-9}$ H
intensity of magnetization	1 e.m.u.	$= 10^3$ A/m
magnetic field strength	1 e.m.u.	$= (1/4\pi) \times 10^3$ A/m
magnetic flux	1 e.m.u.	$= 10^{-8}$ Wb
magnetic flux density	1 e.m.u.	$= 10^{-4}$ T
magnetic moment	1 e.m.u.	$= 10^{-3}$ Am^2
magnetomotive force	1 e.m.u.	$= (10/4\pi)$ A
mass susceptibility	1 e.m.u/g	$= 4\pi \times 10^{-3}$/kg
potential	1 e.m.u.	$= 18^{-8}$ V
resistance	1 e.m.u.	$= 10^{-9}$ Ω

APPENDIX 2   PHYSICAL CONSTANTS

Avogadro's number	N	$= 6 \cdot 023 \times 10^{23}/\text{mol}$
Bohr magneton	β	$= 9 \cdot 27 \times 10^{-24} \text{ A m}^2$
Boltzmann's constant	k	$= 1 \cdot 380 \times 10^{-23} \text{ J/K}$
electron volt	eV	$= 1 \cdot 602 \times 10^{-19} \text{ J}$
electron charge	e	$= 1 \cdot 602 \times 10^{-19} \text{ C}$
electronic rest mass	m_e	$= 9 \cdot 109 \times 10^{-31} \text{ kg}$
electronic charge to mass ratio	e/m_e	$= 1 \cdot 759 \times 10^{-11} \text{ C/kg}$
energy for $T = 290\text{K}$	kT	$= 4 \times 10^{-21} \text{ J}$
energy of ground state H atom		
(Rydberg energy)		$= 13 \cdot 60 \text{ eV}$
Faraday constant	F	$= 9 \cdot 65 \times 10^7 \text{ C/mol}$
permeability of free space	μ_0	$= 4\pi \times 10^{-7} \text{ H/m}$
permittivity of free space	ϵ_0	$= (1/36\pi) \times 10^{-9} \text{ F/m}$
Planck's constant	h	$= 6 \cdot 626 \times 10^{-34} \text{ Js}$
proton mass	m_p	$= 1 \cdot 672 \times 10^{-27} \text{ kg}$
proton to electron mass ratio	m_p/m_e	$= 1,836 \cdot 1$
Average radius of wave function of H		
atom in ground state		$= 0 \cdot 529 \times 10^{-10} \text{ m}$
		$= 0 \cdot 529 \text{ Å}$
standard gravitational acceleration	g	$= 9 \cdot 807 \text{ m/s}^2$
		$= 32 \cdot 17 \text{ ft/s}^2$
Stefan–Boltzmann constant	σ	$= 5 \cdot 67 \times 10^{-8} \text{ J/m}^2 \text{ s K}^4$
universal constant of gravitation	G	$= 6 \cdot 67 \times 10^{-11} \text{ N m}^2/\text{kg}^2$
		$= 3 \cdot 32 \times 10^{-11} \text{ lbf ft}^2/\text{lb}^2$
universal gas constant	R	$= 8 \cdot 314 \text{ J/mol K}$
velocity of light in vacuo	c	$= 2 \cdot 9979 \times 10^8 \text{ m/s}$
volume of 1 mol of ideal gas at N.T.P.		$= 22 \cdot 42 \text{ l}$

APPENDIX 3 PHYSICAL PROPERTIES OF ELEMENTS

Atomic number	Element	Density at 20°C (10^3 kg/m³)	Melting point (°C)	Boiling point (°C)	Young's modulus (10^{10} N/m²)	Shear modulus (10^{10} N/m²)	Poisson's ratio
1	Hydrogen (H)	0·00009	−259·2	−252·7			
2	Helium (He)	0·00018		−268·9			
3	Lithium (Li)	0·534	180·5	1,330	1·15		
4	Beryllium (Be)	1·848	1,277	2,770	29·65	14·48	0·08
5	Boron (B)	2·34	(2,100)	(2,550)			
6	Carbon (C) Graphite	2·25	3,700[a]	4,830	0·69 (Polycrystalline)		
	Diamond	3·52			82·74	34·48	0·25
7	Nitrogen (N)	0·00125	−210	−196			
8	Oxygen (O)	0·00143	−219	−183			
9	Fluorine (F)	0·0017	−219·6	−188			
10	Neon (Ne)	0·0009	−249	−246			
11	Sodium (Na)	0·971	97·8	892	0·9		
12	Magnesium (Mg)	1·74	650	1,105	1·72		
13	Aluminium (Al)	2·699	660	2,450	6·9	2·62	0·3
14	Silicon (Si)	2·33	1,410	2,680	11·0		0·34
15	Phosphorus (P) (white)	1·83	44·2	280			
16	Sulphur (S) (yellow)	2·07	119	445			
17	Chlorine (Cl)	0·0032	−101	−34·7			
18	Argon (A)	0·0018	−189·4	−186			
19	Potassium (K)	0·86	63·7	760	0·34		
20	Calcium (Ca)	1·55	838	1,440	2·07	0·69	0·31
21	Scandium (Sc)	2·99	1,540	2,730			
22	Titanium (Ti)	4·507	1,670	3,260	11·58	4·14	0·34
23	Vanadium (V)	6·1	1,860	3,400	13·79	5·03	0·36
24	Chromium (Cr)	7·19	1,875	2,665	24·82		
25	Manganese (Mn)	7·43	1,245	2,150	15·86		

26	Iron (Fe)	7·87	1,536	3,000	19·65		7·93	0·28
27	Cobalt (Co)	8·85	1,495	2,900	20·68			0·31
28	Nickel (Ni)	8·90	1,453	2,730	21·37		7·93	0·31
29	Copper (Cu)	8·96	1,083	2,600	12·41		4·62	0·35
30	Zinc (Zn)	7·13	419·5	906	9·65		3·45	0·35
31	Gallium (Ga)	5·91	29·8	2,240	0·97			
32	Germanium (Ge)	5·32	937	2,830	7·58			
33	Arsenic (As)	5·72		613[a]				
34	Selenium (Se)	4·79	217	685				
35	Bromine (Br)	3·12	−7·2	58				
36	Krypton (Kr)	0·0037	−157	−152				
37	Rubidium (Rb)	1·53	38·9	688	0·023		0·69	0·28
38	Strontium (Sr)	2·60	768	1,380	1·72			
39	Yttrium (Y)	4·47	1,510	3,030			3·45	0·34
40	Zirconium (Zr)	6·49	1,852	3,580	9·65		3·72	0·38
41	Niobium (Nb) [Columbium (Cb)	8·57	2,470	4,900	10·34			
42	Molybdenum (Mo)	10·22	2,610	5,550	34·48			
43	Technetium (Tc)		(2,100)	(3,900)	40·68			
44	Ruthenium (Ru)	12·2	(2,500)	(4,900)	41·37		18·61	0·25
45	Rhodium (Rh)	12·44	1,965	4,500	28·96			
46	Palladium (Pd)	12·02	1,552	4,000	11·72		4·83	0·39
47	Silver (Ag)	10·49	960·8	1,761	7·58		2·76	0·38
48	Cadmium (Cd)	8·65	320·9	765	5·52		2·76	0·29
49	Indium (In)	7·31	156·2	2,000	1·10			
50	Tin (Sn)	7·298	231·9	2,270	4·69		1·72	0·36
51	Antimony (Sb)	6·62	630·5	1,380	7·58			
52	Tellurium (Te)	6·24	449·5	990	4·14			
53	Iodine (I)	4·94	113·7	183				
54	Xenon (Xe)	0·0059	−112	−108				
55	Caesium (Cs)	1·903	28·7	690			0·17	
56	Barium (Ba)	3·5	714	1,640	1·24			

Atomic number	Element	Density at 20°C (10^3 kg/m³)	Melting point (°C)	Boiling point (°C)	Young's modulus (10^{10} N/m²)	Shear modulus (10^{10} N/m²)	Poisson's ratio
57	Lanthanum (La)	6·19	930	3,470	6·90	3·45	0·37
58–71	(Rare earth elements)						
72	Hafnium (Hf)	13·09	2,250	5,400	13·79	18·62	0·17
73	Tantalum (Ta)	16·6	2,980	5,400	18·62	15·17	0·26
74	Tungsten (W)	19·3	3,410	5,900	34·48	20·69	0·25
75	Rhenium (Re)	21·04	3,170	5,900	48·27	23·44	0·26
76	Osmium (Os)	22·57	(3,000)	5,500	55·16	22·06	0·39
77	Iridium (Ir)	22·5	2,455	5,300	51·71	5·52	0·42
78	Platinum (Pt)	21·45	1,769	4,530	14·48	2·76	
79	Gold (Au)	19·32	1,063	2,970	8·27		
80	Mercury (Hg)	13·55	−38·6	357			
81	Thallium (Tl)	11·85	303	1,457	0·69	0·28	0·45
82	Lead (Pb)	11·36	327·4	1,725	1·79	0·55	0·45
83	Bismuth (Bi)	9·80	271·3	1,560	3·17		
84	Polonium (Po)		250				
85	Astatine (At)		(300)				
86	Radon (Rn)	0·01	(−70)	−61·8			
87	Francium (Fr)		(27)				
88	Radium (Ra)	5·0	700				
89	Actinium (Ac)		(1,000)				
90	Thorium (Th)	11·66	1,750	(3,850)	7·58	2·76	0·30
91	Protactinium (Pa)	15·4	(1,200)				
92	Uranium (U)	19·07	1,132	3,820	17·24	7·58	0·24

(Numbers in parentheses are uncertain)   ª Sublimes.

ANSWERS TO PROBLEMS

Chapter 1

1·3 820 eV
1·4 5.92×10^{-29} m
1·5 3.5 Å
1·6 0.078 Å
1·7 1.25×10^{14} cm^{-2}
1·8 2.5×10^{14} cm^{-2} s^{-1}

1·9 $y = \dfrac{\alpha}{4} - \dfrac{2\alpha}{\pi^2} \left(\cos \omega t + \dfrac{1}{3^2} \cos 3\omega t \right.$

$\left. + \dfrac{1}{5^2} \cos 5\omega t + \ldots \right) + \dfrac{\alpha}{\pi}$

$\left(\sin \omega t - \dfrac{1}{2} \sin 2\omega t + \dfrac{1}{3} \sin \right.$

$\left. 3\omega t - \ldots \right)$

Chapter 2

2·2 8.64×10^9
2·3 2.3×10^5 ms^{-1}; 2.0×10^9 m^{-1}

Chapter 3

3·1 $3.36 \times 10^{-24} n^{2/3}$ J ($n = 1, 2, 3 \ldots$)
3·2 1.06 Å; 0.6 Å
3·3 -2.17×10^{-18} J; -5.42×10^{-19} J;
-2.41×10^{-19} J; 2.46×10^{15} Hz;
2.91×10^{15} Hz; 4.55×10^{14} Hz;
$n = 3$ to $n = 2$
3·4 5, 5, 7
3·5 $-e^4 m / 2h^2 \epsilon_0^2 n^2$
3·9 0.1 eV

Chapter 5

5·1 10.6 eV
5·2 109° 28'
5·3 1.03×10^{-11} N, 2.32×10^{-10} N
5·6 1.33 eV, 4.8 eV

Chapter 6

6·1 $(\sqrt{3} - 1)$, 1.04 Å
6·4 9° 36', 28° 6'
6·5 2, 4

6·7 2.25×10^3 kgm^{-3}
6·8 8.50×10^{28} m^{-3}, 8.50×10^{28} m^{-3}
6·10 1.83 Å, 1.95 Å, 1.27 Å, 1.44 Å
6·11 {111}

Chapter 7

7·2 1.5 km s^{-1}
7·3 7×10^{-6}, 2.1×10^{-5}
7·6 0.921; Pb: 2×10^{12} Hz;
Au: 3.5×10^{12} Hz;
NaCl: 5.8×10^{12} Hz;
Fe: 7.5×10^{12} Hz;
Si: 1.4×10^{13} Hz;
Diamond: 3.9×10^{13} Hz.
7·7 3 translational, 3 rotational, 2 vibrational, $4R$ per mole.
7·10 591K, 4.49×10^{-9} kg m^{-2}s^{-1}.

Chapter 8

8·1 9.2×10^6; 7.2×10^{-21}
8·2 2×10^{-19} J/atom
8·4 8.75×10^{-8} N/m
8·5 5.7×10^4 J/m^3

Chapter 9

9·1 17.9 cm^2
9·2 0.527
9·6 116 MN/m^2
9·9 206 GN/m^2, 80.5 GN/m^2

Chapter 10

10·1 (a) 62 wt % Sn
(b) $\alpha_e = 18$ wt % Sn,
$\beta_e = 97$ wt % Sn
(c) none
(d) $\approx 20\%$
(e) $\alpha \approx 95\%$, $\beta \approx 5\%$ (90 wt % Sn alloy)
$\alpha \approx 42\%$, $\beta \approx 58\%$ (eutectic alloy)
(f) ≈ 20

10·3 (b) 100% γ, 0·4 wt % C
     (c) $\alpha \approx 88\%$, cementite $\approx 12\%$
     (d) eutectic mixture of $\gamma \approx 49\%$
         and cementite $\approx 51\%$

Chapter 11

11·2 (a) $\approx 63\%$
     (b) liquid 5·5 wt %, Cristobalite
         $\approx 0\%$, Mullite ≈ 72 wt %
         Al_2O_3
11·5 1·5 mm, 1,500 MN/m²
11·7 ΔT_c for A and B are 719°C and
     63°C respectively. Select A as
     higher ΔT_c and higher K

Chapter 12

12·1 0·9 μm
12·2 6×10^7, $2·8 \times 10^7$
12·5 $1·6 \times 10^5$, $2·5 \times 10^5$
12·6 24·7

Chapter 13

13·2 $8·5 \times 10^{28}$ electrons/m³
     $7·4 \times 10^{-7}$ m/s
13·4 4·09, 2·48, 8·54, 6·76 eV
13·5 $4·1 \times 10^{-3}$, $6·7 \times 10^{-3}$,
     $4·4 \times 10^{-3}$, $9·0 \times 10^{-4}$,
     $7·9 \times 10^{-4}$ m² V^{-1} s^{-1}
13·6 $2·3 \times 10^{-14}$, $3·8 \times 10^{-14}$,
     $2·5 \times 10^{-14}$ s
13·7 $1·58 \times 10^6$ m s^{-1}, $5·28 \times 10^{-8}$ m
     $1·39 \times 10^6$ m s^{-1}, $3·68 \times 10^{-8}$ m
     $1·39 \times 10^6$ m s^{-1}, $3·48 \times 10^{-8}$ m
13·8 $2·6 \times 10^{-8}$ Ω m, 79 nm, 1·06 nm

Chapter 14

14·1 0·58, 0·26, 0·074, $2·84 \times 10^{26}$ m²
     V^{-1} s^{-1}
14·2 $3·48 \times 10^7$ m^{-3}
14·3 $9·47 \times 10^{17}$ m^{-3}, 76 Ω m
     $2·19 \times 10^{20}$ m^{-3}, 0·24 Ω m
14·4 $1·77 \times 10^{29}$ m^{-3}; 1 in 2×10^{10}
14·5 12·3 Ω^{-1} m^{-1}, $21·7 \times 10^{20}$ m^{-3}
14·6 $1·59 \times 10^{-6}$ at %
14·7 2,060K
14·8 0·51 mA
14·9 $1·37 \times 10^3$ Ω^{-1} m^{-1}
14·10 $1·06 \times 10^{-4}$ W
14·11 $1·57 \times 10^{14}$ Hz
14·12 $2·54 \times 10^{-8}$ m

Chapter 15

15·1 18·6 kW
15·2 $E_D = 3·5$ eV, $D_0 = 1$ cm²/s
15·3 $t = 1,271$ s, $n(O, t) = 4·67 \times 10^{23}$
     m^{-3}
15·4 Linear: Dry 0·288 μm/h, wet 4·94
     μm/h
     Parabolic: Dry 0·024 μm²/h, wet
     0·532 μm²/h

Chapter 18

18·1 (a) 22 μm^{-1}
     (b) 11·5 m^{-1}
18·8 (a) 0·8 pm
     (b) 2·4 nm

INDEX